Advances in Intelligent Systems and Computing

189

For further volumes:
http://www.springer.com/series/11156

Álvaro Herrero, Václav Snášel, Ajith Abraham,
Ivan Zelinka, Bruno Baruque, Héctor Quintián,
José Luis Calvo, Javier Sedano,
and Emilio Corchado (Eds.)

International Joint Conference CISIS'12 ICEUTE'12 SOCO'12 Special Sessions

Editors
Álvaro Herrero
Department of Civil Engineering
University of Burgos
Burgos
Spain

Václav Snášel
VŠB-TU Ostrava
Ostrava
Czech Republic

Ajith Abraham
Machine Intelligence Research Labs
(MIR Labs)
Scientific Network for Innovation
and Research Excellence
Auburn, Washington
USA

Ivan Zelinka
VŠB-TU Ostrava
Ostrava
Czech Republic

Bruno Baruque
Department of Civil Engineering
University of Burgos
Burgos
Spain

Héctor Quintián
Universidad de Salamanca
Salamanca
Spain

José Luis Calvo
University of Coruña
Ferrol, A Coruña
Spain

Javier Sedano
Instituto Tecnológico de Castilla y León
Lopez Bravo 70, Pol. Ind. Villalonquéjar
Burgos
Spain

Emilio Corchado
Universidad de Salamanca
Salamanca
Spain

ISSN 2194-5357 e-ISSN 2194-5365
ISBN 978-3-642-33017-9 e-ISBN 978-3-642-33018-6
DOI 10.1007/978-3-642-33018-6
Springer Heidelberg New York Dordrecht London

Library of Congress Control Number: 2012945518

Printed on acid-free paper

Springer is part of Springer Science+Business Media (www.springer.com)

Preface

This volume of Advances in Intelligent and Soft Computing contains accepted papers presented at CISIS 2012 and ICEUTE 2012, both conferences held in the beautiful and historic city of Ostrava (Czech Republic), in September 2012.

CISIS aims to offer a meeting opportunity for academic and industry-related researchers belonging to the various, vast communities of **Computational Intelligence**, **Information Security**, and **Data Mining**. The need for intelligent, flexible behaviour by large, complex systems, especially in mission-critical domains, is intended to be the catalyst and the aggregation stimulus for the overall event.

After a through peer-review process, the CISIS 2012 International Program Committee selected 30 papers which are published in these conference proceedings achieving an acceptance rate of 40%.

In the case of ICEUTE 2012, the International Program Committee selected 4 papers which are published in these conference proceedings.

The selection of papers was extremely rigorous in order to maintain the high quality of the conference and we would like to thank the members of the Program Committees for their hard work in the reviewing process. This is a crucial process to the creation of a high standard conference and the CISIS and ICEUTE conferences would not exist without their help.

CISIS 2012 and ICEUTE 2012 enjoyed outstanding keynote speeches by distinguished guest speakers: Prof. Ponnuthurai Nagaratnam Suganthan, Prof. Jeng-Shyang Pan, Prof. Marios M. Polycarpou, Prof. Fanny Klett and Mr. Milan Kladnicek.

For this special edition, as a follow-up of the conference, we anticipate further publication of selected papers in a special issue of the prestigious Logic Journal of the IGPL Published by Oxford Journals.

Particular thanks go as well to the Conference main Sponsors, IT4Innovations, VŠB-Technical University of Ostrava, IEEE.- Systems, Man and Cybernetics Society CzechoSlovakia, IEEE.- Systems, Man and Cybernetics Society Spain, MIR labs, Spanish Association for Artificial Intelligence and IFCOLOG.

We would like to thank all the special session organizers, contributing authors, as well as the members of the Program Committees and the Local Organizing Committee for their hard and highly valuable work. Their work has helped to contribute to the success of the CISIS 2012 and ICEUTE 2012 events.

September, 2012

Álvaro Herrero
Václav Snášel
Ajith Abraham
Ivan Zelinka
Bruno Baruque
Héctor Quintián
José Luis Calvo
Javier Sedano
Emilio Corchado

CISIS 2012

Organization

General Chair

Emilio Corchado	University of Salamanca, Spain

International Advisory Committee

Ajith Abraham	Machine Intelligence Research Labs - MIR Labs, Europe
Antonio Bahamonde	President of the Spanish Association for Artificial Intelligence, AEPIA
Michael Gabbay	Kings College London, UK

Program Committee Chair

Álvaro Herrero	University of Burgos, Spain
Václav Snasel	VSB-Technical University of Ostrava, Czech Republic
Ajith Abraham	VSB-Technical University of Ostrava, Czech Republic
Ivan Zelinka	VSB-Technical University, Ostrava, Czech Republic
Emilio Corchado	University of Salamanca, Spain

Program Committee

Alain Lamadrid Vallina	Department of High Education, Cuba
Alberto Peinado Domínguez	University of Malaga, Spain
Alessandro Zanassi	Bocconi University, Italy
Amparo Fúster Sabater	CSIC, Spain
Ana Isabel González-Tablas	Carlos III of Madrid University, Spain
Andre CPLF de Carvalho	University of São Paulo, Brazil
Andrés Correal Cuervo	University of Boyacá, Colombia
Ángel Arroyo	University of Burgos, Spain

Ángel Grediaga Olivo	University of Alicante, Spain
Ángel Martín del Rey	Univsersity of Salamanca, Spain
Ángela I. Barbero Díez	University of Valladolid, Spain
Antoni Bosch Pujol	Autonomous University of Barcelona, Spain
Antonino Santos de Riego	University of La Coruña, Spain
Antonio J. Tomeu Hardasmal	University of Cadiz, Spain
Antonio Maña Gómez	University of Malaga, Spain
Antonio Zamora Gómez	University of Alicante, Spain
Araceli Queiruga Dios	Univsersity of Salamanca, Spain
Arturo Ribagorda Garnacho	Carlos III of Madrid University, Spain
Belén Vaquerizo	University of Burgos, Spain
Benjamín Ramos Alvarez	Carlos III of Madrid University, Spain
Bernardete Ribeiro	University of Coimbra, Portugal
Borja Sanz Urquijo	University of Deusto, Spain
Bruno Baruque	University of Burgos, Spain
Candelaria Hernández Goya	University of La Laguna, Spain
Carlos Laorden	University of Deusto, Spain
Carlos Marcelo Martínez Cagnazzo	University of La República, Uruguay
Carlos Munuera Gómez	University of Valladolid, Spain
Carlos Pereira	Universidade de Coimbra, Portugal
César Hervás-Martínez	University of Córdoba, Spain
Chelo Malagón Poyato	CSIC, Spain
Claudio Jorge Felizia	University of Buenos Aires, Argentina
Constantino Malagón Luque	University Antonio de Nebrija, Spain
Cristina Alcaraz Tello	University of Malaga, Spain
Daniel Sadornil Renedo	University of Cantabria, Spain
Danilo Pástor Ramírez	Politechnic High School of Chimborazo, Ecuador
Dario Forte	University of Milano Crema, Italy
David García Rosado	University of Castilla la Mancha, Spain
Davide Anguita	University of Genova, Italy
Davide Leoncini	University of Genova, Italy
Debasis Giri	Haldia Institute of Technology, India
Diego Avila Pesantez	Higher Polytechnic School of Chimborazo, Ecuador
Domingo Gómez Pérez	University of Cantabria, Spain
Edgar Martínez Moro	University of Valladolid, Spain
Eduardo Carozo Blumsztein	University of Montevideo, Uruguay
Eduardo Fernández-Medina Patón	University of Castilla la Mancha, Spain
Edwin Mamani Zeballos	Military School of Engineering, Bolivia
Enrico Appiani	Elsag Datamat, Italy
Enrique Daltabuit	National Autonomous of México University, México
Enrique De la Hoz	University of Alcalá, Spain

Enrique González Jiménez	Autonomous University of Madrid, Spain
Eva Volna	University of Ostrava, Czech Republic
Fabián Velásquez Clavijo	University of los Llanos, Colombia
Fausto Montoya Vitini	Institute of Applied Physics, CSIC, Spain
Federico García Crespí	Miguel Hernández University, Spain
Felipe Andrés Corredor Chavarro	University of los Llanos, Colombia
Fernando Piera Gómez	Computer Technicians Association, ATI, Spain
Fernando Tricas García	University of Zaragoza, Spain
Filip Merhaut	Getmore s.r.o.
Francisco Fernández-Navarro	University of Córdoba, Spain
Francisco Herrera	University of Granada, Spain
Francisco José Navarro Ríos	University of Granada, Spain
Francisco Plaza	University of Salamanca, Spain
Francisco Rodríguez Henríquez	CINVESTAV IPN, México
Francisco Valera Pintor	Carlos III of Madrid University, Spain
Gabriel Díaz Orueta	UNED, Spain
Gabriel López Millán	University of Murcia, Spain
Gerald Schaefer	Loughborough University, UK
Gerardo Rodríguez Sánchez	University of Salamanca, Spain
Gonzalo Alvarez Marañón	CSIC, Spain
Gonzalo Martínez Ginesta	Alfonso X University, Spain
Guillermo Morales-Luna	CINVESTAV, Mexico
Gustavo Adolfo Isaza Echeverry	University of Caldas, Colombia
Hector Alaiz	University of León, Spain
Héctor Quintián	University of Salamanca, Spain
Hugo Pagola	University of Buenos Aires, Argentina
Hujun Yin	University of Manchester, UK
Ignacio Arenaza	University of Mondragon, Spain
Ignacio Luengo Velasco	Complutense de Madrid University, Spain
Igor Santos Grueiro	University of Deusto, Spain
Isaac Agudo Ruiz	University of Malaga, Spain
Ivonne Valeria Muñoz Torres	ITESM, México
Janusz Kacprzyk	Systems Research Institute Polish Academy of Sciences, Poland
Javier Areitio Bertolín	University of Deusto, Spain
Javier Carbó Rubiera	Carlos III of Madrid University, Spain
Javier Fernando Castaño Forero	University of los Llanos, Colombia
Javier Nieves Acedo	University of Deusto, Spain
Javier Sánchez-Monedero	University of Córdoba, Spain
Javier Sedano	Instituto tecnológico de Castilla y León, Spain
Jesús Esteban Díaz Verdejo	University of Granada, Spain
Jesús María Minguet Melián	UNED, Spain
Jirí Dvorsky	VSB-Technical University of Ostrava, Czech Republic

Joan Borrel Viader	Autonomous University of Barcelona, Spain
Joan-Josep Climent	University of Alicante, Spain
Joaquín García-Alfaro	Carleton University, Canada
Jordi Herrera Joancomarti	Autonomous University of Barcelona, Spain
Jorge Diéz Peláez	University of Oviedo, Spain
Jorge Eduardo Sznek	Nacional del Comahue University, Argentina
Jorge López Hernández-Ardieta	Carlos III of Madrid University, Spain
Jorge Ramió Aguirre	Universidad Politécnica de Madrid, Spain
Jose A. Onieva	University of Málaga, Spain
José Ángel Domínguez Pérez	University of Salamanca, Spain
José Antonio Montenegro Montes	University of Malaga, Spain
José Antonio. Onieva	University of Málaga, Spain
José Daniel Britos	National University of Córboda, Argentina
José de Jesús Angel Angel	CINVESTAV IPN, México
José Esteban Saavedra López	University of Oruro, Bolivia
José Francisco Martínez	INAOE, Mexico
José Francisco Vicent Francés	University of Alicante, Spain
Jose L. Salmeron	Pablo Olavide University, Spain
José Luis Calvo	University of A Coruña, Spain
José Luis Ferrer Gomila	University of Islas Baleares, Spain
José Luis Imaña	Complutense de Madrid University, Spain
José Luis Rivas López	University of Vigo, Spain
José Luis Salazar Riaño	University of Zaragoza, Spain
José Manuel Benitez	University of Granada, Spain
José Manuel Molina	Carlos III of Madrid University, Spain
Jose María Gómez Hidalgo	Optenet, Spain
Josep Conde Colom	Lleida University, Spain
Josep Rifá Coma	Autonomous University of Barcelona, Spain
Jouni Lampinen	University of Vaasa, Finland
Juán Á. Muñoz	University of Almería, Spain
Juan C. Fernández	University of Córdoba, Spain
Juan Guillermo Lalinde-Pulido	EAFIT University, Colombia
Juan Jesús Barbarán Sánchez	University of Granada, Spain
Juan José del Coz Velasco	University of Oviedo, Spain
Juan José Ortega Daza	University of Malaga, Spain
Juan Manuel Corchado	University of Salamanca, Spain
Juan Manuel García García	Morelia Technologic Institute, Mexico
Juán Manuel González Nieto	Queensland University of Technologic, Australia
Juan Pedro Hecht	University of Buenos Aires, Argentina
Juan Tapiador	University of York, UK
Juan Tena Ayuso	University of Valladolid, Spain
Judith Redi	Delft University of Technology, The Netherlands
Juha Karhunen	Helsinki University of Technology, Finland
Julio Cesar Hernandez Castro	University of Portsmouth, UK

Petro Gopych	V.N. Karazin Kharkiv National University, Ukraine
Pino Caballero Gil	University of La Laguna, Spain
Rafael Álvarez	University of Alicante, Spain
Rafael Calzada Pradas	Carlos III of Madrid University, Spain
Rafael Corchuelo	University of Sevilla, Spain
Rafael Martínez Gasca	University of Sevilla, Spain
Ramón Rizo Aldeguer	University of Alicante, Spain
Ramón Torres Rojas	Marta Abreu de Las Villas Central University, Cuba
Raquel Redondo	University of Burgos, Spain
Raúl Durán Díaz	University of Alcala, Spain
Reinaldo Nicolás Mayol Arnao	University of Los Andes, Venezuela
Ricardo Contreras Arriagada	University of Concepción, Chile
Ricardo Llamosa-Villalba	Industrial University of Santander, Colombia
Richard Duro	University of La Coruña, Spain
Roberto Gómez Cárdenas	ITESM, México
Roberto Hiribarne Guedes	Marta Abreu de Las Villas Central University, Cuba
Roberto Uribeetxeberria	University of Mondragon, Spain
Rodolfo Zunino	University of Genova, Italy
Rodrigo Adolfo Cofré Loyola	Católica del Maule University, Chile
Roman Jašek	Tomas Bata University in Zlin, Czech Republic
Roman Senkerik	Tomas Bata University in Zlin, Czech Republic
Rosanna Costaguta	National University of Santiago del Estero, Argentina
Rosaura Palma Orozco	CINVESTAV IPN, México
Rubén Vázquez Medina	National Polytechnic Institute, Mexico
Salvador Alcaraz Carrasco	Miguel Hernández University, Spain
Santiago Martín Acurio Del Pino	Católica del Ecuador Pontificial University, Ecuador
Seema Verma	Banasthali University, India
Sergi Robles Martínez	Autonomous University of Barcelona, Spain
Sergio Bravo Silva	University of Bío Bío, Chile
Sergio Decherchi	University of Genova, Italy
Sergio Pozo Hidalgo	University of Sevilla, Spain
Sorin Stratulat	University Paul Verlaine - Metz, France
Tomás Jímenez García	University of Murcia, Spain
Tzai-Der Wang	Cheng Shiu University, Taiwan
Urko Zurutuza Ortega	University of Mondragon, Spain
Valentina Casola	University of Naples Federico II, Italy
Vincenzo Mendillo	Central University of Venezuela, Venezuela
Vladimir Cobarrubias Llanten	University of Chile, Chile
Wei Wang	University of Luxembourg, Luxembourg
Wenjian Luo	University of Science and Technology of China, China
Wilfred Torres Moya	Andrés Bello Catholic University, Venezuela
Wiltord Pedrycz	University of Alberta, Canada

Xiangliang Zhang	King Abdullah Univ. of Science and Technology, Saudi Arabia
Xiuzhen Chen	ParisTech, France
Zuzana Oplatkova	Tomas Bata University in Zlin, Czech Republic

Special Sessions

Intelligent Network Security and Survivability

Andrzej Kasprzak	Wroclaw University of Technology, Poland
Krzysztof Walkowiak	Wroclaw University of Technology, Poland
Leszek Koszałka	Wroclaw University of Technology, Poland
Michał Woźniak	Wroclaw University of Technology, Poland
Bartosz Krawczyk	Wroclaw University of Technology, Poland
Bartosz Kurlej	Wroclaw University of Technology, Poland
Jacek Rak	Gdansk University of Technology, Poland
Marcin Zmyslony	Wroclaw University of Technology, Poland
Mariusz Koziol	Wroclaw University of Technology, Poland
Michal Choras	University of Technology and Life Sciences, Poland
Michal Kucharzak	Wroclaw University of Technology, Poland
Miroslaw Klinkowski	National Institute of Telecommunications, Poland
Robert Burduk	Wroclaw University of Technology, Poland
Teresa Gomes	University of Coimbra, Portugal
Tomasz Kajdanowski	Wroclaw University of Technology, Poland
Wojciech Kmiecik	Wroclaw University of Technology, Poland

Text Processing Applications to Secure and Intelligent Systems

Pablo G. Bringas	University of Deusto, Spain
Igor Santos	University of Deusto, Spain
Xabier Ugarte-Pedrero	University of Deusto, Spain
Carlos Laorden	University of Deusto, Spain
David Barroso Berrueta	
Felix Brezo	University of Deusto, Spain
Igor Ruiz-Agúndez	University of Deusto, Spain
Jose Maria Gómez Hidalgo	Optenet
Juan Álvaro Muñoz Naranjo	University of Almería, Spain
Paolo Fosci	Università di Bergamo, Italy
Pedram Hayati	University of Western Australia, Australia
Simone Mutti	Università degli Studi di Bergamo, Italy
Vidyasagar Potdar	University of Western Australia, Australia

Organising Committee

Václav Snášel - Chair	VSB-Technical University of Ostrava, Czech Republic, (Chair)
Jan Platoš	VSB-Technical University of Ostrava, Czech Republic (Co-chair)
Pavel Krömer	VSB-Technical University of Ostrava, Czech Republic (Co-chair)
Katerina Kasparova	VSB-Technical University of Ostrava, Czech Republic
Hussein Soori	VSB-Technical University of Ostrava, Czech Republic
Petr Berek	VSB-Technical University of Ostrava, Czech Republic

ICEUTE 2012

Organization

International Advisory Committee

Jean-Yves Antoine	Université François Rabelais, France
Reinhard Baran	Hamburg University of Applied Sciences, Germany
Fernanda Barbosa	Instituto Politécnico de Coimbra, Portugal
Bruno Baruque	University of Burgos, Spain
Emilio Corchado	University of Salamanca, Spain
Wolfgang Gerken	Hamburg University of Applied Sciences, Germany
Arnaud Giacometti	Université François Rabelais, France
Helga Guincho	Instituto Politécnico de Coimbra, Portugal
Álvaro Herrero	University of Burgos, Spain
Patrick Marcel	Université François Rabelais, France
Gabriel Michel	University Paul Verlaine - Metz, France
Viorel Negru	West University of Timisoara, Romania
Jose Luis Nunes	Instituto Politécnico de Coimbra, Portugal
Salvatore Orlando	Ca' Foscari University, Italy
Veronika Peralta	Université François Rabelais, France
Carlos Pereira	Instituto Politécnico de Coimbra, Portugal
Teppo Saarenpää	Turku University of Applied Sciences, Finland
Sorin Stratulat	University Paul Verlaine - Metz, France

Program Committee Chair

Václav Snasel	VSB-Technical University of Ostrava, Czech Republic
Bruno Baruque	University of Burgos, Spain
Álvaro Herrero	University of Burgos, Spain
Ajith Abraham	VSB-Technical University of Ostrava, Czech Republic
Emilio Corchado	University of Salamanca, Spain

Program Committee

Agostino Cortesi	Ca' Foscari University, Italy
Ana Borges	Instituto Politécnico de Coimbra, Portugal
Anabela de Jesus Gomes	Instituto Politécnico de Coimbra, Portugal
Anabela Panão Ramalho	Instituto Politécnico de Coimbra, Portugal

Ángel Arroyo	University of Burgos, Spain
Antonio José Mendes	University of Coimbra, Portugal
Arnaud Giacometti	Université François Rabelais, France
Begoña Prieto	University of Burgos, Spain
Belén Vaquerizo	University of Burgos, Spain
Carlos López	University of Burgos, Spain
Carlos Pereira	Instituto Politécnico de Coimbra, Portugal
César Hervás-Martínez	University of Córdoba, Spain
Claudio Silvestri	Ca' Foscari University, Italy
David Fairen-Jimenez	The University of Edinburgh, UK
Dominique Laurent	Cergy-Pontoise University, France
Dragan Simic	Novi Sad University, Serbia
Elisabeth Delozanne	Université Pierre et Marie Curie Paris, France
Fernanda Barbosa	Instituto Politécnico de Coimbra, Portugal
Fernando López	University of La Coruña, Spain
Francisco Duarte	Instituto Politécnico de Coimbra, Portugal
Francisco Fernández-Navarro	University of Córdoba, Spain
Francisco Leite	Instituto Politécnico de Coimbra, Portugal
François Bret	Université François Rabelais, France
Gabriel Michel	University Paul Verlaine - Metz, France
Héctor Quintián	University of Salamanca, Spain
Helga Guincho	Instituto Politécnico de Coimbra, Portugal
Hujun Yin	University of Manchester, UK
Ivan Zelinka	VSB-Technical University of Ostrava, Czech Republic
Javier Bajo	Pontifical University of Salamanca, Spain
Javier Barcenilla	University Paul Verlaine - Metz, France
Javier Sánchez-Monedero	University of Córdoba, Spain
Javier Sedano	Instituto tecnológico de Castilla y León, Spain
Jean-Yves Antoine	Université François Rabelais, France
Jiří Dvorský	VSB-Technical University of Ostrava, Czech Republic
Joaquín Pacheco	University of Burgos, Spain
Jon Mikel Zabala	Lund University, Sweden
Jorge Barbosa	Instituto Politécnico de Coimbra, Portugal
José Luis Calvo	University of A Coruña, Spain
Jose Luis Nunes	Instituto Politécnico de Coimbra, Portugal
Juan C. Fernández	University of Córdoba, Spain
Juan Manuel Corchado	University of Salamanca, Spain
Juan Pavón	Complutense University of Madrid, Spain
Konrad Jackowski	Wroclaw University of Technology, Poland
Leonel Morgado	University of Trás-os-Montes e Alto Douro, Portugal
Leticia Curiel	University of Burgos, Spain
Lourdes Sáiz	University of Burgos, Spain

María José Marcelino	University of Coimbra, Portugal
Michał Wozniak	Wroclaw University of Technology, Poland
Miguel Ángel Manzanedo	University of Burgos, Spain
Milos Kudelka	VSB-Technical University of Ostrava, Czech Republic
Nicoletta Cocco	Ca' Foscari University, Italy
Nuno Ferreira	Instituto Politécnico de Coimbra, Portugal
Patrick Marcel	Université François Rabelais, France
Pavel Krömer	VSB-Technical University of Ostrava, Czech Republic
Pedro A. Gutiérrez	University of Córdoba, Spain
Rafael Corchuelo	University of Seville, Spain
Raúl Marticorena	University of Burgos, Spain
Reinhard Baran	Hamburg University of Applied Sciences, Germany
Richard Duro	University of La Coruña, Spain
Robin Vivian	University Paul Verlaine - Metz, France
S. Hariharan	JJ College of Engineering and Technology, India
Salvatore Orlando	Ca' Foscari University, Italy
Shanmugasundaram Hariharan	Bharathiar University, India
Sorin Stratulat	University Paul Verlaine - Metz, France
Teppo Saarenpää	Turku University of Applied Sciences, Finland
Thomas Connolly	University of the West of Scotland, United Kingdom
Veronika Peralta	Université François Rabelais, France
Viorel Negru	West University of Timisoara, Romania
Wolfgang Gerken	Hamburg University of Applied Sciences, Germany
Zdenek Tronicek	Czech Technical University in Prague, Czech Republic

Organising Committee

Václav Snášel - Chair	VSB-Technical University of Ostrava, Czech Republic (Chair)
Jan Platoš	VSB-Technical University of Ostrava, Czech Republic (Co-chair)
Pavel Krömer	VSB-Technical University of Ostrava, Czech Republic (Co-chair)
Katerina Kasparova	VSB-Technical University of Ostrava, Czech Republic
Hussein Soori	VSB-Technical University of Ostrava, Czech Republic
Petr Berek	VSB-Technical University of Ostrava, Czech Republic

SOCO 2012

Organization

General Chair

Emilio Corchado	University of Salamanca, Spain

International Advisory Committee

Ashraf Saad	Armstrong Atlantic State University, USA
Amy Neustein	Linguistic Technology Systems, USA
Ajith Abraham	Machine Intelligence Research Labs - MIR Labs, Europe
Jon G. Hall	The Open University, UK
Paulo Novais	Universidade do Minho, Portugal
Antonio Bahamonde	President of the Spanish Association for Artificial Intelligence, AEPIA
Michael Gabbay	Kings College London, UK
Isidro Laso-Ballesteros	European Commission Scientific Officer, Europe
Aditya Ghose	University of Wollongong, Australia
Saeid Nahavandi	Deakin University, Australia
Henri Pierreval	LIMOS UMR CNRS 6158 IFMA, France

Industrial Advisory Committee

Rajkumar Roy	The EPSRC Centre for Innovative Manufacturing in Through-life Engineering Services, UK
Amy Neustein	Linguistic Technology Systems, USA
Jaydip Sen	Innovation Lab, Tata Consultancy Services Ltd., India

Program Committee Chair

Emilio Corchado	University of Salamanca, Spain
Václav Snášel	VSB-Technical University of Ostrava, Czech Republic
Ajith Abraham	VSB-Technical University of Ostrava, Czech Republic

Special Sessions

Soft Computing Models for Control Theory & Applications in Electrical Engineering

Pavel Brandstetter	VSB - Technical University of Ostrava, Czech Republic (Co-chair)
Emilio Corchado	University of Salamanca, Spain (Co-chair)
Daniela Perdukova	Technical University of Kosice, Slovak Republic
Jaroslav Timko	Technical University of Kosice, Slovak Republic
Jan Vittek	University of Zilina, Slovak Republic
Jaroslava Zilkova	Technical University of Kosice, Slovak Republic
Jiri Koziorek	VSB - Technical University of Ostrava, Czech Republic
Libor Stepanec	UniControls a.s., Czech Republic
Martin Kuchar	UniControls a.s., Czech Republic
Milan Zalman	Slovak University of Technology, Slovak Republic
Pavel Brandstetter	VSB - Technical University of Ostrava, Czech Republic
Pavol Fedor	Technical University of Kosice, Slovak Republic
Petr Palacky	VSB - Technical University of Ostrava, Czech Republic
Stefan Kozak	Slovak University of Technology, Slovak Republic

Soft Computing Models for Biomedical Signals and Data Processing

Lenka Lhotská	Czech Technical University, Czech Republic (Co-chair)
Martin Macaš	Czech Technical University, Czech Republic (Co-chair)
Miroslav Burša	Czech Technical University, Czech Republic (Co-chair)

Chrysostomos Stylios	TEI of Epirus, Greece
Dania Gutiérrez Ruiz	Cinvestav, Mexico
Daniel Novak	Czech Technical University, Czech Republic
George Georgoulas	TEI of Epirus, Greece
Michal Huptych	Czech Technical University, Czech Republic
Petr Posik	Czech Technical University, Czech Republic
Vladimir Krajca	Czech Technical University, Czech Republic

Advanced Soft Computing Methods in Computer Vision and Data Processing

Irina Perfilieva	University of Ostrava, Czech Republic (Co-chair)
Vilém Novák	University of Ostrava, Czech Republic (Co-chair)
Antonín Dvořák	University of Ostrava, Czech Republic
Marek Vajgl	University of Ostrava, Czech Republic
Martin Štěpnička	University of Ostrava, Czech Republic
Michal Holcapek	University of Ostrava, Czech Republic
Miroslav Pokorný	University of Ostrava, Czech Republic
Pavel Vlašanek	University of Ostrava, Czech Republic
Petr Hurtik	University of Ostrava, Czech Republic
Petra Hodáková	University of Ostrava, Czech Republic
Petra Murinová	University of Ostrava, Czech Republic
Radek Valášek	University of Ostrava, Czech Republic
Viktor Pavliska	University of Ostrava, Czech Republic

Organising Committee

Václav Snášel - Chair	VSB-Technical University of Ostrava, Czech Republic (Chair)
Jan Platoš	VSB-Technical University of Ostrava, Czech Republic (Co-chair)
Pavel Krömer	VSB-Technical University of Ostrava, Czech Republic (Co-chair)
Katerina Kasparova	VSB-Technical University of Ostrava, Czech Republic
Hussein Soori	VSB-Technical University of Ostrava, Czech Republic
Petr Berek	VSB-Technical University of Ostrava, Czech Republic

CISIS 12

ICEUTE

Contents

CISIS 2012

General Track

Applications of Intelligent Methods for Security

Infrastructure and Network Security

ICEUTE 2012

General Track

SOCO 2012

Special Sessions

Soft Computing Models for Control Theory & Applications in Electrical Engineering

Automatic Analysis of Web Service Honeypot Data Using Machine Learning Techniques

Abdallah Ghourabi, Tarek Abbes, and Adel Bouhoula

Higher School of Communication of Tunis SUP'COM, University of Carthage, Tunisia
{abdallah.ghourabi,adel.bouhoula}@supcom.rnu.tn,
tarek.abbes@isecs.rnu.tn

Abstract. Over the past years, Honeypots have proven their efficacy for understanding the characteristics of malicious activities on the Internet. They help security managers to collect valuable information about the techniques and motivations of the attackers. However, when the amount of collected data in honeypots becomes very large, the analysis performed by a human security administrator tends to be very difficult, tedious and time consuming task. To facilitate and improve this task, integration of new methods for automatic analysis seems to be necessary. We propose in this paper a new approach based on different machine learning techniques to analyze collected data in a Web Services Honeypot. The aim of this approach is to identify and characterize attacks targeting Web services using three classifiers (SVM, SVM Regression and Apriori) depending on the nature of collected data.

Keywords: data analysis, Honeypot, machine learning, Web service attacks.

1 Introduction

A honeypot [9] can be defined as a computer system voluntarily vulnerable to one or more known threats, deployed on a network for the purpose of logging and studying attacks on the honeypot. Honeypots are very useful for collecting different types of attack traffic. However, the identification and the characterization of attacks from the honeypot log files can be challenging due to the high dimensionality of the data and the large volume of collected traffic. A large amount of the collected data is related to normal activities performed by users who test or scan the Honeypot. Another large part of this data is generated by frequent and repetitive attacks that are already seen. The presence of these two parts complicates the analysis task of log files since it hides new abnormal activities which require immediate attention from the security personnel.

The large amount of data collected by the honeypot makes its analysis by a human analyst a very difficult and fastidious task. To resolve this problem and facilitate the analysis of collected data, some researchers are looking for automatic analysis solutions based on techniques like statistics, machine learning and data mining. In this paper, we propose a new approach based on machine learning techniques to analyze data collected from Web Service Honeypot. The purpose of this approach is to automatically process collected data using the most suitable among three possible classifiers (SVM, SVM Regression and Apriori). Each classifier analyzes a set of

Á. Herrero et al. (Eds.): Int. JointConf. CISIS'12-ICEUTE'12-SOCO'12, AISC 189, pp. 1–11.
springerlink.com

parameters extracted from captured SOAP requests on the honeypot, and then classifies the current activity as normal or suspicious. The advantage of our approach is to exploit the complementarily of three different classifiers in order to identify various types of attacks.

The remaining parts of the paper are organized as follows: Section 2 reviews related work. Section 3 describes our approach for analyzing collected data in a Web Service Honeypot. Section 4 reports our experimental results. Finally, we conclude the paper in Section 5.

2 Related Work

With the continuing evolution of honeypots in recent years, and the deployment of new distributed honeypot platforms around the world, the amount of collected information is becoming extremely large. To facilitate the analysis of these data, several research works suggest the use of automated analysis solutions such as statistics, machine learning and data mining.

For example, Pouget and Dacier [6] proposed a simple clustering approach to analyze data collected from the honeypot project "Leurre.com". Their objective is to characterize the root causes of attacks targeting their Honeypots. The aim of this algorithm is to gather all attacks presenting some common characteristics (duration of attack, targeted ports, number of packets sent, etc...) based on generalization techniques and association-rules mining. Resulting clusters are further refined using "Levenshtein distance". The final goal of their approach is to group into clusters all attacking sources sharing similar activity fingerprints, or attack tools.

In [2], the authors presented some results obtained from their project CADHo (Collection and Analysis of Data from Honeypots). The purpose of this project was to analyze data collected from the environment "Leurre.com" and to provide models for the observed attacks. They proposed simple models describing the time-evolution of the number of attacks observed on different honeypot platforms. Besides, they studied the potential correlations of attack processes observed on the different platforms taking into account the geographic location of the attacking machines and the relative contribution of each platform in the global attack scenario. The correlation analysis is based on a linear regression models.

In [10], Thonnard and Dacier proposed a framework for attack patterns' discovery in honeynet data. The aim of this approach is to find, within an attack data set, groups of network traces sharing various kinds of similar patterns. In the paper, the authors applied a graph-based clustering method to analyze one specific aspect of the honeynet data (the time series of the attacks). The results of the clustering applied to time-series analysis enable to identify the activities of several worms and botnets in the traffic collected by the honeypots.

In another paper, Seifert et al. [7] proposed static heuristics method to classify malicious Web pages. To implement this method, they used the machine learning to build a decision tree in order to classify Web pages as normal or malicious. Subsequently, malicious Web pages are forwarded to a high-interaction client honeypot for a second inspection. The purpose of this method is to reduce the number of Web pages that will be inspected by the honeypot.

In [5], Herrero et al. proposed a neural intelligent system to visualize and analyze network traffic data collected by a network of honeypots (Honeynets). Their approach is based on the use of different neural projection and unsupervised methods to visualize honeypot data by projecting the high-dimensional data points onto a lower dimensional space. The authors applied this neural-visualization in a big volume of traffic data in order to project it onto a lower dimensional subspace plotted in two or three dimensions, which makes it possible to examine the structure with the naked eye.

As we can conclude from these works, the use of Computational Intelligence to analyze data collected by honeypots represents a very interesting idea given the simplicity and the efficiency that can be added to the analysis task. On our side, we propose in this paper an intelligent system based on machine learning techniques to facilitate the analysis of data obtained from our WS Honeypot. In this approach, we exploit the complementarily of three different classification techniques (SVM, SVM Regression and Apriori). The advantage of our approach compared to works described above is that it can detect several types of attacks due to the diversity of implemented machine learning techniques.

3 Analysis of Collected Data from a Web Service Honeypot

Web services are becoming more and more an integral part of next-generation web applications. A Web service is defined as a software system designed to enable data exchange between heterogeneous systems and applications across the Internet. Exchanged data are SOAP messages based on a set of XML standards and using HTTP as a transport protocol. Web services are vulnerable to many types of attacks such as DoS attack, SQL and XML injection, parameters tampering and still others [3]. We demonstrated in a previous work [4] the usefulness of Honeypot in order to study attacks against Web services and discover the employed techniques. In this paper, we try to demonstrate the utility of using machine learning techniques to facilitate the analysis of collected data from a Web service Honeypot.

3.1 Description of WS Honeypot

WS Honeypot is a solution based on honeypot in order to detect and study attacks against Web Services. It is a honeypot which acts as a Web service. The purpose of its deployment is to attract attackers targeting Web services to study and understand their techniques and motivations.

The role of WS Honeypot is to simulate the behavior of a Web service. It must convince the attackers that they are interacting with a real Web service. It incorporates mechanisms of traffic capture and monitoring tools to intercept and analyze requests sent to the simulated Web service. Captured data is then used to extract the characteristic parameters of each SOAP message captured by the WS Honeypot. These parameters are useful to classify activities generated by these messages. The extraction process creates three categories of characteristic parameters:

- *SOAP message content:* IP source, message length, request preamble, response preamble, invoked operations in each request, input parameters of each operation.
- *Resource consumption:* response time, CPU usage, memory usage.

- *Operations list:* the Web service offers several operations that can be called differently depending on user. For each user, we extract a list of all invoked operations to build a profile.

This diversity of extracted parameters is very useful for detecting different types of attacks. The captured data is then analyzed in an automatic way to identify and characterize attacks.

3.2 Analysis of Data Captured by WS Honeypot

Data collected by WS Honeypot should be analyzed by a human expert in order to identify and characterize captured attacks. However, when the collected information becomes very large, the human expert will be overwhelmed with a great amount of data in the audit trails and may fail to notice severe attacks. To facilitate the analysis task, we append to the WS Honeypot automatic solutions for data processing based on machine learning methods. These techniques contribute to learn the normal behavior of activities in the honeypot, and consequently, detect any significant deviation from this normal behavior. In this case, the human analyst can decide whether the activity constitutes a true threat or not.

The operating principle of our approach is described in Figure 1. WS Honeypot captures all SOAP messages in order to extract some characteristic parameters as mentioned above. The extracted parameters are stored in datasets in order to be analyzed. During this phase, we classify incoming messages according to invoked operations in the request. The classification refers to a learning dataset built progressively during the deployment of the Honeypot. If a classification error occurs, for example due to an abnormal activity, this activity is transmitted to the human expert to check if it represents an attack.

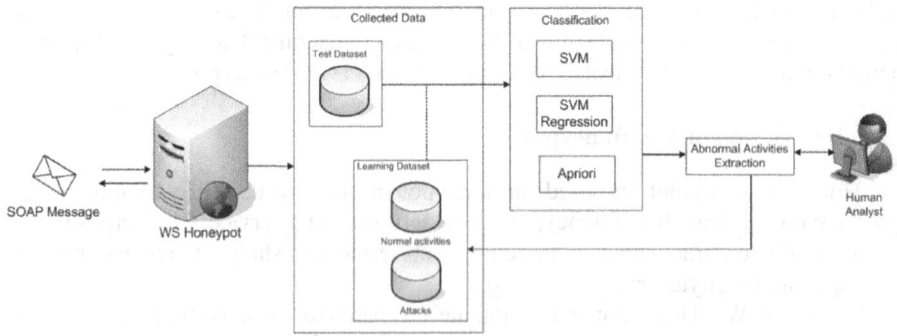

Fig. 1. Analysis process of data collected by WS Honeypot

The classification process is divided into three parts. For each group of parameters extracted from the requests, we associate a classifier based on a machine learning method (SVM, regression or association rules):

- For "SOAP message content" category, we use an SVM (support vector machine) classifier. This solution is useful to detect attacks like SQL injection, XML injection and parameters tampering.

- For "Resource consumption" category, we employ an SVM implementation for regression. This classifier is useful to detect several types of DoS attacks.
- For "Operations list", we use the association rule algorithm "Apriori". This solution is useful to detect unauthorized accesses to the Web service.

Collected Data
The characteristic parameters extracted from the captured SOAP messages on the WS Honeypot, are stored in two datasets (learning dataset and test dataset). The learning dataset contains two subsets: information characterizing the normal activities carried out in the WS Honeypot and information describing the attacks detected progressively during the deployment of WS Honeypot and validated by the human analyst. The test dataset contains the characteristic information of new captured SOAP messages. These data must be classified in order to make a decision about the nature of the activity (normal or attack).

Classification
The classification process is based on three machine learning algorithms: Support Vector Machine (SVM) [11], an SVM implementation for regression [8] and the association rules algorithm "Apriori" [1]. Each algorithm is interested in a group of extracted parameters. During this phase, the data stored in the test dataset are classified periodically based on the learning dataset. If a classification error is generated due to an abnormal activity, this activity will be transmitted to the human analyst to check if it represents an attack. Upon the analyst decision, the activity is added to the "learning dataset" either in normal activities or attacks data.

SVM Classifier
We employ SVM method in order to classify the first category of extracted features "SOAP message content". This analysis allows us the detection of malicious code injection and tampered parameters. The purpose of classification is to group the SOAP messages according to invoked operations in the web service requests, i.e., classes are assigned according to the operations provided by the WS Honeypot.

The classification of SOAP messages can lead to several classes; however SVM is basically a binary classifier. To resolve this problem, we use an approach based on "one-versus-one" method for adopting SVM to classification problems with multi-class (three or more classes). The multi-class classification refers to assigning each of the SOAP messages into one of the k classes provided by the Web service.

We denote by x_i, a SOAP message to classify. Each instance x_i contains the characteristic parameters of this message ($x_i = x_{i1}, x_{i2}, \ldots, x_{in}$). Besides, we denote by C, $C \in \{1,2,..k\}$ the k classes representing the k-operations provided by the Web service. The goal of SVM classifier is to achieve the following two tasks:

1. Assigning each instance x_i to one possible class C, $C \in \{1,2,..k\}$. According to the input parameters and the type of the sent request, SVM classifier tries to find to which class the instance x_i belongs and the distance of x_i from the separating hyper-plane.

2. Knowing the distance of x_i from the separating hyper-plane, and the class to which it belongs, the second task of SVM classifier is to calculate the probability p_i with which x_i belongs to class j.

$$p_i = P(C = j|x_i), j \in \{1,2,..k\}$$

The probability p_i of the instance x_i, allows determining the nature of the captured SOAP message. If the probability is high, then the message is considered as normal and legitimate. However, if the probability is below a certain threshold, then it is considered as an anomaly and has to be sent to the human analyst to check if it represents an attack.

SVM Regression classifier

SVM Regression classifier is useful for the second category of extracted features "Resource consumption". These features are related to the resources consumption during the SOAP request processing such as CPU usage, memory usage and response time. The analysis of these variables is very useful to detect several DoS (denial of service) attacks.

Regression analysis is applied when we have more than one predictor variable to be analyzed with the response variable at the same time [12]. For example, in our case, we want to examine that certain characteristics of SOAP message processing, such as, response time, CPU usage and memory usage are in accordance with the size of the SOAP message. To do this, we used an SVM implementation for regression based on polynomial kernel [8].

Given a training dataset $D = \{(x_i, y_i)|i = 1,2, ..., n\}, x \in \mathbb{R}^n, y \in \mathbb{R}$. The x_i instance represents the processing characteristics of a SOAP request such as CPU usage, memory usage and response time. The y_i parameter contains the size of the SOAP request. The goal of SVM Regression classifier is to achieve the following two tasks:

1. Find a function $f(x)$ which approximates the relation between the dataset points and it can be used afterwards to infer the predicted output for a new input instance x. The function $f(x)$ must have at most ε deviation from the actually obtained targets y for all the training data. This problem can be solved using a loss function $\mathcal{L}(y, f(x))$, which describes how the estimated function deviates from the true one. For our case, we use a so called ε–insensitive loss function [8], described by:

$$\mathcal{L}(y, f(x)) = \begin{cases} 0 & \text{if } |y - f(x)| \leq \varepsilon \\ |y - f(x)| - \varepsilon & \text{otherwise.} \end{cases}$$

2. For each new instance x, determine the predictive value $f(x)$ which represents the predictive size of the corresponding SOAP request. Then calculate the following score:

$$error = |f(x) - y|$$

Where y is the actual size of the processed SOAP request.

The classification process is done as follows: If the error value is low, then the SOAP request is considered normal. However, if the error value is upper a certain threshold, then it is considered as an anomaly and has to be sent to the human analyst to check if it matches a DoS attack.

Apriori Classifier

The Apriori classifier examines the third category of extracted features "Operations list". This classifier is based on association rules to detect if the same user has conducted an abnormal sequence of activities. For example, if a large amount of repeated messages are sent from a specific source address over a very short time period, most likely it is a flooding attack. Furthermore, a bad scenario is detected if a user accesses the Web service without being successfully authenticated during the login step, although the latter is necessary. Another suspect situation is discovered when a client fails to login several times using the same identifier.

The first step of this classifier is to build, using the training data, a profile describing the sequence of activities performed by a normal user. We note by S_i the sequence of activities performed by a user i on the Web service. $S_i = \{item_1, item_2, \ldots, item_n\}$, where $item_k$ represents the kth operation performed by the user i. Consider a dataset D of a set of sequences S_i. The Apriori classifier generates a set F of all frequent items taking into account the two parameters of association rules "minimum support" and "minimum confidence".

$$F = \{S_i \subseteq D \mid support(S_i) \geq minSupport\}$$

In other words, F contains a set of frequent sequences of activities that a user can achieve in the Web service. The next step of the Apriori classifier is to determine for each new user if he has conducted a normal sequence of activities or not, by checking if his sequence of activities is included in F or not.

4 Experimental Tests

The purpose of a honeypot is to attract attackers by offering vulnerable applications that contain interesting information. For this reason, we prepared our WS Honeypot to simulate a Web service for online shopping. The simulated Web service contains eight operations as mentioned in Figure 2.

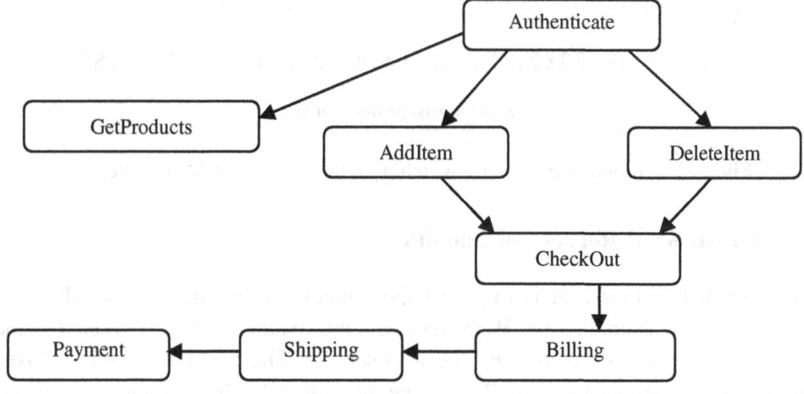

Fig. 2. An example of legitimate transitions of Web service operations in the WS Honeypot

To test the efficacy of our approach, we prepared three experiments. Each experience is intended for one of the three classifiers (SVM, SVM Regression and Apriori). However, before starting the experiments, firstly, the WS Honeypot must undergo a training phase to build a learning dataset which describes the normal operation of the simulated Web service.

4.1 1st Test: SVM Classifier

In this test, we are interested in the "Authenticate" operation of the Web service described in Figure 2. The purpose of this operation is to authenticate users trying to access the Web service by issuing a username and a password. Some attackers attempt to exploit this operation in order to obtain illegal access. To achieve their goal, attackers rely on an unexpected input leading to SQL injection or XML injection. During our experiment, we sent to the WS Honeypot 85 SOAP messages requesting the "Authenticate" operation, among them, there were 12 malicious requests.

After the classification phase, the SVM classifier assigns to each SOAP message (x_i) a probability p_i which calculates the probability of its membership to the class "Authenticate". If the value of p_i is below a threshold equal to 0.25, then the SOAP message is abnormal and it is probably an attack. Our experiment shows that all of the 12 malicious messages have a probability less than 0.25 (see Figure 3).

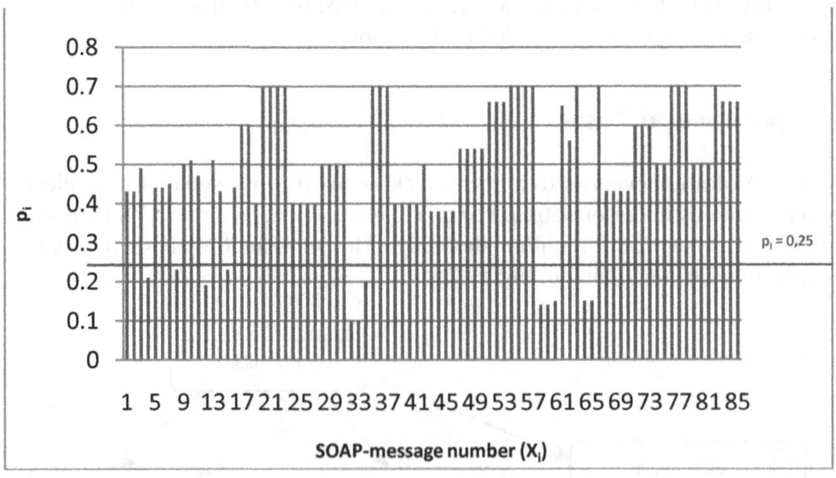

Fig. 3. The probability p_i with which x_i belongs to class "Authenticate"

4.2 2nd Test: SVM Regression Classifier

The purpose of this classifier is to detect DoS attacks by identifying SOAP requests that are trying to damage the Web service and exhaust the resources (memory, processor, etc.) of the server hosting the web service. The SVM Regression classifier determines the nature of the SOAP message according to the calculated error value. When the characteristic values of the SOAP-requests processing (such as CPU usage, memory usage and response time) become very large compared to the actual values, the error value becomes also greater.

In our test, we sent 50 SOAP messages to the WS Honeypot, among them, there were three malicious messages. The purpose of these three messages is to perform an Oversize Payload attack against the Web service. Since that SOAP message is an XML-based document, the principle key of this attack is to send a very large SOAP request, containing a long list of elements, in order to exhaust the server resources during the XML document parsing. After the classification process, the values of calculated errors associated to the three malicious messages were very large compared to the rest of messages (see Figure 4).

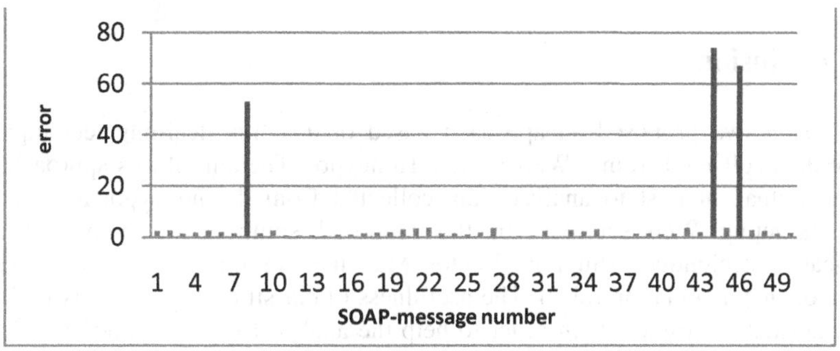

Fig. 4. Calculated error for each SOAP-message processing

4.3 3rd Test: Apriori Classifier

The role of Apriori Classifier is to identify users who have conducted an abnormal sequence of activities by using association rules technique. This classifier begins, firstly, by constructing a set F which contains a list of normal sequences of activities that a user can conduct in the Web service. Afterwards, for each new user, it monitors all activities achieved and arranges them in a sequence S called sequence of activities. If S is included in the list F, then it is considered as a normal sequence. Otherwise the user will be considered as suspect.

In Figure 5, we present some normal sequences of activities found in F. In our test, we queried the WS Honeypot in order to generate two suspicious situations. In the first situation, we tried several times to authenticate using fake identifiers. In the second one, we stole the session of another user by using "session hijacking" technique to access the Web service without authenticate. The two sequences related to these operations are:

- S1 = Authenticate → Authenticate → Authenticate → Authenticate → Authenticate
- S2 = getProduct → AddItem → CheckOut → Billing → Shipping → Payment

As the Apriori classifier didn't find these two sequences in the set F, so it declared them as suspicious.

Authenticate → getProduct → AddItem → CheckOut → Billing → Shipping → Payment
Authenticate → getProduct → getProduct
Authenticate → getProduct → AddItem → DeleteItem → AddItem →CheckOut → Billing → Shipping → Payment
Authenticate → CheckOut → Billing → Shipping → Payment
Authenticate → getProduct → AddItem → DeleteItem → AddItem
Authenticate → AddItem → CheckOut → Billing → Shipping → Payment
Authenticate → getProduct → AddItem → AddItem
…..

Fig. 5. Some normal sequences of activities achieved in WS Honeypot

5 Conclusion

In this paper, we proposed an approach based on machine learning techniques to analyze data collected from a Web Service Honeypot. The aim of this approach is to help the human analyst to analyze data collected from the honeypot and identify attacks targeting Web services. In the proposed solution, we employed three classification techniques (Support Vector Machine, an implementation SVM for Regression and association rules). The usefulness of classifiers is to classify collected data as normal or suspicious in order to help the analyst to identify and characterize attacks on the honeypot.

Experimental tests presented in this paper have shown the effectiveness of our approach to detect several types of Web service attacks. However, to better evaluate and validate our approach, the WS Honeypot should be interacting with real attackers. Therefore, we prepare its deployment on the Internet in order to study and characterize current attacks. As future work, we envisage adding a clustering-based technique in order to regroup detected attacks and determine their common proprieties.

References

1. Agrawal, R., Srikant, R.: Fast algorithms for mining association rules in large databases. In: 20th International Conference on Very Large Data Bases, VLDB, Santiago, Chile, pp. 487–499 (1994)
2. Alata, E., Dacier, M., Deswarte, Y., Kaâniche, M., Kortchinsky, K., Nicomette, V., Pham, V.H., Pouget, F.: Collection and analysis of attack data based on honeypots deployed on the Internet. In: First Workshop on Quality of protection, Security Measurements and Metrics, Milan, Italy (2005)
3. Ghourabi, A., Abbes, T., Bouhoula, A.: Experimental analysis of attacks against web services and countermeasures. In: 12th International Conference on Information Integration and Web based Applications & Services (iiWAS 2010), Paris, France (2010)
4. Ghourabi, A., Abbes, T., Bouhoula, A.: Design and implementation of web service honeypot. In: 19th International Conference on Software, Telecommunications and Computer Networks, Split, Croatia (2011)

5. Herrero, Á., Zurutuza, U., Corchado, E.: A Neural-Visualization IDS for Honeynet Data. Int. J. Neural Syst. 22(2) (2012)
6. Pouget, F., Dacier, M.: Honeypot-based Forensics. In: AusCERT Asia Pacific Information Technology Security Conference (AusCERT 2004), Brisbane, Australia (2004)
7. Seifert, C., Komisarczuk, P., Welch, I.: Identification of malicious web pages with static heuristics. In: Austalasian Telecommunication Networks and Applications Conference, Adelaide (2008)
8. Smola, A.J., Schölkopf, B.: A tutorial on support vector regression. Statistics and Computing 14(3), 199–222 (2004)
9. Spitzner, L.: Definitions and value of honeypots (2003), http://www.tracking-hackers.com/papers/honeypots.html
10. Thonnard, O., Dacier, M.: A framework for attack patterns discovery in honeynet data. Digital Investigation 8, S128–S139(2008)
11. Vapnik, V.N.: The nature of statistical learning theory. Springer-Verlag New York, Inc., New York (1995)
12. Wang, Y.: Statistical techniques for network security: modern statistically based intrusion detection and protection. IGI Global (2009)

A Security Pattern-Driven Approach toward the Automation of Risk Treatment in Business Processes

Angel Jesus Varela-Vaca[1], Robert Warschofsky[2], Rafael M. Gasca[1], Sergio Pozo[1], and Christoph Meinel[2]

[1] Computer Languages and Systems Department, Quivir Research Group
ETS. Ingeniería Informática, Avd. Reina Mercedes S/N, University of Seville, Seville, Spain
{ajvarela,gasca,sergiopozo}@us.es
[2] Hasso-Plattner-Institute
Prof.-Dr.-Helmert Str. 2-3
14482 Potsdam, Germany
{robert.warschofsky,meinel}@hpi.uni-potsdam.de

Abstract. Risk management has become an essential mechanism for business and security analysts, since it enable the identification, evaluation and treatment of any threats, vulnerabilities, and risks to which organizations maybe be exposed. In this paper, we discuss the need to provide a standard representation of security countermeasures in order to automate the selection of countermeasures for business processes. The main contribution lies in the specification of security pattern as standard representation for countermeasures. Classical security pattern structure is extended to incorporate new features that enable the automatic selection of security patterns. Furthermore, a prototype has been developed which support the specification of security patterns in a graphical way.

Keywords: Business Process Management, Security, Pattern, Risk Treatment, Automation.

1 Introduction

New technologies have emerged into the business process scene providing applications to automate the generation of IT products based on business processes management. Most of the recognized business process products, such as Intalio BPM suite, Bonita Soft, BizAgi, and AuraPortal BPMS, are capable to generate automatically entire applications from definitions of business process diagrams and interactions with web services. These applications do not pay attention on security risks and less in the treatment of these. In general, security treatments are applied in a second thought. It is desirable to provide the business analyst with tools that enable the risk assessment of business process designs and also

Á. Herrero et al. (Eds.): Int. JointConf. CISIS'12-ICEUTE'12-SOCO'12, AISC 189, pp. 13–23.
springerlink.com

the specification of necessary security countermeasures to apply. Nevertheless, in general the selection and configuration of security countermeasures is a human, manual, complex, time consuming and error-prone task that involves many security stakeholders p(managers and administrator). Security countermeasures can vary from technical controls to management controls; such as list of procedures, backup policies, and the specification of access-control policies). In order to understand the complexity of security countermeasures an example could be network Intrusion Detection Systems (IDS) where two different approaches [5] [9] apply machine learning methods for intrusion detection. The question is which specific approach is adequate for the requirements of the organization.

The main problems regarding to countermeasures are: (1) how to describe the countermeasures in business processes; (2) countermeasures are very heterogeneous; (3) countermeasures are described in natural language or informal way (4) in general the selection of countermeasures is carried out in manual way without criteria. A derived problem emerges from the complexity in the selection of a set of countermeasures that comply with several organizational constraints. This complexity increase when there exist multiple objectives to achieve such as reduce the return of investment (ROI) until certain value, reduce risks until ten percent, among other.

In this work, we propose a formalization based on security patterns templates in order to standardize the representation of security countermeasures in business processes. Furthermore, we propose an extension of these patterns with new features that enables to include organizational metrics and constraints for the automatic selection of countermeasures. In order to overcome the complexity of selection we propose to apply automatic algorithms based on artificial intelligence techniques. To support the proposals, OPBUS framework [14] has been improved by including a risk treatment stage which support the agile specification and automatic selection of security patterns as countermeasures.

The rest of the paper is organized as follows: in Section 2 security patterns, and the need to model and extend security patterns is detailed; in Section 3, an application example scenario is presented; in Section 4 the prototype developed is described; Section 5 gives an overview about related work in the domain of security patterns and model-driven security; the last section concludes this paper and outlines the future work.

2 Specifying Security Patterns

Security patterns [12] are widely recognized way for the description of security solutions. Security patterns are based on the idea of design patterns that has been introduced by Christopher Alexander in 1977: *"A pattern describes a problem which occurs over and over again our environment, and then describes the core of the solution to that pattern"*.

In general, security patterns [11] are defined in an informal way, usually using the natural language. Patterns are described in documents that have a specific structure: Name, Context, Forces, Problem, Solution. Other sections; such as Implementation details, Structure, Examples, can be incorporated to the security patterns in order to improve its information.

There exist relevant approaches where security patterns has been formalized such as [11] [12] [7]. The formalization in [12] is focused on the definition of ontologies to map security concepts within security patterns in order to enable search engines and query capabilities. In [7], security patterns have been formalized as profiles to automate the generation of security policies for SOA environments. In [11], a catalogue of security patterns is defined using natural language, and UML diagrams. Nevertheless, these formalizations are unsuitable to automate the application of security patterns.

In our approach, the security patterns are utilized for the selection of treatments for certain risks. Hence, security patterns have to be equipped with information that enable the evaluation how much adequate is a pattern among others. Furthermore, in the selection security patterns should comply with certain business objectives (cost, acceptable risk levels) pre-stated. We propose the modelling of security pattern as shown in Table 1. The table presents, regarding to classic security patterns, four new sections: Security goal, Security intention, Risk type and Attributes. This extension is based on the UML Profile for Modeling QoS and Fault Tolerance (hereinafter UML profile) [1]. In the next sections, security pattern structured is detailed.

Table 1. Template of extended security patterns

<table>
<tr><td rowspan="2">Name</td><td colspan="2">Description</td><td colspan="2">is a label that identifies and summarize of the pattern</td></tr>
<tr><td colspan="2">UML QoS & FT Profile</td><td colspan="2">::QoSValue</td></tr>
<tr><td rowspan="2">Security Intentions</td><td colspan="2">Description</td><td colspan="2">is a label to describe security intentions to cover</td></tr>
<tr><td colspan="2">UML QoS & FT Profile</td><td colspan="2">::QoSCharateristic</td></tr>
<tr><td rowspan="2">Security Goals</td><td colspan="2">Description</td><td colspan="2">list of indicates the security goals to fullfil</td></tr>
<tr><td colspan="2">UML QoS & FT Profile</td><td colspan="2">::QoSCharateristic</td></tr>
<tr><td rowspan="3">Context</td><td rowspan="2">Risk Type</td><td colspan="2">Description</td><td>indicates el type of risk of the pattern</td></tr>
<tr><td colspan="2">UML QoS & FT Profile</td><td>::RiskAssessment ::Treatment ::TreatmentOption</td></tr>
<tr><td>Attributes</td><td colspan="2">Description</td><td>describes the attributes concerning to the context related to risk</td></tr>
<tr><td rowspan="2">Forces</td><td colspan="2">Description</td><td colspan="2">describe the constraint that exist in the business process, they affect the problem</td></tr>
<tr><td colspan="2">UML QoS & FT Profile</td><td colspan="2">::QoSConstraint</td></tr>
<tr><td rowspan="2">Problem</td><td colspan="2">Description</td><td colspan="2">indicates the problem that occurs within the context</td></tr>
<tr><td colspan="2">UML QoS & FT Profile</td><td colspan="2">::QoSCharateristic</td></tr>
<tr><td rowspan="2">Solution</td><td colspan="2">Description</td><td colspan="2">indicates the solution for the problem within the context</td></tr>
<tr><td colspan="2">UML QoS & FT Profile</td><td colspan="2">::QoSCharateristic</td></tr>
</table>

2.1 Security Goals and Intentions in Security Patterns

Security patterns as countermeasure specify well-known solutions to common security problems, hence there should exist a direct relation between countermeasures and the security goals the pattern is enforcing. In fact, security pattern templates as shown in [12] this relation is fuzzy. In other approaches, such as the approaches in [7] [6], the authors define a relation with security intentions attaching security intentions in the problem section. Although, a security intention could enforce various security goals this relation is also fuzzy. For instance, in [7], the authors define the security patterns: 'Secure Pipe' and 'Information Protection' as intention. However it is not clear which security goals the pattern is enforcing. We propose the enhancement of security pattern templates with security goal and intention concepts in order to provide with new criteria for the selection of security patterns.

For a better understanding of the problem and due to the heterogeneity of all concepts, we have formalized all concepts by means of an ontology as shown in Figure 1. The ontology includes various examples to illustrate each concept. For instance, we can observe how risk concepts in the context are related to threats and vulnerabilities. Threats and vulnerabilities represent problems from the point of view of a security pattern. There exist various databases, such as NIST Vulnerability Database [3], and Common Weakness Enumeration (CWE) [2], that provide information referring to technological vulnerabilities. Identifiers utilized by CWE or NIST can be adopted for the specification of security patterns in our approach. For instance, Figure 1 shows a SQL injection attached the identifier CWE-89. This is used by CWE dictionary to identify a specific SQL-Injection vulnerability. In addition, CWE provides a particular section (Common Consequences) to indicate which security goal is affected. In the same way, other databases such as CRAMM database can be adopted by means of defining the alignment of its concepts.

2.2 Risk Type and Attributes for Selection

Following the UML Profile, there exist four categories of risk treatments: avoid, reduce, transfer and retain. Security patterns as countermeasures should be categorized inside of one of the four categories. This classification could be useful to rule out countermeasures for in-suitable risks and to do more efficient searches. We therefore propose to include a risk type property in order to support the classification of the pattern. For instance, an organization has decided to transfer one risk making outsourcing, if there exist a database with one hundred security patterns splitted in four different categories, it would be desirable that rule out several patterns for its category and focus on the patterns which category is transfer.

On the other hand, metrics are necessary in order to evaluate which countermeasure is better than others. We propose to incorporate a new section within the context to gather metrics that enable the measurement of the security pattern. In general, these attributes are related to metrics that the organization

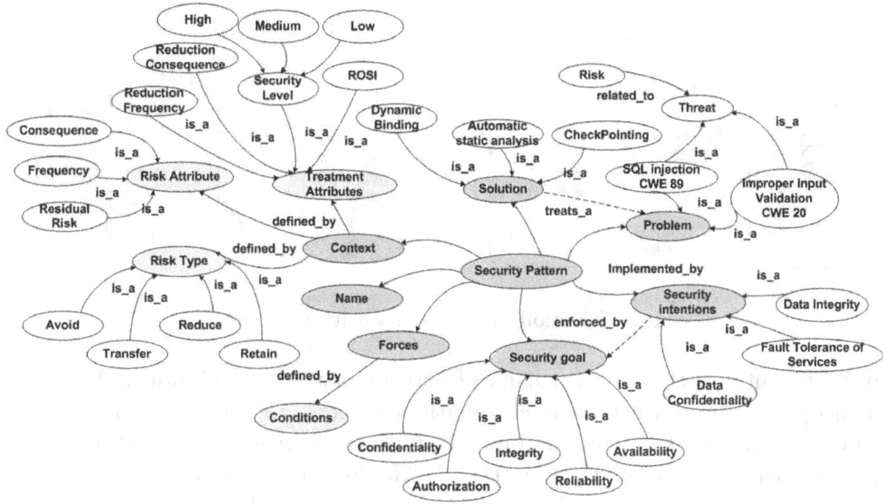

Fig. 1. Ontology for problem, solution, security goals, and security intentions

need to carry out an evaluation of countermeasures in monetary terms; such as annual cost, cost of implementation, annual loss expectancy (ALE). Otherwise, from the point of view of the risk management it would be useful to introduce metrics related to the level of risk allowed, and security level (High, Medium and Low), risk reduction, impact, priority of application of the pattern, etc. In Figure 1 various examples of risk attributes are given. For instance, an organization has detected a vulnerability related to SQL injections since the information introduced in a web form is not validated. In most of the cases, SQL injections are prevented by using information input validation. Other mechanisms such as data analysis methods can exist as well. Different mechanisms are adequate however which one is the best for the organization.

2.3 Forces and Selection of Security Patterns

As mentioned earlier, forces section indicates conditions in the context for the application of the solution. These conditions reflects restrictions of application relative to the context, thus to the metrics previously defined in the context. An example of constraint could be for instance $NeutralizationSQLInjection.Cost \leq TotalCost$ that indicates the cost of the implementation of 'Neutralization of SQL Injection' must be less than the total cost. These concepts (Cost and TotalCost) has previously been defined in the context. Another example could be related the level of security achieved in the application of this pattern, *SecurityLevel=Medium*.

The selection aim consist of finding a list of security patterns (Figure 2) that comply with the objectives specified in the business process, the constraint of application indicated in the forces of security patterns, and treat risks in the activities that need to be treated. This selection could be carried out in manual

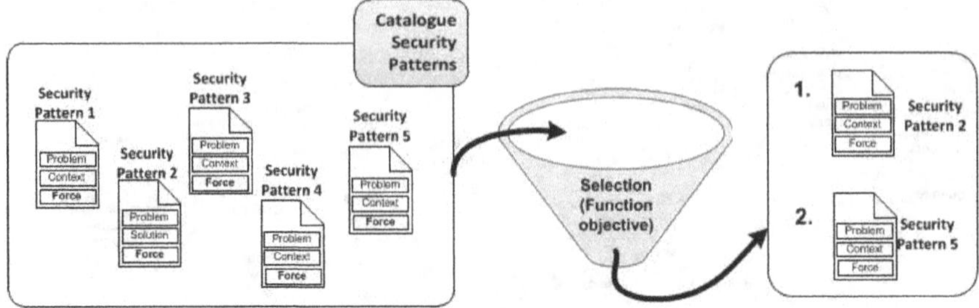

Fig. 2. Selection process of security patterns

way or in automatic way by means of certain reasoning technique. As proof of concept, an approach for the automatic selection of countermeasures could be an algorithm that receives as input: a business process, and a catalogue of countermeasures (set of security patterns). The business process composed by a set of activities (that need to be treated), risks associated to activities and the business process, a set of threats, and objectives of the business process. Firstly, the catalogue is reduced by means of security goal and risk type stated in the business process. After that, the algorithm strives for the selection of security patterns that treat risks of activities, comply with cost of the business process, and satisfy the constraints in the pattern. Finally, the algorithm returns a set of security patterns associated to activities of the business process.

2.4 Problem and Solution Items in Security Patterns

Security patterns already include a section to describe problems within the context. In [12], the authors use the problem section to describe the problem by natural language. However, in [7], the problem section is utilized to specify the security intention as a objective to achieve and not to describe problem. For our propose, the problem section is related to a specific threat or vulnerability.

In our particular use of security pattern as countermeasures, solutions are related to solve security risks concerning threats or vulnerabilities. That is, the solution field of security pattern template has been used to gather information about particular solutions to certain kind of vulnerabilities. For example, a solution such as 'Automatic static analysis' treat the vulnerability of 'CWE-20 - Improper Input Validation'. This solution has been obtained from the CWE database. CWE state three different types of detection methods: automated static analysis, manual static analysis and fuzzy. These methods could be used as identifiers to specify solutions in the security patterns at this level. The identifiers used in this field are general and do not describe explicit mechanism. In the same way, we can define our own identifiers according to the solutions that are not picked up in the CWE database. For instance, Figure 1 shows a solution named 'Checkpointing' concern to the 'Fault Tolerance Services' intention.

Although, this solution is not in CWE database. However, it can be applied in business processes to enforce the reliability even the integrity as demonstrated in [13]. For a better understanding, the description of security patterns and the automatic selection of these are illustrated in an example in section below.

3 Illustrative Example

An example scenario where a hosting organization would like to provide a service to publish entries in a blog system. The organization has decided to automate this using business process management systems and developed a business process model which contains three basic activities: (1) Login, to enter username and password information to authenticate customers; (2) Request, to request information (title and content) about a new entry for blogs; (3) Publish: the entry information is registered in the database and published in a blog system.

The organization wishes to increase the security where the priority is customers integrity and confidentiality. In manual way, a security analysis of the system can detect possible vulnerabilities referring to: (1) Passwords integrity and confidentiality due to use an insecure channel in public networks; and (2) Data integrity and confidentiality due to an insecure channel in public networks.

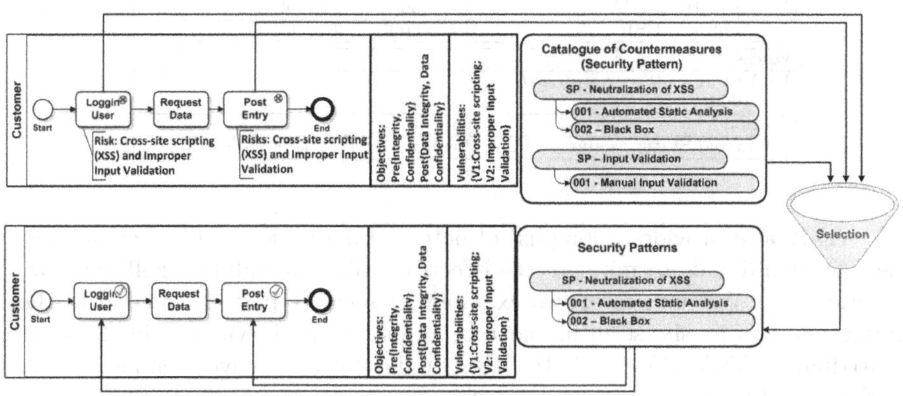

Fig. 3. Example of automatic risk treatment of a business process

The organization wish to automate the customer security by including countermeasures in the business processes. In order to ensure a security-aware development of business process a risk assessment is done through OPBUS [14]. After diagnosis, OPBUS indicates which vulnerabilities and activities have to be treated as shown by red marks in Figure 3. In that case vulnerabilities 'V1' concerns to 'Cross-site scripting (XSS)' vulnerability and 'V2' concerns to 'Improper input validation' produces a risk in activities 'Loggin User' and 'Post Entry' that exceed the risk levels allowed. Hence, the exploiting of vulnerabilities or the materialization of threats can produce effects that cannot be assumed and they has to be treat. After a first filter, OPBUS provide two types of security patterns to treat these vulnerabilities as shown in Figure 3. These countermeasures are defined based on CWE database which indicates some detection methods for this kind of vulnerabilities. Detection methods are mechanisms to prevent the threat of vulnerabilities. In this particular case countermeasures are stated as follows: (1) CWE-79 solutions: Automated static analysis based on data flow analysis, and Black box based on automated test; (2) CWE-20 solution: Manual input validation.

Table 2. Neutralize XSS based on data analysis

Name		
Neutralization of XSS		
Problem		
CWE-79: Improper Neutralization of Input During Web Page Generation		
Security Intention		
Data integrity		
Security goal		
Integrity,Confidentiality		
Context		
Risk Type		
Reduce		
Attributes		
Risk Reduction Freqpuency (RRF)	[30,60]%	[50,70]%
Risk Reduction Consequence (RRC)	[10,30]%	[10,30]%
Annual Number Attacks (ANA)	30	30
Cost per Attack (CA)	[60-100]	[60-100]
Cost Solution (CS)	[500-600]	[900-1000]
Forces		
$ANA * CA + CS < TotalCost$ $CS < 600$		
Solution		
Automated Static Analysis - Data Flow Analysis	Black Box - Testing	

CWE-79 also provides a long list of potential mitigation even specifying implementation details to mitigate the effects of this vulnerability. Following descriptions of CWE database, two examples of security patterns related to the mitigation of cross-site scripting have been defined as shown in Table 2. A set of attributes (ANA, CA, CS, RRF, RRC) and constraints have been included in attribute and forces sections respectively.

The main difference between both countermeasures are the values in the attributes. Attributes included within context can be stated by a single value and intervals when there is no certainty of the value. The main question is which countermeasure is better for the organization. After selection, OPBUS has proposed 'Neutralization of XSS' as solution to treat the problems identified in the activities. If we observe carefully the main differences between both countermeasures are the cost of implementation (CS) and the risk reduction in terms of frequency (RRF). For instance, 'RRF' in the first countermeasure has a interval of [30,60]%, and the second one has a interval of [50,70]%. Taking into consideration the constraints included in forces section: $CS < 600$ and

$ANA * CA + CS < TotalCost$. It is not clear which countermeasure is better in plain sight a comparative process is necessary. However, in other cases with a large number of vulnerabilities the list of countermeasure could multiply and to select the adequate countermeasures become a very complex task. The complexity in the selection of adequate countermeasure makes it mandatory to include mechanisms that enable the automatic selection treatment.

After applying the security patterns the model can be re-diagnosed obtaining residual risks. The results of these residual risks are lower than the previous. Figure 3 shows the results of the diagnosis after including security patterns as countermeasures.

4 OPBUS Prototype

Security patterns has been integrated as part of the OPBUS model. The model is composed of three main sub-models: risk model, business process model, and security policy model. Security patterns have been modelled as part of the security policy model, and according to the structure listed on Table 1. In general, security countermeasures are gathered by means of security policies. The risk model provide an extension of business process models, and security policy model is used to define policies by means of security patterns as specific countermeasures. Security patterns can be defined as security constraints that enforce certain security goals within a security policies. As we can observe in Figure 4 risk model includes 'Treatment' and 'Countermeasure' by means of security patterns to enable the specification of risk treatments into threat scenarios.

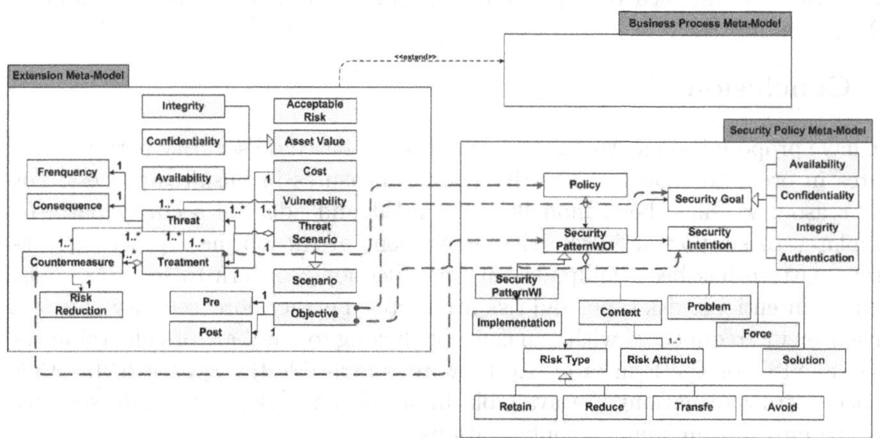

Fig. 4. Business process extensions

A prototype has been developed as part of the OPBUS tools [4]. The prototype have been developed as Eclipse plug-in which integrates a business process modeller that enable the automatic risk assessment of the model. The main novelty is to support the specification of security patterns. A specific properties tab has been enable in order to set up the security pattern attributes. Security patterns can also be linked to threats as countermeasures related to certain vulnerabilities.

5 Related Work

Various approaches have emerged in the context of business process management in order to bridge the gap between security and business domains [15] [8] [7]p [10]. Most of these initiatives are focused on providing new domain-specific languages (DSLs) for the annotation of assets, security requirements, and threats into the business process models. Nevertheless, most of the studies avoid the introduction of risk treatments.

In [8], Menzel et al. propose an approach to adequate the generation of security requirement of authorization and trust in accordance with specifics risk levels. In the same way, in [15], Wolter et al. provides security annotations for graphical business processes that enable the set up of authorization directly over the model. However, these approaches are only focused on the definition of authorization and trust in service-based environments.

Concerning to security pattern, in [7], Menzel provide a new application for security patterns to automate the generation of security policies in SOA environments. Menzel provides different security patterns in order to fulfill certain security intentions such as User Authentication, Identity Provisioning, Data Confidentiality and Data Authenticity. Furthermore, security patterns has been formalized by means of ontologies. The main problem in the approach is the coupling to the solution. That is, security model, formalization and security patterns have been defined to support the specification of solutions only valid to SOA environments.

6 Conclusion

We have proposed to incorporate a formalization based on security pattern templates in order to represent security countermeasures in business process. For this reason, we have been modelled, extended and adapted security patterns templates to support it in the risk assessment of the business processes. Security pattern has been equipped with new sections concern to security goals, security intentions, risk type, and risk attributes. Furthermore, security patterns have been interconnected with real concepts belong to database of vulnerabilities such as NIST or CWE in order to do more compatible the approach. In other respects, transformations are available in the framework to translate security pattern into certain security configurations.

To the best of our knowledge, this is the first proposal that define a specific format to define countermeasures to automate the risk treatment in business processes. To future work, we propose to define algorithms to automate the selection of optimal countermeasures following security pattern templates proposed in this paper. Furthermore, we propose to extend the prototype including these algorithms for the automatic selection of countermeasure.

Acknowledgements. This work has been partially funded by the Department of Innovation, Science and Enterprise of the Regional Government of Andalusia project under grant P08-TIC-04095, by the Spanish Ministry of Science and Education project under grant TIN2009-13714, and by FEDER (under the ERDF Program). The authors would like to thank Dr. Michael Menzel and Hasso-Plattner-Institute for the valuable information that has contributed to develop the ideas in this paper.

References

1. UML Profile for Modeling QoS and Fault Tolerance Characteristics and Mechanisms (2009), `http://www.omg.org/spec/QFTP/1.1`
2. Common Weakness Enumeration (2011), `http://cwe.mitre.org/index.html`
3. NIST National Vulnerability Database (2011), `http://nvd.nist.gov/`
4. OPBUS tools (2012), `http://estigia.lsi.us.es/angel/OPBUS/`
5. Corchado, E., Herrero, L.: Neural visualization of network traffic data for intrusion detection. Applied Soft Computing 11(2), 2042–2056 (2011)
6. Menzel, M., Warschofsky, R., Meinel, C.: A pattern-driven generation of security policies for service-oriented architectures. In: 2010 IEEE International Conference on Web Services (ICWS), pp. 243–250 (July 2010)
7. Menzel, M.: Model-driven Security in Service-oriented Architectures. Ph.D. thesis. Hasso-Plattner - University of Potsdam (2010)
8. Menzel, M., Thomas, I., Meinel, C.: Security requirements specification in service-oriented business process management. In: International Conference on Availability, Reliability and Security (ARES), pp. 41–48. IEEE Computer Society (2009)
9. Mrutyunjaya, Abraham, A., Das, S., Patra, M.R.: Intelligent Decision Technologies 5(4), 347–356 (2011)
10. Rosemann, M., zur Muehlen, M.: Integrating risks in business process models. In: 16th Australasian Conference on Information Systems (ACIS 2005), Paper 50, pp. 1–10 (2005)
11. Schumacher, M. (ed.): Security Engineering with Patterns - Origins, Theoretical Models, and New Applications. LNCS, vol. 2754. Springer, Heidelberg (2003)
12. Schumacher, M., Fernandez-Buglioni, E., Hybertson, D., Buschmann, F., Sommerlad, P.: Security Patterns: Integrating Security and Systems Engineering. John Wiley and Sons, Ltd (2006)
13. Varela-Vaca, A.J., Gasca, R.M.: OPBUS: Fault Tolerance against integrity attacks in business processes. In: 3rd International Conference on Computational Intelligence in Security for Information Systems, CISIS 2010 (2010)
14. Varela-Vaca, A., Gasca, R., Jimenez-Ramirez, A.: A model-driven engineering approach with diagnosis of non-conformance of security objectives in business process models. In: 2011 Fifth International Conference on Research Challenges in Information Science (RCIS), pp. 1–6 (May 2011)
15. Wolter, C., Menzel, M., Schaad, A., Miseldine, P., Meinel, C.: Model-driven business process security requirement specification. Journal of Systems Architecture - Embedded Systems Design 55(4), 211–223 (2009)

Neural Network Ensembles Design with Self-Configuring Genetic Programming Algorithm for Solving Computer Security Problems

Eugene Semenkin, Maria Semenkina, and Ilia Panfilov

Department of System Analysis and Operation Research, Siberian State Aerospace University, Krasnoyarsky rabochy avenue, 31, 660014, Krasnoyarsk, Russia
eugenesemenkin@yandex.ru, {semenkina88,crook_80}@mail.ru

Abstract. Artificial neural networks based ensembles are used for solving the computer security problems. Ensemble members and the ensembling method are generated automatically with the self-configuring genetic programming algorithm that does not need preliminary adjusting. Performance of the approach is demonstrated with test problems and then applied to two real world problems from the field of computer security – intrusion and spam detection. The proposed approach demonstrates results competitive to known techniques.

Keywords: self-configuring genetic programming, artificial neural networks, ensembles, intrusion detection, spam detection.

1 Introduction

Today, computers become more powerful and interconnected that makes their security one of the most important concerns. Attacks are more danger and diverse ranging from unsolicited email messages that can trick users to dangerous viruses that can erase data or damage computer systems. Conventional security software requires a lot of human effort to identify and work out threats. This human labor intensive process can be more efficient by applying machine learning algorithms [1]. There is much research related to the data mining techniques application for solving specific computer security problems, e.g. the intrusion detection [2-4, 5] or spam detection [6]. Artificial neural networks (ANN) are one of the most widely used data mining techniques here ([6, 7-10]).

The highly increasing computing power and technology made possible the use of more complex intelligent architectures, taking advantage of more than one intelligent system in a collaborative way. This is an effective combination of intelligent techniques that outperforms or competes to simple standard intelligent techniques.

One of the hybridization forms, the ensemble technique, has been applied in many real world problems. It has been observed that the diversity of members, making up a "committee", plays an important role in the ensemble approach [11]. Different techniques have been proposed for maintaining the diversity among members by running on the different feature sets [12] or training sets (e.g., bagging [13] and boosting [14]). Some techniques, such as neural networks, can be run on the same

Á. Herrero et al. (Eds.): Int. JointConf. CISIS'12-ICEUTE'12-SOCO'12, AISC 189, pp. 25–32.
springerlink.com

feature and training sets producing the diversity by different structures [15]. Simple averaging, weighted averaging, majority voting, and ranking are common methods usually applied to calculate the ensemble output.

Johansson et al. [16] used genetic programming (GP) [17] for building an ensemble from the predefined number of the ANNs where the functional set consisted of the averaging and multiplying and the terminal set included the models (i.e., ANNs) and constants. In [18], a similar approach was proposed where first a specified number of the neural networks is generated and then a GP algorithm is applied to build an ensemble making up the symbolic regression function from partial decisions of the specific members.

In this paper, we apply the self-configuring genetic programming technique to construct formula that shows how to compute an ensemble decision using the component ANN decisions. The algorithm involves different operations and math functions and uses the models providing the diversity among the ensemble members. Namely, we use neural networks, automatically designed with our GP algorithm, as the ensemble members. The algorithm automatically chooses component ANNs which are important for obtaining an efficient solution and doesn't use the others.

With the approach developed an end user has no necessity to be an expert in the computational intelligence area but can implement reliable and effective data mining tool. It makes the approach very useful for computer security experts making them free from extra efforts on the intellectual information technology (IIT) algorithmic core implementation and allowing them to concentrate their attention in the area of their expertise, i.e. the computer security as such.

The rest of the paper is organized as follows. Section 2 describes the method for the GP self-configuring and it's testing results confirming the method usefulness. Section 3 describes the method of the ANN automated design and its performance evaluation. In Section 4 we describe the GP-based approach to the ANN ensembles automated integration and the results of the performance comparative analysis. In Section 5 we apply developed approach to solving benchmark problems from the area of the computer security. In Conclusion section we discuss the results and directions of the further research.

2 Operator Rates Based Self-Configuration of GP Algorithm

Before suggesting the GP use to end users, e.g., computer security specialists, for application in data mining tools development, we have to save them from main troubles which are the problem even for evolutionary computation experts. It is really hard job to configure GP settings and tune its parameters, i.e., we have to suggest a way to avoid this problem.

We apply the operator probabilistic rates dynamic adaptation on the level of population with centralized control techniques [19, 20]. To avoid the issues of precision caused while using real parameters, we used setting variants, namely types of selection, crossover, population control and level of mutation (medium, low, high). Each of these has its own probability distribution, e.g., there are 5 settings of selection: fitness proportional, rank-based, and tournament-based with three tournament sizes. During initialization all probabilities are equal to 0.2 and they will be changed according to a special rule through the algorithm execution in such a way that the sum of probabilities should be equal to 1 and no probability could be less than predetermined minimum balance.

When the algorithm creates the next off-spring from the current population it first has to configure settings, i.e. to form the list of operators with using the probability operator distributions. The algorithm then selects parents with the chosen selection operator, produces an off-spring with the chosen crossover operator, mutates off-spring with the chosen mutation probability and puts off-spring into the intermediate population. When the intermediate population is filled, the fitness evaluation is computed and the operator rates (probabilities to be chosen) are updated according to operator productivities. Then the next parent population is formed. The algorithm stops after a given number of generations or if the termination criterion is met.

The productivity of an operator is the ratio of the average off-spring fitness obtained with this operator and the average fitness of the overall off-spring population. Winning operator increases its rate obtaining portions all other operators. We call our algorithm as self-configuring genetic programming (SelfCGP).

As a commonly accepted benchmark for GP algorithms is still an "open issue" [21], we used the symbolic regression problem with 17 test functions borrowed from [22] for preliminary evaluation. Experiments settings were 100 individuals, 300 generations and 100 algorithm runs for each test function. Having no place here for detailed description of experiments and results, we summarize it briefly [23]. We evaluated the algorithm reliability, i.e. proportion of 100 runs when the approximation with sufficient precision was found. The worse reliability (for the most hard problem) averaged over 100 runs was equal to 0.42. The best reliability was equal to 1.00. SelfCGP reliability averaged over 17 test function is better than averaged best reliability of conventional GP. The computation consumption is also better. It gives us a possibility to recommend SelfCGP for solving symbolic regression problems as better alternative to conventional GP. Main advantage of SelfCGP is no need of algorithmic details adjustment without any losses in the performance that makes this algorithm useful for many applications where terminal users being no experts in evolutionary modeling intend to apply the GP for solving these problems.

3 ANN Automated Design with Self-Configuring GP Algorithm

Usually, the GP algorithm works with tree representation, defined by functional and terminal sets, and exploit the specific solution transformation operators (selection, crossover, mutation, etc.) until termination condition will be met [17].

For the ANN automated design, the terminal set of our GP includes 16 activation functions such as bipolar sigmoid, unipolar sigmoid, Gaussian, threshold function, linear function, etc. The functional set includes specific operation for neuron placement and connections. The first operation is the placing a neuron or a group of neurons in one layer. There will no additional connections appeared in this case. The second operation is the placing a neuron or a group of neurons in sequential layers in such a way that the neuron (group of neurons) from the left branch of tree preceded by the neuron (group of neurons) from the right branch of tree. In this case, new connections will be added that connect the neurons from the left tree's branch with the neurons from the right tree's branch. Input neurons cannot receive any signal but have to send a signal to at least one hidden neuron.

The GP algorithm forms the tree from which the ANN structure is derived. The ANN training is executed to evaluate its fitness that depends on its performance in solving problem in hand, e.g., the approximation precision or the number of misclassified instances. For training this ANN, connection weights are optimized with self-configuring genetic algorithm (SelfCGA) [24] that similarly to SelfCGP does not need any end user efforts to be the problem adjusted doing it automatically. When GP finishes giving the best found ANN structure as the result, this ANN is additionally trained with again SelfCGA hybridized with local search.

We compared the performance of the ANNs designed with our SelfCGP algorithm with the alternative methods on the set of problems from [25]. Materials for the comparison we have taken from [26] where together with results of authors' algorithm (CROANN) the results of 15 other approaches are presented on three classification problems (Iris, Wisconsin Breast Cancer, Pima Indian Diabetes) from [25].

Analyzing comparison results, we observed that the performance of our approach is high enough comparing to alternative algorithms (1st, 3rd and 4th positions, correspondingly). However, the main benefit from our SelfCGP algorithm is the possibility to be used by the end user without expert knowledge in ANN modeling and evolutionary algorithm application. Additional dividend is the size of designed ANNs. The ANNs designed with SelfCGP contain few hidden neurons and connections and use not all given inputs although perform well. Now we can conclude that the self-configuring genetic programming algorithm is the suitable tool for ANN automated design. We may use it for the design of ANN ensembles.

4 Integration of ANN Ensembles with Self-CGP Algorithm

Having the developed appropriate tool for ANN automated design that doesn't require the effort for its adjustment, we applied our SelfCGP algorithm to construct formula that shows how to compute an ensemble decision using the component ANN decisions. The algorithm involves different operations and math functions and uses the ANN models providing the diversity among the ensemble members. In our numerical experiments, we used neural networks, automatically designed with our SelfCGP algorithm, as the ensemble members. The algorithm automatically chooses the component ANNs which are important for obtaining an efficient solution and doesn't use the others. The ensemble component ANNs are taken from the preliminary ANN pool that includes 20 ANNs generated beforehand with SelfCGP. For the designing every ANN, corresponding data set was randomly divided into two parts, i.e., the training sample (70%) and the validation sample (30%).

The first experiment was conducted for comparing the performance of the ensembling method based on the SelfCGP with the others. We used the same three problems from [25] and the real world problem of the simulation of the spacecraft solar arrays degradation (SAD) process [27] as the test bed. Results were averaged over 20 runs. In Table 1 below we present our results. The second row contains the performance of non-ensembling ANN model designed with the SelfCGP. Numbers in the first three columns are the error measure calculated as it was given in [26] and the numbers in the last column are relative deviation from the correct value.

Table 1. Ensembling methods comparison

Classifier	Iris (% of mistakes)	Cancer (% of mistakes)	Diabetes (% of mistakes)	SAD (relative deviation)
SelfCGP+ANN+Ensemble	**0**	**0**	**17.18**	**0.0418**
SelfCGP+ANN	0.0133	1.05	19.69	0.0543
ANN ensemble with weighted averaging	0.0267	1.03	19.03	0.0503
ANN ensemble with simple averaging	0.0267	1.09	19.75	0.0542

Results in Table 1 demonstrate that the SelfCGP based ensembling method used the ANNs integration outperforms conventional ensembling methods and the single best ANN designed with SelfCGP.

5 ANN Ensemble Design for Solving Computer Security Problems

Having proved the high workability of the SelfCGP designed ANN ensembles, we can verify our approach solving hard problems from the area of computer security.

The first problem is the detection of PROBE attacks. Corresponding dataset "KDD'99 Cup" is hosted in the Machine Learning Repository [25]. For the approach effectiveness evaluation, all patterns relevant to PROBE attacks were marked as referring to the first class, the others were marked as belonging to the second class. We used the following attributes in our experimental study: 1, 3, 5, 8, 33, 35, 37, 40. The choice of these attributes has been made empirically based on the analysis of related works and their description can be found in [28]. The results were compared with other approaches collected in [29]. The comparison results are shown in Table 2 below.

From Table 2 we can conclude that the classifier automatically designed with SelfCGP as the ANN-based ensemble demonstrates the high performance compared with the best known results (PSO-RF and RF). The classifier based on the single ANN designed with SelfCGP also demonstrates competitive performance.

The second problem is one of the e-mail spam detection. Corresponding data set was also taken from [25]. This data set includes 4600 instances of spam and non-spam e-mail messages each described with 57 attributes two of which are integers and the others are real. The classifier has to separate spam and non-spam messages.

Results for comparison were taken from [30] where authors compared their reinforcement learning controlled classifiers mixture (RL) with a single multilayer perceptron (MLP) and AdaBoost ensembling technique using MLPs (Boost). We have conducted 20 runs of our algorithm and averaged results. Comparison results are given in Table 3.

Table 2. Performance comparison for PROBE attack detectors

Applied technique	Detection rate, %	False positive rate, %
PSO-RF	**99.92**	**0.029**
SelfCGP+ANN+Ensemble	**99.79**	**0.027**
Random Forest	**99.80**	0.100
SelfCGP+ANN	98.78	0.097
Bagging	99.60	0.100
PART (C4.5)	99.60	0.100
NBTree	99.60	0.100
Jrip	99.50	0.100
Ensemble with majority voting	99.41	0.043
Ensemble with weighted averaging	99.17	0.078
Ensemble with simple averaging	99.18	0.122
BayesNet	98.50	1.000
SMO (SVM)	84.30	3.800
Logistic	84.30	3.400

Table 3. Performance comparison for spam detectors

Applied technique	Error (%)
SelfCGP+ANN+Ensemble	5.04
Ensemble with majority voting	5.23
Ensemble with weighted averaging	5.33
Ensemble with simple averaging	5.43
SelfCGP+ANN	5.43
Boost	6.48
RL	7.41
MLP	8.33

From Table 3 we can see that ANN-based ensemble automatically designed with SelfCGP outperforms all other techniques, second best is the majority voting ensemble based on ANNs automatically designed with SelfCGP. Moreover, the single SelfCGP designed ANN outperforms competing ensembling methods (boosting, RL).

6 Conclusion

The SelfCGP based automatic designing the ensembles of ANNs allows improving the effectiveness of the data analysis. The obtained results are approved by solving two real-world problems from the area of computer security.

Computational efforts for the implementation of the described approach and the model complexity increase comparing to any single learning model. However, it is the usual drawback of any ensembling technique that is compensated with higher performance. Only additional computational consumption is because of the necessity to run the genetic programming algorithm that combines single ANN outputs into an output of the ensemble. However, it is less than the efforts for evolutionary generating of one single ANN, i.e., cannot be considered as a serious disadvantage when we design tens of ANNs. At the same time, our experiments show that the SelfCGP never includes all possible single models into an ensemble taking usually a few of them. As the greater part of the ensemble computational complexity is given by the computational efforts needed for calculating the output for each model, our approach has the advantage upon usual ensembling methods that take all available single models for the following averaging or voting. Remind also that the end user has no necessity to be an expert in ANN design and evolutionary computations but can implement reliable and effective data mining tool. All above makes the approach developed in this study very useful for computer security experts making them free from extra efforts on the intellectual information technology algorithmic core implementation.

The further development of the approach is aimed to the expansion of its functionality by including the other types of IITs (fuzzy logic systems, decision trees, neuro-fuzzy systems, other kinds of ANNs, multiobjective selection, etc.).

Acknowledgments. The research is partially supported through the Governmental contracts № 16.740.11.0742 and 11.519.11.4002.

References

1. Maloof, M. (ed.): Machine Learning and Data Mining for Computer Security. Springer (2006)
2. Victoire, T.A., Sakthivel, M.: A Refined Differential Evolution Algorithm Based Fuzzy Classifier for Intrusion Detection. European Journal of Scientific Research 65(2), 246–259 (2011)
3. Bloedorn, E.E., Talbot, L.M., DeBarr, D.D.: Data Mining Applied to Intrusion Detection: MITRE Experiences. In: Machine Learning and Data Mining for Computer Security: Methods and Applications. Springer, London (2006)
4. Julisch, K.: Intrusion Detection Alarm Clustering. In: Machine Learning and Data Mining for Computer Security Methods and Applications. Springer, London (2006)
5. Patcha, A., Park, J.-M.: An Overview of Anomaly Detection Techniques: Existing Solutions and Latest Technological Trends. Computer Networks (2007)
6. Özgür, L., Güngör, T., Gürgen, F.: Spam Mail Detection Using Artificial Neural Network and Bayesian Filter. In: Yang, Z.R., Yin, H., Everson, R.M. (eds.) IDEAL 2004. LNCS, vol. 3177, pp. 505–510. Springer, Heidelberg (2004)
7. Han, C., Li, Y., Yang, D., Hao, Y.: An intrusion detection system based on neural network. In: Proceedings of Mechatronic Science, Electric Engineering and Computer (MEC), pp. 2018–2021 (2011)
8. Saravanakumar, S., Mohanaprakash, T.A., Dharani, R., Kumar, C.J.: Analysis of ANN-based Echo State Network Intrusion Detection in Computer Networks. International Journal of Computer Science and Telecommunications 3(4), 8–13 (2012)
9. Panda, M., Abraham, A., Das, S., Patra, M.R.: Network intrusion detection system: A machine learning approach. Intelligent Decision Technologies 5(4), 347–356 (2011)
10. Pervez, S., Ahmad, I., Akram, A., Swati, S.U.: A Comparative Analysis of Artificial Neural Network Technologies in Intrusion Detection Systems. In: Proceedings of the 6th WSEAS International Conference on Multimedia, Internet & Video Technologies, Lisbon, Portugal, pp. 84–89 (2006)
11. Dietterich, T.G.: An experimental comparison of three methods for constructing ensembles of decision trees: bagging, boosting, and randomization. Machine Learning 40(2), 139–158 (2000)

12. Ho, T.K., Hull, J.J., Srihari, S.N.: Decision combination in multiple classifier systems. IEEE Transactions on Pattern Analysis and Machine Intelligence 16(1), 66–75 (1994)
13. Breiman, L.: Bagging predictors. Machine Learning 24(2), 123–140 (1996)
14. Friedman, J.H., Hastie, T., Tibshirani, R.: Additive logistic regression: a statistical view of boosting. Annals of Statistics 28(2), 337–374 (2000)
15. Navone, H.D., Granitto, P.M., Verdes, P.F., Ceccatto, H.A.: A learning algorithm for neural network ensembles. In: Inteligencia Artificial, Revista Iberoamericana de Inteligencia Artificial, vol. 12, pp. 70–74 (2001)
16. Johansson, U., Lofstrom, T., Konig, R., Niklasson, L.: Building Neural Network Ensembles using Genetic Programming. In: International Joint Conference on Neural Networks (2006)
17. Poli, R., Langdon, W.B., McPhee, N.F.: A Field Guide to Genetic Programming (2008), http://lulu.com, http://www.gp-field-guide.org.uk
18. Bukhtoyarov, V., Semenkina, O.: Comprehensive evolutionary approach for neural network ensemble automatic design. In: Proceedings of the IEEE World Congress on Computational Intelligence, pp. 1640–1645 (2010)
19. Gomez, J.: Self Adaptation of Operator Rates in Evolutionary Algorithms. In: Deb, K., Tari, Z. (eds.) GECCO 2004. LNCS, vol. 3102, pp. 1162–1173. Springer, Heidelberg (2004)
20. Meyer-Nieberg, S., Beyer, H.-G.: Self-Adaptation in Evolutionary Algorithms. In: Lobo, F., Lima, C., Michalewicz, Z. (eds.) Parameter Setting in Evolutionary Algorithm, pp. 47–75 (2007)
21. O'Neill, M., Vanneschi, L., Gustafson, S., Banzhaf, W.: Open issues in genetic programming. In: Genetic Programming and Evolvable Machines, vol. 11, pp. 339–363 (2010)
22. Finck, S., et al.: Real-parameter black-box optimization benchmarking 2009. Presentation of the noiseless functions. Technical Report Researh Center PPE (2009)
23. Semenkin, E., Semenkina, M.: Self-configuring genetic programming algorithm with modified uniform crossover. In: IEEE Congress on Evolutionary Computation (CEC 2012), Brisbane, Australia (accepted for publication, 2012)
24. Semenkin, E., Semenkina, M.: Self-configuring Genetic Algorithm with Modified Uniform Crossover Operator. In: Tan, Y., Shi, Y., Ji, Z. (eds.) ICSI 2012, Part I. LNCS, vol. 7331, pp. 414–421. Springer, Heidelberg (2012)
25. Frank, A., Asuncion, A.: UCI Machine Learning Repository. School of Information and Computer Science. University of California, Irvine (2010), http://archive.ics.uci.edu/ml
26. Yu, J.J.Q., Lam, A.Y.S., Li, V.O.K.: Evolutionary Artificial Neural Network Based on Chemical Reaction Optimization. In: IEEE Congress on Evolutionary Computation (CEC 2011), New Orleans, LA (2011)
27. Bukhtoyarov, V., Semenkin, E., Shabalov, A.: Neural Networks Ensembles Approach for Simulation of Solar Arrays Degradation Process. In: Corchado, E., Snášel, V., Abraham, A., Woźniak, M., Graña, M., Cho, S.-B. (eds.) HAIS 2012, Part III. LNCS, vol. 7208, pp. 186–195. Springer, Heidelberg (2012)
28. Stolfo, S., Fan, W., Lee, W., Prodromidis, A., Chan, P.: Cost-based Modeling for Fraud and Intrusion Detection: Results from the JAM Project. In: Proceedings of the 2000 DARPA Information Survivability Conference and Exposition, DISCEX 2000 (2000)
29. Malik, A.J., Shahzad, W., Khan, F.A.: Binary PSO and random forests algorithm for PROBE attacks detection in a network. In: IEEE Congress on Evolutionary Computation, pp. 662–668 (2011)
30. Dimitrakakis, C., Bengio, S.: Online Policy Adaptation for Ensemble Classifiers. IDIAP Research Report 03-69 (2006)

Clustering for Intrusion Detection: Network Scans as a Case of Study

Raúl Sánchez[1], Álvaro Herrero[1], and Emilio Corchado[2]

[1] Department of Civil Engineering, University of Burgos, Spain
C/ Francisco de Vitoria s/n, 09006 Burgos, Spain
{ahcosio,rsarevalo}@ubu.es
[2] Departamento de Informática y Automática, Universidad de Salamanca
Plaza de la Merced, s/n, 37008 Salamanca, Spain
escorchado@usal.es

Abstract. MOVICAB-IDS has been previously proposed as a hybrid intelligent Intrusion Detection System (IDS). This on-going research aims to be one step towards adding automatic response to this visualization-based IDS by means of clustering techniques. As a sample case of study for the proposed clustering extension, it has been applied to the identification of different network scans. The aim is checking whether clustering and projection techniques could be compatible and consequently applied to a continuous network flow for intrusion detection. A comprehensive experimental study has been carried out on previously generated real-life data sets. Empirical results suggest that projection and clustering techniques could work in unison to enhance MOVICAB-IDS.

Keywords: Network Intrusion Detection, Computational Intelligence, Exploratory Projection Pursuit, Clustering, Automatic Response.

1 Introduction

The ever-changing nature of attack technologies and strategies is one of the most harmful issues of attacks and intrusions, which increases the difficulty of protecting computer systems. For that reason, among others, Intrusion Detection Systems (IDSs) [1-3] have become an essential asset in addition to the computer security infrastructure of most organizations.

In the context of computer networks, an IDS can roughly be defined as a tool designed to detect suspicious patterns that may be related to a network or system attack. Intrusion Detection (ID) is therefore a field that focuses on the identification of attempted or ongoing attacks on a computer system (Host IDS - HIDS) or network (Network IDS - NIDS).

MOVICAB-IDS (MObile VIsualisation Connectionist Agent-Based IDS) has been proposed [4, 5] as a novel IDS comprising a Hybrid Artificial Intelligent System (HAIS). It monitors the network activity to identify intrusive events. This hybrid intelligent IDS combines different AI paradigms to visualise network traffic for ID at packet level. Its main goal is to provide security personnel with an intuitive and informative visualization of network traffic to ease intrusion detection. The proposed

Á. Herrero et al. (Eds.): Int. JointConf. CISIS'12-ICEUTE'12-SOCO'12, AISC 189, pp. 33–45.
springerlink.com

MOVICAB-IDS applies an unsupervised neural projection model to extract interesting traffic dataset projections and to display them through a mobile visualisation interface

A port scan may be defined as a series of messages sent to different port numbers to gain information on their activity status. These messages can be sent by an external agent attempting to access a host to find out more about the network services the host is providing. A port scan provides information on where to probe for weaknesses, for which reason scanning generally precedes any further intrusive activity. This work focuses on the identification of network scans, in which the same port is the target for a number of computers. A network scan is one of the most common techniques used to identify services that might then be accessed without permission [6]. Because of that, the proposed extension of MOVICAB-IDS is faced up with this kind of simple but usual situations.

Clustering is the unsupervised classification of patterns (observations, data items, or feature vectors) into groups (clusters). The clustering problem has been addressed in many contexts and by researchers in many disciplines; this reflects its broad appeal and usefulness as one of the steps in exploratory data analysis.

The remaining sections of this study are structured as follows: section 2 introduces the proposed framework and applied models and techniques. Experimental results are presented in section 3 while the conclusions of this study are discussed in section 4, as well as future work.

2 On the Network Data Visualization and Analysis

The general framework for the proposed projection-based intrusion detection taking part in MOVICAB-IDS is depicted in Fig. 1.

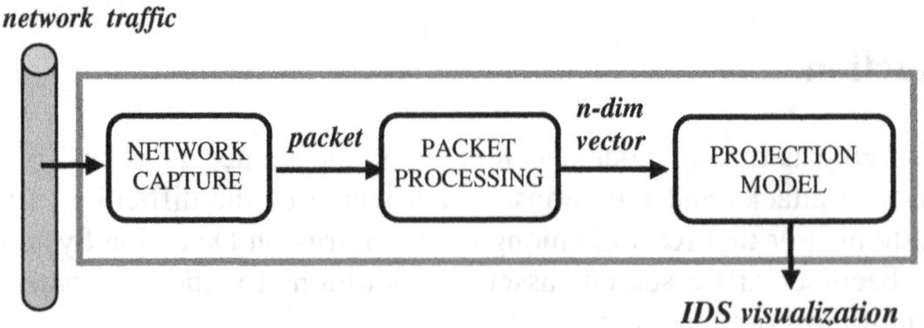

Fig. 1. MOVICAB-IDS general architecture

This framework could be described as follows:

- packets traveling through the network are intercepted by a **capture device**;
- traffic is **coded by a set of features** spanning a multidimensional vector space;
- a **projection model** operates on feature vectors and yields as output a suitable representation of the network traffic. The projection model clearly is the actual core of the overall IDS. That module is designed to yield an effective and intuitive representation of network traffic, thus providing a powerful tool for the security staff to visualize network traffic.

Present work focuses on the upgrading of the previous framework, to incorporate now new facilities as depicted in Fig. 2. It is now required and enhanced visualization by combining projection and clustering results to ease traffic by personnel. By doing so, further information on the nature of the travelling packets could be compressed in the visualization. On the other hand automatic response is an additional feature of some IDSs that could be incorporated in MOVICAB-IDS. Additionally to some classifiers for the automatic detection, clustering is proposed for those cases in which classifiers do usually fail (0-day attacks for example).

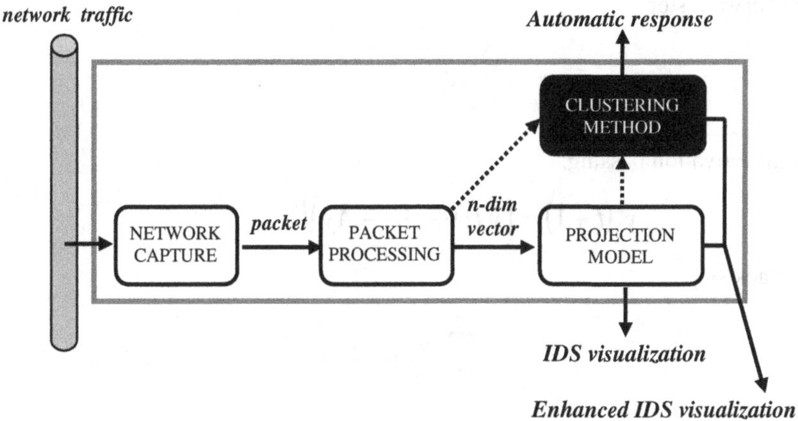

Fig. 2. Clustering extension of MOVICAB-IDS

The following subsections describe the different techniques that take part in the proposed solution. For the dimensionality reduction as a projection method, Cooperative Maximum Likelihood Hebbian Learning [7] is explained as it proved to be the most informative one among many considered [5]. It is described in section 2.1. On the other hand, to test clustering performance some of the standard methods have been tested, namely: *k*-means and agglomerative clustering. They are described in sections 2.2 and 2.3 respectively.

2.1 Cooperative Maximum Likelihood Hebbian Learning

The standard statistical method of Exploratory Projection Pursuit (EPP) [8] provides a linear projection of a data set, but it projects the data onto a set of basis vectors which best reveal the interesting structure in data; interestingness is usually defined in terms of how far the distribution is from the Gaussian distribution.

One neural implementation of EPP is Maximum Likelihood Hebbian Learning (MLHL) [9], [10]. It identifies interestingness by maximising the probability of the residuals under specific probability density functions which are non-Gaussian.

One extended version of this model is the Cooperative Maximum Likelihood Hebbian Learning (CMLHL) [7] model. CMLHL is based on MLHL [9], [10] adding lateral connections [7], [11] which have been derived from the Rectified Gaussian Distribution [12]. The resultant net can find the independent factors of a data set but does so in a way that captures some type of global ordering in the data set.

Considering an N-dimensional input vector (x), and an M-dimensional output vector (y), with W_{ij} being the weight (linking input j to output i), then CMLHL can be expressed [7], [11] as:

1. Feed-forward step:

$$y_i = \sum_{j=1}^{N} W_{ij} x_j, \forall i \ . \tag{1}$$

2. Lateral activation passing:

$$y_i(t+1) = \left[y_i(t) + \tau(b - Ay) \right]^{+} \ . \tag{2}$$

3. Feedback step:

$$e_j = x_j - \sum_{i=1}^{M} W_{ij} y_i, \forall j \ . \tag{3}$$

4. Weight change:

$$\Delta W_{ij} = \eta . y_i . sign(e_j) | e_j |^{p-1} \ . \tag{4}$$

Where: η is the learning rate, τ is the "strength" of the lateral connections, b the bias parameter, p a parameter related to the energy function [9], [10], [7] and A a symmetric matrix used to modify the response to the data [7]. The effect of this matrix is based on the relation between the distances separating the output neurons.

2.2 Clustering

Cluster analysis [13] is the organization of a collection of data items or patterns (usually represented as a vector of measurements, or a point in a multidimensional space) into clusters based on similarity. Hence, patterns within a valid cluster are more similar to each other than they are to a pattern belonging to a different cluster. This notion of similarity can be expressed in very different ways.

Pattern proximity is usually measured by a distance function defined on pairs of patterns. A variety of distance measures are in use in the various communities [14], [15], [16]. A simple distance measure such as the Euclidean distance is often used to reflect dissimilarity between two patterns, whereas other similarity measures can be used to characterize the conceptual similarity between patterns [17], it depends on the type of data we want to analyse. Furthermore, the clustering output can be hard (allocates each pattern to a single cluster) or fuzzy (where each pattern has a variable degree of membership in each of the output clusters). A fuzzy clustering can be converted to a hard clustering by assigning each pattern to the cluster with the largest measure of membership.

There are different approaches to clustering data [13], [15], but given the high number and the strong diversity of the existent clustering methods, we have focused on the ones shown in Figure 3 based on the suggestions in [13].

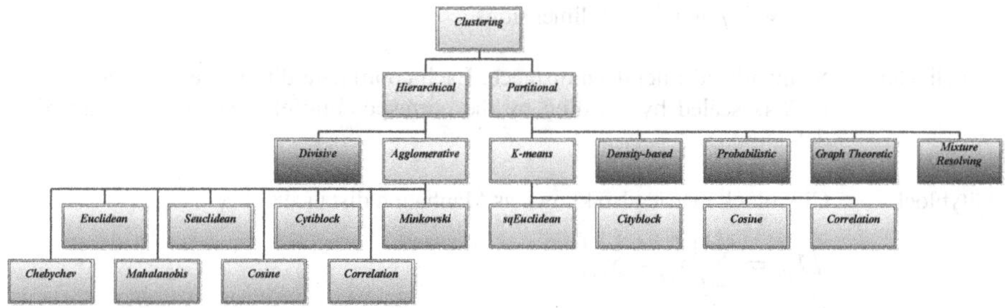

Fig. 3. Clustering methods used on this paper: one hierarchical (Agglomerative) and other partitional method (K-means)

Hierarchical methods generally fall into two types:

1. Agglomerative: an agglomerative approach begins with each pattern in a distinct cluster, and successively joins clusters together until a stopping criterion is satisfied or until a single cluster is formed.
2. Divisive: a divisive method begins with all patterns in a single cluster and performs splitting until a stopping criterion is met or every pattern is in a different cluster. This method is neither applied nor discussed in this paper.

Partitional clustering aims to directly obtain a single partition of the data instead of a clustering structure, such as the dendrogram produced by a hierarchical technique.

There is no clustering technique that is generally applicable in the clustering of the different structures presented in multidimensional data sets. Humans can competitively and automatically cluster data in two dimensions, but most real problems involve clustering in a higher dimensional space, that is the case of network security data sets. It is difficult to obtain an intuitive interpretation of data in those spaces.

Since similarity is fundamental to the definition of a cluster, a measure of the similarity is essential to most clustering methods and it must be chosen carefully. We will focus on the well-known distance measures used for patterns whose features are all continuous:

Table 1. Some of the well-known distance measures that are usually employed in clustering methods

Metric	**Description**
Euclidean	Euclidean distance: $$D_{ab} = \sqrt{\sum_{j=1}^{p} \left(x_{aj} - x_{bj}\right)^2}$$ Where: • x_{aj} , x_{bj} values taken by the jth variable for the objects a and b, respectively in the multi-variable space. • p number of dimensions.
sEuclidean	Standardized Euclidean distance. Each coordinate difference between rows in X is scaled by dividing by the corresponding element of the standard deviation.
Cityblock	City block metric also known as Manhattan distance: $$D_{ab} = \sum_{j=1}^{p} \left\| x_{aj} - x_{bj} \right\|$$ Where: • x_{aj} , x_{bj} values taken by the jth variable for the objects a and b, respectively in the multi-variable space. • p number of dimensions.
Minkowski	Minkowski distance: $$D_{ab} = \sqrt[\lambda]{\sum_{j=1}^{p} \left\| x_{aj} - x_{bj} \right\|^{\lambda}}$$ • x_{aj} , x_{bj} values taken by the jth variable for the objects a and b, respectively in the multi-variable space. • p number of dimensions. • λ=1 Cityblock distance. • λ=2 Euclidean distance.
Chebychev	Chebychev distance (maximum coordinate difference).
Mahalanobis	Mahalanobis distance, using the sample covariance of X: $$D_{ab} = \sqrt{\left(x_a - x_b\right)^T S^{-1} \left(x_a - x_b\right)}$$ • x_a , x_b values of the objects a and b, respectively in the multi-variable space. • S covariance matrix.
Cosine	One minus the cosine of the included angle between points (treated as vectors).
Correlation	One minus the sample correlation between points (treated as sequences of values).

The most popular metric for continuous features is the Euclidean distance which is a special case of the Minkowski metric (p =2). It works well when a data set has compact or isolated clusters [18]. The problem of using directly the Minkowski metrics is the tendency of the largest-scaled feature to dominate the others. Solutions to this problem include normalization of the continuous features (sEuclidean distance).

Linear correlation among features can also distort distance measures, it can be relieved by using the squared Mahalanobis distance that assigns different weights to different features based on their variances and pairwise linear correlations. The regularized Mahalanobis distance was used in [18] to extract hyperellipsoidal clusters.

2.2.1 *K*-Means

K-means is the simplest and most commonly used partitional algorithm employing a squared error criterion [19], but it also can be used with other distance measures. It starts with a random initial partition of k clusters (centroids: is the point to which the sum of distances from all objects in that cluster is minimized) and assign the patterns to clusters based on the similarity between the pattern and the centroid until a convergence criterion is met (e.g. minimize the sum of point-to-centroid distances, summed over all k clusters). The k-means algorithm is popular because it is easy to implement, and its time complexity is O(n), where n is the number of patterns to cluster. The main problem of this algorithm is that it is sensitive to the selection of the initial partition and may conclude with a local minimum (not a global minimum) depending on the initial partition.

This study uses four different distance measures, the method have been tested on all of them and the best result can be seen on the results section.

Table 2. Distance measures employed for K-means in this study

Metric	Description
sqEuclidean	Squared Euclidean distance. Each centroid is the mean of the points in that cluster.
Cityblock	Sum of absolute differences. Each centroid is the component-wise median of the points in that cluster.
Cosine	One minus the cosine of the included angle between points (treated as vectors). Each centroid is the mean of the points in that cluster, after normalizing those points to unit Euclidean length.
Correlation	One minus the sample correlation between points (treated as sequences of values). Each centroid is the component-wise mean of the points in that cluster, after centering and normalizing those points to zero mean and unit standard deviation.

2.2.2 Agglomerative

A hierarchical algorithm produces a dendrogram representing the nested grouping of patterns and similarity levels at which groupings change. The dendrogram can be broken at different levels to produce different clusterings of the data. The hierarchical agglomerative clustering algorithm has three phases:

1. *First phase*: compute the proximity matrix containing the distance between each pair of patterns. Treat each pattern as a cluster.
2. *Second phase*: find the most similar pair of clusters using the proximity matrix. Merge these two clusters into one cluster. Update the proximity matrix to reflect this merge operation.
3. *Third phase*: if all patterns are in one cluster, stop. Otherwise, go to step 2.

Based on the way the proximity matrix is updated in the second phase, a variety of linking methods can be designed (this study has been developed with the linking methods shown in Table 3).

Table 3. Linkage functions employed for agglomerative clustering in this study

Method	Description
Single	Shortest distance. $d'(k,\{i,j\}) = \min\{d(k,i), d(k,j)\}$
Complete	Furthest distance. $d'(k,\{i,j\}) = \max\{d(k,i), d(k,j)\}$
Ward	Inner squared distance (minimum variance algorithm), appropriate for Euclidean distances only.
Median	Weighted center of mass distance (WPGMC: Weighted Pair Group Method with Centroid Averaging), appropriate for Euclidean distances only.
Average	Unweighted average distance (UPGMA: Unweighted Pair Group Method with Arithmetic Averaging).
Centroid	Centroid distance (UPGMC: Unweighted Pair Group Method with Centroid Averaging), appropriate for Euclidean distances only.
Weighted	Weighted average distance (WPGMA: Weighted Pair Group Method with Arithmetic Averaging).

3 Experimental Results

This section describes the dataset used for evaluating the proposed clustering methods and how they were generated. Then, the obtained results are also detailed.

3.1 Datasets

Five features were extracted from packet headers to form the data set:

- **Timestamp**: the time difference in relation to the first captured packet. Sequential integer nonlinear [0:262198].
- **Source Port**: the port of the source host from where the packet is sent. Discrete integer values {53, ..., 5353}.

- **Destination Port**: the port of the destination host to where the packet is sent. Discrete integer values {53, ..., 36546}.
- **Size**: total packet size (in Bytes). Discrete integer values {60, ..., 355}.
- **Protocol ID**: we have used values between 1 and 35 to identify the packet protocol. Discrete integer values {65, ..., 112}.

As an initial experiment to enhance MOVICAB-IDS detection capabilities, clustering techniques have been applied to a simple dataset containing three network scans aimed at port numbers 161, 162 and 3750. Additionally, it contains a great background of network traffic that may be considered as "normal".

As previously used in other experiments, further details on the data can be found in [4, 5].

3.2 Results

The best results obtained by applying the previously introduced techniques to the described datasets are shown in this section. The results are projected through CMLHL and further information about the clustering results is added to the projections, mainly by the glyph metaphor (colors and symbols). All the projections comprise a legend that states the color and symbol used to depict each packet, according to the original category: normal (Cat. 1), scan #1 (Cat. 2), scan #2 (Cat. 3) or scan #3 (Cat. 4), and the assigned cluster (Clust.).

Initially, the well-known *k*-means algorithm has been applied several times to the data by combining the different algorithm options. Best results are shown in Fig. 4.

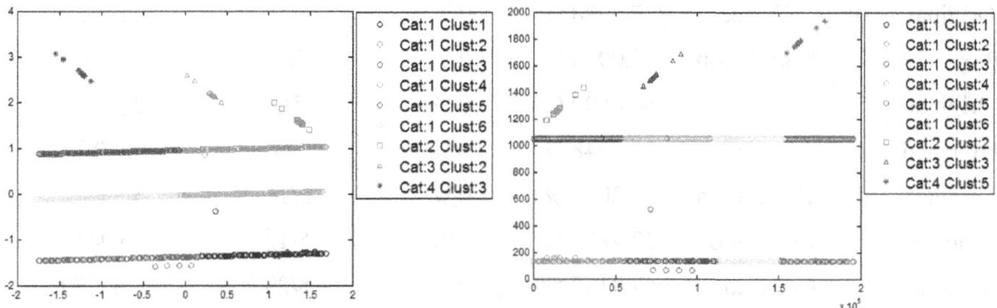

4.a *k*-means on projected data (*k*=6 and sqEuclidean distance)

4.b *k*-means on original data (*k*=6 and sqEuclidean distance)

Fig. 4. Best clustering result through *k*-means under the frame of MOVICAB-IDS

From Fig. 4 it can be seen that all the packets in each one of the scans (represented as non-horizontal small bars) are clustered in the same group. However, some other packets, regarded as normal, have been also included in those clusters. Apart from these two projections, some more experiments have been conducted, whose details (performance, true positive and false positive rates, values of *k* parameter, etc.) can be seen in Table 4.

Table 4. K-means experiments with different conditions.

Data	k	Distance criteria	False Positive	False Negative	Replicates/ Iterations	Sum of Distances
Projected	2	sqEuclidean	48,0186 %	0 %	5/4	1705,77
Original	2	sqEuclidean	46.6200 %	2.0979 %	5/5	9,75E+11
Projected	4	sqEuclidean	22,9604 %	0 %	5/8	643,352
Original	4	sqEuclidean	69,1143 %	0 %	5/8	4,38E+11
Projected	6	sqEuclidean	22,9604 %	0 %	5/8	301,218
Original	6	sqEuclidean	45,4545 %	0 %	5/24	2,91E+11
Projected	2	Cityblock	46,2704 %	0 %	5/7	1380,1
Original	2	Cityblock	49.6503 %	2.0979 %	5/9	3,50E+07
Projected	4	Cityblock	22,9604 %	0 %	5/8	710,545
Original	4	Cityblock	72,0249 %	0 %	5/15	2,15E+07
Projected	6	Cityblock	22,9604 %	0 %	5/14	526,885
Original	6	Cityblock	48,0187 %	0 %	5/10	1,41E+07
Projected	2	Cosine	47,9021 %	0 %	5/3	316,193
Original	2	Cosine	78,5548 %	0 %	5/5	15,4214
Projected	4	Cosine	22,9604 %	0 %	5/7	86,2315
Original	4	Cosine	46,8531 %	0 %	5/12	3,79324
Projected	6	Cosine	22,9604 %	0 %	5/5	35,9083
Original	6	Cosine	47,2028 %	0 %	5/24	2,51022
Projected	2	Correlation	52,0979 %	0 %	5/3	273,91
Original	2	Correlation	80,0699 %	0 %	5/6	20,6143
Projected	4	Correlation	51,8648 %	0 %	5/7	46,7877
Original	4	Correlation	47,2028 %	0 %	5/16	5,53442
Projected	6	Correlation	27,4876 %	0,3497 %	5/12	16,9416
Original	6	Correlation	47,3193 %	0 %	5/29	3,69279

The setting of the *k* parameter is one of the key points in applying *k*-means. For this experimental study, different values of *k* parameter were tested; the best of them (in terms of false positive and negative rates) are the ones in Table 4. One of the harmful points in computer security, in general terms, and intrusion detection, in particular, is the false negative rate (FNR). It can be easily seen in Table 4 that only in few of the experiments the FNR is not zero. For those cases, it keeps as a very low value as the number of packets in the network scans is much lower than those from normal traffic. On the other hand, there is not a clear difference (in terms of clustering error) between the experiments on original and projected data, although for a certain number of clusters, the results on projected data are better. Additionally, the number of needed iterations is lower for the projected data, as the dimensionality of the data has been previously reduced through CMLHL. By looking at the sum of distances (sum of point-to-centroid distances, summed over all *k* clusters), a clear conclusion can not be drawn as it depends on the distance method.

Given that *k*-means did not achieved satisfactory results on projected/original data, agglomerative clustering has been also used. Comprehensive details of the run experiments with no clustering error are shown in Table 5. Some of the results, with different values for distance criteria, linkage and number of clusters, are depicted in Fig. 5.

Table 5. Experimental setting of the agglomerative method

Data	Distance	Linkage	Cutoff	Range	Cluster
Projected	Euclidean	Single	0,37	0,307 - 0,3803	9
Projected	sEuclidean	Single	0,37	0,3087 - 0,3824	9
Projected	Cityblock	Single	0,42	0,4125 - 0,443	9
Projected	Minkowski	Single	0,38	0,307 - 0,3803	9
Projected	Chebychev	Single	0,35	0,2902 - 0,366	9
Projected	Mahalanobis	Single	0,35	0,3084 - 0,3824	9
Original	sEuclidean	Single	1,80	1,533 - 1,813	5
Original	sEuclidean	Complete	4,62	4,62 - 4,628	4
Original	sEuclidean	Average	3,00	2,696 -3,271	4
Original	sEuclidean	Weighted	3,20	3 - 3,261	4
Original	Mahalanobis	Single	2,40	2,289 - 2,438	4
Original	Mahalanobis	Complete	6,00	5,35 - 6,553	3
Original	Mahalanobis	Average	4,00	3,141 - 4,624	3
Original	Mahalanobis	Weighted	4,00	3,504 - 4,536	3

As previously stated, Table 5 contains those results whit no clustering error. It can be seen that in the case of projected data, the minimum number of clusters without error is 9, while in the case of original data, it could be lowered to 3 with appropriate distance method. From the intrusion detection point of view, a higher number of clusters does not mean a higher error rate because more than one cluster can be assigned to both normal and attack traffic.

In the case of original data, the sEuclidean and Mahalanobis distances are minimizing the number of clusters without error. On the contrary, some other distances are applicable in the case of projected data with same performance regarding clustering error.

The results of one of the experiments from Table 5 are depicted on Fig. 5: traffic visualization and the dendrogram associated to agglomerative clustering. It has been selected to show how clustering results improve the visualization capabilities of MOVICAB-IDS. The following sample has been chosen: Euclidean distance, linkage single, cutoff: 0.37, 9 groups without error. It is shown that clusters 1, 4 and 6 are associated to the three network scans and the remaining ones are associated to normal traffic.

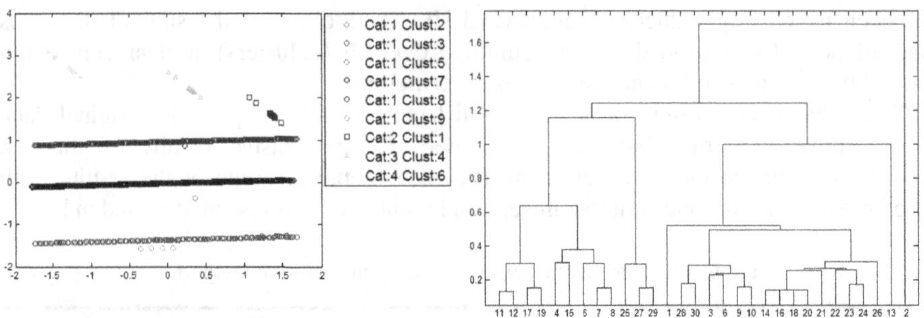

5.a Agglomerative clustering on projected data **5.b** Corresponding dendrogram

Fig. 5. Best results of agglomerative clustering under the frame of MOVICAB-IDS

4 Conclusions and Future Work

This paper has proposed the use of clustering technics to perform ID on numerical traffic data sets. Experimental results show that some of the applied clustering methods, mainly hierarchical ones, perform a good clustering in the analysed data, according to false positive and negative rates. It can then be concluded that the applied methods are able to properly detect new attacks when projected together with normal traffic. As an unsupervised process is proposed as a whole, the projections ease the task of labelling each one of the clusters as normal or attack traffic.

Future work will be based on the analysis of some other attack situations and the broadening of considered clustering methods. Moreover, new distance metrics would be developed to improve clustering results on projected data. By doing so, the automatic detection facilities of MOVICAB-IDS would be greatly improved.

Acknowledgments. This research has been partially supported through the project of the Spanish Ministry of Science and Innovation TIN2010-21272-C02-01 (funded by the European Regional Development Fund). The authors would also like to thank the vehicle interior manufacturer, Grupo Antolin Ingenieria S.A., within the framework of the MAGNO2008 - 1028.- CENIT Project also funded by the MICINN, the Spanish Ministry of Science and Innovation PID 560300-2009-11 and the Junta de Castilla y León CCTT/10/BU/0002.□

References

1. Computer Security Threat Monitoring and Surveillance. Technical Report. James P. Anderson Co. (1980)
2. Denning, D.E.: An Intrusion-Detection Model. IEEE Transactions on Software Engineering 13, 222–232 (1987)
3. Chih-Fong, T., Yu-Feng, H., Chia-Ying, L., Wei-Yang, L.: Intrusion Detection by Machine Learning: A Review. Expert Systems with Applications 36, 11994–12000 (2009)

4. Herrero, Á., Corchado, E.: Mining Network Traffic Data for Attacks through MOVICAB-IDS. In: Abraham, A., Hassanien, A.-E., de Carvalho, A.P. (eds.) Foundations of Computational Intelligence Volume 4. SCI, vol. 204, pp. 377–394. Springer, Heidelberg (2009)
5. Corchado, E., Herrero, Á.: Neural Visualization of Network Traffic Data for Intrusion Detection. Applied Soft Computing 11, 2042–2056 (2011)
6. Abdullah, K., Lee, C., Conti, G., Copeland, J.A.: Visualizing Network Data for Intrusion Detection. In: Sixth Annual IEEE Information Assurance Workshop - Systems, Man and Cybernetics, pp. 100–108 (2005)
7. Corchado, E., Fyfe, C.: Connectionist Techniques for the Identification and Suppression of Interfering Underlying Factors. International Journal of Pattern Recognition and Artificial Intelligence 17, 1447–1466 (2003)
8. Friedman, J.H., Tukey, J.W.: A Projection Pursuit Algorithm for Exploratory Data-Analysis. IEEE Transactions on Computers 23, 881–890 (1974)
9. Corchado, E., Corchado, J.M., Sáiz, L., Lara, A.M.: Constructing a Global and Integral Model of Business Management Using a CBR System. In: Luo, Y. (ed.) CDVE 2004. LNCS, vol. 3190, pp. 141–147. Springer, Heidelberg (2004)
10. Fyfe, C., Corchado, E.: Maximum Likelihood Hebbian Rules. In: 10th European Symposium on Artificial Neural Networks (ESANN 2002), pp. 143–148 (2002)
11. Corchado, E., Han, Y., Fyfe, C.: Structuring Global Responses of Local Filters Using Lateral Connections. Journal of Experimental & Theoretical Artificial Intelligence 15, 473–487 (2003)
12. Seung, H.S., Socci, N.D., Lee, D.: The Rectified Gaussian Distribution. In: Advances in Neural Information Processing Systems, vol. 10, pp. 350–356 (1998)
13. Jain, A.K., Murthy, M.N., Flynn, P.J.: Data Clustering: A Review. ACM Computing Surveys 31 (1999)
14. Anderberg, M.R.: Cluster Analysis for Applications. Academic Press, Inc., New York (1973)
15. Jain, A.K., Dubles, R.C.: Algorithms for Clustering Data. Prentice-Hall Advanced Reference Series. Prentice-Hall, Inc., Upper Saddle River (1988)
16. Diday, E., Simon, J.C.: Clustering Analysis. In: Fu, K.S. (ed.) Digital Pattern Recognition, pp. 47–94. Springer, Secaucus (1976)
17. Michalski, R., Stepp, R.E., Diday, E.: Automated construction of classifications: conceptual clustering versus numerical taxonomy. IEEE Trans. Pattern Anal. Mach. Intell. PAMI-5(5), 396–409 (1983)
18. Mao, J., Jones, A.K.: A self-organizing network for hyperellipsoidal clustering (HEC). IEEE Trans. Neural Netw. 7, 16–29 (1996)
19. McQueen, J.: Some methods for classification and analysis of multivariate observacions. In: Proceedings of the Fifth Berkeley Symposium on Mathematical Statistics and Probability, pp. 281–297 (1967)

Sensor Networks Security Based on Sensitive Robots Agents: A Conceptual Model

Camelia-M. Pintea and Petrica C. Pop

Tech Univ Cluj-Napoca, North Univ Center Baia Mare,
76 Victoriei, 430122 Baia Mare, Romania
cmpintea@yahoo.com, petrica.pop@ubm.ro

Abstract. Multi-agent systems are currently applied to solve complex problems. From this class of problem the security of networks is a very important and sensitive problem. We propose in this paper a new conceptual model *Hybrid Sensitive Robot Metaheuristic* for *Intrusion Detection*. The proposed technique could be used with machine learning based intrusion detection techniques. Our novel model uses the reaction of virtual sensitive robots to different stigmergic variables in order to keep the tracks of the intruders when securing a sensor network.

Keywords: intrusion detection, sensor network, intelligent agents.

1 Introduction

Prevention and detection of intruders in a secure network is nowadays a challenging issue. The intrusion detection system based on computational intelligence (*CI*) has proved in time to have huge advantages over traditional detection systems due to characteristics of *CI* methods: adaptation, fault tolerance, high computational speed etc. It is essential to design efficient *Intrusion Detection Systems (IDS)* especially for open medium networks as wireless sensor devices.

The intrusions can be divided into two major categories: missue intrusions and anomaly intrusions. Missue intrusions are the attacks knowing the weak points of a system. Anomaly intrusions are based on observations of normal system usage patterns and detecting deviations from the given norm. The mentioned intrusions are hard to quantify because there are no fixed patterns that can be monitored and as a result a more fuzzy approach is often required.

The *Intrusion Preventing Systems (IPS)* are network security appliances that monitor network and/or system activities for malicious activities. *IPS* is a device used to block all the unwanted access to the targeted host, to remove malicious part of packets and as well it may reconfigure the network device where an attack is detected [3].

Social autonomic cooperative colonies as ants, bees and others have the capability to coordinate and construct complex systems [4]. Using their behaviour, engineers have built real collective robotic systems. The metaheuristics based on classes of specialized robots provide feasible solutions for nowadays complex problems. One of these techniques is *Sensitive Robot Metaheuristic* developed

Á. Herrero et al. (Eds.): Int. JointConf. CISIS'12-ICEUTE'12-SOCO'12, AISC 189, pp. 47–56.
springerlink.com

by Pintea et al. [19,21]. The sensitive model was introduced and explained in [5,6,21] and used to solve complex problems in [7,20,21]. The *SRM* model was implemented first to solve a large drilling problem but it has the potential to solve other *NP*-hard problems including intrusion detection. The model ensure a balance between diversification and intensification in searching.

The aim of the current paper is to provide an effective stigmergic-based technique for *IDS* in a sensor network graph, that consist of multiple detection stations called sensor nodes. The new *Hybrid Sensitive Robot Metaheuristic for Intrusion Detection (HSRM-ID)* model uses a collection of robots endowed with a stigmergic sensitivity level. The sensitivity of robots allow them to detect and react to different stigmergic variables involving the attacks into a secure network. The hybrid model combines elements from *Sensitive Robot Metaheuristic (SRM)* [19] as *Ant Colony System (ACS)* [10], autonomous mobile robots and the intrusion detection based on emotional ants for sensors *(IDEAS)* [2].

2 Sensitive Stigmergic Robots

The metaheuristic *Sensitive Robot Metaheuristic (SRM)* [19] combining the concepts of stigmergic communication and autonomous robot search is used to solve *NP*-hard optimization problems. The basic concepts are defined and described further in this section, see for more details [4,5,6,7,19,20,21].

Definition 1. *Stigmergy occurs when an action of an insect is determined or influenced by the consequences of the previous action of another insect.*

Definition 2. *Sensitive robots refers to artificial entities with a Stigmergic Sensitivity Level (SSL) expressed by a real number in the unit interval [0, 1].*

Definition 3. *Environment explorers'robots are sensitive robots with small Stigmergic Sensitivity Level (sSSL) with the potential to autonomously discover new promising regions of the search space.*

Definition 4. *Environment exploiters robots are sensitive robots with high Stigmergic Sensitivity Level (hSSL) emphasizing search intensification.*

An important characteristic of stigmery is that individual behaviour modifies the environment, which in turn modifies the behaviour of other individuals [11]. The *SRM* technique attempts to address the coupling between perception and action as direct as possible in an intelligent stigmergic manner.

As it is known, *robot communication* relies on local environmental modifications that can trigger specific actions. The set of the rules defining actions (stimuli pairs) used by a homogeneous group of stigmergic robots defines their behaviour and determines the type of structure the robots will create [4,26]. Robot stigmergic communication does not rely on chemical deposition as it is for artificial ant-based colonies [10]. A stigmergic robot action is determined by the environmental modifications caused by prior actions of other robots. The value of quantitative stigmergy modify the future actions of robots. Discrete stimulus are involved in qualitative stigmergy and the action is switched to a different action [4,26].

Some real-life applications of the behaviour-based approach, including autonomous robots, are in data mining, military applications, industry and agriculture, waste management, health care.

3 Intrusion Detection Techniques Using Artificial Intelligence

In the current section we start with the introduction of the main concepts of *IDS* and as well we present a survey of *Artificial Intelligence*-based existing models for computer security.

3.1 Intrusion Detection System

Due to increasing incidents of computer attacks, it is essential to build efficient intrusion detection mechanisms. The definitions of the main concepts related to this domain are given in what it follows, see for example [8,13].

Definition 5. *Intrusion detection technology is a technology designed to monitor computer activities for the purpose of finding security violations.*

Definition 6. *Intrusion detection system (IDS) is a system that implements intrusion detection technology.*

Definition 7. *A security violation of a system is any deliberate activity that is not wanted including denial of service attacks, port scans, gaining of system administrator access and exploiting system security holes.*

Definition 8. *Intrusion Prevention System (IPS) is active, in-line device in the network that can drop packets or stop malicious connection before reaching the targeted system.*

IPS is able to detect and prevent attacks but it has not deeper detection capabilities of *IDS*. Neither of *Intrusion Detecting System* and *Intrusion Prevention System* is capable to provide in depth security. *Intrusion Detecting and Prevention System (IDPS)*, a combinations of *IDS* and *IPS*, is a more effective system capable of detection and prevention [22]. Based on the placement, the *IDPS* is divided into four classes as follows:

1. a *network-based system*, which is able to monitor traffic of network or its particular segment and identify different network attacks. Fig 1 is showing an example of architecture of a network-based system, the *In-line Network-Based IDPS Sensor Architecture* according to [22].
   An example of network-based system is Snort [14]. *Snort* is an open source network intrusion prevention and detection system - nowadays a standard for IPS - that combines the benefits of signature, protocol and anomaly-based inspection. A number of problems associated with *Network-based system* according to [17] are:

- they cannot fully detect novel attacks;
- variations of known attacks are not fully detected;
- they generate a large amount of alerts, as well as a large number of false alerts;
- the existing *IDS* is focus on low-level attacks or anomalies and do not identify logical steps or strategies behind these attacks.

2. *host-based systems* describe the class of software able to monitor a single system, analyse characteristics and log to at one host. These systems are deployed on critical hosts.
3. *wireless-based systems* analyse wireless traffic to monitor intrusion or any suspicious activity. They scan traffic but are not able to identify attack in the application layer or higher layer network protocols as UDP and TCP. It may be deployed at the point where unauthorized wireless network could be accessed.
4. *behaviour-based systems* are used for examining network traffic in order to identify attacks (e.g. Denial of Service attacks). These systems are deployed to monitor flow of network or flow between internal and external network.

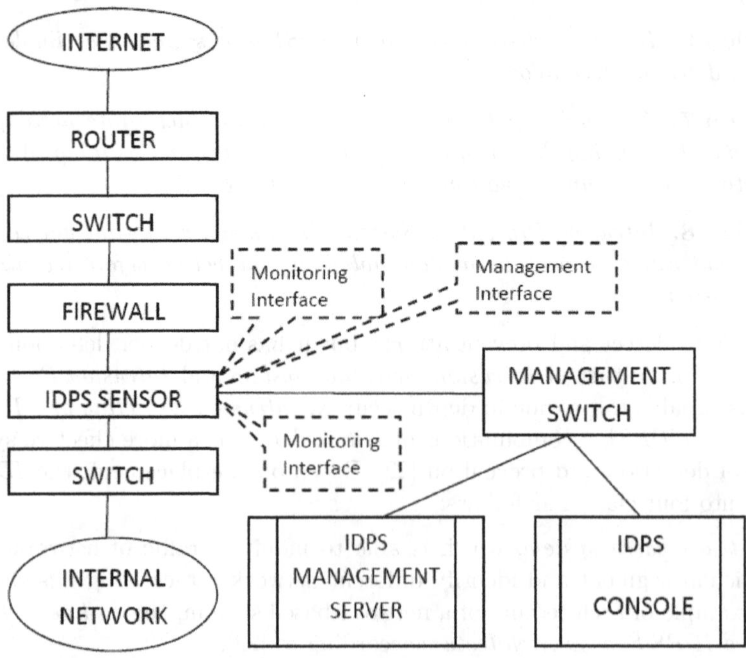

Fig. 1. Inline Network-Based IDPS Sensor Architecture based on [22]

3.2 Artificial Intelligence in Intrusion Detection System

The current paper deals with an artificial intelligent approach for intrusion detections. A short review of the main *AI* techniques already used and their benefits for detecting intrusion in network systems follows.

According to Beg et al. [3], the intrusion detection classical algorithms have the following disadvantages: false alarm rate and constant updates of database with new signatures.The network administrator responds to alarms and updates the signatures that increases in time. For example, in the already mentioned *Snort* signatures increased from 1500 to 2800 over two years [14]. In order to improve the administrator work, reducing the number of false alarms and better intrusion detection are introduced artificial intelligence mechanisms [23]. Some of *AI* techniques used in intrusion detection are data mining, genetic algorithm, neural network, multi-agents, ant-net miner, etc.

Lee et al. [15] introduced a data mining classification mechanism with association rules from the audit data - knowledge present in a knowledge base - providing gaudiness for data gathering and feature selection. In order to detect abnormal behaviour one can use genetic algorithms, see for example [1]. In [18], neural networks use back propagation *MLP* for a small network in order to detect anomalies and identify user profiles after end of each log session.

It shall also be remarked that several of the leading methods for detecting intrusions and detecting intrusions are hybrid artificial approaches, which combine different *AI* solution techniques [9,16,25]. Some hybrid methods used in the literature are data mining and fuzzy logic techniques [16], data mining and genetic algorithm selecting the best rules for the system [9]. In the future could be implemented hybrid models involving intelligent evolutionary agents [12] and dynamic decision boundary using Support Vector Machine [24] for handle a large number of features.

Banerjee et al. [2] introduced an intrusion detection based on emotional ants for sensors *(IDEAS)*, which could keep track of the intruder trials. This technique is able to work in conjunction with the conventional machine learning based intrusion detection techniques to secure the sensor networks.

4 Hybrid Sensitive Robot Metaheuristic for Intrusion Detection

In this section we introduce a new hybrid metaheuristic in order to detect the intruders in a sensor network. The new model is called *Hybrid Sensitive Robot Metaheuristic* for *Intrusion Detection (HSRM-ID)*, is based on *Sensitive Robot Metaheuristic (SRM)* introduced in [19] and uses a specific rule in order to generate a state of thinking or the choice of an intruder [2].

The proposed *(HSRM)* can be modelled using two distinct groups of sensitive stigmergic robots. The first group of robots-agents is endowed with small sensitive values *SSL* and they are sensitive-explorers (*sSSL: small SSL-robots*).

They can sustain diversification in intruders searching. In the second group are the robots-agents with high sensitive stigmergic values (*hSSL: high SSL-robots*). They are sensitive-exploiters and could exploit intensively the regions already identified with attacks from intruders. In time, based on the experience of robots-agents, the sensitive stigmergic level *SSL* can increase or decrease.

The pseudo-code description of the *Hybrid Sensitive Robot Metaheuristic* for *Intrusion Detection* is described in what it follows.

```
Hybrid Sensitive Robot Algorithm for Intrusion Detection
Begin
  Set parameters; initialize stigmergic values of the trails;
  For k=1 to m do
    Place robot k on a randomly chosen node of a sensor network;
  For i=1  to Niter do
      Each robot incrementally builds a solution
      based on the autonomous search sensitivity;
      The sSSL robots choose the next node
      based on the attack probability (1);
      A hSSL-robot uses the information supplied by
      the sSSL robots to chose the new node (2);
      Apply a local stigmergic updating rule (3);
      Apply the rule generating a state of thinking
      or the choice of an intruder (4):
    A global updating rule is applied (5);
    Validate the path and detect intruder;
        Endfor
End
```

The stigmergic value of an edge is τ and the visibility value is η. A *tabu list* with the already visited nodes is maintained, see [10] for more details. In order to divide the colony of m robots in two groups it is used a random variable uniformly distributed over $[0, 1]$. Let q be a realization of this random variable and q_0 a constant $0 \leq q_0 \leq 1$. If the inequality $q > q_0$ stands the robots are endowed with small sensitive stigmergic value *sSSL* robots and otherwise they are highly sensitive stigmergic robots (*hSSL*). A *hSSL-robot* uses the information supplied by the *sSSL* robots.

In order to define the rule to generate a state of thinking or the choice of an intruder we use the same notations as in Banerjee et al. [2]:

- $A(I, s, t)$ denotes the tendency of an intruder I to be assigned to the sensor node s at moment t.
- $I_1(intruder1)\_C(I, s, t)$ is the potential to generate the state of choice to a particular path in the network sensor graph.
- $I\_C(I, s, t)$ is the intensity of the attack,
- $f\_C(.)$ is a function specific of the thinking of intruder
- $T\_c(I, t)$ is the threshold value.

The new hybrid model *(HSRM-ID)* for identifying the affected path of a sensor network graph is described further.

- Initially the *SSL* robots are placed randomly in the network space. The parameters of the algorithm are initialized.
- A *SSL* robot chooses the next move with a probability based on the distance to the candidate node and the stigmergic intensity on the connecting edge. In order to stop the stigmergic intensity increasing unbounded each time unit evaporation takes place.
- Let i be the current node. The next node is chosen probabilistically. Let $J^k{}_i$ be the unvisited successors of node i by robot k and $u \in J^k{}_i$. As in *Ant Colony System* technique [10] the probability of choosing the next node u, possible to be attacked, is shown in (1).

$$p^k{}_{iu}(t) = \frac{[\tau_{iu}(t)][\eta_{iu}(t)]^\beta}{\Sigma_{o \in J^k{}_i}[\tau_{io}(t)][\eta_{io}(t)]^\beta}, \tag{1}$$

  where β is a positive parameter, $\tau_{iu}(t)$ is the stigmergic intensity and $\eta_{iu}(t)$ is the inverse of the distance on edge (i, u) at moment t.
- The new node j is choose by *hSSL* robots using (2):

$$j = argmax_{u \in J_i^k}\{\tau_{iu}(t)[\eta_{iu}(t)]^\beta\}, \tag{2}$$

  where β determines the relative importance of stigmergy versus heuristic information.
- Update trail stigmergic intensity by local stigmergic rule (3):

$$\tau_{ij}(t+1) = q_0^2\tau_{ij}(t) + (1-q_0)^2 \cdot \tau_0. \tag{3}$$

  where (i, j) are the edges belonging to the most successful traversing across sensor nodes.
- Equation (4) illustrates the rule to generate a state of thinking or the choice of an intruder [2].

$$\text{If } I\_C(I,s,t) = I\_C(I,s,t) - T\_C(I,t) \text{then } l\_C(l,s,t) > I\_C(l,t) \\ \text{else } I\_C(I,s,t) = 0. \tag{4}$$

- A global updating rule is applied [2] as in (5) and is used a tabu list where to store the track and edge details.

$$\tau_{ij}(t+1) = q_0^2\tau_{ij}(t) + (1-q_0)^2 \cdot \sum_{j=1}^{k} \Delta s^j \tau_{ij}(t), \tag{5}$$

  where

$$\Delta s^j t_{ij} = \begin{cases} f(s^j) \text{ if } s^j \text{ contributes to } \tau_{ij} \\ 0 \qquad \text{otherwise} \end{cases} \tag{6}$$

  and where q_0 is the evaporation rate, $\Delta s^j \tau_{ij}$ is the combination of a solution s^j with the update for pheromone value τij; $f(s^j)$ is the function specific to the thinking of the intruder and k is the number of solution used for updating the pheromones.

– Update the intensity of attack value $I\_C(I, s, t)$ through validating the path and detect intruder.

The output of the algorithm is the most affected path of a sensor network with n nodes. Termination criteria is given by the number of iterations, denoted by N_{iter}. The complexity of the proposed algorithm is $O(n^2 \cdot m \cdot N_{iter})$.

5 The Analyse of the New Concept

In the following is performed an analyse of the *Hybrid Sensitive Robot Algorithm for Intrusion Detection*. The artificial pheromone from the edges of the sensor network graph reveals as the attacked zone within the network. Each bio-inspired robot uses his one specific properties as his level of sensitivity in order to detect the intruders and the artificial stigmergy in order to find the attacked edges.

Table 1. Analyse the action of agents-robots based on the pheromone level on the edges of the sensor network graph

Agents	Intruders searching type	Pheromone Level	Detecting intrusion	Action Type
sSSL robots	explorers	low	no	continue to explore
		high	possibly intruders	notify the hSSL-robots
hSSL robots	exploiters	low	the attack is not certified	update pheromone trails
		high	attack is highly present	identify the affected path

Table 1 illustrates the behaviour of different groups of sensitive bio-inspired robots when investigate the sensor network in search of intrusion. As a concept, the introduced model *Hybrid Sensitive Robot Algorithm for Intrusion Detection* has more chances to improve the intrusion detection systems comparing with the existing approaches from the literature, due to the sensitivity property of the bio-inspired robots. As well the diversity of robots groups implies also different values of virtual pheromone trail values. The robots with small stigmergic value are constantly sustaining diversification in intruders searching and as a complementary action, the robots with high sensitive stigmergic values are testing the already identified networks attacked regions. In the future we will perform numerical experiments to assess the performance of the proposed algorithm.

6 Conclusions

Nowadays the networks are threatened by security attacks and resource limitations. In order to deal with this security network problem efficient intruders detection and prevention systems are used. Within this paper we introduce a new concept *Hybrid Sensitive Robot Algorithm for Intrusion Detection* based on bio-inspired robots. It is used a qualitative stigmergic mechanism, each robot is endowed with a stigmergic sensitivity level facilitating the exploration and exploitation of the search space. In the future some computational tests will be proposed and further hybrid AI techniques will be involved for securing the networks.

Acknowledgement. This research is supported by Grant PN II TE 113/2011, New hybrid metaheuristics for solving complex network design problems, funded by CNCS Romania.

References

1. Alhazzaa, L.: Intrusion Detection Systems using Genetic Algorithms (2007)
2. Banerjee, S., Grosan, C., Abraham, A.: IDEAS: Intrusion Detection based on Emotional Ants for Sensors. In: Intelligent Systems Design and Applications, pp. 344–349. IEEE C.S. (2005)
3. Beg, S., Naru, U., Ashraf, M., Mohsin, S.: Feasibility of Intrusion Detection System with High Performance Computing: A Survey. Int. J. for Advances in Computer Science 1(1), 26–35 (2010)
4. Bonabeau, E., Dorigo, M., Tehraulaz, G.: Swarm intelligence from natural to artificial systems. Oxford Univ. Press, Oxford (1999)
5. Chira, C., Pintea, C.-M., Dumitrescu, D.: Sensitive stigmergic agent systems: a hybrid approach to combinatorial optimization. In: Innovations in Hybrid Intelligent Systems, Advances in Soft Computing, vol. 44, pp. 33–39 (2008)
6. Chira, C., Pintea, C.-M., Dumitrescu, D.: Cooperative learning sensitive agent system for combinatorial optimization. In: NICSO 2007. SCI, vol. 129, pp. 347–355 (2008)
7. Chira, C., Dumitrescu, D., Pintea, C.-M.: Learning sensitive stigmergic agents for solving complex problems. Computing and Informatics 29(3), 337–356 (2010)
8. Crothers, T.: Implementing Intrusion Detection Systems. Wiley (2003)
9. Dhanalakshmi, Y., Babu, I.R.: Intrusion detection using data mining along fuzzy logic and genetic algorithms. Int. J. of Computer Science and Network Security 8(2), 27–32 (2008)
10. Dorigo, M., Gambardella, L.M.: Ant Colony System: A cooperative learning approach to the Traveling Salesman Problem. IEEE Trans. Evol. Comp. 1, 53–66 (1997)
11. Grassé, P.-P.: La Reconstruction du Nid et Les Coordinations Interindividuelles Chez Bellicositermes Natalensis et Cubitermes. Insect Soc. 6, 41–80 (1959)
12. Iantovics, B., Enachescu, C.: Intelligent Complex Evolutionary Agent-based Systems. In: Development of Intelligent and Complex Systems, AIP, vol. 124, pp. 116–124 (2009)
13. Ierace, N., Urrutia, C., Bassett, R.: Intrusion Prevention Systems. In: Ubiquity, p. 2. ACM (2005)
14. Kim, B., Yoon, S., Oh, J.: Multi-hash based Pattern Matching Mechanism for High-Performance Intrusion Detection. Int. J. of Computers 1(3), 115–124 (2009)
15. Lee, W., Stolfo, S., Mok, K.: Mining audit data to build intrusion detection models, Knowledge Discovery and Data Mining, pp. 66–72. AAAI Press, New York (1998)

16. Luo, J.: Integrating Fuzzy Logic and Data Mining Methods for Intrusion detection, Master thesis, Mississippi State University (1999)
17. Northcutt, S.: Network Intrusion Detection. New Riders Publishers (2002)
18. Ryan, J., Lin, M.-J., Miikkulainen, R.: Intrusion Detection with Neural Networks. In: Advances in Neural Information Processing Systems, vol. 10. MIT Press (1998)
19. Pintea, C.-M., Chira, C., Dumitrescu, D., Pop, P.C.: A Sensitive Metaheuristic for Solving a Large Optimization Problem. In: Geffert, V., Karhumäki, J., Bertoni, A., Preneel, B., Návrat, P., Bieliková, M. (eds.) SOFSEM 2008. LNCS, vol. 4910, pp. 551–559. Springer, Heidelberg (2008)
20. Pintea, C.-M., Chira, C., Dumitrescu, D.: Sensitive Ants: Inducing Diversity in the Colony. In: Krasnogor, N., Melián-Batista, M.B., Pérez, J.A.M., Moreno-Vega, J.M., Pelta, D.A. (eds.) NICSO 2008. SCI, vol. 236, pp. 15–24. Springer, Heidelberg (2009)
21. Pintea, C.M.: Combinatorial optimization with bio-inspired computing. PhD Thesis, Babes-Bolyai University (2008)
22. Scarfone, K., Mell, P.: Guide to Intrusion Detection and Prevention Systems, IDPS (2007), `http://csrc.nist.gov/publications/nistpubs/800-94/SP800-94.pdf`
23. Selvakani, S., Rajes, R.S.: Genetic algorithm for forming rules for intrusion detection. Int. J. of Computer Science and Network Security 7(11), 285–290 (2007)
24. Stoean, R., et al.: Support Vector Machine Learning with an Evolutionary Engine. J. Operational Research Society 60(8), 1116–1122 (2009)
25. Yao, J.T., Zhao, S.L., Saxton, L.V.: A study on fuzzy intrusion detection. In: SPIE, Data Mining, Intrusion Detection, Information Assurance and Data Networks Security, vol. 5812, pp. 23–30 (2005)
26. White, T.: Expert Assessment of Stigmergy: A Report for the Department of National Defence, `http://www.scs.carleton.ca/~arpwhite/stigmergy-report.pdf`

Comments on a Cryptosystem Proposed by Wang and Hu

R. Durán Díaz[1], L. Hernández Encinas[2], and J. Muñoz Masqué[2]

[1] Universidad de Alcalá, 28871-Alcalá de Henares, Spain
raul.duran@uah.es

[2] Instituto de Seguridad de la Información, CSIC, 28006-Madrid, Spain
{luis,jaime}@iec.csic.es

Abstract. In this paper, we analyze a new proposal for a knapsack-type cryptosystem, recently published by Wang and Hu ([1]), along with two cryptanalyses of it, carried out by Youssef ([2]) and Lee ([3]). The cryptosystem proves to be safe only if the keys have very large sizes, but this severely impacts the use of the system from a practical point of view.

Keywords: Equivalent keys, Public-key cryptography, Quadratic knapsack problem.

1 Introduction

The heavy computational cost associated to cryptosystems based on problems such as integer factorization or discrete logarithm poses a daunting challenge when coming to implementing them over low-resource devices such as, for example, smart cards, RFID tags, and the like. This fact has renewed the interest for cryptosystems based on other types of mathematical problems, as it is the case for those based on the subset sum problem or on knapsacks, for which a number of revisions and new proposals have come to light as of late ([1]–[6]). Security shows also a strong link with computation, and hence other techniques, such as those from computational intelligence (see, for example, [7]–[10]) remain helpful as well.

In this paper, we analyze a new proposal of a knapsack-type cryptosystem, recently published by Wang and Hu ([1]), along with two cryptanalyses of it, carried out by Youssef ([2]) and Lee ([3]). Our main contribution is a discussion on the strengths and the weaknesses of both the cryptosystem and the cryptanalyses.

This paper is organized as follows: Section 2 summarizes the main phases in Wang-Hu cryptosystem operation along with some comments about its security followed by Youssef's and Lee's cryptanalyses. Section 3 comments on the issue of the bit lengths of the keys, comparing the ones for the present cryptosystem with others'. Section 4 is devoted to describe a possible Shamir-like attack on the proposed cryptosystem. Section 5 describes the numerical experiments conducted and, finally, section 6 summarizes the results obtained from the previous discussion and experiments.

Á. Herrero et al. (Eds.): Int. JointConf. CISIS'12-ICEUTE'12-SOCO'12, AISC 189, pp. 57–65.
springerlink.com

2 Wang-Hu Cryptosystem and Cryptanalyses

The security of the cryptosystem proposed by Wang and Hu is based on the difficulty of solving the quadratic knapsack problem; for this reason, this cryptosystem is a particular case of the so-called knapsack cryptosystems.

In the next paragraphs, the cryptosystem will be very briefly described. We will present the key generation, the encryption, and the decryption processes.

2.1 Key Generation

A user computes its public key as follows:

1. Chooses two cargo vectors $A = (a_1, \ldots, a_n)$ and $B = (b_1, \ldots, b_n)$ at random satisfying [1, Theorem 1].
2. Randomly chooses a 2-dimensional square matrix $C = (c_{ij})$ with determinant equal to 1. The bit length of its entries is upper bounded by a constant, i.e., $|c_{ij}| = O(1)$, where $|x|$ stands for the binary length of the integer x.
3. Computes
$$\begin{pmatrix} \hat{A} \\ \hat{B} \end{pmatrix} = \begin{pmatrix} \hat{a}_1 \cdots \hat{a}_n \\ \hat{b}_1 \cdots \hat{b}_n \end{pmatrix} = C \begin{pmatrix} A \\ B \end{pmatrix}$$
4. Chooses at random two prime numbers p, q verifying
$$p > 15^2 \sum_{i=1}^{n} \hat{a}_i, \quad q > 15^2 \sum_{i=1}^{n} \hat{b}_i, \tag{1}$$
and computes $N = p \cdot q$.
5. Uses the Chinese Remainder Theorem to generate a cargo vector $E = (e_1, \ldots, e_n)$, where $e_i = \hat{a}_i \pmod p$, $e_i = \hat{b}_i \pmod q$.
6. Randomly chooses an element $v \in \mathbb{Z}_N^*$.
7. Computes the set $F = \{f_1, \ldots, f_n\}$, where $f_i = e_i \cdot v \pmod N$.

Finally, the secret key is N, p, q, C^{-1}, and $v^{-1} \pmod N$; whereas the public key is simply the set F (with an optional permutation on the indices).

2.2 Encryption

Any plain text M must be first represented by a tuple, $M = (m_1, \ldots, m_n)$, with $|m_i| = 4$. To encrypt M the sender computes $c = \sum_{i=1}^{n} f_i \cdot m_i^2$.

2.3 Decryption

For decrypting a given ciphertext, c, the receiver performs the following:

1. Computes the value $t = c \cdot v^{-1} \pmod N = \sum_{i=1}^{n} e_i \cdot m_i^2 \pmod N$.
2. Determines
$$\begin{pmatrix} s_A \\ s_B \end{pmatrix} = C^{-1} \begin{pmatrix} t \pmod p \\ t \pmod q \end{pmatrix}.$$
3. Solves the following simultaneous quadratic compact knapsack problem to recover the plaintext by using [1, Algorithm 1] $s_A = \sum_{i=1}^{n} a_i \cdot m_i^2$, $s_B = \sum_{i=1}^{n} b_i \cdot m_i^2$, where $m_i^2 \in I$, and $I = \{i^2 : i = 0, \ldots, 15\}$.

3 Security

The security of this cryptosystem is reviewed in [1] by considering a number of possible attacks and proving that the system is secure against any and all of them. The attacks considered are the following:

1. Solving the quadratic compact knapsack problem.
2. Relinearization attack.
3. Diophantine approximation attack.
4. Orthogonal lattice attack.
5. Known N attack.
6. Known p and q attack.

Nevertheless, there are at least two cryptanalyses, described in [2] and [3], aimed at breaking Wang-Hu cryptosystem.

Cryptanalysis by Youssef. Youssef's cryptanalysis ([2]) is characterized by the following:

- It is heuristic.
- It requires the knowledge of a very large percentage of the plaintext (larger than 60%).
- It uses LLL algorithm ([11],[12]), which, though runs in polynomial time, requires large amounts of computation even for moderate-sized parameters.

However, in practice this cryptanalysis can be thwarted by simply adding a random padding to the plaintext.

Cryptanalysis by Lee. Moreover, Lee's cryptanalysis ([3]) is more effective. In fact, it is a heuristic key-recovery attack, based on the fact that there exist some specific relations among the four last elements, $f_n, f_{n-1}, f_{n-2}, f_{n-3}$, of the public key, F. These relations permit one to recover the private key.

Lee's cryptanalysis has the following properties:

- It is heuristic.
- It also uses LLL algorithm ([11],[12]).
- It attacks the system by using only the public key, without resorting to any plaintext or ciphertext.

This attack strongly depends on the shortest vector determined by the heuristic LLL algorithm. The paper does not provide a bound for the number of shortest vectors that must be tried. So, the algorithm may lead to no solutions or it may require a large computational time. Moreover, the author himself suggests how to thwart this attack, though his method implies an increase of a 50% in the ciphertext length.

These reasons help to understand why we deem these cryptanalyses not practical for real-life implementations of Wang-Hu cryptosystem.

4 Bit Length of the Keys

In [1, section 5.2], the authors state that the bit length of the public key is around 6157 bits for $n = 100$, but this computation is flawed and the real value is ten times as large. Actually, the size of the public key for such value of n is about 61600, thus rendering that cryptosystem not competitive against other well-known cryptosystems. For example, in Chor-Rivest cryptosystem defined over the field $\mathbb{F}_{q^h}$, the public key size is 14986 bits for $q = 127$ and $h = 17$ with a density of 1.068; for $q = 233$ and $h = 17$, the public key size is 30989 bits and the density is $d = 1.742$ (see [13, Table 1]). Besides, the cryptosystems associated to these sets of parameters are not vulnerable neither to the attack described in [14] nor to any other type of attack we know of.

Moreover, if a key size similar to the one in [1, section 5.2] is to be taken for a Chor-Rivest cryptosystem, then it is required to select parameters such as $q = 409$ and $h = 17$. This particular case provides a density of 2.773 and a public key size of 60123 bits (see [13, Table 1]).

5 A Shamir-Like Attack

As it is well known, Shamir designed an attack against the additive knapsack cryptosystem proposed by Merkle and Hellman ([15]), which runs in polynomial time ([16]).

Shamir's attack considers a public knapsack, be it A, and tries to find any other trapdoor equivalent to the one employed by the user. In other words, Shamir's attack finds a 'new' (partial) key that is equivalent to the user's private key in the sense that it solves the knapsack problem associated to A in polynomial time. It is important to keep in mind that the original knapsack indeed has a solution. This 'new' key may or may not be a part of the user's actual private key.

Shamir is able to prove that general knapsack cryptosystems, such as Merkle-Hellman's, have infinite many trapdoors allowing one to decrypt any given message, no matter which was the actual ciphering key used to encrypt that message. In this sense, Shamir's attack computes an "equivalent" private key.

Following this line of thought, we searched for values p', q', providing cargo vectors associated to the same public key of a particular instance of Wang-Hu cryptosystem. We have considered the case $n = 3$, which is small and very far from $n = 100$, as in the example proposed in the original paper. We arrived at the following conclusions:

1. Only in very few cases, there do exist primes $r_i < N$, distinct from the actual cryptosystem parameters that are equivalent to the original ones. As Tables 1–4 show, only in 4 out of a total of 51 randomly analyzed cases, a prime r_1 has been obtained that is distinct from the original ones; in 3 cases, a pair of primes, r_1, r_2 has been obtained, with identical properties. In a particular case, up to 7 new primes have been obtained, verifying the condition in (1).

2. The public key associated to the parameters p', $q' \in \{r_i, p, q\}$ is the same as the original one, though its computation is very costly since it requires an exhaustive search along with one execution of the probabilistic Miller-Rabin test per candidate, thus eventually amounting (see [17, Theorem 9.4.5]) to a computational cost of
$$O(N \cdot (\log_2 N)^3). \tag{2}$$
3. While we have been able to find several p', q', distinct from p and q that verify the necessary conditions required to decipher, none of the corresponding cargo vectors, A', B', satisfy the conditions in [1, Theorem 1].

From a theoretical standpoint, it is possible to justify the scarcity of equivalent parameters as follows. Let r be another prime such that

$$r > 15^2 \sum_{i=1}^{n} c_i, \qquad c_i = f_i \pmod{r}, \tag{3}$$

then we have that $f_i = c_i + h_i r$. Hence, r divides $\gcd(f_1 - c_1, \ldots, f_n - c_n)$. Now, it is also well known (see [18, solution to exercise 10, p. 641] and the references therein) that the probability that n integers be coprime is $1/\zeta(n) = 1/\sum_{k\geq 1} k^{-n}$. Moreover, there is another result ([19, Theorem 11.2]) stating that $\lim_{n\to\infty} \zeta(n) = 1$.

As a consequence, the probability of obtaining integers r verifying the condition (3) is asymptotically null. For example, for $n = 10$, we have $\sum k^{-n} \cong 0.9990064130690307817 5$, when 10^4 terms are summed up. Obviously, this probability is much closer to 1 for $n = 100$. Therefore, we can safely conclude that an attack "à la Shamir" in order to obtain equivalent keys is infeasible.

6 Numerical Experiments

In order to determine primes $r < N$ verifying the condition (3), an algorithm has been developed and executed using the symbolic algebraic system MAPLE.

The algorithm has two phases: First, two cargo vectors $A = (a_1, a_2, a_3)$ and $B = (b_1, b_2, b_3)$ are generated at random satisfying [1, Theorem 1]. Then, primes p and q are generated satisfying equation (1), and $N = p \cdot q$. The knapsack $F = (f_1, f_2, f_3)$ is computed such that $f_i = a_i \pmod{p}$ and $f_i = b_i \pmod{q}$, $i = 1, 2, 3$, by applying the Chinese Remainder Theorem. In this way, we have a 'simplified' version of the cryptosystem (we have skipped steps 3, 6–7 in the key generation since they are immaterial for the search of the primes).

The second phase performs an exhaustive search, which is shown next using MAPLE pseudo-code:

```
r ← 2
while true do
  if r >= 225 ∑_{i=1}^{n} f_i (mod r) then
    print (r, f (mod r)):
```

```
      r ← nextprime (r)
   end if
   r ← nextprime (r)
   if r > N then break
   end if
end do
```

Observe that the search is limited to primes below N so that the computational cost in (2) is justified. The results are collected and displayed in Tables 1–4. Table 1 shows an excerpt of the 68 random cases analyzed: The columns p and q display the value of the two original primes and the column F shows the original knapsack.

The algorithm always finds the original primes p and q, but in some cases, it is able to also find primes other than the original ones. Tables 2–4 show different results, depending on the number of extra primes found. For each of them, the corresponding knapsack is also displayed (when sufficient space is available).

Table 1. Knapsacks with both primes

Case No.	p	q	F
1	33083	241207	[2565719013, 6495946386, 7638302318]
...			
5	88211	88883	[1750106353, 1096727571, 4818614157]
...			
12	80329	20261	[154794062, 1308880912, 546317621]
...			
19	37811	76079	[1554750563, 1170779877, 1340740290]
...			
32	2707	2402329	[5522956753, 1513472879, 3896580324]
...			
40	2742317	2707	[6888703221, 3323695212, 2012862941]
...			
42	90227	82129	[4043703568, 5785084778, 4179134259]
...			
45	98561	65027	[2562388976, 2562389121, 3934653778]
...			
49	38039	122651	[3922011138, 2825879350, 941959804]
...			
52	907	3046501	[2022880287, 2022883009, 2638273438]
...			
68	82591	101701	[3669270462, 958633913, 6775930908]

Table 2. Knapsacks with an extra prime

Case No.	r_1	$F \pmod{r_1}$
12	2276311	$[4914, 2087, 2981]$
32	11294383	$[3466, 25557, 18189]$
40	5484631	$[6685, 8826, 3364]$
42	19347841	$[4799, 80319, 603]$

Table 3. Knapsacks with two extra primes

Case No.	r_1	$F \pmod{r_1}$	r_2	$F \pmod{r_2}$
5	11667281	$[14203, 3157, 27104]$	11667301	$[11203, 1277, 18844]$
19	6294499	$[9310, 3063, 12003]$	6294511	$[6346, 831, 9447]$
49	17126491	$[44699, 8335, 2799]$	17126519	$[38287, 3715, 1259]$
52	3155819	$[308, 3030, 8754]$	6092993	$[6611, 9333, 7469]$

Table 4. Knapsacks with more than two extra primes

Case No.	r_1	r_2	r_3	r_4	r_5	r_6	r_7
45	1570091	4047997	4235351	6071993	12143939	12143981	14002117

7 Conclusions

- The idea of Wang and Hu is to propose yet another knapsack-type cryptosystem in an original and creative way. In this sense, [1, Theorem 1] is a very interesting result even from a pure mathematical viewpoint.
- A strength of the system is the fact that the existence of equivalent keys has an asymptotically null probability.
- As weaknesses, two cryptanalyses against the system have been published.
- The first one is the attack by Youssef. This attack is successful only if more than 60% of the plaintext is known. But this requirement imposes a strong limitation, so that the cryptanalysis is of little practical use.
- Lee's attack is more successful but it has also several drawbacks. One of them is the high computational cost, $O(n^6)$. A second one is the strong dependence on the shortest vector given by the LLL algorithm. Third, the author himself suggests how to circumvent the attack, though his solution implies a serious increase, more than 50%, in the ciphertext length.
- As a summary, existing attacks are successful for key lengths smaller than approximately 60,000 bits. Lee's attack is able to completely break Wang-Hu's cryptosystem unless we adopt the modification suggested by Lee himself,

but this implies again a key length on the order of 60,000 bits. These figures make it apparent that the cryptosystem is not competitive with respect to other common knapsack-type cryptosystems such as, for example, Chor-Rivest cryptosystem, that exhibits a good security level for key lengths as small as 30,000 bits, as explained in section 4.

Acknowledgment. This work has been partially supported by Ministerio de Ciencia e Innovación (Spain) under the Grant TIN2011-22668.

References

1. Wang, B., Hu, Y.: Quadratic compact knapsack public-key cryptosystem. Comput. Math. Appl. 59(1), 194–206 (2010)
2. Youssef, A.M.: Cryptanalysis of a quadratic knapsack cryptosystem. Comput. Math. Appl. 61(4), 1261–1265 (2011)
3. Lee, M.S.: Cryptanalysis of a quadratic compact knapsack public-key cryptosystem. Comput. Math. Appl. 62, 3614–3621 (2011)
4. Kate, A., Goldberg, I.: Generalizing cryptosystems based on the subset sum problem. Int. J. Inf. Secur. 10(3), 189–199 (2011)
5. Wang, B., Wu, Q., Hu, Y.: A knapsack-based probabilistic encryption scheme. Inform. Sci. 177(19), 3981–3994 (2007)
6. Youssef, A.M.: Cryptanalysis of a knapsack-based probabilistic encryption scheme. Inform. Sci. 179(18), 3116–3121 (2009)
7. Herrero, Á., Zurutuza, U., Corchado, E.: A Neural-Visualization IDS for Honeynet Data. International Journal of Neural Systems 22(2), 1–18 (2012)
8. Liu, H., Abraham, A., Snášel, V., McLoone, S.: Swarm scheduling approaches for work-flow applications with security constraints in distributed data-intensive computing environments. Information Sciences 192, 228–243 (2012)
9. Corchado, E., Herrero, Á.: Neural visualization of network traffic data for intrusion detection. Applied Soft Computing 11(2), 2042–2056 (2011)
10. Panda, M., Abraham, A., Das, S., Patra, M.R.: Network intrusion detection system: A machine learning approach. Intelligent Decision Technologies 5(4), 347–356 (2011)
11. Lenstra, A., Lenstra Jr., H., Lovász, L.: Factoring polynomials with rational coefficients. Math. Ann. 261, 515–534 (1982)
12. Nguyen, P.Q., Vallée, B. (eds.): The LLL Algorithm. Survey and Applications. Information Security and Cryptography. Springer, Heidelberg (2010)
13. Hernández Encinas, L., Muñoz Masqué, J., Queiruga Dios, A.: Analysis of the efficiency of the Chor-Rivest cryptosystem implementation in a safe-parameter range. Inform. Sci. 179, 4219–4226 (2009)
14. Vaudenay, S.: Cryptanalysis of the Chor-Rivest cryptosystem. J. Cryptology 14, 87–100 (2001)
15. Merkle, R., Hellman, M.: Hiding information and signatures in trap-door knapsacks. IEEE Trans. Inform. Theory 24(5), 525–530 (1978)
16. Shamir, A.: A polynomial time algorithm for breaking the basic Merkle-Hellman cryptosystem. IEEE Trans. Inform. Theory 30(5), 699–704 (1984)

17. Bach, E., Shallit, J.: Algorithmic Number Theory, Vol. I: Efficient Algorithms. The MIT Press, Cambridge (1996)
18. Knuth, D.: The Art of Computer Programming, 3rd edn. Addison-Wesley Series in Computer Science, vol. 2 - Seminumerical Algorithms. Addison-Wesley Publishing Co., Reading (1998)
19. Apostol, T.: Introduction to Analytic Number Theory, 4th corrected edn. Undergraduate Texts in Mathematics. Springer, NY (1976)

Equivalent Inner Key Recovery Attack to NMAC

Fanbao Liu[1,2], Changxiang Shen[2], and Tao Xie[3]

[1] School of Computer, National University of Defense Technology, Changsha, 410073, Hunan, P.R. China
[2] School of Computer, Beijing University of Technology, Beijing, 100124, P.R. China
[3] The Center for Soft-Computing and Cryptology, National University of Defense Technology, Changsha, 410073, Hunan, P.R. China
liufanbao@gmail.com

Abstract. We propose a general equivalent inner key recovery attack to the NMAC (Nested Message Authentication Code) instantiated with secure hash function in a related key setting, by applying a generalized birthday attack with two groups. We can recover the equivalent inner key of NMAC in about $2^{n/2+1}$ on-line MAC queries. The assumption of that the underlying hash function must be collision resistant is dropped in the security proof of NMAC. However, our result shows that NMAC, even instantiated with a collision resistant Merkle-Damgård hash function, is not secure as its designer claimed.

Keywords: NMAC, Equivalent Key Recovery, Verifiable Forgery, Birthday Attack.

1 Introduction

HMAC (Hash-based Message Authentication Code)[2,1], a derivative of NMAC (Nested Message Authentication Code), is a practically and commonly used, widely standardized MAC construction nowadays. HMAC has two advantages. First, HMAC can make use of current hash functions, the most widely used ones are based on Merkle-Damgård construction [12,4], without modification. Second, it is provable secure under two assumptions that the keyed compression function of the underlying hash function and the key derivation function in HMAC are pseudo random functions (PRFs) [1].

All in all, NMAC is the base of HMAC. For an iterated hash function H with Merkle-Damgård construction, NMAC is defined by

$$\mathrm{NMAC}_{(K_{\mathrm{out}}, K_{\mathrm{in}})}(M) = H(K_{\mathrm{out}}, H(K_{\mathrm{in}}, M))$$

where M is the input message, K_{in} and K_{out} are two random and independent secret n-bit keys.

Á. Herrero et al. (Eds.): Int. JointConf. CISIS'12-ICEUTE'12-SOCO'12, AISC 189, pp. 67–75.
springerlink.com

After some prevalent iterated hash functions had been broken [20,8,22,24,21], the security of NMAC and HMAC instantiated with those hash functions were analysed [3,6,19,23], which emphasized that NMAC and HMAC instantiated with broken hash functions are weak.

We analyse the security of NMAC based on the assumption that the underlying hash function is collision resistant (CR), instead of broken one. The security property CR is dropped to prove the security of NMAC [1]. Here, we notice that CR is a stronger notion than the origin assumption of that the underlying compression function is a PRF.

We propose the first and general equivalent key recovery attack to NMAC instantiated with secure hash function, which needs about $O(2^{n/2})$ on-line MAC queries to recover the equivalent inner key, in a related key setting. The attack is based on the generalized birthday attack with two groups. Our result shows that NMAC is not secure to some extent, since an attacker needs only $2^{n/2+1}$ calls to the NMAC oracle to recover the equivalent inner key, instead of the claimed 2^n calls.

This paper is divided into five sections. Section 2 presents the related definitions and background. Section 3 recalls some related work about key recovery attack to NMAC and HMAC. Section 4 proposes an equivalent key recovery attack to NMAC, in a related key setting. We conclude the paper in the last section.

2 Preliminaries

2.1 Notations

n	The length of hash result
b	The length of a message block
K	A secret key with n bits
$\|M\|$	The length of the string M
H	A concrete hash function with n-bit result
$x\|\|y$	The concatenation of two bit strings x and y
$pad(M)$	The padding bits of M with length information
ICV_i	The intermediate chaining variables for the i-th iteration of H

2.2 Birthday Paradox

A generalized variant. Given two groups G_1 with r elements, G_2 with s elements drawn uniformly and independently at random from $\{0,1\}^n$, find $x_1 \in G_1$ and $x_2 \in G_2$, such that $x_1 = x_2$.

The probability $\Pr(|G_1 \cap G_2| = i)$ that there are i distinct elements in the intersection of the two groups is denoted by $P(2^n, r, s, i)$. $P(2^n, r, s, i)$ converges towards a Poisson distribution $\wp_\lambda(i)$ with parameter λ, where $r \times s/2^n \to \lambda$, $r, s, 2^n \to +\infty$ [7].

A solution x_1, x_2 exists with great probability once $r \times s \geq 2^n$ holds, and if the list sizes are favourably chosen, the complexity of the optimal algorithm is $O(2^{n/2})$ [7,18].

The birthday problem has numerous applications throughout cryptography and cryptanalysis, and the direct one is collision searching.

2.3 Brief Description of Merkle-Damgård Hash Function

Cryptographic hash functions with Merkle-Damgård structure compress message M of arbitrary length to a fixed length output $H(M)$. MD5 [16] and SHA-1 [5] are two typical Merkle-Damgård structure hash functions in use, which take a variable-length message M (actually, $|M| < 2^{64}$) as input and outputs a 128-bit and 160-bit hash values, respectively.

M is first padded to be multiples of b bits, a '1', added at the tail of M, followed by '0's, until the bit length becomes $(b-64)$ on modulo b, and finally, the length of the unpadded message M is added to the last 64 bits. The padded M' is further divided into chunks of $(M_0, M_1, \ldots, M_{|M'|/b-1})$, each is a b-bit block.

The compression function h takes a b-bit block M_i and a n-bit chaining variable ICV_i, initialized to IV, as input, and outputs ICV_{i+1}. For example, $ICV_1 = h(IV, M_0)$, and $H(M) = ICV_N = h(ICV_{N-1}, M_{N-1})$. For the details of the concrete compression functions, please refer [15,16,5].

Extension Attack. Let $pad(M)$ denote the padding bits of M. For arbitrary unknown M_0, let $R = H(M_0)$, then for $M_1 = M_0||pad(M_0)||x$, where x is randomly generated. We can generate the hash of M_1 by computing $H(M_1) = h(R, x||pad(M_1))$, with no knowledge about M_0 except its length.

Security Properties of Hash Functions. Cryptographic hash functions need to satisfy the following security properties [11,13]:

1. pre-image resistance: it should be computation infeasible to find a x that satisfies $y = H(x)$, for any hash result y;
2. 2nd pre-image resistance: it should be computation infeasible to find a x' that satisfies $H(x') = H(x)$, for a known x;
3. collision resistance: it should be computation infeasible to find any two different inputs x and x', satisfying $H(x') = H(x)$.

For an ideal hash function with n-bit result, finding a pre-image requires approximately 2^n hash operations. On the other hand, finding a collision requires only $2^{n/2}$ hash operations; this follows from the birthday paradox [7]. In this paper, we assume that the underlying hash functions of all MACs are secure, which means collision resistance (CR).

2.4 NMAC

NMAC [2,1], proposed by Bellare et al., is the basis of the most widely used cryptographic algorithm HMAC. NMAC is built from iterated hash function H, where the IV of H is replaced with a secret n-bit key K. Based on the secret prefix approach [17], the NMAC algorithm is defined as:

$$\mathrm{NMAC}(M) = \mathrm{NMAC}_{(K_{\mathrm{out}},K_{\mathrm{in}})}(M) = H(K_{\mathrm{out}}, H(K_{\mathrm{in}}, M))$$

where keys K_{in}, $K_{\mathrm{out}} \in \{0,1\}^n$ in NMAC are to replace the IV of hash function H before further process. In practice, both keys are random and independently generated [2].

2.5 Security Notion of MAC

A universal forgery attack results in the ability to forge MACs for any message. A selective forgery attack results in a MAC tag on a message of the adversary's choice. An existential forgery merely results in some valid message/MAC pair not already known to the adversary.

3 Related Work

In AsiaCrypt 2006, Contni et al. presented distinguishing, forgery, and partial key recovery attacks on HMAC and NMAC instantiated with security broken hash function such as MD4, using collisions [3]. Their results demonstrated that the strength of NMAC and HMAC can be greatly weakened by the insecurity of the underlying hash functions.

In Crypto 2007, Fouque et al. explored the differential path of already broken hash function MD4 and presented the first key-recovery attack on NMAC and HMAC instantiated with MD4 after about 2^{88} MAC queries and 2^{95} hash computations.

In EuroCrypt 2008, Wang et al. showed how to use near collision, instead of collision of MD4, to launch key recovery attack to NMAC/HMAC-MD4, and they successfully reduced the MAC queries from 2^{88} to 2^{72}, and the number of hash computations from 2^{95} to 2^{77} [19].

In EuroCrypt 2009, Wang et al. presented a distinguishing attack on NMAC and HMAC instantiated with MD5 without related key, which needs 2^{97} MAC queries with success probability 0.87 [23].

4 Equivalent Key Recovery Attack to NMAC

As pointed out by Bellare et al., the on-line birthday attack for verifiable forgery attack is also applicable to NMAC [1]. In this section, we focus on how to recover the equivalent inner key of NMAC instantiated with secure hash functions, by using a generalized birthday attack in a related key setting.

Since any Merkle-Damgård hash function is vulnerable to extension attack, we may explore this property to launch selective forgery attack to NMAC. To achieve this target, we need to complete an equivalent inner key recovery attack to NMAC, and the only thing to do is to get the intermediate chaining variable ICV_i of the inner hashing of NMAC. However, we cannot launch such an attack without related key setting, since our attack is based on the assumption that the underlying hash function of NMAC is a secure one.

4.1 Related Key Setting

To recover the n-bit equivalent inner key K_{ei}, we have the following related key setting for our equivalent key recovery attack to NMAC.

There are two oracles $\text{NMAC}_{(K_{\text{out}},K_{\text{in}})}$ and $\text{NMAC}_{(K'_{\text{out}},K'_{\text{in}})}$. The relation between $(K_{\text{out}}, K_{\text{in}})$ and $(K'_{\text{out}}, K'_{\text{in}})$ is set as follows:

$$K_{\text{out}} = K'_{\text{out}} \quad \text{and} \quad K'_{\text{in}} \in \{Constants\}$$

where these two oracles share the same outer key, and the inner key of the oracle $\text{NMAC}_{(K'_{\text{out}},K'_{\text{in}})}$ can be any known constant, such as the IV.

4.2 Equivalent Inner Key Recovery

The overall strategy of the equivalent inner key recovery attack to NMAC, in a related key setting, is shown as follows.

1. Query $\text{NMAC}_{(K_{\text{out}},K_{\text{in}})}$ oracle for the corresponding values of $2^{n/2}$ different M_is, store their values in group one G_1.
2. Query $\text{NMAC}_{(K'_{\text{out}},K'_{\text{in}})}$ oracle for the corresponding values of $2^{n/2}$ different M'_js, store their values in group two G_2.
3. There will be a pair (M_i, M'_j) that not only satisfies $\text{NMAC}_{(K_{\text{out}},K_{\text{in}})}(M_i) = \text{NMAC}_{(K'_{\text{out}},K'_{\text{in}})}(M'_j)$ (the application of generalized birthday attack with two groups), but also satisfies $H(K_{\text{in}}, M_i) = H(K'_{\text{in}}, M'_j)$ (an inner collision happens).
4. Since $H(K_{\text{in}}, M_i) = H(K'_{\text{in}}, M'_j)$, and we know the value of K'_{in} and M'_j, hence we can calculate the very value of the equivalent inner key $K_{\text{ei}} = H(K_{\text{in}}, M_i) = H(K'_{\text{in}}, M'_j)$.

In the above attack, an inner collision between two oracles $\text{NMAC}_{(K_{\text{out}},K_{\text{in}})}$ and $\text{NMAC}_{(K'_{\text{out}},K'_{\text{in}})}$ must be found first. The problem is why an inner collision must happen. If we remove the outer hashing of NMAC, we can directly observer that a collision pair (M_i, M'_j) can be found with great probability, after querying both oracles of $H(K_{\text{in}}, M_i)$ and $H(K_{\text{in}'}, M'_j)$ with enough times. Moreover, the application of outer hashing of NMAC can't hide the existence of such inner collision.

How to Judge the Inner Collision. After a collision pair (M_i, M'_j) is found, we first generate the padding bits pad_0 for M_i and M'_j, where $pad_0 = pad(M'_j)$. Further, we randomly generate a message x, and append x to $M_i||pad_0$ and $M'_j||pad_0$, respectively. We query the corresponding MAC value on-line to both NMAC oracles for these two messages. After that, we further check whether $\text{NMAC}_{(K_{\text{out}},K_{\text{in}})}(M_i||pad_0||x) = \text{NMAC}_{(K'_{\text{out}},K'_{\text{in}})}(M'_j||pad_0||x)$ still holds. If so, (M_i, M'_j) is an exact inner collision pair between these two oracles, the attack succeeds. Otherwise, (M_i, M'_j) is an outer collision pair, which will be simply discarded.

Success Probability. We calculate the success probability of the above attack. We notice that $r = s = 2^{n/2}$ (hence $\lambda = r{\cdot}s/2^n = 1$), the probability sp of that at least one inner collision happens is computed as

$$sp = 1 - P(2^n, r, s, 0) = 1 - \wp_\lambda(0) + \varepsilon = 1 - e^{-1} + \varepsilon \geq 0.632$$

where $\varepsilon \leq 10^{-5}$ [7].

The elements of group G_1 and G_2 computed by NMAC need $2^{n/2}$ on-line NMAC queries, respectively. We can store these values of both groups in hash tables. The above algorithm will require $O(2^{n/2})$ time and space to be completed.

We point out that the equivalent inner key of NMAC is totally dependent on the generalized birthday attack (WCR), not the strength of the used inner key, in the related key setting.

However, if the outer key K_{out} of NMAC is leaked, then, it just needs a generalized birthday attack to recover the equivalent inner key to break the entire system [10].

We also point out that such an equivalent key recovery attack is not applicable to the outer key of NMAC, in a similar related key setting, since the outer hashing of NMAC, with fixed length input, is resistant to the extension attack[1].

4.3 Enlarging the Success Probability

In the above attack, the success probability of that at least one inner collision happens is greater than 0.632, which is acceptable sometimes. However, we can enlarge the success probability through doing more queries, since sp is determined by $P(2^n, r, s, 0)$, which converges towards a Poisson distribution with parameter λ.

For example, if we now want the success probability sp to be $sp \geq 1 - 10^{-4}$, by changing only r and s (but preserving $r = s$ both powers of 2), we can choose $r = s = 2^{n/2+2}$, and then $\lambda = r \cdot s/2^n = 16$, finally, we have

$$sp = 1 - P(2^n, 2^{n/2+2}, 2^{n/2+2}, 0) = 1 - \wp_\lambda(0) + \varepsilon = 1 - e^{-16} + \varepsilon \geq 1 - 10^{-4}$$

where $\varepsilon \leq 10^{-5}$.

4.4 Selective Forgery Attack to NMAC

After an inner collision pair (M_i, M'_j) is found, we can apply $H(K'_{\text{in}}, M'_j)$ to compute the equivalent inner key k_{ei} of NMAC. However, we can not use k_{ei} to launch selective forgery attack to NMAC directly, since we do not know the outer key of NMAC. But we notice that, in such a situation, NMAC is reduced to the no inner key version of $\text{NMAC}_{K_{\text{out}}}$, which is shown in (1). If we launch additional off-line birthday attack to $H(K_{\text{ei}}, x)$ further, which needs about $2^{n/2}$ hash computations, we can launch any selective forgery attack to NMAC through additional one on-line query.

---

[1] In fact, the outer hashing of NMAC is a PRF, for it applies the compression function directly, instead of the hash function.

$$\mathrm{NMAC}_{K_{\mathrm{out}}}(M) = H(K_{\mathrm{out}}, H(K_{\mathrm{ei}}, M)) \qquad (1)$$

However, it is interesting to notice that NMAC is provable secure under the assumption of that the underlying compression function h is a PRF [25], which means that collision resistance of the underlying hash function can be dropped.

As pointed out by the editors of Cryptology ePrint Archive in our preliminary version of this paper [9], the equivalent key recovery attack to NMAC is not applicable to practical HMAC, since the HMAC keys are derived from a base key, and then there exists no related key.

4.5 A Note on the Extension Attack

We conclude directly from our results that the security of NMAC is dependent on the secrecy of the outer key, instead of the secrecy of the both random and independent keys[2].

We further point out that the security of the envelop MAC with two different and independent keys K_1 and K_2, shown as (2), is totally dependent on collision resistance of the underlying hash function, not the strength of both keys. Since the outer key can be recovered through "Slice-by-Slice" attack [14] using hash collisions (collision resistance, with complexity of $O(2^{n/2})$), after the outer key is recovered, such MAC system can be attacked directly with extension attack to perform selective forgery attack[3], hence, the overall complexity is $O(2^{n/2})$.

$$Envelop_{(K_1,K_2)}(M) = H(K_1||M||K_2) \qquad (2)$$

5 Conclusion

We propose a general equivalent key recovery attack to NMAC, which can recover the equivalent inner key in about $2^{n/2+1}$ on-line MAC queries, in a related key setting. The attack is based on the generalized birthday attack, and is applicable to any concrete NMAC instantiated with Merkle-Damgård hash function. Moreover, this attack is based on the assumption that the underlying hash function is collision resistant, which is dropped in the security proof of NMAC [1], instead of broken hash function. Hence, our attack shows that the NMAC is not secure as it claimed to some extent, and confirms that the extension attack is a real threat to the Merkle-Damgård hash function.

Acknowledgement. We thank the anonymous reviewers for their valuable comments. This work was partially supported by the program Core Electronic Devices, High-end General Purpose Chips and Basic Software Products in China (No. 2010ZX01037-001-001), and supported by the 973 program of China under contract 2007CB311202, and by National Science Foundation of China through the 61070228 project.

[2] Once the outer key is leaked, the equivalent inner key can be recovered without a related key setting.

[3] The recovered outer key K_2 is just another kind of padding bits to the attacker.

References

1. Bellare, M.: New Proofs for NMAC and HMAC: Security Without Collision-Resistance. In: Dwork, C. (ed.) CRYPTO 2006. LNCS, vol. 4117, pp. 602–619. Springer, Heidelberg (2006)
2. Bellare, M., Canetti, R., Krawczyk, H.: Keying Hash Functions for Message Authentication. In: Koblitz, N. (ed.) CRYPTO 1996. LNCS, vol. 1109, pp. 1–15. Springer, Heidelberg (1996)
3. Contini, S., Yin, Y.L.: Forgery and Partial Key-Recovery Attacks on HMAC and NMAC Using Hash Collisions. In: Lai, X., Chen, K. (eds.) ASIACRYPT 2006. LNCS, vol. 4284, pp. 37–53. Springer, Heidelberg (2006)
4. Damgård, I.: A Design Principle for Hash Functions. In: Brassard, G. (ed.) CRYPTO 1989. LNCS, vol. 435, pp. 416–427. Springer, Heidelberg (1990)
5. Eastlake, D.E., Jones, P.: US secure hash algorithm 1 (SHA1). RFC 3174, Internet Engineering Task Force (September 2001), `http://www.rfc-editor.org/rfc/rfc3174.txt`
6. Fouque, P.-A., Leurent, G., Nguyen, P.Q.: Full Key-Recovery Attacks on HMAC/NMAC-MD4 and NMAC-MD5. In: Menezes, A. (ed.) CRYPTO 2007. LNCS, vol. 4622, pp. 13–30. Springer, Heidelberg (2007)
7. Girault, M., Cohen, R., Campana, M.: A Generalized Birthday Attack. In: Günther, C.G. (ed.) EUROCRYPT 1988. LNCS, vol. 330, pp. 129–156. Springer, Heidelberg (1988)
8. Leurent, G.: MD4 is Not One-Way. In: Nyberg, K. (ed.) FSE 2008. LNCS, vol. 5086, pp. 412–428. Springer, Heidelberg (2008)
9. Liu, F., Shen, C., Xie, T., Feng, D.: Cryptanalysis of HMAC and Its Variants (2011) (unpublished)
10. Liu, F., Xie, T., Shen, C.: Equivalent Key Recovery Attack to H^2-MAC. International Journal of Security and Its Applications 6(2), 331–336 (2012)
11. Menezes, A.J., Vanstone, S.A., Oorschot, P.C.V.: Handbook of Applied Cryptography, 1st edn. CRC Press, Inc., Boca Raton (1996)
12. Merkle, R.C.: One Way Hash Functions and DES. In: Brassard, G. (ed.) CRYPTO 1989. LNCS, vol. 435, pp. 428–446. Springer, Heidelberg (1990)
13. Preneel, B.: Cryptographic Primitives for Information Authentication - State of the Art. In: Preneel, B., Rijmen, V. (eds.) State of the Art in Applied Cryptography. LNCS, vol. 1528, pp. 49–104. Springer, Heidelberg (1998)
14. Preneel, B., van Oorschot, P.C.: On the Security of Two MAC Algorithms. In: Maurer, U.M. (ed.) EUROCRYPT 1996. LNCS, vol. 1070, pp. 19–32. Springer, Heidelberg (1996)
15. Rivest, R.: The MD4 Message-Digest algorithm. RFC 1320, Internet Engineering Task Force (April 1992), `http://www.rfc-editor.org/rfc/rfc1320.txt`
16. Rivest, R.: The MD5 Message-Digest algorithm. RFC 1321, Internet Engineering Task Force (April 1992), `http://www.rfc-editor.org/rfc/rfc1321.txt`
17. Tsudik, G.: Message authentication with one-way hash functions. SIGCOMM Comput. Commun. Rev. 22, 29–38 (1992)
18. Wagner, D.: A Generalized Birthday Problem. In: Yung, M. (ed.) CRYPTO 2002. LNCS, vol. 2442, pp. 288–304. Springer, Heidelberg (2002)
19. Wang, L., Ohta, K., Kunihiro, N.: New Key-Recovery Attacks on HMAC/NMAC-MD4 and NMAC-MD5. In: Smart, N.P. (ed.) EUROCRYPT 2008. LNCS, vol. 4965, pp. 237–253. Springer, Heidelberg (2008)

20. Wang, X., Lai, X., Feng, D., Chen, H., Yu, X.: Cryptanalysis of the Hash Functions MD4 and RIPEMD. In: Cramer, R. (ed.) EUROCRYPT 2005. LNCS, vol. 3494, pp. 1–18. Springer, Heidelberg (2005)
21. Wang, X., Yin, Y.L., Yu, H.: Finding Collisions in the Full SHA-1. In: Shoup, V. (ed.) CRYPTO 2005. LNCS, vol. 3621, pp. 17–36. Springer, Heidelberg (2005)
22. Wang, X., Yu, H.: How to Break MD5 and Other Hash Functions. In: Cramer, R. (ed.) EUROCRYPT 2005. LNCS, vol. 3494, pp. 19–35. Springer, Heidelberg (2005)
23. Wang, X., Yu, H., Wang, W., Zhang, H., Zhan, T.: Cryptanalysis on HMAC/NMAC-MD5 and MD5-MAC. In: Joux, A. (ed.) EUROCRYPT 2009. LNCS, vol. 5479, pp. 121–133. Springer, Heidelberg (2009)
24. Xie, T., Liu, F., Feng, D.: Could The 1-MSB Input Difference Be The Fastest Collision Attack For MD5? In: Eurocrypt 2009, Poster Session, Cryptology ePrint Archive, Report 2008/391 (2008), `http://eprint.iacr.org/`
25. Yasuda, K.: HMAC without the "Second" Key. In: Samarati, P., Yung, M., Martinelli, F., Ardagna, C.A. (eds.) ISC 2009. LNCS, vol. 5735, pp. 443–458. Springer, Heidelberg (2009)

Hybrid Compression of the Aho-Corasick Automaton for Static Analysis in Intrusion Detection Systems

Ciprian Pungila

West University of Timisoara,
Blvd. V. Parvan 4, Timisoara 300223, Timis, Romania
cpungila@info.uvt.ro
http://info.uvt.ro/

Abstract. We are proposing a hybrid algorithm for constructing an efficient Aho-Corasick automaton designed for data-parallel processing in knowledge-based IDS, that supports the use of regular expressions in the patterns, and validate its use as part of the signature matching process, a critical component of modern intrusion detection systems. Our approach uses a hybrid memory storage mechanism, an adaptation of the Smith-Waterman local-sequence alignment algorithm and additionally employs path compression and bitmapped nodes. Using as a test-bed a set of the latest virus signatures from the ClamAV database, we show how the new automata obtained through our approach can significantly improve memory usage by a factor of times compared to the unoptimized version, while still keeping the throughput at similar levels.

Keywords: static analysis, aho-corasick, smith-waterman, high compression, pattern fragmentation, parallel algorithm.

1 Introduction

The static analysis process implemented in intrusion detection systems nowadays aims at identifying malicious patterns of code inside an executable's binary code. This is achieved through multiple-pattern matching algorithms, such as the Aho-Corasick [1] method, and is still accepted as being the best method of employing almost (if not entirely) zero false positives in the process. The primary issue for deployment is the high-amount of memory such an approach uses, since often the automata contains several millions of states. We will relate to ClamAV [2], an open-source antivirus with a large malware signature database.

We discuss related work in section II of this paper. We then describe our algorithm and methodology in section III, while section IV contains the experimental results achieved during our tests.

2 Related Work

Multiple pattern matching algorithms are commonly used in intrusion detection systems (IDS), especially in antivirus engines [3] [4] or network IDS [5] [6].

Á. Herrero et al. (Eds.): Int. JointConf. CISIS'12-ICEUTE'12-SOCO'12, AISC 189, pp. 77–86.
springerlink.com

An adaptive, unsupervised learning Aho-Corasick-based approach for analyzing program behavior through system call analysis was also presented in [7]. Path compression for the Aho-Corasick automata was introduced by Tuck et al and applied to the Snort signatures database [8] and improved slightly by Zha et al [9]. A method for efficiently serializing very large automata used in virus signature matching was first proposed in [10]. In [11] a model is shown for finding the severity of a threat in the cloud, with the model proposed supporting system call analysis and a detection engine backed by an attack database.

Our main paper's contributions are: a) proposing an approach based on dynamic programming (the Smith-Waterman algorithm in our case) that reduces memory usage of large automata used in knowledge-based IDS by extracting their common subsequences and storing them only once, without significantly affecting throughput, as well as b) implementing and testing our approach on viral signatures, along with the help of other techniques, while maintaining the local hit ratio. To the best of our knowledge, there are no studies showing the memory efficiency of large automata used in knowledge-based IDS for malware detection, for instance when using the ClamAV antivirus signatures, nor any studies showing how much this can be reduced through path-compression, bitmapped nodes or pattern fragmentation by employing dynamic programming methods, such as the approach we discuss in this paper. Our approach can be successfully used for modeling the detection engine shown in [11] (as well as the system call analysis handler, using a pattern matching approach as shown in [7]). While unsupervised learning methods based on pattern recognition have been proposed for network IDS in [12] (using ANNs) and [13] (using decision trees and several stochastic models), our approach is aimed at an approach based on one of the fastest algorithms for multiple pattern matching, along with the dynamic programming aspect of extracting common subsequences from the patterns for minimizing the storage space required and studying the impact on the overall performance.

2.1 The Aho-Corasick Algorithm

The Aho-Corasick (A-C) algorithm [1] is a widely-spread approach for performing multiple pattern matching and involves building a deterministic finite state machine (DFSM) that, in the unoptimized version, parses input characters based on success pointers and a failure pointer, which points to another state in the automaton in case a mismatch occurs at the current state. The failure pointer is computed as follows (Figure 1a): a) for the root and all children at the first level of the automaton, the failure always points back to the root; b) for all other states, the failure pointer is computed as follows: i) we find the largest suffix of the word at the current level of the tree that is also matched by the automata, and we point back to it, or ii) if no such suffix of the word exists, the failure pointer links back to the root.

Path-compression is achieved by grouping multiple consecutive states in the automata that only have one single possible transition pointer, into a single, compressed node which stores the transitions as a linked list, along with a list of failure pointers for each character of them. This introduces additional processing complexity since it may happen that a failure pointer of another state will point to a certain character inside a compressed node, therefore each failure pointer now requiring an offset for this particular situation. Bitmapped nodes have also been used for reducing the data structure memory size, therefore reducing the overall memory required for processing the Snort patterns that were processed.

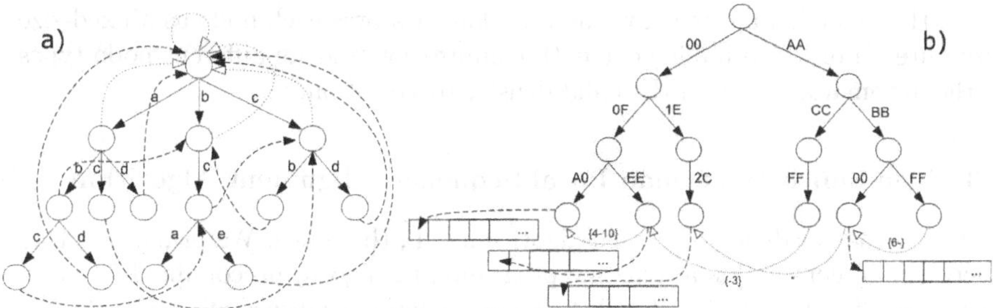

Fig. 1. a) An Aho-Corasick automaton for the input set {abc, abd, ac, ad, bca, bce, cb, cd}. Dashes transitions are failure pointers. b) Our extended Aho-Corasick automaton (failures not included) for the regular expressions 001E2C*AABB00{6-}AABBFF and 000FA0{4-10}000FEE{-3}AACCFF.

2.2 Memory Efficiency

A highly efficient automata serialization technique, aimed at depth-limited or smaller trees and supporting one single node type, was proposed in [10], where a hybrid CPU-GPU-based approach was used to transform very large A-C and Commentz-Walter automata into smaller stacks, by compressing the data structures in a single stack-based model. In this approach, all children of a given node are not stored at random pointers in memory, but at consecutive offsets in the stack, avoiding therefore the memory gaps created by memory allocation routines (Figure 2b).

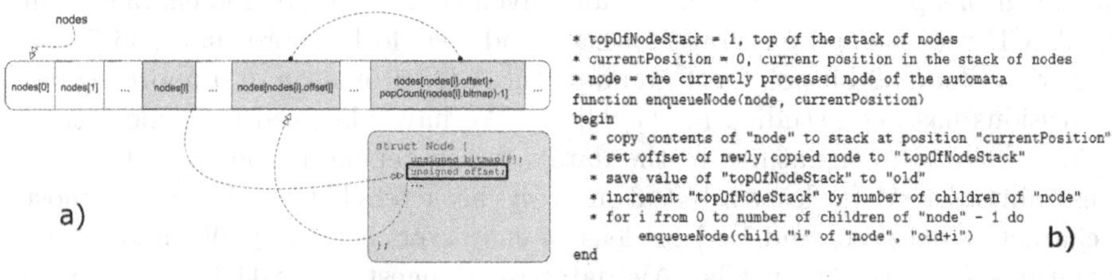

Fig. 2. a) The stack-based representation of very large Aho-Corasick automata; b) The modified algorithm for building a compact, stack-based representation of very large Aho-Corasick automata.

We have extended the model in [10] to support bitmapped nodes, as well as path-compressed nodes and added support for the different types of signature constraints existing in the ClamAV database. In the original approach, there were two stacks created: a *node* stack (used to store all nodes, with the single constraint of specifying that all children of a node must be stored as consecutive memory cells in the stack) and an *offset* stack (used to specify the starting offsets of the children of all nodes in the *node* stack); to reduce memory usage even further, we have completely discarded the *offset* stack (Figure 2a). Furthermore, the extended approach we used no longer stores each node as a fixed-size structure, in order to implement path-compression and its different node types in the automata, based on the child density distribution.

2.3 The Smith-Waterman Local Sequence Alignment Algorithm

Based on the Needleman-Wunsch algorithm [14], the Smith-Waterman local sequence alignment [15] is a dynamic programming algorithm, commonly used in bioinformatics [16] for identifying the longest common subsequences in two DNA strands. It makes use of a two-dimensional matrix, where each cell corresponds to a pairing of a letter from each sequence, and starts from the upper-left corner, following a mostly diagonal path, moving down or to the right. In case of an alignment of the two letters, the path traverses the matrix diagonally. There are two steps for computing the alignment: compute the similarity score and backtrace into the similarity matrix and search for the optimal alignment. A value $v_{i,j}$ in the matrix for sequences A (of length n) and B (of length m) is being computed as below in *O(mn)*, where *sub*, *ins* and *del* are penalties for substitution, insertion and deletion. We used $ins = 1, del = 1$, $sub = 2$. An important application is in the problem of finding the longest common subsequence (LCS) and finding all common subsequences between two given sequences.

3 Implementation

A signature from the ClamAV database is *00687474{10-40}2E70{-14}00*0053* and has several subsequences that are inserted into the automata along with their constraints. Since the unoptimized automata needs to store a pointer for all transitions possible at each node, and given that there are 256 characters in the ASCII set, the space storage for each node would be more than $256 * 4 =$ 1,024 bytes for a 32-bit pointer. Extending the A-C automaton to support regular expressions has been studied in [17] [4] [7]. We have observed than more than 96.5% of the viral signatures in the database are depending on three types of constraints: {*n-m*} (at least n but at most m characters between), {*-n*} (at most n characters between) and {*n-*} (at least n characters between). We used 62,302 signatures from the main ClamAV database (longest was 804 bytes, average length was 45 bytes), and we extended the A-C automata to support regular expression matching based on these three types of constraints, extending an idea described in [7]. In the preprocessing part, we split a signature that is a regular expression into several subexpressions, each being added individually into the automaton, then storing for each of these leaves a pointer to the previous one, including a list of positions in the file where previous matches occurred (Figure 1b). Later on, this list gets updated dynamically as the input file is being parsed and matches occur - this information is then being used for matching a regular expression by analyzing the previous occurrences of all its subexpressions.

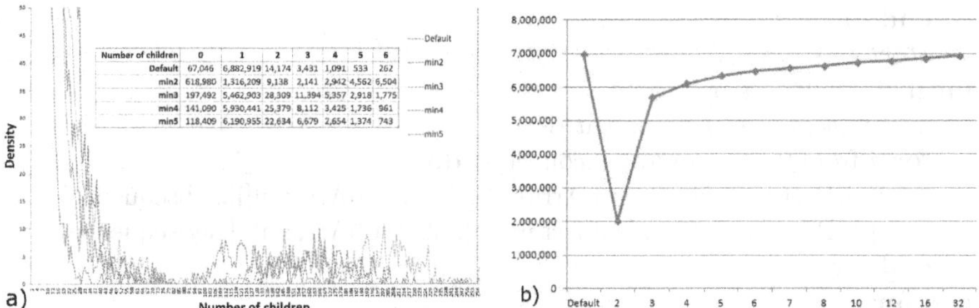

Fig. 3. a) Children density distribution in the automaton built from the ClamAV database. Similar distributions (not shown) as for min were also obtained for max. b) The total number of nodes obtained for our automata, when using the ClamAV signatures and also when using our **min**-based common signature extraction approach. For **max**, a closely similar distribution was obtained.

Since many viruses share common areas of code, which leads to common sequences in the malware signatures, our idea to compress the Aho-Corasick automaton is based on the ability to fragment patterns in the database in the preprocessing stage by identifying these, along with the path compression technique. Given the large number of signatures (after sorting them upfront), we have applied the Smith-Waterman algorithm for identifying all common subsequences of code in the signatures, in groups of 1,000, progressively, up to the maximum tested. Assuming the common subsequence is C, and that the words in the tree are $P_1 = S_1CS_2$ and $P_2 = T_1CT_2$, the resulting modification will imply transforming the patterns into $P_1 = S_1\{0-0\}C\{0-0\}S_2$ and $P_2 = T_1\{0-0\}C\{0-0\}T_2$, so that C is stored only once in the automaton. We have used minimum subsequence lengths of 2 to 8, 10, 12, 16 and 32 and also two variations of the algorithm, one that performs splits starting with the longest common subsequences first (referenced as *max*), while the other begins with the shortest (referenced as *min*):

$start \leftarrow 0$, $LIMIT \leftarrow 1000$, $LENGTH \leftarrow \{2,3,4,5,6,7,8,10,12,16,32\}$
$beginFrom \leftarrow 2, endAt \leftarrow maxKeywordLength$ (for min, reversed for max)
repeat
    $lowerLimit = start \times LIMIT$
    $upperLimit = lowerLimit \times LIMIT - 1$
    for i from $lowerLimit$ to $upperLimit$ **do**
        for j from $i+1$ to $upperLimit$ **do**

```
            pArray ← getAllCommonSubsequences(pattern[i], pattern[j])
            for k from 0 to pArray.count()-1 do
                if pArray[k].length() > LENGTH then
                    insertSubSeqInArray(pArray[k].length(), pArray[k], i, j)
                    incrementOccurences(pArray[k])
                end if
            end for
        end for
    end for
    start ← start + 1
until all sequences processed
for i from beginFrom to endAt do
    for j from 0 to subArray[i].count()-1 do
        splitPattern(pattern[subArray[i][j].id1], subArray[i][j].subsequence)
        splitPattern(pattern[subArray[i][j].id2], subArray[i][j].subsequence)
    end for
end for
```

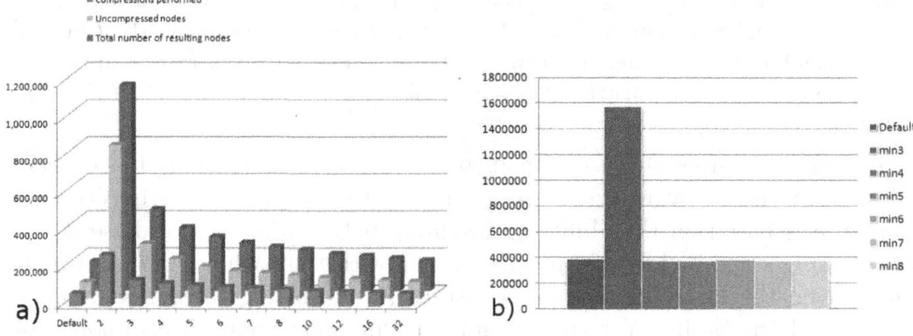

Fig. 4. a) The resulting number of nodes after applying path-compression in our automata, for the **min** approach. A highly similar distribution was obtained for the **max** approach also. b) Local hits during test scenario.

The number of nodes resulted is shown in Figure 3b, while path-compression results are shown in Figure 4a. We implemented two variants: the first uses, for each state in the automaton, a sorted list with all the possible transitions and the binary search approach to achieve locating a transition in $O(log_2 N)$ (which is a good compromise for reducing the memory required for storing the automaton by an important factor), and a bitmapped-version, where popcounts are computed faster than performing a binary search, and uses even less memory, starting from an idea proposed by Zha [9] for Snort signatures. Based on the observations in Figure 3a, the nodes having the most transitions are the ones corresponding to child node densities from 0 to 15, followed by the ones corresponding to densities from 16 to 48. We have setup therefore four types of bitmapped nodes:

type I, for nodes having from 0 to 15 transitions, type II for nodes having from 16 to 48 transitions, type III, for generic bitmapped nodes and type IV for compressed nodes. In fact, in the $O(log_2N)$ implementation, to even further reduce memory usage, types I, II and III all have two sub-types: the first includes the *bitmapSequences, pointer* and *matches* fields (which only apply to nodes that are "leaves" of a subexpression), while the second sub-type does not have these fields (this only applies to nodes that are not leaves, so we do not need that specific information for them). In conclusion, the bitmapped implementation has 4 types of nodes, while the non-bitmapped implementation uses 7 types of nodes (4 primary types and 3 sub-types).

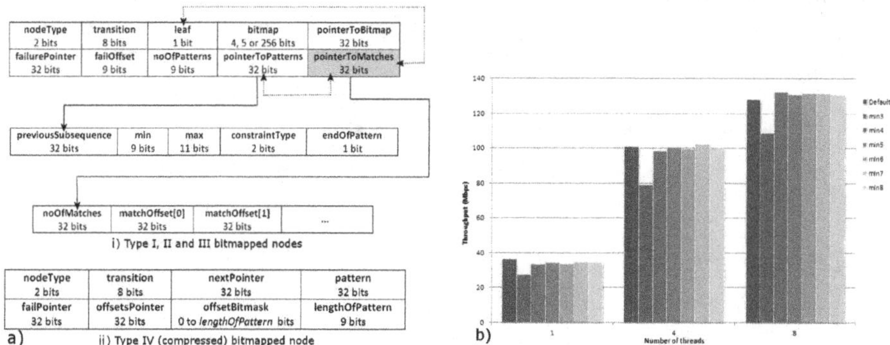

Fig. 5. a) Compressed nodes. b) Throughput of the automaton in multi-threaded implementations.

Our path-compression method applies only to nodes that have a single transition possible, so all leaf nodes are uncompressed nodes. Also, the lists used for storing match positions are dynamically modified: when a match is found, if the match is part of a regular expression, a check is being made to determine if the previous subexpression node has been matched before, and if the constraints are being met (there are four types of constraints, but constraints of types $\{n\text{-}\}$ are the only ones that, when matched, permanently store the match position), store the match position at the current node in the list; if constraints are not met, remove all permitted previous matches. In Figure 5a, the *transition* field is used for nodes whose parent node is a type I or II node. To locate if at the current node there is a match for the current input character, we are storing all children as a sorted list with the key *transition* representing the character for a match. We are using a binary search approach for locating a match, that performs a search in $O(log_2N)$ (where N is the total number of children of that particular node). For all other nodes, this field is ignored. The *leaf* field ensures the presence (or absence) of the *pointerToMatches* and *pointerToPatterns* fields. In order to avoid sparse memory locations, type IV (bitmapped) nodes in Figure 5a are followed in the actual node stack by the pattern itself, stored as a normal sequence of characters, after which the next node structure follows. We implemented a version of the algorithm that performs data-parallel scanning on the automaton using different pointers and different linked lists in each thread, that implements overlapped scanning at the boundaries of the data allocated for each thread (of length *d-1*, where *d* is the maximum possible depth of the automaton from the current level) [7] and using a hashtable for all matches found to avoid duplicates.

4 Experimental Results

We have conducted our experiments on a quad-core Intel i7 2600K CPU, clocked at 4.3 GHz supporting Hyper-Threading and the memory usage obtained is shown in Figure 6.

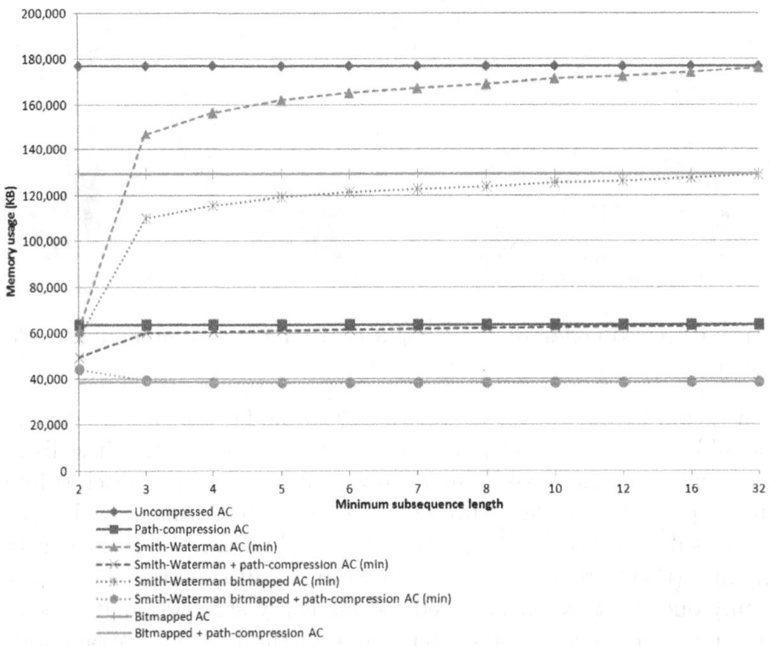

Fig. 6. Memory usage for different implementations of the static analysis automaton

Results have shown that for the default database of signatures from ClamAV, applying path-compression to the unoptimized automaton had achieved a drop in memory usage of more than 64%, a simple bitmap implementation (that does not employ path-compression) had reduced the memory usage by almost 27%, while the bitmapped implementation employing path-compression has reduced memory usage by more than 78%. By pre-processing the patterns and applying the pattern fragmentation algorithm presented earlier, we were able to achieve even better compression ratios, the best occurring (by creating a highly dense tree on the first few levels, as shown earlier) for a minimum subsequence length of 2 (Figure 6), but at the very high expense of processing time. The pattern fragmentation approach becomes however very efficient for a high number of patterns to be matched containing a large number of common subsequences of minimum lengths 3 or 4 (since the probability of a local hit decreases as the minimum length increases).

We tested the throughput of the automaton on a file comprised of 50 MB random binary data. The results (Figure 5b) show that for a minimum subsequence length of 2 (for both *min* and *max*), the throughput decreases significantly (to about 0.1 Mbps for the single-threaded implementation and 0.4 Mbps for the 8-threads version, which is why it was not included in the graphic) compared to the default one (which is explained by the high-density of nodes at the first three levels in the automaton, that highly increases the probability of a match), however for a minimum subsequence length of 3 the throughput decreases by about 25%, while for the next lengths it is kept at the same level as the one in the default database (because of lower match probabilities). The local-hit ratio, which determines the number of matches achieved in the leaf nodes of the tree, shows that for a minimum pattern length of 2 the number of hits is very high (about 31 hits per byte), for length 3 the local-hit ratio is about 4 times bigger than in the default approach, while for the next lengths (Figure 4b) they are about the same as when using the default database (the highest throughput was achieved for a minimum subsequence length of 4 when using 8 threads).

5 Conclusion

We have proposed a hybrid method for compressing the Aho-Corasick automata used in knowledge-based IDS and tested it for virus signature matching, and have shown how we can reduce the memory used by the unoptimized automata by as much as 78% using our own approach based on pattern fragmentation through dynamic programming, combined with an efficient stack-based representation supporting bitmapped nodes along with a modified path-compression technique. Throughputs of the automata from various scenarios tested were compared and we have determined that for long-enough subsequence lengths, the throughput matches that of the original automaton, also decreasing slightly the local hit ratio.

We believe that our approach could be also applied to a wider spectrum of domains that employ exact pattern-matching using a high number of patterns and regular expressions, such as DNA sequencing, natural language processing and others. Future work includes implementing additional support in the automaton to further reduce the processing complexity through constraints, such as dependencies on the executable's entry point.

Acknowledgements. This work was partially supported by the grant of the European Commission FP7-REGPOT-CT-2011-284595 (HOST), and Romanian national grant PN-II-ID-PCE-2011-3-0260 (AMICAS).

References

1. Aho, A., Corasick, M.: Efficient string matching: An Aid to blbiographic search. CACM 18(6), 333–340 (1975)
2. Clam AntiVirus, http://www.clamav.net
3. Cha, S.K., Moraru, I., Jang, J., Truelove, J., Brumley, D., Andersen, D.G.: Split Screen: Enabling Efficient, Distributed Malware Detection. In: Proc. 7th USENIX NSDI (2010)
4. Lee, T.H.: Generalized Aho-Corasick Algorithm for Signature Based Anti-Virus Applications. In: Proceedings of 16th International Conference on Computer Communications and Networks, ICCN (2007)
5. Snort, http://www.snort.org/
6. Paxson, V.: Bro: A system for detecting network intruders in real-time. Computer Networks 31, 2435–2463 (1999)
7. Pungila, C.: A Bray-Curtis Weighted Automaton for Detecting Malicious Code Through System-Call Analysis. In: 11th International Symposium on Symbolic and Numeric Algorithms for Scientific Computing (cisis), pp. 392–400 (2009)
8. Tuck, N., Sherwood, T., Calder, B., Varghese, G.: Deterministic memory-efficient string matching algorithms for intrusion detection. In: 23rd Annual Joint Conference of the IEEE Computer and Communications Societies (INFOCOM), vol. 4, pp. 2628–2639 (2004)
9. Zha, X., Sahni, S.: Highly Compressed Aho-Corasick Automata For Efficient Intrusion Detection. In: IEEE Symposium on Computers and Communications (ISCC), pp. 298–303 (2008)
10. Pungila, C., Negru, V.: A Highly-Efficient Memory-Compression Approach for GPU-Accelerated Virus Signature Matching. In: Information Security Conference, ISC (2012)
11. Arshad, J., Townend, P., Xu, J.: A novel intrusion severity analysis approach for Clouds. Future Generation Computer Systems (2011), doi:10.1016/j.future.2011.08.009
12. Corchado, E., Herrero, A.: Neural visualization of network traffic data for intrusion detection. Applied Soft Computing 11(2), 2042–2056 (2011)
13. Panda, M., Abraham, A., Patra, M.R.: Hybrid Intelligent Approach for Network Intrusion Detection. Procedia Engineering 30, 1–9 (2012), doi:10.1016/j.proeng.2012.01.827
14. Needleman, S.B., Wunsch, C.D.: A general method applicable to the search for similarities in the amino acid sequence of two proteins. J. Mol. Biol. 47, 443–453 (1970)
15. Smith, T.F., Waterman, M.S.: Identification of Common Molecular Subsequences. J. Mol. Biol. 147, 195–197 (1981)
16. Bioinformatics Explained: Smith-Waterman, http://www.clcbio.com/sciencearticles/Smith-Waterman.pdf
17. Scarpazza, D.P.: Top-Performance Tokenization and Small-Ruleset Regular Expression Matching. Int. J. Parallel Prog. 39, 3–32 (2011)

Tracking Traitors in Web Services via Blind Signatures

J.A. Álvarez-Bermejo[1] and J.A. López-Ramos[2]

[1] Dpt. Arquitectura de Computadores
Universidad de Almería
jaberme@ual.es

[2] Dpt. Álgebra y Análisis Matemático
Universidad de Almería
jlopez@ual.es

Abstract. This paper presents a method and its implementation proposal, built on the blind signatures protocol, to trace and report users sharing a valid license illegally when accessing services provided through Internet (Web services, Streaming, etc). The method devised is able to identify the legitimate user from those users who are illegally accessing services with a shared key. This method is robust when detecting licenses built with no authorization. An enhancement of the protocol to identify the last usage of a certain license is also provided, allowing to detect a traitor when an unauthorized copy of a license is used.

1 Introduction

The distribution of software, protected by licenses, has always been an issue for the intellectual creators and their business' models. Implementing countermeasures for software piracy is a must for the software industry even before the boom of the Internet era, the *RojaDirecta* saga should make clear why the software industry -and the content industry- is looking for new enforcement tools [1]. When the Internet was not so popular, sharing licenses was hard and the methods employed to protect software were merely based in built-in passwords. Reverse engineering, decompilation and other not so advanced techniques made it possible; a further step was then adopted, hash functions and cryptography. As the code to check licenses was embedded in the program, the illegal usage and duplication of licenses were an easy task. Support to avoid the discovery of built-in hash functions was also provided from the Operating Systems in order to harness possible abuses. As the Internet began to be used as a medium to distribute licensed software (p2p networks are the proof), the licensing protection mechanism began to be weaker, those who cracked software posted their tricks or distributed software packages to let any user unprotect software.

As the Internet acquired more relevance, protecting licenses and software became harder. Software is illegally cracked and then redistributed, in a high percent of the distributions *malware* is included in the distributed packages,

Á. Herrero et al. (Eds.): Int. JointConf. CISIS'12-ICEUTE'12-SOCO'12, AISC 189, pp. 87–96.
springerlink.com

therefore creating ways to protect software is necessary not only for the sake of the creators' economy but for the integrity of the users, that might be compromised. As a solution, many authentication services were moved to Internet servers, using licensed software means that it might be necessary establishing a first connection to a server in order to authenticate the software and let it run. Examples of such software range from antivirus software to games. Implementing a license protection is a hot spot in software factories like the antivirus industry. The antiviruses need a license to retrieve updated virus signatures databases in order to keep the system protected. The strength of an antivirus, among others, relies on its databases. Antivirus companies invest so much effort in building such databases. A duplicated license harness this useful effort.

This paper proposes a method and a feasible architecture to access services through the Internet, using licenses that cannot be replicated, duplicated, shared, or even cracked. Concurrent usage of licenses would trigger the alarm system. When licenses are not used concurrently, the method is able to detect the abuse and report it to the legitimate user of the original license. The licenses do not add information on users' identity as opposed to other contributions like the one developed in [2] and protects identity of legitimate users as in [3]. This work not only detects false licenses but also abuses (like duplications, stolen keys, etc) of original licenses. Comparing our method with the one developed in [2] ours do not act on content itselft but on the permission of users to access the media. This way our protocol is faster and therefore data can be cyphered.

1.1 Agent Based Computing

Traditional methods to establish connections to Internet servers do not offer the dynamism that a robust licensing mechanism may need so we based our method on a layered multiagent architecture [4].Distributed systems based on agents are very attractive because of their inherent scalability and autonomy. Viewing software agents (usually in the form of objects) as brokers and the interactions among agents as the exchange of services, a multiagent system closely resembles a community of human beings doing business with each other, and are widely envisioned to be able to perform many commercial activities on behalf of human beings. Useful properties of agents (see [5]) that distinguish them from traditional processes or threads:

- reactive: responds fast to changes in the environment.
- autonomous: controls its own actions.
- proactive: does not simply act in response to the environment.
- temporally continuous: continually running process.
- communicative/sociallyable: communicates with other agents
- learning/adaptive: changes its behaviour based on its previous experiences.
- mobile : able to migrate from one machine to another.
- flexible: the answer to an event is not predefined.

A multiagent system is a collection of agents that work in conjunction with one another. They may cooperate to achieve a common goal or compete to achieve individual goals. In this system the use of agents enhances communication and allows for speedier transfer of data. Agent oriented computing differs from centralized object based computing in several ways and as such requires different analysis and design methodologies [6]. Devices such as smartphones have seen a major upsurge in popularity. Using a multiagent layered architecture to stream media to those devices is efficient, as shown in [7].

1.2 Multiagent Layered Architecture Proposal

A multiagent layered architecture, as shown in figure 1, is the implementation proposed for the method developed in this paper to protect media/content streaming, software or whatever that is covered by a license restriction.

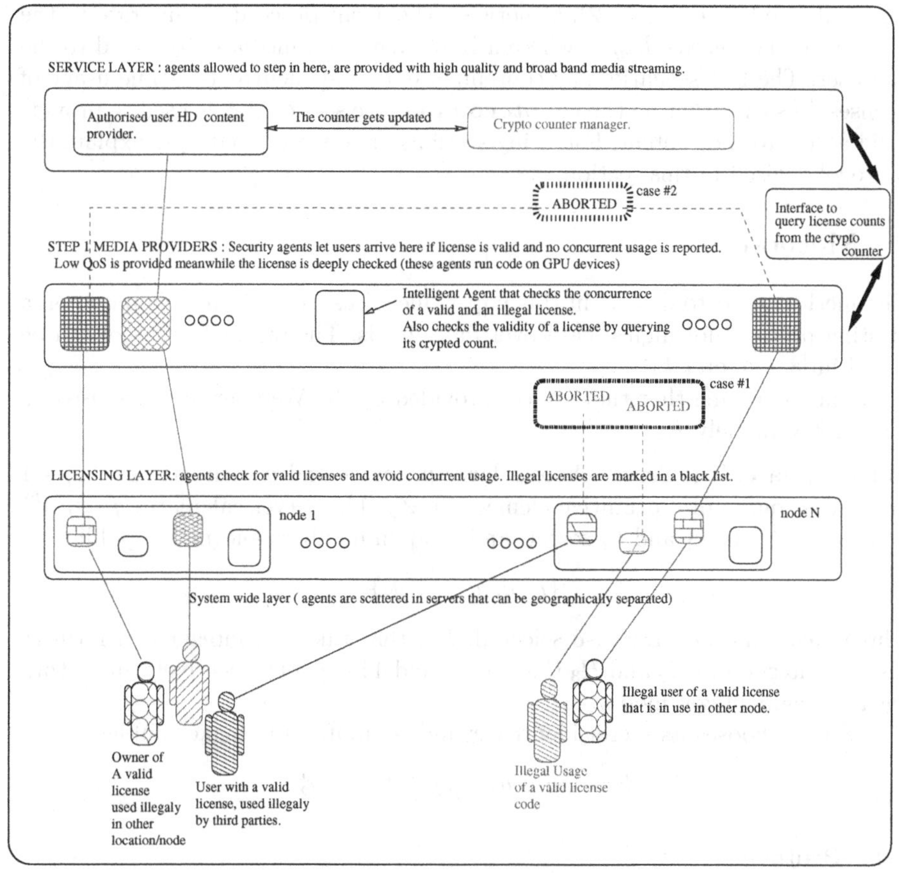

Fig. 1. Multiagent layered architecture proposal

As sketched in figure 1, our proposal establishes several layers of cooperating and intelligent agents to solve the problem of whether a license is valid or not. An agent can migrate to the upper layer taking with itself all the necessary information about the owner of the license. When an user connects to the server, an agent is selected to manage the licensing procedure and the connection at the first stage, the *Licensing Layer*, here the agents verify whether there is another license running at the same server using the protocol devised in section 2. In figure 1 it can be seen that if there exist several copies (see figure 1, case 1) then the connection is aborted and the owner of the license is reported (the license is then blacklisted). It may happen that the connection with an illegal copy of a license (while the valid one is active) is taking place in a different server, in this case the agent would proceed to the next step, the *Media Providers* layer where the user is provided with a low quality streaming, meanwhile the agent is verify globally if the license corresponds to a legitimated user. If another copy is running, then as in case 1, the licenses are blacklisted and the users are reported (see figure 1, case 2). Otherwise, the agent proceeds to migrate to the next layer, the *Service Layer*, where a High Definition media is streamed to the end user. The last scenario may contemplate a non-concurrent-in-time usage of licenses, in such situation the **cryptocounter**, see section 2.2, is used to provide valid access to the content. Following sections are devoted to deeply explain the protocol devised in this section.

2 Method

The method used to detect, in this paper, illegal usage of licenses is based on a existing protocol for digital cash over finite fields. The original protocol can be found in [8, Chapter 11].

We are assuming that the services provided by the Web Server make use of a session key, namely K_S.

Set up phase: The Server chooses large primes p and q such that $p = 2q + 1$ and g the square of a primitive element in $\mathbb{Z}_p$. The server calculates $g_1 \equiv_p g^{k_1}$ and $g_2 \equiv_p g^{k_2}$ for k_1 and k_2 secret, and random numbers and makes public

$$\{p,\ q,\ g,\ g_1,\ g_2\}$$

Three hash functions are also selected: H_1, that takes an integer and returns, also, an integer and H_2 and H_3 that take 5 and 4 integers respectively and return each, a single integer mod q.
The Server chooses its secret identity x and computes and makes public

$$\{h_0 \equiv_p g^x,\ h_1 \equiv_p g_1^x,\ h_2 \equiv_p g_2^x\}$$

2.1 Protocol

1. When an user logs into a content Server by the first time, the agent selected to guide him through the connection, chooses u randomly and sends $I \equiv_p g_1^u$ to the content Server.

2. The Server stores I along with some information to identify the user and sends, to the agent, a private ticket that will be used to get the corresponding session key K_S every time the user wishes to access the service.
3. The Server chooses w and computes and sends to the agent the triple

$$\{z' \equiv_p (Ig_2)^x,\ h \equiv_p g^w,\ k \equiv_p (Ig_2)^w\}$$

4. The agent then chooses seven random numbers $(s, x_1, x_2, n_1, n_2, n_3, n_4)$ and computes:
   - $A \equiv_p (Ig_2)^s$.
   - $B \equiv_p g_1^{x_1} g_2^{x_2}$.
   - $z \equiv_p z'^s$.
   - $a \equiv_p h^{n_1} g^{n_2}$.
   - $b \equiv_p k^{sn_1} A^{n_2}$.
   - $a' \equiv_p h^{n_3} g^{n_4}$.
   - $c \equiv_q n_1^{-1} H_2(A, B, Z, a, b)$ and $d \equiv_q n_3^{-1} H_2(A, B, z, a, H_1(T))$, where $H_1(T)$ denotes the hash value of the user's ticket T.
5. The agent sends to the Server, the pair $\{c,\ d\}$
6. The Server computes and sends back to the agent the pair

$$\{c_1 \equiv_q cx + w,\ d_1 \equiv_q dx + w\}$$

7. The agent computes $r \equiv_q n_1 c_1 + n_2$ and $r' \equiv_q n_3 d_1 + n_4$. The license corresponding to the ticket T will be then

$$L = \{A,\ B,\ z,\ a,\ b,\ r,\ a',\ r'\}$$

Theorem 1. *The following equalities hold for a* $L = \{A,\ B,\ z,\ a,\ b,\ r,\ a',\ r'\}$ *corresponding to the ticket* T*:*

$$g^r \equiv_p a h_0^{H_2(A,B,z,a,b)},\quad A^r \equiv_p b z^{H_2(A,B,z,a,b)},\quad g^{r'} \equiv_p a' h_0^{H_2(A,B,z,a,H_1(T))}$$

Proof. Firstly we recall that $r' \equiv_q n_3 d_1 + n_4$ and thus $r' \equiv_q n_3 dx + n_3 w + n_4$. Then $d \equiv_q n_3^{-1} H_2(A, B, z, a, H_1(T))$

$$\begin{aligned} a' h_0^{H_2(A,B,z,a,H_1(T))} &\equiv_p h^{n_3} g^{n_4} h_0^{H_2(A,B,z,a,H_1(T))} \\ &\equiv_p (g^w)^{n_3} g^{n_4} (g^x)^{H_2(A,B,z,a,H_1(T))} \\ &\equiv_p g^{wn_3} g^{n_4 + xH_2(A,B,z,a,H_1(T))} \\ &\equiv_p g^{wn_3 + n_4 + xH_2(A,B,z,a,H_1(T))} \\ &\equiv_p g^{wn_3 + n_4 + xn_3 d} \\ &\equiv_p g^{r'} \end{aligned}$$

Analogously we get $g^r \equiv_p a h_0^{H_2(A,B,z,a,b)}$

Let us show now that $A^r \equiv_p b z^{H_2(A,B,z,a,b)}$.

$$\begin{aligned}
bz^{H_2(A,B,z,a,b)} &\equiv_p k^{sn_1} A^{n_2} z'^{sH_2(A,B,z,a,b)} \\
&\equiv_p (Ig_2)^{wsn_1} (Ig_2)^{sn_2} (Ig_2)^{xsH_2(A,B,z,a,b)} \\
&\equiv_p (Ig_2)^{wsn_1+sn_2+xsH_2(A,B,z,a,b)} \\
&\equiv_p (Ig_2)^{wsn_1+sn_2+xn_1sc} \\
&\equiv_p ((Ig_2)^s)^{wn_1+n_2+xn_1c} \\
&\equiv_p A^{wn_1+n_2+xn_1c} \\
&\equiv_p A^{n_1(w+cx)+n_2} \\
&\equiv_p A^r
\end{aligned}$$

since $H_2(A, B, z, a, b) \equiv_q n_1c$ and $c_1 \equiv_q w + cx$.

8. The user accesses the content Server and demands the license.
9. The agent in charge of driving the validation procedure, sends to the Server the pair $(h(T), L)$ corresponding to the currently logged user.
10. The Server checks that the equalities of Theorem 1. If these holds, then the Server stores the license and computes the hash $H_3(A, B, H_1(T), t)$ where t denotes a time-stamp.
11. The agent now computes

$$s_1 = H_3(A, B, H_1(T), t)us + x_1 \quad \text{and} \quad s_2 = H_3(A, B, H_1(T), t)s + x_2$$

where u is the private information generated by the agent in step 1 and s, x_1 and x_2 are the random integers generated also by the agent in step 4.
12. The Server checks that $g_1^{s_1} g_2^{s_2} \equiv_p A^{H_3(A,B,H_1(T),t)} B$ and if so, then the client is sent the session key K_S encrypted using T and the pair $(h(T), L)$ is stored along with the 4-tuple $(s_1, s_2, H_3(A, B, H_1(T), t))$ until the user leaves/logs out the system.

Theorem 2. *With the above notation, the following equality holds*

$$g_1^{s_1} g_2^{s_2} \equiv_p A^{H_3(A,B,H_1(T),t)} B$$

Proof. Since $I \cong_p g_1^u$, $A \equiv_p (Ig_2)^s$ and $B \equiv_p g_1^{x_1} g_2^{x_2}$ we get that

$$\begin{aligned}
g_1^{s_1} g_2^{s_2} &\equiv_p g_1^{H_3(A,B,H_1(T),t)us+x_1} g_2^{H_3(A,B,H_1(T),t)s+x_2} \\
&\equiv_p (g_1^u)^{H_3(A,B,H_1(T),t)s} g_1^{x_1} g_2^{H_3(A,B,H_1(T),t)s} g_2^{x_2} \\
&\equiv_p (g_1^u g_2)^{H_3(A,B,H_1(T),t)s} g_1^{x_1} g_2^{x_2} \\
&\equiv_p ((Ig_2)^s)^{H_3(A,B,H_1(T),t)} g_1^{x_1} g_2^{x_2} \\
&\equiv_p A^{H_3(A,B,H_1(T),t)} B
\end{aligned}$$

Proposition 1. *The above protocol avoids the usage of a certain license by two different users simultaneously, it protects and avoids the usage of a stolen license and the reuse of a recorded message previously submitted for a requested service.*

Proof. If a legitimated user shares all his private information (which comprises the ticket, the license and all the private numbers generated through all the steps of the protocol) with someone else and this fake license is used to authenticate $(H_1(T), L)$ along with another triple $(s'_1, s'_2, H_3(A, B, H_1(T), t')$ while the legitimated user (or a third person holding the information) is still logged in the system, then the Server, as it is shown in [8, 11.1.9], the private information u that identifies the legal user by means of $I \equiv_p g_1^u$ is derived from $u \equiv_q (s_1 - s_1')(s_2 - s_2')^{-1}$.

Note that if someone steals a license $(H_3(T), L)$, in order to get the service, the user needs to know, also, the private information u to produce s_1 and s_2. Trying to forge these numbers implies satisfying the Theorem 2 which is a hard problem since it involves the discrete logarithm.

The reuse of a recorded message $(H_1(T), L)$ is equivalent to the previous comment.

Proposition 2. *The above protocol provides anonymity to users.*

Proof. It easily observed that through this protocol there is no possibility for anybody, including the Server, to compute u, which is the only information that identifies the user.

Proposition 3. *Only the corresponding authority (the server) is able to provide valid licenses.*

Proof. Computing numbers, that verifies the Theorem 1, involves discrete logarithms, for instance even a legal user knowing A, B, z, a and T will need to solve a discrete logarithm in order to produce r' such that $g^{r'} \equiv_p a' h_0^{H_2(A,B,z,a,H_1(T))}$

However, since an user might share his key with somebody else and due to anonymity, this unauthorized usage could be detected only in the case in which the legal and the unauthorized users make use of it simultaneously as noted in Proposition 1.

2.2 Detecting Copies of Licenses

The purpose of this section is to settle an extension to the method in order to detect when a copy of a license, when no other copy of it is active at the moment, is used and determine who is responsible of the corresponding sharing (violation).

We now introduce a new parameter in the license L that is nothing but a cryptocounter where the only one able to update values of the counter is the owner of a certain private key. A version also based on the discrete logarithm is easily obtained as follows:

Let $f \in \mathbb{Z}_p$ the initial value of the counter and let k_1 be random. Then the first value of the counter is $(a_1, b_1) = (g^{k_1}\ mod\ p, (g^c)^{k_1}\ mod\ p)$. To increase the counter we make $(a_2, b_2) = (a_1 g^{k_2}\ mod\ p, b_1(g^c)^{k_2} g\ mod\ p)$, for c some secret value owned by the Server.
Now given $(a_n, b_n) = (a_{n-1}^{k_n}\ mod\ p, b_{n-1}(g^c)^{k_n} g\ mod\ p)$, then $a_n^{p-1-a} b_n = g^n$ and so only the Server, who is the only knowing a can get g^n and thus, via a simple search, n.

The application is now clear. Thus a license is defined as

$$L = \{A,\ B,\ z,\ a,\ b,\ r,\ a',\ r',\ c'\}$$

for c' a cryptocounter as above. The Server has to store a copy of the license and the 4-tuple appearing in Step 12. Then each time a license is used, the cryptocounter is increased by the Server. When the user leaves the system, the Server sends the user L' with the corresponding value c' increased.

Thus, the protocol is the same until step 9. In step 10, the Server checks the same equalities as above and that the corresponding c' equals the stored value. In case the value that c' gives rise is less than the value that provides the stored one, this means that a copy of a license is being used to access the service. Then the protocol follows the same Steps 10, 11 and 12 and operates as in Proposition 1 to detect the legal user that shared all his private information.

Thus, the only way to avoid fraud detection when using this protocol is that both legal and unauthorized users share the information that the Server sends back after leaving the service each time. This attack could be carried out when the group of people which the legal user shares his license with is formed by few and trusted and people, since in other case it is quite probable that one of the members of the group shares the information to others and the situation could become one of those that the protocol detects as fraudulent.

3 Implementation

The implementation of such a system need to be scalable. Common agents' platforms tend to be inefficient to handle thousands of agents concurrently operating on a same server with different calculations (checking licenses' validity). The implementation we propose here is one built on high performance concurrent and migratable objects provided by the charm++ [9] runtime. Implementations for high performance and costly applications are proven to be efficient in terms of concurrent usage of resources, scalability and load adaptivity issues when using message-driven computing. To build our model, high-level abstractions are useful to target the model of the implementation without having to deal with details of the architecture underneath [10]. A farm cpan based model (see Fig.2) was used to have fully concurrent charm objects per processor. Load balancing is executed asynchronously as exposed in [11]. Figure 2 shows an implementation proposal for the licensing server. To instrument the agents layer, the platform proposed in [12] was used as a model for our implementation so both, agent communications and overloaded computing nodes were efficiently reduced.

As figure 2 shows agents are labeled as *slave objetcts.* Whenever a new server is added to the infrastructure, then a new *node object* is created to host and monitor the computation. Slaves can be sent to the new node when the node is set to available. Slave objects communicate using messages (remote invocation of their methdos, accesible via proxy objects). In order to avoid that the monitoring of the computation affects performance, agents are able to send *signals* to a special object that do not belong to the computation. This object

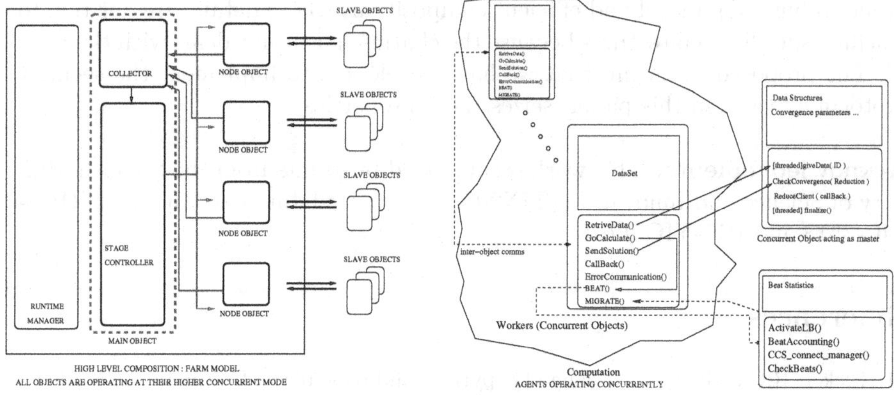

Fig. 2. Multiagent layered implementation proposal

gathers signaling information from agents and is able to detect a decrease in performance [11] so load balancing is invoked asynchronously and agents are moved to other computing nodes. This implementation hides network latencies for when the users are not operating concurrently at the same node. When the system needs to check if the user is operating concurrently in other nodes, then the main characteristic of the charm++ objects (agents) is that of being able to hide latencies and concurrently provide content to the user. As many agents may be operating concurrently to prove the validity of a number of user licenses, the implementation proposal due to the nature of the cited objects is able to exploit the network communication gaps *agent-server* to concurrently operate with a number of agents. Following times for checking a license in a node are shown: preliminary tests show that the time (in ms) needed to provide access to a unique user is *942 ms* in average. Whereas it raises to *1922 ms* when ten users are concurrently logging in. *5129 ms* for one hundred users.

4 Conclusions

A new protocol to protect services provided through the Internet has been shown and proved to work efficiently when illegal usage of licenses is taking place, even when the licenses are not in use at the same time. Also, a layered multiagent architecture was sketched in order to show a feasible implementation. The agents' platform proposed is built by composing high performance and concurrent objects that communicate using asynchronous messages. Such issue enables the proposed platform to interleave computation and communication phases, so when an agent is communicating, then another is efficiently scheduled to do its CPU related computations. We have, also, used and enhanced our proposal by following the implementation sketched in [12] to relieve the effect of unmaskarable latencies and to efficiently move agents from one processor to another. These latency hiding issues and the migratable characteristic of the cited objects are prone to build high level and efficient compositions whose details are not tied to machine specific architectures because the charm++ runtime deals with them for us. The proposed platform, that we have developed, to implement the security protocol exposed in this paper, seizes these properties.

Acknowledgements. This work is supported by grants from the Spanish Ministry of Science and Innovation (TIN2008-01117), and Junta de Andalucía (P08-TIC-3518 and P11-TIC7176).

References

1. Picker, R.C.: The yin and yang of copyright and technology. Commun. ACM 55(1), 30–32 (2012)
2. Lou, X., Hwang, K.: Collusive piracy prevention in p2p content delivery networks. IEEE Transactions on Computers 58(7), 970–983 (2009)
3. Ding, Y., Fan, L.: Traitor tracing and revocation mechanisms with privacy-preserving. In: Wang, Y., Ming Cheung, Y., Guo, P., Wei, Y. (eds.) CIS, pp. 842–846. IEEE (2011)
4. Kosuga, M., Yamazaki, T., Ogino, N., Matsuda, J.: Adaptive qos management using layered multi-agent system for distributed multimedia applications. In: ICPP, pp. 388–394 (1999)
5. Franklin, S., Graesser, A.C.: Is it an agent, or just a program?: A taxonomy for autonomous agents. In: Jennings, N.R., Wooldridge, M.J., Müller, J.P. (eds.) ECAI-WS 1996 and ATAL 1996. LNCS, vol. 1193, pp. 21–35. Springer, Heidelberg (1997)
6. Silva, D., Braga, R., Reis, L., Oliveira, E.: A generic model for a robotic agent system using gaia methodology: Two distinct implementations. In: IEEE Conference on Robotics Automation and Mechatronics, RAM, pp. 280–285 (2010)
7. Leetch, G., Mangina, E.: A multi-agent system to stream multimedia to handheld devices. In: Sixth International Conference on Computational Intelligence and Multimedia Applications, pp. 2–10. IEEE (2005)
8. Trappe, W., Washington, L.: Introduction to cryptography: with coding theory. Pearson Prentice Hall (June 2006)
9. Kale, L., Arya, A., Bhatele, A., Gupta, A., Jain, N., Jetley, P., Lifflander, J., Miller, P., Sun, Y., Venkataraman, R., Wesolowski, L., Zheng, G.: Charm++ for productivity and performance: A submission to the 2011 HPC class II challenge. Technical Report 11-49. Parallel Programming Laboratory (November 2011)
10. Capel Tunon, M., Lopez, M.: A parallel programming methodology based on high level parallel compositions (cpans). In: 14th International Conference on Electronics, Communications and Computers, CONIELECOMP 2004., pp. 242–247 (February 2004)
11. Alvarez-Bermejo, J.A., Roca-Piera, J.: A Proposed Asynchronous Object Load Balancing Method for Parallel 3D Image Reconstruction Applications. In: Hsu, C.-H., Yang, L.T., Park, J.H., Yeo, S.-S. (eds.) ICA3PP 2010, Part I. LNCS, vol. 6081, pp. 454–462. Springer, Heidelberg (2010)
12. Jang, M.-W., Agha, G.: Adaptive Agent Allocation for Massively Multi-agent Applications. In: Ishida, T., Gasser, L., Nakashima, H. (eds.) MMAS 2005. LNCS (LNAI), vol. 3446, pp. 25–39. Springer, Heidelberg (2005)

C&C Techniques in Botnet Development

Félix Brezo[1], José Gaviria de la Puerta[1], Igor Santos[1], David Barroso[2], and Pablo Garcia Bringas[1]

[1] DeustoTech Computing - University of Deusto, Bilbao BI 48007, Spain
{felix.brezo,jgaviria,isantos,pablo.garcia.bringas}@deusto.es
http://www.s3lab.deusto.es
[2] Telefonica I+D, Madrid MA, Spain
dbarroso@lostinsecurity.com

Abstract. Botnets are one of the most important threats towards nowadays users of the Internet. The joint of malware capabilities to be exploited in the network services and the increasing number of daily transactions performed in the cloud, makes them an attractive target for cybercriminals who have evolved their old IRC-based communication channels, into decentralized P2P networks, HTTP/S botnets and even Twitter-controlled networks. Against this background, this article analyses the threat that will affect computer networks in the upcoming years by going through these different Command & Control channels used by botmasters to keep the control of their hijacked networks.

Keywords: botnets, crimeware, cyberfraud, C&C, source analysis.

1 Introduction

More often than ever, computer security is gaining its place in human being's ordinary life. The Anonymous attacks on different e-platforms [1,2] have brought closer to ordinary people the reality of DDoS, relegating to the memories of the elder the security experts appearing in the media to speak about punctual and almost funny threats collapsing thousands of computers.

The scientific community is concerned about the problems that will involve the increasing complexity of traditional communication channels in the next generation of botnets. This is a hot topic that has concerned security agencies in the world for a long time. One of the first important ones was the "Operation Bot Roast" [3] in 2007, by which the FBI detected a botnet compounded by more than a million compromised computers. In 2010, an action leaded by the Spanish *Guardia Civil* in cooperation with Panda Labs dismantled the "Botnet Butterfly" with 12.7 million computers committed [4]. Their capabilities have not been misjudged by governments and armies. In fact, the DDoS attacks received by Estonia in 2007 lead to the creation of Cooperative Cyber Defence Centre of Excellence [5], while their functionalities to achieve information about economical and military targets [6] had already being suggested by Colonel Charles W. Williamson III [7].

Á. Herrero et al. (Eds.): Int. JointConf. CISIS'12-ICEUTE'12-SOCO'12, AISC 189, pp. 97–108.
springerlink.com

This document is structured as follows. Section 2 explains briefly how a botnet works and which are some of the most well-known functionalities. Section 3 analyses the characteristics of some of the most important botnet tipologies going through particular cases. Section 4 delves into the different approaches that exist for detecting and tracking these networks, together with the advance of some outlines to work on in the future. Finally, Section 5 brings down the conclusions after analysing the menaces commented all along this document.

2 The Botnet Threat

Microsoft's End to End Trust defines botnets as "networks of dedicated and remoted controlled computers by one or more cybercriminals" [8]. A botnet infection will begin whenever a concrete software program, called a bot, is downloaded by the victim exploiting any of the traditional infection vectors used by common malware instances. In this regard, the main difference between other types of malware and botnets is the fact of getting connected to a public server run by the botmaster (the botnet operator) for issuing the commands to be executed by the infected machines. As described in the pages that follow, this control panel can be implemented using a bunch of different techniques depending on the needs and experience of its final user.

Botnets are attractive to malware writers for two reasons: they are easily managed and can produce direct or indirect economical benefit for the botnet controller. Nowadays bots are a mixture of threats closely related to other previous' areas of malware: they can spread themselves like worms, remain hidden as many viruses and allow remote control of the infected machine by third parties. These circumstances, together with other evidence related to the writing of code by means of cooperative efforts (as happened with SDBot, whose code is commented by different authors), allow the proliferation of a wide range of variations and mutations based on the specific purpose for which they are sought.

These functionalities can vary depending on the complexity of the final target, but we have to assume that may include a combination of the following:

- **DDoS support.** A denial of service attack, also called DoS, is an attack on a computer system or network that causes a service or resource being inaccessible for legitimate users. A Distributed DoS (DDoS) will cause the loss of network connectivity by consuming the bandwidth of the victim network or overloading its computational resources using different attack sources. In this way, Roman Studer states that DDoS attacks cause various types of financial costs to a company [9]: for those depending on online functionalities like e-banking and ecommerce businesses, the revenue out of online functions will be lost during system unavailability, while "they may also suffer monetary penalties as a result of losing the ability to meet service level agreements and by failing to provide an agreed service the company could cause damage to a third party and be faced litigation and charged accordingly". Furthermore, while some investments on recovering tools and backup policies have already been performed [10,11], an in-depth analysis of the menaces is needed to prevent them from happening again, as they may also affect, not only the companies image, but also the stock price in the extremely sensible markets of new technologies.

- **Remote execution of protected programs.** With very different targets, botnets may only be the first step into a deeper infection in the system. Thus, attackers may get full control of the infected machines to perform different covert operations without the victim noticing it.
- **Keylogging.** A keylogger is a software (or even hardware) device that handles specific record keystrokes on the keyboard to store them in a file or to send them via Internet. Often used as a daemon malware type, the objective of this feature is to open a door to capture login and password details, credit card information and other sensible data.
- **Click fraud.** Click fraud refers to a type of Internet fraud, in which the target of the attack are the pay-per-click services (such as Google AdSense). Though click fraud can also be performed manually, with the help of certain kinds of software its automatication makes the fraud more profitable. By means of these tools, fraudsters can manipulate billing systems lying behind target by increasing the traffic and the amount of clicks.
- **Spam.** Practically all the spam sent worldwide, is mailed through machines under direct control of spam operators using botnets to accomplish such objectives. According to SpamHaus[1], amongst 300 and 400 spammers were responsible for themselves of the 80% of global traffic of such malicious content. Microsoft [8] estimates that botnets are responsible for 87 percent of all the e-mail unsolicited, equivalent to about 151,000 million e-mail messages a day. At the same time, on February 25, 2010, Microsoft, industry partners, academic and legal communities announced a successful collaborative effort to disable the spam-specialised win32/Waledac[2] botnet [12].

3 Botnet Samples

In this section we are going to analyse some aspects in the procedure of coding a botnet. The reader may take into account that there exist different topologies for implementing a Command & Control Channel as shown in figure 1:

- **Centralized (fig. 1(a)).** In centralized topologies all the infected nodes will be connected to a given server directly commanded by the botmaster, creating a critical failure point.
- **Uncentralized (fig. 1(b)).** An uncentralized topology provides an additional protection and increasing versatility for the botmaster, who deploys his commands through different server machines.

[1] ROKSO: Register Of Known Spam Operations `http://www.spamhaus.org/rokso`

[2] Win32/Waledac is a trojan that is used to send spam. It also has the ability to download and execute arbitrary files, harvest email addresses from the local machine, perform denial of service attacks, proxy network traffic and sniff passwords.

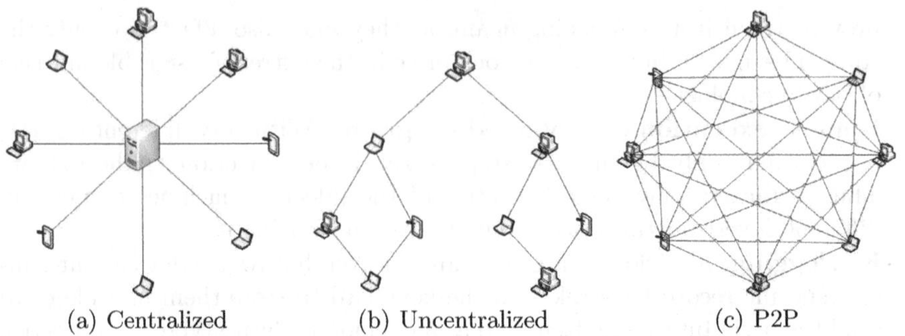

(a) Centralized (b) Uncentralized (c) P2P

Fig. 1. Different topologies in botnet management

- **Peer-to-peer (fig. 1(c)).** In a peer-to-peer topology each and every node of the network can act either as a client or a server. The main advantage of this approach derives in its flexibility and capability to face connection-related issues and detection.

With recent attacks on several credit card servers as a retaliatory measure by the constraints brought by these companies to Wikileaks [13,14], it has been shown that even in 2011 IRC remains being a channel used to transmit commands in Anonymous botnets [1,2]. However, more modern techniques should be developed in the design of a reliable botnet detection system to fight the proliferation of new C&C techniques such as chat (or private messages) in the P2P servers, static HTML pages connections, more complex HTTP-based botnets and more modern Twitter or Pastebin based ones. To achieve this objective, the following Command & Control techniques have been studied.

3.1 IRC

IRC-based botnets compounded the initial stage of the botnet threat with an impact between 2005 and 2008 [15]. Those networks were connected periodically to a specified IRC channel awaiting further instructions, being capable of retransmitting the commands in real time, either through private messages (PRIVMSG IRC) or through the publication of thematic messages (TOPIC).

Sdbot. Sdbot was a backdoor that allows hackers to gain remote access to the affected computer to carry out malicious actions that compromise user confidentiality and impede the tasks. Sdbot also uses an own IRC client as C&C channel. In spite of requiring the user intervention for being propagated, amongst his malicious functionalities it can launch denial of service attacks against websites and download and execute files on the infected computer.

Agobot. Agobot, also frequently known as Gaobot, is a family of computer worms used for botnet creation and released under version 2 of the GNU General Public License. The Agobot source code describes it as: “a modular IRC bot for Win32/Linux” and its first implementation is attributed to Axel “Ago” Gembe, a German programmer. Agobot is a multi-threaded and mostly object oriented program written in C++ as well as a small amount of assembly. Browsing the project, anyone will be able to surf easily inside a forest of modular .cpp files. Agobot was another example of a botnet that required little or no programming knowledge to be used thanks to its modularity.

Rbot. Rbot is a worm that spreads through network shares. It searches and makes a list of the shared folders used for application downloads (eg. P2P), where it will release a copy of itself. It has some features that allow it to act as a backdoor developing botnet functionalities using IRC as channel. The worm uses its engine to connect to a preset IRC channel to wait for commands from a remote attacker. Some of the commands that an attacker may have sent to any victim were launching DDoS attacks; getting registration keys (Product ID and CD Keys); adding and deleting shares, groups and users; amongst others.

3.2 P2P

Traditionally propagated through shared binary in other P2P networks by making use of the P2P protocol, they have the advantage of being more difficult to destabilize as they do not have a unique core from which issuing orders and/or sharing resources and information. They make use of the facilities of traditional P2P networks which allow high connection and disconnection ratios. Some examples of P2P network architectures are the aforementioned Trojan.Peacomm and Stormnet [16].

Special mention must be *parasitic botnets* [17,18], whose communication channel makes use of third party services: first, a bot is selected to perform a specific search with a specific title (which can be fixed or calculated by some sort of algorithm) that would allow each bot to identify a) what other file must be found to reveal the commands to execute or b), in which searching response the command to be sent will be coded not needing to download any file. Thus, the botmaster will only receive a positive response if a member of the same botnet contains it, initiating then the communication protocol. This philosophy uses *in-band-messages* (common P2P traffic) which permits the sending and reception of commands [19]. The pro is that this traffic might be easily confused with legitimate traffic within the network, difficulting its detection and complicating the analysis of traffic as this approach will sue a pre-filtering process.

However, Shishir Nagaraja et al. [20] have worked on P2P network detection. Concretely, they have developed some research to find whether ISPs can detect those efficient communication structures of P2P botnets so as to use this basis for botnet defense. The point is that ISPs, enterprise networks, and IDSs have significant visibility into these communication patterns due to the potentially large number of paths between bots that traverse their routers, making them a good starting point to detect this traffic.

IMMONIA. IMMONIA is a P2P bot designed to be spread by its own means on IRC platforms. It is written in AutoIt v3[3]. Being coded using a scripting programming language, whose code[4] was attributed to someone named Hypoz in 2011, is fully documented and pretty detailed commented. The most interesting point here is its decentralized P2P structure running through IRC platforms developing greater structural complexity because all of them can act as both, client and server, being more difficult to intercept and study.

3.3 HTTP/HTTPS

In the HTTP-based, the botmaster informs the infected computers from the new list of commands to execute, updating the contents of a web page that bots are to periodically visit. Some modern examples are Flu and Zeus, described below.

Zeus. Zeus botnet is an evolution of a Trojan horse that stole banking information by Man-in-the-browser keystroke logging and Form Grabbing, being mai spread through drive-by downloads and phishing schemes. First identified in July 2007 when it was used to steal information from the United States Department of Transportation [3], it became more widespread in March 2009. By that time, Zeus had infected over 3.6 million computers in the United States. Binsaleeh et al. presented in 2010 [21] a reverse engineering analysis on the Zeus crimeware toolkit, which they defined as "one of the recent and powerful crimeware tools that emerged in the Internet underground community to control botnets".

Its source code was released in May 2011 after being sold in different underground forums [22]. Pretty well balanced, the attackers C&C channel is written in PHP, while the client side of the Malware is coded C++. According to its documentation, Zeus developers were working on more than 15 updates and modifications only in 2010 [23].

- **DDBB connection.** The Zeus control panel was developed on PHP 5.2.6, so there are PHP.ini settings included that will need to be configured, as well as MySQL setting recommendations.
- **Server-side capabilities.** Zeus includes Socks 4/4a/5 support with IPv4 and IPv6 via UDP. This will allow the botmaster to connect to an infected host even if there is a NAT in place or a restrictive firewall.
- **C&C channels.** The botmaster can do a number of things with the stolen data and the infected systems. One interesting aspect of the Zeus control panel is the ability to search for targeted information: IP, NAT, number of active bots online, country, operating system, and other key demographics. The botmaster can have alerts delivered via IM on Jabber if they wish to let the botmaster know when a victim is using a given bank, allowing them to trigger a session capture on demand.

---

[3] AutoIt v3 is a freeware BASIC-like scripting language designed for automating the Windows GUI and other general scripting. Its default extension is .au3.

[4] By this publication available on `http://www.mediafire.com/?1wdq1o076amww03`

The following are some of the evil functionalities implemented on Zeus: traffic interception from the web browsers, modification of pages on the fly, inject forms, or redirect the system to a fake page; and data harvesting, monitoring all of the stored information on a system that is not immediately protected (eg. browser cookies) and certificates, intercepting FTP-Login and POP3 data and keylogging capabilities.

Flu. Flu[5] is a reverse Trojan designed to permit the construction of a botnet. It is designed with a client-server architecture, where, the server, is a small executable programmed in C# which will infect any Windows operating system to get control of the machine that is hosting. The client is developed in PHP and runs on an Apache web server. It aims to provide commands to all those individual servers that Flu has spread across the Internet to obtain information from the machines where they are installed and to store it in the web server. The user information may be accessed at any time from a graphical interface developed in HTML and PHP using the browser. The purpose of the project is to allow all those users who want to learn in the development of remote management tools, to take advantage of the expertise of others. Also, the project also aims to raise awareness among users about the dangers of malware, showing the operation of a simple way through and showing Flu protection and disinfection.

The source is quite well commented to match the educational purposes. At the same time, there exists a fully accessible document ready to be checked by occasional writers at the official webpage. The basic deployment of the botnet is fully explained by Juan Antonio Calles and Pablo González [24] in the *Flu Trojan's User Guide* with very brief explanations.

- **DDBB connection.** The DB structure in the free version of Flu was not professionallly designed: it included a pair of tables, one for botnet users (with users and passwords stored in plain text) and another one to store the details of the machines belonging to it.
- **C&C channels.** The bidirectional communications in Flu are performed by means of an Internet browser. On the one hand, the infected victim requests the .xml file that contains the list of commands to be executed to the web server. Once the victim has executed the requested commands, it returns its answers to the web server. All the information is encrypted by default using the AES algorithm and a key of 128 bits. In the web interface, the botmaster will be able to track all the infected nodes, together with an option to load, in each and every node, a remote shell to give personalised orders to that node. In another tab, group orders can also be dispatched, by selecting predefined-attacks or launching commands in every victim's console.
- **Bot program creation.** The bot generation takes place by means of an automated generator which receives as parameter the URL where the botnet server will be allocated, together with the absolute path to the .xml commands file.

[5] `www.-flu-project.com`

Though Flu b0.4 is conceived as an educational project, it includes some interesting features even for real malware writers: individual and group remote shell execution; information gathering, screen printing; user account creation with full control capabilities; file stealing (though for design limitations, up to 3.5 MB only); Windows firewall deactivation and keylogging functionalities amongst others. In spite of this, there also exists a more complex version of Flu oriented to fight pederasts called Flu-AD (standing for *Flu-Anti Depredadores* or Flu-CounterPredator in English) presented in No cON Name 2011 with some fully implemented characteristics as rootkit capabilities, webcam and Sound capture, crypter, so as not to be detected by antivirus software and different utilities to recover usernames and passwords of *predators* accounts.

Prablinha. Prablinha[6] is another reverse Trojan designed to permit the construction of a botnet, made open-source in December 2010. It is also designed with a client-server architecture, where, the zombies in designed in C# are generated by a bot generator also designed in C#. The management utilities are developed in PHP and run on an Apache web server. Prablinha was also thought to teach users about how easy DDoS could be made. The system is conceived to perform HTTP, UDP and TCP Denial of service attacks, as well as to provide the attacker a simple remote shell in the victim's system.

3.4 New Trends: Social Malware

The widespread of social networks is also attractive for malware writers. In fact, very recently, botmasters have begun to exploit social network websites such as Twitter [25] with pretty good results. As social networks existence depends on being online 24/7, some programmers have found in them a good starting point to successfully host their Command & Control platforms [26]. Additionally, if we add that users trust in their contacts suggestions to click and visit the links they receive, much more than on anonymous webpages, social networks will undoubtedly gain more and more attention in short-term malware evolution.

4 Discussion on Generic Countermeasures

We can divide the efforts performed to fight botnets in two different main lines which complement each other: detection and deactivation. In the section that follows we briefly describe part of the current work on botnet fighting.

4.1 Detection

At the time of detection and tracking botnets, there are two different approaches: the active one, usually based on *honeypots*[7] as the one developed in *The Honeynet Project* [27] and a more passive one, monitoring the network traffic, for

[6] http://www.indetectables.net/foro/viewtopic.php?t=29086

[7] A honeypot is an entity created only to be attacked in order to detect and analyse new threats and the effects of these connections.

example, using a neural visualization IDS for Honeynet Data as developed by Herrero et al. [28,29]. Nevertheless, one of the main obstacles when dealing with botnets arises when trying to simulate a real environment under controlled circumstances, as some authors have stated when developing tools to generate simulated botnet traffic for researching purposes in large-scale networks [30].

By means of signatures. Detection using signatures is currently facing mayor challenges posed by modern techniques capable of developing polymorphic families. This is especially remarkable when facing malware instances detection [31], as the use of obfuscation techniques to hide bots methods and attributes[8] is increasing the complexity of finding the executables in the endpoint suggesting that semantic-based detection techniques may be needed.

Cooperative behaviour. Typically, studies going on this path are oriented towards the pursuit of activities relating the infected machines, with the ultimate goal of using this information to optimize the tracking techniques.The target is to make statistical attacks seeking behaviour profiles that differ from that performed by a non-infected user.

Offensive behaviour Assuming that botnets send massive amounts of information in relatively short periods of time, Xie et al. [32] uses the volume of data sent along with the information obtained from spam servers for tracking such contents before they cause real damage.

End-user tracking. Authors like Ormerod et al. sustained a very different approach to face the problem: they defame botnet toolkits through discouraging or prosecuting the end-users of the stolen credentials by combatting identity theft toolkits [33]. In the same line, Riccardi et al. have performed some efforts on mitigating the impact of financial botnets by identifying and giving a *malicious score* to any node connecting to the system [34]. To accomplish their objective, however, they need boosted information sharing policies amongst all the peers taking part in the process.

4.2 Deactivation

Nevertheless, detecting a *botnet* is never the final step. Botwriters know these issues and are working on defacing them providing the botnet security measures similar to those in benign commercial network tools. Once detected the existence of a botnet, there are different outlines of work for terminating its activities.

- Closing them by placing them in quarantine at an early stage of development to disinfect compromised nodes one by one.
- Closing the communication C&C channels to prevent them from spreading even more commands thereby, limiting its operational capabilities first and

[8] Obfuscation may be performed using free tools such as EazFuscator.NET for C# applications.

their propagation ways then. For example, Binsaleeh et al. injected falsified information into the botnet communications to defame the toolkit [21]. Another implementation of this technique are the Index Poisoning Attack techniques, used in the past by companies to prevent the redistribution of software, video and other content protected by copyright by including false records [35,36] to make it more difficult for the nodes to communicate effectively with their pals.

- Hijacking of the network by sending commands to deactivate it. The downside is that many botnets currently use asymmetric encryption systems [37] to prevent the detection of any of the nodes making pretty more difficult to apply Sybil attack techniques such as the ones proposed by Vogt et al. to infiltrate a node in the list of bots and then sabotage the communications from inside by altering hashes, files or commands [38].

Thus, because of its complexity, collaborative efforts are needed to take down this menace as conceived in The German Anti-Botnet Initiative [39] or as proven in the operation performed in March 2012 by Kaspersky Lab, CrowdStrike Intelligence Team, Dell SecureWorks and members of the Honeynet Project, entities involved in the takedown of Hlux/Kelihos botnet [40].

5 Conclusion

The communication between bots has undergone major changes since the use of IRC clients. Moreover, to facilitate the exponential growth of such networks, it is necessary to ensure that those commands sent by the botmaster are not lost without being received, interpreted and/or executed by any of the peers infected.

In the face of the alarming growth of malicious applications, which numbers have been beating every year [41], strong efforts will be needed to carry out more researching work in this field. In this way, a report presented by the computer security firm Kaspersky, as part of Technology Day 2011 showed that botnets are one of the most powerful threats in nowadays cyberspace, with perspectives of going on evolving "dramatically" in 2020 as they will incorporate more and more mobile devices with Internet connection. Thus, because of its complexity, collaborative efforts are needed to take down certain botnets as the one performed in March 2012, amongst Kaspersky Lab, CrowdStrike Intelligence Team, Dell SecureWorks and members of the Honeynet Project, entities involved in the takedown of Hlux/Kelihos botnet [40].

At the same time, a new battlefield is appearing. The power and capabilities of mobile devices, tablets and smartphones are arising new security concerns as stated by Kok et al. [42]. The amount of personal data stored in them and the increasing daily transactions that will be performed with such devices in the future, highlights them as a very attractive target for malware writers.

Although being dealing with a field whose expertise is still under development, we have proposed in this paper some outlines of work to cope with a phenomenon that could jeopardize all services connected to the network in the middle/long term. Thus, each and every organisation with access to the Internet must be prepared proactively, assuming, as part of the computer security protocols, that our systems may suffer in the future from hypothetical massive attacks linked to this new form of organized crime.

References

1. Lillington, K.: Time to talk: Anonymus speaks outs
2. InfoSecurity: Anonymus hacking group uses IRC channles to co-ordinate DDoS attacks (2011)
3. Office, F.N.P.: Over 1 Million Potential Victims of Botnet Cyber Crime (2007)
4. Corrons, L.: Mariposa botnet (2010)
5. NATO/OTAN: Tackling new security challenges. Technical report (2011)
6. Lemos, R.: U. S. military to build botnets? 737 (2008)
7. Williamson, C.W.: Carpet bombing in cyberspace: Why America needs a military botnet
8. Trust, E.T.E.: Desactivando redes de ordenadores controlados por ciberdelincuentes para crear un internet ms seguroy fiable (2010)
9. Studer, R.: Economic and Technical Analysis of BotNets and Denial-of-Service Attacks. In: Communication Systems IV. University of Zurich, Department of Informatics (2011)
10. Bleaken, D.: Botwars: the fight against criminal cyber networks. Computer Fraud & Security 2010(5), 17–19 (2010)
11. Smith, K., Lin, P.: Keeping internet marketing up and running: potential disasters and how to plan for them. International Journal of Electronic Marketing and Retailing 4(1), 1–15 (2011)
12. Cranton, T.: Cracking Down on Botnets (2010)
13. Seiiler, J.: Entrance of Wikileaks Into Fourth Estate Creates Perils, Opportunities
14. Bloxham, A., Swinford, S.: WikiLeaks cyberwar: hackers planning revenge attack on Amazon.
15. Zhuge, J., Holz, T., Han, X., Guo, J., Zou, W.: Characterizing the irc-based botnet phenomenon. In: Reihe Informatik. Pace University, White Plains (2007)
16. Grizzard, J., Sharma, V., Nunnery, C., Kang, B., Dagon, D.: Peer-to-peer botnets: Overview and case study. In: Proceedings of the First USENIX Workshop on Hot Topics in Understanding Botnets (2007)
17. Wang, P., Wu, L., Aslam, B., Zou, C.: An advanced hybrid peer-to-peer botnet. In: USENIX Workshop on Hot Topics in Understanding Botnets (HotBots 2007) (2007)
18. Wang, P., Wu, L., Aslam, B.: C. Zou, C.: A systematic study on peer-to-peer botnets. In: Proceedings of 18th Internatonal Conference on Computer Communications and Networks, ICCCN 2009 (2009)
19. Naoumov, N., Ross, K.: Exploiting p2p systems for ddos attacks (2009)
20. Nagaraja, S., Mittal, P., Hong, C.Y., Caesar, M., Borisov, N.: Botgrep: Finding p2p bots with structured graph analysis (2010)
21. Binsalleeh, H., Ormerod, T., Boukhtouta, A., Sinha, P., Youssef, A., Debbabi, M., Wang, L.: On the analysis of the zeus botnet crimeware toolkit. In: Eighth Annual International Conference on Privacy Security and Trust, PST (2010)
22. Seltzer, L.: Zeus Source Code Released
23. Ragan, S.: Overview: Inside the Zeus Trojans source code
24. Calles, J.A., Gonzàlez, P.: Troyano Flu b0.4 Windows. Manual de Usuario (2011)

25. Nazario, J.: Twitter-based Botnet Command Channel (2009)
26. Kartaltepe, E., Morales, J., Xu, S., Sandhu, R.: Social Network-Based Botnet Command-and-Control: Emerging Threats and Countermeasures. In: Zhou, J., Yung, M. (eds.) ACNS 2010. LNCS, vol. 6123, pp. 511–528. Springer, Heidelberg (2010)
27. Spitzner, L.: The honeynet project: Trapping the hackers. IEEE Security & Privacy 1(2), 15–23 (2003)
28. Herrero, L., Zurutuza, U., Corchado, E.: A neural-visualization ids for honeynet data. International Journal of Neural Systems 22(2), 1250005 (2012)
29. Corchado, E., Herrero, Á.: Neural visualization of network traffic data for intrusion detection. Applied Soft Computing 11(2), 2042–2056 (2011)
30. Massi, J., Panda, S., Rajappa, G., Selvaraj, S., Swapana, R.: Botnet detection and mitigation. In: Student-Faculty Research Day, CSIS. Pace University, White Plains (2010)
31. Goebel, J., Holz, T.: Rishi: Identify bot contaminated hosts by irc nickname evaluation. In: Proceedings of the USENIX Workshop on Hot Topics in Understanding Botnets, HotBots (2007)
32. Xie, Y., Yu, F., Achan, K., Panigrahy, R., Hulten, G., Osipkov, I.: Spamming botnets: Signatures and characteristics. ACM SIGCOMM Computer Communication Review 38(4), 171–182 (2008)
33. Ormerod, T., Wang, L., Debbabi, M., Youssef, A., Binsalleeh, H., Boukhtouta, A., Sinh, P.: Defaming botnet toolkits: A bottom-up approach to mitigating the threat. In: eCrime Researchers Summit, eCrime (2010)
34. Riccardi, M., Oro, D., Luna, J., Cremonini, M., Vilanova, M.: A framework for financial botnet analysis. In: eCrime Researchers Summit, eCrime (2010)
35. Liang, J., Naoumov, N., Ross, K.: The index poisoning attack in p2p file sharing systems. In: IEEE INFOCOM, Citeseer, vol. 6 (2006)
36. Lou, X., Hwang, K.: Prevention of index-poisoning DDoS attacks in peer-to-peer file-sharing networks. Submitted to IEEE Trans. on Multimedia, Special Issue on Content Storage and Delivery in P2P Networks (2006)
37. Staniford, S., Parxson, V., Weaver, N.: How to own the internet in your spare time. In: Proceedings of the 11th USENIX Security Symposium (2002)
38. Vogt, R., Aycock, J., Jacobson, M.: Army of botnets. In: Proceedings of the 2007 Network and Distr. System Sec. Symposium (NDSS 2007), Citeseer, pp. 111–123 (2007)
39. Karge, S.: The german anti-botnet initiative. In: OECD Workshop: The Role of Internet Intermediaries in Advancing Public Policy Objectives, Organization for Economic Co-Operation and Development (2011)
40. Ashford, W.: Collaborative strike takes down second hlux/kelihos botnet (2012)
41. Gostev, A.: Kaspersky Security Bulletin. Malware Evolution 2010. Technical report, Karspersky Labs (February 2011)
42. Kok, J., Kurz, B.: Analysis of the botnet ecosystem. In: 10th Conference of Telecommunication, Media and Internet Techno-Economics (CTTE). VDE, pp. 1–10 (2011)

How Political Illusions Harm National Stability: Fiscal Illusion as a Source of Taxation

Paulo Mourao[1] and José Pedro Cabral[2]

[1] Department of Economics/NIPE, University of Minho,
4700 Braga, Portugal
[2] Câmara Municipal de Vila Real,
5000 Vila Real, Portugal

Abstract. Fiscal illusion is the most common form of political illusions. This article introduces an agent-based model developed for testing the Fasiani model of fiscal illusion. The Fasiani model introduces fiscal illusion as a source of extra taxation that can lead to an impoverishment of citizens and to null national production after a several periods of increasing levels of fiscal illusion. We modeled this strategy as a "dictator game," in which we show that a state that constantly deceives its citizens is a source of national instability, deeply harming its national security.

Keywords: Fiscal Illusion, Taxation, Taxes Programs Codes.

1 Introduction

This article is the first to program fiscal illusion using program codes. The advantages of programming such phenomena are various. First, programming leads to extending the debate around public finance issues to other fields of science, such as computational engineering, multi-agent systems, centralized control systems, and performance modeling (applications and systems). Second, programming can generate didactic tools that are useful for illustrating the strategies discussed in different games (debated in subjects like games theory, public finances, or public economics). Finally, programming offers a new stimulus for testing the claims of models that have not yet been tested, as is the case with the Fasiani model.

Mauro Fasiani was the first author to model fiscal illusion based on the study of Amilcare Puviani's [1] analysis. According to Mauro Fasiani [2], if the level of fiscal illusion is higher, then it is easier to raise taxes, diminishing the final amount of disposable income received by taxpayers. Consequently, fiscal illusion is a source of (increasing) taxation, which leads to social instability and to higher number of riots. Therefore, we are going to focus on these fiscal and macroeconomic threats to national security. These threats cannot be neglected in spite of an increasing attention on more recently identified issues, like Corchado and Herrero [3] or Panda et al [4].

However, until now no work about this issue has discussed program codes. This work is clearly the first to do that.

For this purpose, the article has four sections: section 2, introducing the model of fiscal illusion according to Fasiani; section 3, introducing the codes and the figures derived from running our program on running; section 4, a conclusion.

Á. Herrero et al. (Eds.): Int. JointConf. CISIS'12-ICEUTE'12-SOCO'12, AISC 189, pp. 109–115.
springerlink.com

2 Fiscal Illusion According to Mauro Fasiani

Fiscal illusion happens when the taxpayer does not realize how much she or he pays to the state. The state tries to collect the maximum amount of taxes with the minimum amount of socially exhibited cost. As Colbert argued [5], "The art of taxation consists in so plucking a goose as to obtain the largest amount of feathers with the least possible amount of hissing." Therefore, all of us suffer from fiscal illusion.

The first author to study the specific phenomenon of fiscal illusion was Amilcare Puviani [1]. Later, Mauro Fasiani [2] gave an enlarged focus to this theme and developed the first analytical model to study fiscal illusion.

According to Fasiani [2], the final result of fiscal illusion is a decrease in individual utility with respect to money, as figure 1 suggests. This figure was extracted from the original book by Fasiani [2].

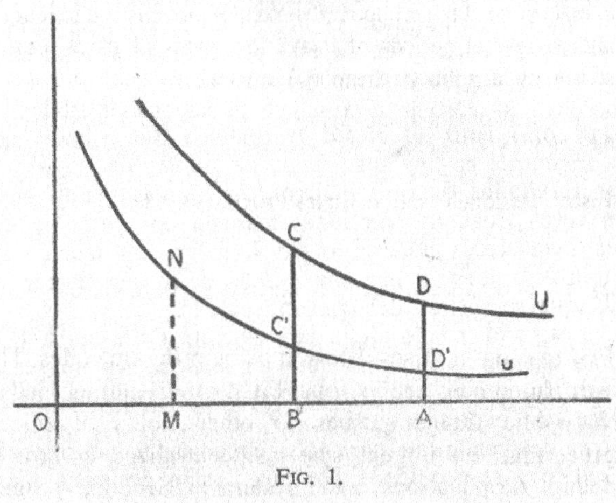

Fig. 1. The Fasiani Fiscal Illusion model. *Source: Fasiani (1941, p. 133)*

Fasiani [2] explained figure 1 as follows:

> If we consider a single individual, we can show the [fiscal illusion] phenomenon in the following way (fig.1). Let us express the monetary units on the absciss axis, and let us express his utility on the ordinates axis. *U* is the utility function before the event [that generates fiscal illusion], and *u* is the utility function after the event. Let us assume that this individual has a fixed income equal to OA, being AB the tax. If this tax is collected when the utility function is *U* then it determines a burden equal to BCDA. If this tax is collected at the moment characterized by the utility function *u* then it determines a burden equal to BC'D'A, and the individual under the effect of the illusion has a smaller burden whose difference is equal to C'CDD.' The State can take advantage of the illusion raising the tax to AM. The increase of

> taxes BM determines a loss of utility equal to MNC'B. It is clear that the advantage of illusion is off when MNC'B is at a maximum, equal to C'CDD'—this is, using other words, when the tax was collected at the moment in which the utility function has been decreased, the individual did not realize the loss of utility C'CDD.' But now, if the State intends to take advantage of the [fiscal illusion] phenomenon, it may collect so many monetary units as those generating a total utility equal to C'CDD.' But beyond these limits, the illusion does not actuate and it is no longer possible to take advantage of it. (p. 133)

We are going to model this strategy of fiscal illusion (that Paulo Mourao [6] considers a kind of "dictator game"). As Mourao [7] also argued, and following Puviani [1] and Fasiani [2], an excessive fiscal illusion can degenerate into fiscal delusion, destroying the value of production (extremely charged by a weighty amount of taxes). This fiscal delusion can lead to riots, popular contestation, and to a final distrust between the state and each citizen, harming the security of nations.

3 Programming Fiscal Illusion

In order to model the behavior described by figure 1, we will consider the U function as the first derivative of the utility of money. Following several authors, a widely generalized function for the utility of money is given by the log of income [ln(y)]. Therefore, $U = 1/y$ and $u = (1-f)/y$, f being a fiscal illusion measure or index between 0 and 1.

Thus, our model will allow the exogeneity of five variables: (initial) income (at monetary units), income growth rate (%/year), explicit taxes (% of income), fiscal illusion (% of money devaluation at the taxpayer perspective), and fiscal illusion growth rate (measured by percentage points). Then, the user can insert a number of years to be studied.

In essence, our model is not a multi-agent model as conventional computing theory usually analyzes. However, following Fasiani's [2] original suggestion, which Mourao [6] claimed to be a kind of "dictator game," we can also think in terms of an alternative multi-agent model with two agents, one (State/proposer) active and the other (taxpayer/respondent) passive.

We have created our model using the language VB.NET. Some of the advantages of object-oriented programming languages are the allowance of a higher level of abstraction/generalization on agents programming and the possibility of developing further extensions.

The program reads:

```
const incomeGrowth
const VarFI
const taxes
const MAXYEARS
var fiscalillusion
var income
```

Then, Run the cycle:

```
For year = 1 To MAXYEARS
      t = Exp(((fiscalillusion / (fiscalillusion - 1)) *
Log(income)) - (1 / (fiscalillusion - 1) * Log(income * (1 -
taxes))))

      ttaxes = income - t
      tfiscalillusion = (income * (1 - taxes)) - t
      net = income - ttaxes
      burden = fiscalillusion / ttaxes
      weight = ttaxes / income

      If tfiscalillusion <= 0 Then
            Exit For
      Else
            DataTable.AddElement(year, income, net, ttaxes,
fiscalillusion, burden, weight)
            income = income + (income * incomeGrowth)
            fiscalillusion = fiscalillusion + VarFI
      End If

End For
```

Figure 2 shows the layout of our program running.

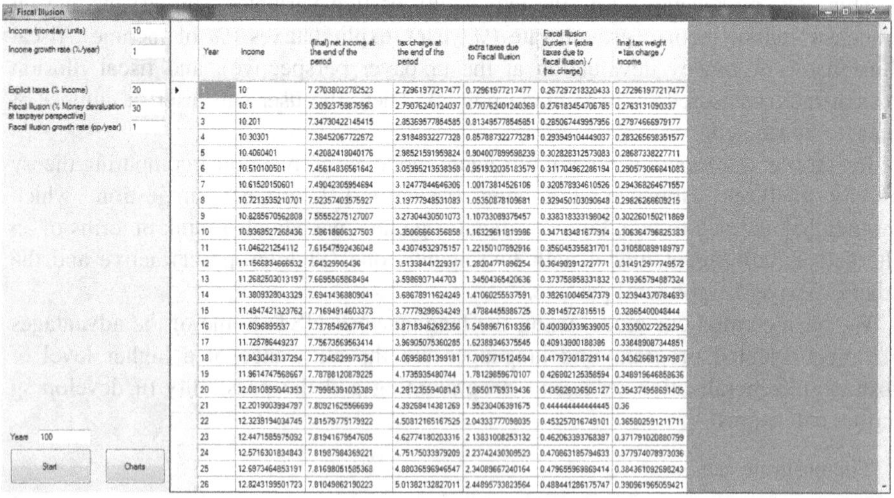

Fiscal Illusion

Income (monetary units) 10
Income growth rate (%/year) 1
Explicit taxes (% income) 20
Fiscal Illusion (% Money devaluation at taxpayer perspective) 30
Fiscal Illusion growth rate (pp/year) 1

Year	Income	(final) net income at the end of the period	tax charge at the end of the period	extra taxes due to Fiscal Illusion	Fiscal Illusion burden = (extra taxes due to fiscal illusion) / (tax charge)	final tax weight = tax charge / income
1	10	7.27038022782523	2.72961977217477	0.72961977217477	0.267297218320433	0.272961977217477
2	10.1	7.30923759875963	2.79076240124037	0.770762401240368	0.276183454706785	0.2763131090337
3	10.201	7.34730422145415	2.85369577854585	0.813495778545851	0.285067449957956	0.27974666979177
4	10.30301	7.38452067722672	2.91848932277328	0.857887322773281	0.293949104449037	0.283265698351577
5	10.4060401	7.42082418040176	2.98521591959824	0.904007899598239	0.302828312573083	0.28687338227711
6	10.510100501	7.45614836561642	3.05395213538358	0.951932035183579	0.311704962286194	0.290573066841083
7	10.61520150601	7.49042305954694	3.12477844646306	1.00173814526106	0.320578934610526	0.294368264671557
8	10.7213535210701	7.52357403575927	3.19777948531083	1.05350878109681	0.329450103090648	0.29826266609215
9	10.8285670562808	7.55552275127007	3.27304430501073	1.10733089375457	0.338318333198042	0.302260150211869
10	10.9368527268436	7.58618606327503	3.35066666356858	1.16329611819986	0.347183481677914	0.306364796825383
11	11.046221254112	7.61547592436048	3.43074532975157	1.22150107892916	0.356045395831701	0.310580899189797
12	11.1566834666532	7.64329905436	3.51338441229317	1.28204771896254	0.364903912727771	0.314912977749572
13	11.2682503013197	7.6695565868494	3.5986937144703	1.34504365420636	0.373758858331832	0.319365794887324
14	11.3809328043329	7.69414368809041	3.68678911624249	1.41060255537591	0.382610046547379	0.323944370784693
15	11.4947421323762	7.71694914603373	3.77779298634249	1.47884455986725	0.39145727815515	0.32865400048444
16	11.6096895537	7.73785492677643	3.87183462692356	1.54989671618356	0.400300339639005	0.333500272252294
17	11.725786449237	7.75673569563414	3.96905075360285	1.62389346375545	0.40913900188386	0.338489087344851
18	11.8430443137294	7.77345829973754	4.06958601399181	1.70097715124594	0.417973018729114	0.343626681297987
19	11.9614747568667	7.78788120879225	4.1735935480744	1.78129859670107	0.426802125358594	0.348919646858636
20	12.0810895044353	7.79985391035389	4.28123559408143	1.86501769319436	0.435628036505127	0.354374958691405
21	12.2019003994797	7.80921625566699	4.39268414381269	1.95230406391675	0.444444444444445	0.36
22	12.3239194034745	7.81579775179922	4.50812165167525	2.04333777098035	0.453257016749101	0.365802591211711
23	12.4471585975092	7.81941679547605	4.62774180203316	2.13831008253132	0.462063393768397	0.371791020880799
24	12.5716301834843	7.81987984369221	4.75175033979209	2.23742430309523	0.470863185794633	0.377974078973036
25	12.6973464853191	7.81698051585368	4.88036596946547	2.34089667240164	0.479655969869414	0.384361092698243
26	12.8243199501723	7.81049862190223	5.01382132827011	2.44895733823564	0.488441286175747	0.390961965059421

Years 100
Start
Charts

Fig. 2. Layout of the Program

How to use the program. After inserting values for the five main parameters and for the number of years to be analyzed, the user presses the "start" button to get the estimation for six columns at each year: income, (final) net income at the end of the period (year); tax charge at the end of the period; extra taxes due to fiscal illusion (for instances, in figure 1, the extra tax is MB); fiscal illusion burden = (extra taxes due to fiscal illusion)/(tax charge); and final tax weight = (tax charge)/income. If, after this step, the user presses the "charts" button, then two set of graphs will appear at a new window: the first set will show fiscal illusion burden and final tax weight; and the second set will show income; final net income; and tax charge.

Testing Fasiani's Claims. Now, using our program, we are able to test several of Fasiani's claims [7], namely:

i) A higher level of fiscal illusion leads to a higher level of taxation, to a higher tax burden and to less disposable income
ii) An increasing level of fiscal illusion not only raises taxes but also may lead to an unsustainable level of taxation (leading to a final situation in which taxpayers are afforded no disposable income);
iii) And if fiscal illusion rises more slowly than income, this strategy can be eternized by governments.

For illustrative purposes, we show next some examples with different values of the parameters.

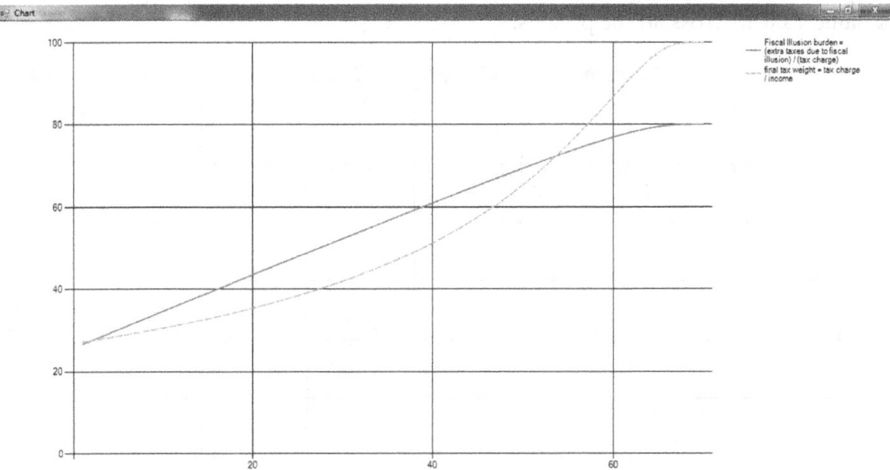

Fig. 3. Graph after Income 10, Income Growth rate 1, Explicit taxes 20, Fiscal Illusion 30, Fiscal Illusion growth rate 1

Observing figure 3, we confirm that an increasing level of fiscal illusion (1 pp/year) actually leads to an increasing tax weight and to an increasing fiscal illusion burden. With the inserted parameters' values, this game would end at year +70 (tax weight = 100%). This is the case of an unsustainable level of taxation that can easily degenerate into riots, and coups d'état!

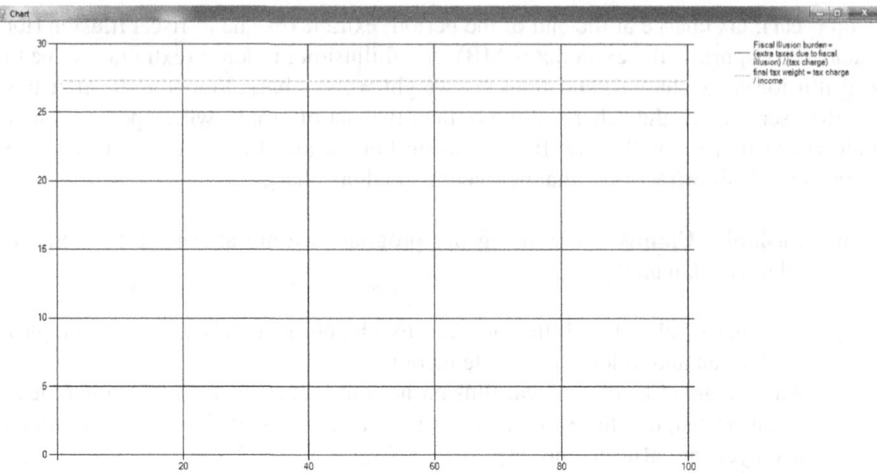

Fig. 4. Graph after Income 10, Income Growth rate 1, Explicit taxes 20, Fiscal Illusion 30, Fiscal Illusion growth rate 0

Observing figure 4, we can check a case in which the level of fiscal illusion is always the same. For this case, the fiscal illusion burden and the tax weight will keep the same values across all the periods.

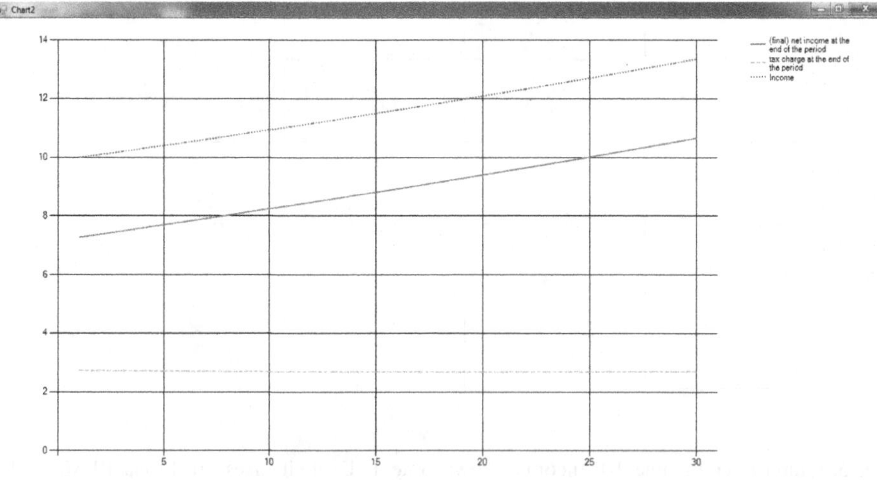

Fig. 5. Graph after Income 10, Income Growth rate 1, Explicit taxes 20, Fiscal Illusion 30, Fiscal Illusion growth rate -1

Finally, figure 5 shows us the (desired) case of a decreasing level of fiscal illusion (-1 pp/year). There we can confirm that when fiscal illusion diminishes, we can expect an increasing trend of income and of net income and a diminishing of tax charges across the periods.

4 Final Remarks and Further Work

This work has tested some of the most important claims of Mauro Fasiani on the dangers of fiscal illusion using a programming language. By using this resource, it has become the first work on fiscal illusion that utilizes the advantages of computing simulation.

Fiscal illusion can be a serious threat to the security of states and citizens. Therefore, studying it and paying attention to it is essential to diminish the probability of fiscal illusion becoming fiscal delusion in the taxpayers' view.

Running our program, we observed the main claims of the authors who have studied fiscal illusion using descriptive strategies. We highlight our findings. First, fiscal illusion increases the tax charge. Second, increasing levels of fiscal illusion can stifle or nullify the national economy. Finally, if fiscal illusion is reduced, net income that is available to taxpayers can be raised.

As a further work, we want to extend this model to a conventional multi-agent model. In this extension, agents will endogenize the parameters and they will optimize objective functions (like production/income functions after optimizing the fiscal illusion parameters).

References

1. Puviani, A.: Teoria della illusione finanziaria, Sandron, Palermo (1903)
2. Fasiani, M.: Principii di Scienza delle Finanze, vol. I, Giappichelli, Torino (1941) [quoted version: Fasiani, M. (1962), Principios de Ciencia de la Hacienda, Aguilar, Madrid; translation by Gabriel de Usera]
3. Corchado, E., Herrero, A.: Neural visualization of network traffic data for intrusion detection. Appl. Soft Comput. 11(2), 2042–2056 (2011)
4. Panda, M., Abraham, A., Das, S., Patra, M.: Network intrusion detection system: A machine learning approach. Intelligent Decision Technologies 5(4), 347–356 (2011)
5. Stiglitz, J.: On Liberty, the Right to Know, and Public Discourse: The Role of Transparency in Public Life. Oxford Amnesty Lecture, Oxford (1999)
6. Mourão, P.: A Case of Dictator Game in Public Finances–Fiscal Illusion between Agents. In: Omatu, S., Paz Santana, J.F., González, S.R., Molina, J.M., Bernardos, A.M., Rodríguez, J.M.C., et al. (eds.) Distributed Computing and Artificial Intelligence. AISC, vol. 151, pp. 249–254. Springer, Heidelberg (2012)
7. Mourao, P.: Fiscal Illusion causes Fiscal Delusion – Please be careful. In: Magalhes, S., et al. (eds.) Global Security, Safety, and Sustainability. CCIS, vol. 92(1), pp. 232–237 (2010)

Software Implementation of Linear Feedback Shift Registers over Extended Fields

O. Delgado-Mohatar[1] and A. Fúster-Sabater[2]

[1] Universidad Internacional de Castilla y León (UNICYL)
Calzadas 5, 09004 Burgos, Spain
oscar.delgado@unicyl.es
[2] Information Security Institute, C.S.I.C.
Serrano 144, 28006 Madrid, Spain
amparo@iec.csic.es

Abstract. Linear Feedback Shift Registers are currently used as generators of pseudorandom sequences with multiple applications from communication systems to cryptography. In this work, design and software implementation of LFSRs defined over extended fields $GF(2^n)$ instead of over the binary field $GF(2)$ are analyzed. The key idea is to take profit of the underlying structure of the processor over which the application is executed. The study has been carried out for diverse extended fields and different architectures. Numerical results prove that extended fields provide speedup factors up to 10.15. The benefits of these fields are clear for LFSR applications included cryptographic applications.

Keywords: software implementation, extended field, LFSR, security.

1 Introduction

Intelligent technologies applied to security are interdisciplinary in nature, bridging computer science with information systems and engineering. Very different disciplines such as intrusion detection systems [1], [6], machine learning [10] or applied cryptography [8] are all included in this broad field that conjugates theory, design, development and evaluation of systems. This work is concerned with a type of a cryptographic technique that guarantees security (confidentiality and integrity) in the frame of information transmission.

Inside cryptographic technologies, stream ciphers are nowadays the fastest among the encryption procedures so they are implemented in many technological applications e.g. RC4 for encrypting Internet traffic [11] or the encryption function E0 in Bluetooth. Stream ciphers try to imitate the mythic *one-time pad cipher* or *Vernam cipher* [8] and are designed to generate a long sequence (the *keystream sequence*) of seemingly random bits. Some of the most recent designs in stream ciphers can be found in [2].

LFSRs have been traditionally designed to operate over the binary Galois field $GF(2)$ [4]. This approach is appropriate for hardware implementations, but its software efficiency is low because of two important drawbacks:

Á. Herrero et al. (Eds.): Int. JointConf. CISIS'12-ICEUTE'12-SOCO'12, AISC 189, pp. 117–126.
springerlink.com

1. In order to update the state of a LFSR, the processor has to spend many clock cycles to perform the register shifting or output generation operations.
2. Binary LFSRs provide only one output bit per clock pulse, which makes the software implementations very inefficient and involves a clear waste of modern processor capabilities.

According to new trends found in the literature [13], the aim of this work is the use of finite fields more adequate to modern processors. More precisely, the natural choice is the Galois extended field with 2^n elements where n is related to the size of the registers in the underlying processor (currently words of 8, 16 or 32 bits). In this sense, the field elements fit perfectly well the storage unit and can be efficiently handled. At the same time, the total throughput of these extended generators is increased by a factor 8, 16 or 32, respectively. Nevertheless, the arithmetic operations over extended Galois fields are computationally very expensive. Therefore, in this work two fundamental features have been considered: the study of calculation techniques for arithmetic operations over extended Galois fields $GF(2^n)$ (in particular the arithmetic multiplication) as well as the adaptation of such techniques to LFSRs of cryptographic application.

By means of this analysis, we try to answer diverse questions such as what is the more efficient calculation technique for LFSRs or what is the maximum improvement that can be obtained or how these arithmetic calculations can be affected by the characteristics of the hardware architecture used in the implementation.

2 LFSRs over $GF(2^n)$

In this section, we study the relative benefit of moving from the binary field to the extended field by evaluating the use of Galois fields $GF(2^n)$ in general-purpose implementations. The validity of this proposal is based on the fact that the fundamental properties of the sequences generated by bit-oriented LFSRs are still preserved when moving to word-oriented LFSRs [9].

2.1 Extended Linear Feedback Shift Registers

Extended LFSRs are defined in the same way as the traditional ones. Next, some basic concepts are introduced.

Definition 1. An extended LFSR of length L is a finite state machine made out of L memory cells $r_{L-1}, \dots, r_1, r_0$, where each of them stores, s_{t+i}, an element of $GF(2^n)$. At time t, the LFSR state is denoted by $S_t = (s_{t+L-1}, s_{t+L-2}, \dots, s_t)$. At each time unit, the register updates its state. In fact, the state S_{t+1} is derived from the state S_t such as follows:

1. The content of the cell r_{L-1} is the corresponding element of the output sequence.
2. The content of the cell r_i is shifted into the cell r_{i+1} for all i, $0 \leq i \leq L-2$.

3. The new content of the cell r_0 is the feedback element computed according to the following feedback function:

$$s_{t+L} = \sum_{i=0}^{L-1} c_i \otimes s_{t+i}, \ c_i \in GF(2^n), \tag{1}$$

where the coefficients $c_0, c_1, \ldots, c_{L-1}$ are called *feedback coefficients* and the symbol $\otimes$ denotes the multiplication in $GF(2^n)$.

These coefficients define the LFSR *characteristic polynomial*:

$$p(x) = x^L + c_{L-1}\, x^{L-1} + \ldots + c_1\, x + c_0\, . \tag{2}$$

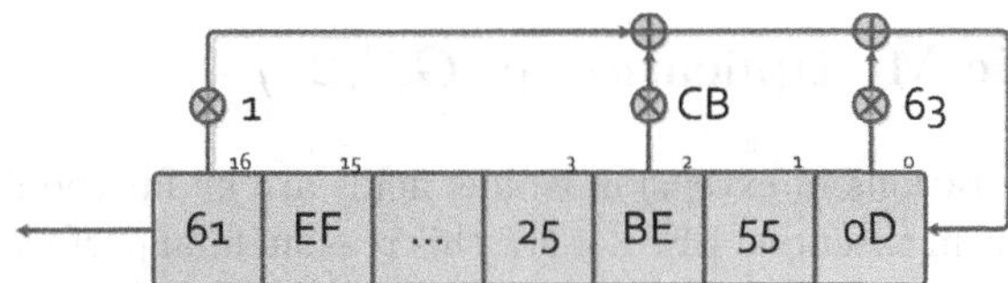

Fig. 1. Extended LFSR of 17 stages defined over $GF(2^8)$

Figure 1 shows an example of extended LFSR with 17 stages defined over $GF(2^8)$. In this new structure, the cell contents, $s_{t+i}\ \ 0 \leq i \leq L-1$, as well as the feedback coefficients (all of them expressed in hexadecimal notation) are elements of $GF(2^8)$. This fact will affect the two arithmetic operations to be performed: addition and multiplication.

2.2 Arithmetic Operations in $GF(2^n)$

The elements of $GF(2^n)$ are represented by means of binary polynomials of degree less than n. For instance, the element a in $GF(2^4)$, notated $a = 0111$, can be represented by means of $a(x) = x^2 + x + 1$. Under this polynomial representation, the arithmetic operations are carried out in terms of an irreducible n-degree polynomial $R(x)$ defined as follows:

$$R(x) = x^n + d_{n-1}x^{n-1} + \ldots + d_1 x + 1, \ d_i \in GF(2). \tag{3}$$

The addition operation for such elements is simply the bit-wise XOR operation among n-dimensional vectors. The multiplication operation for such elements is the polynomial multiplication with the resulting polynomial being reduced mod $R(x)$.

In order to study the extended LFSRs, a set of polynomials $P_8(x)$, $P_{16}(x)$ and $P_{32}(x)$, defined over the fields $GF(2^8)$, $GF(2^{16})$ and $GF(2^{32})$, respectively, has been chosen. In a generic way, the characteristic polynomials have the form:

$$P(x) = x^{17} + x^{15} + c_2 x^2 + c_0, \tag{4}$$

with $c_2, c_0 \in GF(2^n)$ and $c_2, c_0 \neq 0$. The corresponding feedback functions are:

$$s_{t+17} = s_{t+15} \oplus (c_2 \otimes s_{t+2}) \oplus (c_0 \otimes s_t)\,, \quad (5)$$

where $\oplus$ and $\otimes$ are addition and multiplication over $GF(2^n)$, respectively. The specific values of c_2, c_0 for each characteristic polynomial are shown in Table 1.

Table 1. Characteristic polynomials used throughout the work

Polynomial	c_2	c_0
$P_8(x)$	$C6$	67
$P_{16}(x)$	$19B7$	$013C$
$P_{32}(x)$	$F21DA317$	$E28C895D$

3 Arithmetic Multiplication in $GF(2^n)$

Multiplication operations in extension Galois fields are by far the most consuming time operations in extended LFSRs. For this reason, final performance greatly depends on the specific multiplication method used as performed experiments seem to confirm. In next sections, several methods for performing multiplication over the different extended Galois fields are analyzed and compared.

3.1 MulTABLE Method

When the size of n in $GF(2^n)$ is small, the fastest procedure for multiplication is the use of look-up tables of pre-computed results. In general, the table size is $2^{(2n+2)}$ bytes, so that the method is adequate for small values of n. In Table 2, the table size increment for different values of n is depicted. It can be noticed that in practice only the value $n = 8$ is adequate as the table size equals 256 KB. For $n = 16$, the table size would be 16 GB, and for $n = 32$ the table would achieve the astronomic size of 64 EB (!) (1 EB $= 10^{18}$ bytes).

3.2 LogTABLE Method

When multiplication tables cannot be employed, the next fastest way to perform multiplication is to use log and inverse log tables, as described in [5]. Making use of these tables and somewhat of basic arithmetic, two elements can be multiplied by adding their corresponding logs and then taking their corresponding inverse log. In fact,

$$a \cdot b = invlogTable[(logTable[a] + logTable[b])].$$

In this way, the multiplication is reduced to three looking-up in tables plus one addition operation. The first table, *logTable* is defined for indices ranging from 1 up to $2^n - 1$ and stores for each field element its corresponding log. On the other

Table 2. Sizes for a table of pre-computed results

n	Resulting table size	Size in bytes
8	256 KB	2^{18}
10	4096 KB	2^{22}
12	64 MB	2^{26}
16	16 GB	$2^{34} \approx 10^{10}$
...	...	...
32	64 EB	$2^{66} \approx 10^{18}$

hand, the second table *invlogTable* is defined for indices ranging from 0 up to $2^n - 1$ establishing a map between these values and their inverse log. Clearly, $logTable[invlogTable[i]] = i$, and $invlogTable[logTable[i]] = i$. The size of these tables is $2^{(n+2)}$ bytes for each one of them instead of the $2^{(2n+2)}$ bytes of the previous method. Therefore, for $n = 8$ and $n = 16$ the size is 2 KB and 0.5 MB, respectively. Nevertheless for $n = 32$ the approximate size is 32 GB what is out of the capabilities of most modern processors.

3.3 SHIFT Method

When even log tables are unusable, general-purpose multiplication $a \cdot b$ can be implemented as a product of an $n \times n$ matrix and the bit-vector a, where the matrix is a function of operand b and the irreducible polynomial $R(x)$ that defines the field. Although it is significantly slower than the table methods, it is a general-purpose technique that requires no pre-allocation of memory.

3.4 SPLITW8 Method

Finally, for the special case of $n = 32$, a particular method is used that creates seven tables, each of them of 256 KB size. Such tables are employed to multiply 32-bit field elements by breaking them into four eight-bit parts, and then performing sixteen multiplications and XORs operations to calculate the product. With this optimization, the algorithm performs 7 times faster than when using the previous SHIFT method.

4 Software Optimization Techniques

In conjunction with the use of extension Galois fields, which is considered as the major improvement method, other software optimization techniques can be added to improve the final LFSR performance. Three of these software implementation improvements are described and analyzed.

4.1 Circular Buffers

This technique is widely used in many algorithmic applications and is especially suitable for shift register implementations. By using this technique, it is possible to avoid the computationally expensive $L-1$ shifts, where L is the register length. Instead of moving the register contents, several pointers are shifted along the register. From the very moment the pointers reach the buffer end, they come back to the beginning giving the impression of a circular buffer. This technique has been implemented in C language, making use of the same environment and libraries as in previous sections.

4.2 Sliding Windows

In addition to the previous technique, another method, called sliding window, can also be used. Basically, it consists of a buffer twice as long as it should be. When a new value is written into the buffer, in fact it is written in two different positions at the same time. In this way, when new values are added, all the intermediate values are accessible at fixed and known offsets contiguous to the new ones. Unlike the circular buffer method, the sliding window approach allows the shift register to be arbitrarily long and still be efficiently addressed. That is a very important advantage when this method is applied to LFSR implementation.

4.3 Loop Unrolling

Loop unrolling, or loop unwinding, is a well-known code transformation for improving the execution performance of a program [7], [3]. Loop unrolling works by replicating the original loop body multiple times, adjusting the loop termination code and eliminating redundant branch instructions. Thus, loop unrolling is an effective code transformation, which typically improves the execution performance in programs that spend much of their execution time (10 – 30%) in loops. For this reason, this technique is especially suitable for LFSR implementations, as they spend most of their time updating the cells contents by using a loop.

5 Experimental Results

The software implementation performance of the previous methods has been checked in different architectures, whose technical specifications can be found in Table 3. These architectures include from an 8-bit microprocessor, passing through a Pentium III (16-bit microprocessor) up to modern 64-bit processors. The analysis has been carried out by implementing the multiplication methods in C language and making use of a publicly available arithmetic library [12]. The compiler used was gcc version 4.4.1 running in a Linux 2.6.31 kernel, with the -O3 optimization option flag, which activates all the optimization techniques of the underlying microprocessor. The experiments consisted in the generation of a sequence of 10^7 bytes over the different platforms and algebraic fields including the binary field $GF(2)$. Execution time and total throughput were computed.

Table 3. Architectures used for the proposal evaluation

Architecture	CPU Frequency	Cache L1I/L1D/L2	RAM Memory
AMD Athlon 64 Dual Core Processor 4200+	2.2Ghz	64KB/64KB/512KB	2GB
Intel Core 2 Duo	2Ghz	32KB/32KB/4MB	1GB
Intel Pentium III	450Mhz	-	192MB
Micro ATmega 168	16Mhz	-	14KB

Table 4. Execution Time, Throughput and Improvement Factor for different platforms

	$GF(2)$		
Architecture	Time (s)	Throughput (MB/s)	Factor α
ATmega 168	4450	0.002	-
Pentium III	24.37	0.41	-
Dual Core	1.58	6.32	-
Athlon	1.14	8.77	-
	$GF(2^8)$		
Architecture	Time (s)	Throughput (MB/s)	Factor α
ATmega 168	1605	0.006	3
Pentium III	3.89	2.57	6.25
Dual Core	0.56	21.27	2.80
Athlon	0.59	16.94	1.93
	$GF(2^{16})$		
Architecture	Time (s)	Throughput (MB/s)	Factor α
ATmega 168	-	-	-
Pentium III	1.49	6.71	16.35
Dual Core	0.24	41.66	6.60
Athlon	0.33	30.30	3.45
	$GF(2^{32})$		
Architecture	Time (s)	Throughput (MB/s)	Factor α
ATmega 168	-	-	-
Pentium III	1.79	5.58	13.57
Dual Core	0.27	37.03	5.85
Athlon	0.37	27.02	3.08

Table 4 depicts both magnitudes as well as an improvement factor (Factor α) that compares the generation times for extended fields and for the binary field. Factor $\alpha = 2$ means that the extended LFSR is twice faster than its implementation over $GF(2)$. The multiplication operations were performed over random numbers, repeating the measurements ten times, discarding extreme values and calculating the average on the rest. The results in each extended Galois field have been obtained by using the fastest multiplication method (see Table 5) for the corresponding architecture. The results in millions of multiplications per second for the different architectures are depicted in Table 5. The symbol – means that the method is not applicable for that value of n and that particular architecture. The first fact that should be noticed is that multiplication performance goes down as the field degree increases. Indeed, the fastest method in $GF(2^{32})$ is more than 14 times slower than the fastest method in $GF(2^8)$. The reason is that the operand size increases and therefore multiplication operations become more and more expensive. In $GF(2^8)$, the fastest method to perform multiplication is clearly the use of pre-computed tables (MultTABLE method). In $GF(2^{16})$, a table-based method can be used too, although performance drops slightly. In $GF(2^{32})$, due to the impossibility of using any kind of tables, there is an important loss of performance (14 and 8 times slower than in $GF(2^8)$ and $GF(2^{16})$, respectively). The second fact that should be noticed is that larger fields provide larger outputs in each shift. Indeed, one multiplication in $GF(2^{32})$ provides a 32-bit output, while four multiplications in $GF(2^8)$ would be necessary to obtain the same amount of bits. As will be shown in the next sections, the second fact outperforms the first one. As a result, in spite of the increasingly expensive operations, a better performance is always obtained by using extended fields with respect to traditional $GF(2)$.

Table 5. Numerical results for different Multiplication Methods over specific platforms

Architecture	Multiplication Method (M-multiplications/s)			
ATmega168	MulTABLE	LogTABLE	SHIFT	SPLITW8
$GF(2^8)$	-	-	0.00079	-
$GF(2^{16})$	-	-	-	-
$GF(2^{32})$	-	-	-	-
PentiumIII	MulTABLE	LogTABLE	SHIFT	SPLITW8
$GF(2^8)$	31.73	27.77	1.39	-
$GF(2^{16})$	-	4.53	0.55	-
$GF(2^{32})$	-	-	0.18	0.87
Dual Core	MulTABLE	LogTABLE	SHIFT	SPLITW8
$GF(2^8)$	110.03	94.11	4.40	-
$GF(2^{16})$	-	60.25	0.55	-
$GF(2^{32})$	-	-	0.47	8.03
Athlon	MulTABLE	LogTABLE	SHIFT	SPLITW8
$GF(2^8)$	89.62	78.76	5.52	-
$GF(2^{16})$	-	29.05	2.25	-
$GF(2^{32})$	-	-	0.82	3.41

A software implementation of the LFSR by using the optimization techniques described in section 4 has been carried out. Its performance has been analyzed for different combinations of underlying field size (included the binary field) and implementation improvement techniques. Those results for the fastest multiplication method over the Dual Core platform are depicted in Table 6. In addition, the factor β compares execution time with and without improvements.

Table 6. Results of different Improvement Techniques and Factor β

Improvement Technique		$GF(2)$	$GF(2^8)$	$GF(2^{16})$	$GF(2^{32})$
None	Throughput (MB/s)	6.32	21.27	41.66	37.03
	Factor β	-	-	-	-
Circular buffers	Throughput (MB/s)	9.70	22.72	58.82	43.47
	Factor β	1.53	1.06	1.41	1.17
Sliding windows	Throughput (MB/s)	24.39	27.02	62.5	47.61
	Factor β	3.85	1.27	1.50	1.28
Loop unrolling	Throughput (MB/s)	66.67	38.46	76.92	55.55
	Factor β	10.54	1.81	1.84	1.50

6 Conclusions

In this paper an exhaustive study of the LFSR software implementation has been carried out. In fact, we present an in-depth analysis of the impact of the use of extension fields in the construction of more efficient LFSR software implementations. On the other hand, we propose and analyze three additional implementation techniques, in order to determine which is best suited for LFSR implementation. Both extension fields and the proposed techniques can be used simultaneously. Numerical results illustrate how these approaches can provide very significant performance improvements. Compared to traditional methods, speedup factors of up to 10.15 can be easily achieved.

Regarding the evolution of LFSR performance, as the size of the ground field $GF(2^n)$ increases, interesting conclusions have been drawn. In this sense, this research proves that the use of extended fields for LFSR construction and large n is not recommended. Indeed, extended fields greater than $GF(2^{16})$ seem not to be worthwhile, because the internal operation cost is so high that their advantages are lost. In addition, numerical results prove that the optimum implementation technique is the loop unrolling method. The main reason is that it achieves a higher parallelism level than the other proposed techniques and, therefore, it takes greater advantage of the microprocessor pipeline. By combining both results, it can be concluded that the most efficient way to implement LFSRs in software is to operate over the extension field $GF(2^{16})$ and to use the loop unrolling technique.

References

1. Corchado, E., Herrero, A.: Neural visualization of network traffic data for intrusion detection. Appl. Soft Comput. 11(2), 2042–2056 (2011)
2. eSTREAM, the ECRYPT Stream Cipher Project, The eSTREAM Portfolio in 2012, `http://www.ecrypt.eu.org/documents/D.SYM.10-v1.pdf`
3. Dragomir, O., Stefanov, T.P., Bertels, K.: Loop Unrolling and Shifting for Reconfigurable Architectures. In: Proceedings of the 18th International Conference on Field Programmable Logic and Applications, FPL 2008 (September 2008)
4. Golomb, S.W.: Shift Register-Sequences. Aegean Park Press, Laguna Hill (1982)

5. Greenan, K., Miller, E., Schwarz, T.: Optimizing Galois field arithmetic for diverse processor architectures and applications. In: Miller, E., Williamson, C. (eds.) Proc. of MASCOTS, pp. 257–266. IEEE Computer Society (2008)
6. Herrero, A., Zurutuza, U., Corchado, E.: A Neural-Visualization IDS for Noneynet Data. Int. J. Neural Syst. 22(2) (2012)
7. Huang, J.C., Leng, T.: Generalized Loop-Unrolling: A Method for Program Speedup. In: Application-Specific Software Engineering and Technology, IEEE Workshop on Field Programmable Logic, pp. 244–249 (1999)
8. Menezes, A.J., et al.: Handbook of Applied Cryptography. CRC Press, New York (1997)
9. Paar, C.: Efficient VLSI Architectures for Bit-Parallel Computation in Galois Fields. PhD thesis, Institute for Experimental Mathematics. University of Essen, Germany (1994)
10. Panda, M., Abraham, A., Das, S., Patra, M.R.: Network intrusion detection system: a machine learning approach. Intelligent Decision Technologies 5(4), 347–356 (2011)
11. Paul, G., Maitra, S.: RC4 Stream Cipher and Its Variants, Discrete Mathematics and Its Applications. CRC Press, Taylor & Francis Group, Boca Raton, FL (2012)
12. Plank, J.S.: Optimizing Cauchy Reed-Solomon Codes for Fault-Tolerant Storage Applications. Tech. Rep. CS-05-569. University of Tennessee (December 2005)
13. Tsabana, B., Vishne, U.: Efficient Linear Feedback Shift Registers with Maximal Period. Finite Fields and their Applications 8(2), 256–267 (2002)

Simulation Analysis of Static and Dynamic Intermediate Nodes and Performance Comparison of MANETS Routing Protocols

Jahangir Khan[1], Zoran Bojkovic[2], Syed Irfan Hayder[3], Gulam Ali Mallah[4], Abdul Haseeb[1], and Fariha Atta[5]

[1] Faculty of computer science & Information Technology, Sarhad University of Science & IT Peshawar 25000 Pakistan
[2] Faculty of Transport and Traffic Engineering University of Belgrade, Serbia
[3] Graduate School of Science and Engineering, PAF-KIET, PAF Base Korangi Creek Karachi 75190, Pakistan
[4] Department of Computer Science, Shah Abdul Latif University KhairPur Sindh Pakistan
[5] FAST-National University of Computer & Emerging Sciences, Peshawar Pakistan
{Jahangir,haseeb}.csit@suit.edu.pk, z.bojkovic@yahoo.com, hyder@pafkiet.edu.pk, ghulam.ali@salu.edu.pk, fariha.atta.pk@gmail.com

Abstract. Mobile ad hoc networks are self-configuring and self-organizing wireless networks with no infrastructure in which mobile nodes cooperate in a friendly manner for multi-hop forwarding. Due to the mobile nature of the nodes in mobile ad hoc networks, the applications of these networks are diverse, ranging from home networking to intelligent transport, military exercises and campus network. However, this mobility of nodes makes the network vulnerable to frequent topology changes. Consequently, routing in mobile ad hoc networks is much more complex than in other wireless networks. Here in this paper we proposed two reactive protocols i-e AODV and DSR, and to analyse the data delivery ratio of static and dynamic intermediate nodes as well as source to destination in campus networks and also to compare the global and object statistics as well as recent simulated statistics of both reactive protocols. We also compare the global and object simulation statistics to get best results during simulation with different metrics in simulation of DSR and AODV routing protocols. Here we suggest that the combined performance of both protocols make a very strong performance contribution in ad hoc networks using OPNET simulator. Results also shows that 3G ad-hoc can make a very strong contribution towards 4G cellular networks for our future generation.

Keywords: MANETS, DSR, AODV, Data throughput in campus network, Performance, 3G/4G.

1 Introduction

In the fourth generation 4G system concept, the user has freedom and flexibility to select any desired service with reasonably QoS and affordable price, anytime, anywhere. There are many attractive features for 4G which ensure a very high data

Á. Herrero et al. (Eds.): Int. JointConf. CISIS'12-ICEUTE'12-SOCO'12, AISC 189, pp. 127–140.
springerlink.com

rate, global roaming, incorporating the mobile world into the Internet Protocol IP-based core network, establishing an efficient billing system and perfect handoff mechanisms. Improvements over third generation 3G include enhanced multimedia, smooth streaming video, universal access and portability across all types of devices. Migrating current systems to 4G present's enormous challenges. The factors that have led to Long Term Evaluation (LTE) - Advanced, becoming the leading broadband mobile communications' system for 4G, are in large associated with the closed partnership between network operators and equipment vendors in its development. LTE technology provides many enhancements compared its predecessors including high peak data rate, low latency and spectrum flexibility.

The Mobile computing and networking are becoming major drivers of the transition from today's world of personal computers to the future world of ubiquitous/pervasive computing. With an increase in the demand for fast gadgets and computing devices to quickly accomplish tasks in the modern world, communication industry also needs to increase the speed of all types of communication. Wired communication is there for a long time but now users are inclining more towards wireless solutions to provide them faster connectivity anywhere anytime, enabling them to check emails, exchange information during meetings, work from remote locations, and so on. The solution is provided by "Wireless Network" i.e. a network of communicating devices that do not have any physical connectivity among them. However, there are certain situations where the presence of a prior infrastructure for wireless communication is not feasible or is of lesser use. With the advent of technology, user devices are increasingly becoming mobile, thus enabling users to interact and communicate without prior planning and in the environments where there is little or no communication infrastructure. There comes a need for a decentralized network that can quickly be deployed on demand. The potential scenarios of such a decentralized network are emergency relief efforts, disaster recovery, transportation, and military operations etc. Such and other situations cannot afford to have a centralized structure. Instead a decentralized infrastructure with on-demand connectivity is what is needed by these services. The solution is given by Wireless Ad hoc Networks. These networks that do not have a backbone infrastructure of routers for connectivity. Instead, the participating devices play the role of routers to forward information. These devices (nodes) are often mobile and a term *MANET* is coined for such "Mobile" Ad hoc Networks. MANETs are a type of ad hoc networks that are configured on-demand and on-fly. These networks are without any centralized entity for maintaining the network. Since the network is decentralized, the nodes work in collaboration with each other to discover the network topology and route information to each other. These nodes organize themselves arbitrarily and hence the network topology is unpredictable. Mobile nodes can join the network or leave it according to their requirements. The nodes communicate over bandwidth restricted wireless links. Consequently, the network topology is dynamic and changes over time as nodes join and leave the network. However, as nodes move, the strength of connectivity among them may change over time. Despite the fact that nodes may appear, disappear, and re-appear with time, the network connections among nodes should work all the time. That is why wireless networks are more demanding with respect to connectivity as compared to wired networks. The characteristics of MANETs are summarized as: Communication is carried out via radio waves. It lacks proper infrastructure and hence is an autonomous network. Nodes can perform the role of hosts as well as routers. The network topology is dynamic and changes frequently as nodes join and leave the network. The network can be deployed anywhere anytime. A centralized controller does not exist.

This paper is organized as follows. After an Introduction including the paper motivation, Section 2 deals with applications of MANETS, while Section 3 gives an overview of DSR and AODV routing protocols. Statistical comparisons of protocols are carried out in Section 4. Second part of the paper performs the analysis together with the summary of protocols. The last Section concludes the presentation.

2 Applications of MANETS

Wireless ad hoc networks have gained stunning popularity and acceptance in recent years and have become a hot topic of research in the field of wireless communication. The major reasons for the acceptance of this fairly infant technology are low-cost deployment, simple maintenance, fault tolerance, and wireless connectivity. Consequently, the application areas of MANETs are diverse such as class rooms, conference hall, educational campuses, military exercises, communication among soldiers in battlefields, unmanned reconnaissance, public safety, ubiquitous computing, intelligent transportation systems, disaster relief and rescue operations, mine site operations, urgent business meetings, data acquisition in difficult terrains, law enforcement etc. In all of these applications, there is one thing in common. These applications cannot afford to be implemented using the conventional (wired) networking strategies because of the high degree of mobility of the participating nodes. Moreover, it is difficult in these situations to first build up an underlying infrastructure to base the communication on. Therefore, wireless ad hoc network is the network of choice for these applications.

3 Overview of Reactive (On-Demand) Routing Protocols

In source initiated on-demand routing protocols, routes to the destination are discovered on-demand i.e. only as needed. When a node needs to transmit a packet to destination, it initiates a route-discovery procedure. The procedure is complete once the destination is reached and all possible routes to the destination are discovered. This information is maintained by the node until the information gets outdated or is no longer needed. There are a number of source initiated on-demand routing protocols as given in fig. 1.

3.1 Ad-Hoc On-Demand Distance Vector Routing (AODV) Protocol

AODV protocols are based on table-driven proactive DSDV protocols. However AODV protocols discover information about the routes only on demand rather than maintaining the routing tables as in the case of DSDV. Thus the route discovery packets are sent into the network by a source only when it needs to transmit some data to destination. AODV is a destination/hop-by-hop routing protocol i.e. the source

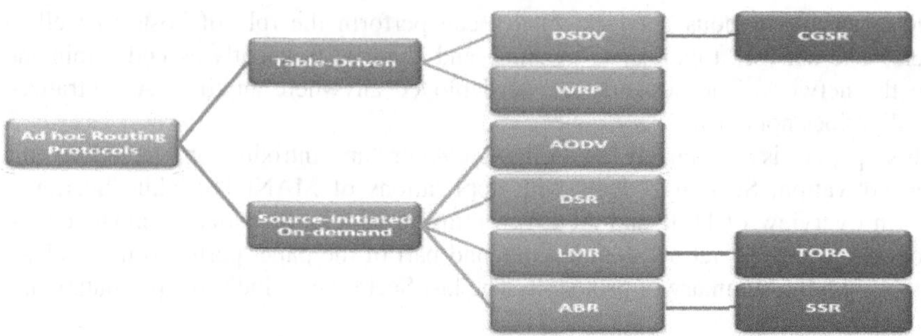

Fig. 1. Proactive and Reactive routing protocols classification

node just specifies the destination node to which the packet has to be sent. Intermediate nodes refer to their routing tables to decide the appropriate next-hop on path to the destination on which to send the packet. AODV protocol therefore uses only the standard IP header without any modification. AODV protocol performs two kinds of activities: route maintenance and route discovery. Route discovery procedure is carried out when a node needs to know the next hop neighbour to forward data to destination. Route maintenance activities are carried out when a next hop neighbour link breaks. When a node needs to send data to another node, it first checks if it has route information about the destination. If it does not have, it initiates a route discovery procedure. A Route Request (RREQ) packet is sent to the neighbouring nodes which in turn send the packet to their neighbours and so on until the packet reaches the destination or to a node that has latest information about the route to the destination. The RREQ packet contains following entries:

1. Source IP address
2. Destination IP address
3. Source Sequence Number: Provides information about the freshness of route to the source.
4. Destination Sequence Number: Indicates how fresh the route to the destination must be before it is accepted by the source.
5. Hop Count

The intermediate nodes between the source and the destination record information about the neighbour from which the RREQ packet is received. This helps in maintaining a reverse path to the source node. This information is maintained by the intermediate nodes for a sufficient amount of time that reflects the transmission of the RREQ packet to the destination and the reply back to the source. This amount of time is dependent on the size of the network. If Route Reply message is not received by the intermediate node within that time, the routing information is removed from the table. Each RREQ packet is uniquely identified by the IP address of the node that generated the packet and a unique broadcast ID. This broadcast ID is incremented every time the node generates a new RREQ packet. Thus the RREQ packet contains information about the node's IP address, broadcast ID, and a sequence number for the destination. When RREQ packet reaches a node that has latest route information to the destination, it compares the destination sequence number in its routing table with that contained in RREQ. If the sequence number in the RREQ is greater than that in the routing table, the RREQ packet is forwarded to neighbours. Otherwise the node sends back a Route Reply (RREP) packet to the node from which the RREQ packet was initiated. For sending back the reply to the source node, the destination/intermediate node does not need to broadcast the RREP packet. Instead the RREP packet is unicast on the reverse path. The intermediate nodes on the reverse path of the RREP packet maintain forward pointers to the node from which RREP is received long with the destination sequence number. This is called Forward Path Setup. After forwarding the RREP towards the source node, if the intermediate node receives another RREP, it checks the destination sequence number and hop count in the new RREP. This new RREP is forwarded towards source node only if

1. The destination sequence number in the new RREP is greater than that the previous RREP, or
2. The destination sequence number is the same but the hop count is small.

3.2 Dynamic Source Routing (DSR) Protocol

Dynamic Source Routing (DSR) is reactive protocol that discovers the routes to destinations on demand. As the name is quite revealing, DSR is a source routing protocol i.e. source node includes the whole path to destination node in data packet. The packet header is extended to include this additional information about the route to destination. DSR protocol carries out two basic routing activities:

a) route discovery
b) route maintenance

When a source has to transmit data to a destination, route discovery procedure is initiated to determine a best possible path to the destination. The route maintenance activities are started when a link in a path to destination is broken down and needs maintenance. When a node needs to send data to destination, it checks its route information if a route to the destination exists. If it cannot find a route to the destination, the source initiates a route-discovery procedure. RREQ packets (containing source and destination IP addresses and a unique packet identifier) are sent to the neighbouring nodes. Each neighbouring node on receiving the RREQ packets checks its routing tables to determine if it has the route information to the destination node. In case of non-availability of this information, the intermediate node too puts its own IP address in the RREQ packet and propagates it further to its neighbouring nodes. In order to prevent the loop back of the RREQ packets, a node does not forward a packet that it has already seen or if the RREQ packet contains the IP address of this node. When the RREQ packet reaches the destination or to an intermediate node that has route information to the destination, a RREP packet is generated by the corresponding node. If the RREQ packet reaches the destination, the destination node puts the route contained in the RREQ packet into the RREP packet. On the other hand, if the RREQ packet reaches an intermediate node that has latest information about the route to the destination, it augments this route information to the route record contained in RREQ packet, places this whole information in an RREP packet, and unicasts it on the reverse path to the initiator of the route request.

Every intermediate node maintains a route cache to store the overheard routes for use in the future route discovery. In case of a link breakage, a Route Error RERR message is sent to source node. All intermediate nodes on the path update their route caches. Source node, on receiving the RERR packet, looks for an alternate route to destination in its routing cache. In case of non-availability of an alternate route, it issues a new RREQ packet for route discovery. Scalability issues are related to DSR protocols i.e. since route discovery packets are broadcast in the network, a larger network leads to a heavier packet storm. In order to prevent packet storm, sequence numbers are associated with packets which helps in duplicate rejection.

4 Statistical Comparisons of AODV and DSR Routing Protocols

In this section we have compared the global statistics and object statistics for DSR and AODV in order to improve performance and efficiency of each protocol.

4.1 Global Simulation Statistics

A plethora of research has taken place and is still being carried out in the field of Mobile Wireless ad hoc networks. The theory of MANETs is comprehensively elaborated by [3, 5] classify routing protocols into table-driven and on demand categories and provide the performance comparison of protocols of both categories. They state that no routing protocol is best for all scenarios. Each protocol has certain advantages and limitation. Based on the situation, the choice of routing protocol may differ. A number of researchers have performed simulations to determine which routing protocol should be used under which network conditions. Since our work is related to the comparison of two on-demand routing protocols i.e. AODV and DSR, we are concerned with the related work done for the performance comparison of AODV and DSR [2] describe the effect of different mobility input parameters on routing and MAC layer protocols. The protocols are evaluated against a number of Quality of Service (QoS) parameters such as throughput, latency, long term fairness, and number of packets received. For routing protocols, AODV and DSR are chosen to be analyzed. It is found that in terms of QoS parameters, no protocol is superior to others in its own class. Moreover, they conclude that it is not possible to analyze the protocols of a single layer in isolation. Instead the whole protocol stack should be treated as a single entity for performance improvement and different combinations of the protocols from each layer should be subject to performance evaluation. Simulations for performance comparison of different routing protocols for wireless ad hoc networks such as AODV, DSDV, DSR, and TORA are carried out by [4]. They conclude that DSDV's performance decreases with an increase in the node mobility TORA's performance degrades with increasing the number of nodes, DSR performance is good for all degrees of mobility rate and movement speed but includes source routing overhead, AODV also performs well for all degrees of mobility rate and movement speed and in addition avoids the source routing overhead, but it includes transmission of a number of routing overhead packets and hence incurs more cost at high mobility as compared to DSR [1] also carry out simulations for the performance analysis of AODV, DSR, and TORA on the basis of average end-to-end delay and packet delivery ratio. They use ns-2 simulator to simulate a network covering an area of 500x500 m2, having a bandwidth of 20Mbps. The 50 nodes of the network move in that rectangular area to simulate the effect of topology changes. The parameters used for performance comparison are packet delivery fraction (throughputl) and end-to-end delay. They find out that increasing the number of nodes in a MANET results in increasing the end-to-end delay and loop detection time. Packet delivery ratio of AODV is smaller than that of DSR because of the higher drop rate for AODV. Packet delay increases with an increase in the size of network. The packet delay time is dominated by the route discovery delay. Source routing protocols, like DSR, have higher route discovery delay than hop-by-hop protocols because intermediate nodes have to extract the next-hop information from the extended packet header. Their simulation results show that DSR performs well for medium mobility of nodes while TORA is suitable for very dense networks having high mobility ratio due to provisions for multipaths and multicast routing. AODV is the winner being suitable for networks of different density and mobility [6] use a simulated network of GloMoSim simulator (a simulator by UCLA) with 50 nodes covering an area of 1000x1000 m2 for CBR traffic. They compare the performance of AODV and DSR for average packet delivery rate under the situation of multiple source -- common destination communication sessions. The performance of both protocols degrades i.e. performance of DSR drops by 10% and that of AODV by 30% under such constrained situation. This performance degradation is due to the fact that both protocols drop packets when they are unable to transmit data on active routes. In case of AODV, the congestion on routes to the same destination result in packet drops which is perceived as link breakage by the protocol. RERR message is sent to the source. Since stale routes are removed from routing tables and only the updated routes are maintained, source node initiates the rout discovery procedure. This degrades the packet delivery ratio. In DSR protocol however, alternate routes in maintained in route caches. The congestion problem is therefore solved locally by referring to alternate routes from route caches. Therefore the performance degradation of DSR is not as severe as that of AODV and hence DSR outperforms AODV in constrained situations [7] compare the performance of DSR and AODV routing protocols. They perform a number of experiments while changing the network size, load, and mobility. The experiments are simulated in ns2 simulator with two configurations i) a network size of 50 nodes in an area of 1500 x 300 m2 and ii) a network size of 100 nodes in an area of 2200 x 600 m2. The two protocols are evaluated against the metrics of average end-to-end delay, packet delivery ratio, and normalized routing load. DSR shows a better performance in terms of delay and throughput for small networks with low mobility and lesser load.

AODV on the other hand, is suitable for larger network or networks with high load or mobility. The reason of performance degradation of DSR for larger networks is due to the fact that DSR maintains information about multiple routes to a destination and there is no mechanism to remove information about stale routes. We use OPNET simulation model for comparison of DSR and AODV routing protocols using the metric of packet delivery ratio. A network of 6 nodes is simulated to cover an area of 4000x3000 m2. We conclude that size of network, mobility of node, and traffic load affect the performance of both protocols. Simulation results show that performance of DSR degrades with an increase in the size of network as well as traffic load. In such situations, AODV shows better performance due to hop-by-hop routing in the light of [11-19].

4.2 Current Research (Object) Simulation Statistics in Campus Networks

Here we propose the simulated analysis for AODV and DSR using OPNET v.12 simulator in order to make a contribution for best and efficient results in the field of MANETS for best performance, data throughput, data delivery ratio from source to destination. Here the statistics on which we are going to improve the performance of each protocols, having Movement space 4000m x 3000m, Maximum speed is 2, 10, and 20 m/s each, Maximum pause time are 0 and 200s and Transmission rates are 2, 5, and 10 packets/s with packet size is 512 bytes for 300s evaluates the performance of important TCP parameters for AODV based 30 nodes in a campus network.

4.2.1 Simulation Statistics of AODV

All nodes in the network are configured to run AODV protocol and FTP sessions. In simulation process different AODV parameters are used as suggested by RFC and WLAN data rate. A single TCP connection is established between wireless nodes of scenario. Two different self mobility trajectories are defined for mobile. The results shows amount of routing traffic generated, route discovery time and the number of hops per route, TCP/IP traffic, no of RREQ and RREP packets, and FTP download and upload time.

The lifetime of a routing table entry is initialized to active route timeout and is defined as 5 seconds for this scenario. If a route is not used and refreshed within defined period of time, AODV marks route as invalid and removes it from IP common table. To establish a route for the first time, originating node uses a TTL start value which is 1 for the first RREQ packet. The individual statistics of source and destination nodes in simulated scenario are shown in fig. 2(a) and 2(b) respectively. The number of hops covered by source node is 2 where destination node is 1. The AODV packets are queued during route establishment, when route is established then packets are transmitted without any waiting in queue. Both nodes routing table contains same number of nodes information. The routing traffic sent and received by both source and destination nodes are also shown in the figures. The routing traffic received by all wireless nodes of ad hoc network during simulation time is shown in the following figure. The traffic received individually by destination node is less than source node. The high traffic received by nodes occur in selected route are mobile node 2 and intermediate node 2. As discussed above that intermediate nodes have high routing traffic because of Hello messages. During simulation time these nodes receive more than 600 bits/sec traffic as shown in the following fig.2.

4.2.2 Object Statistics of DSR

During transmission of data in simulated ad hoc network routing traffic sent by all wireless nodes is shown in following fig. 3. The route selected by DSR protocol in simulated scenario for transmission of data between source and destination nodes consists of four nodes, i.e. source, mobile node 2, intermediate node 2, and destination. Source node sent traffic only once during start of transmission. During reply by destination node a lot of traffic is added to the data as shown in the figures. The intermediate node 2 and mobile node 2 adds nearly same data traffic during simulation time. The source node sent less than 100 bits/sec traffic where a mobile and intermediate node 2 sent more than 500 bits/sec traffic. The traffic sent by all wireless DSR routing nodes in ad hoc network during simulation is shown in the following fig. 3.

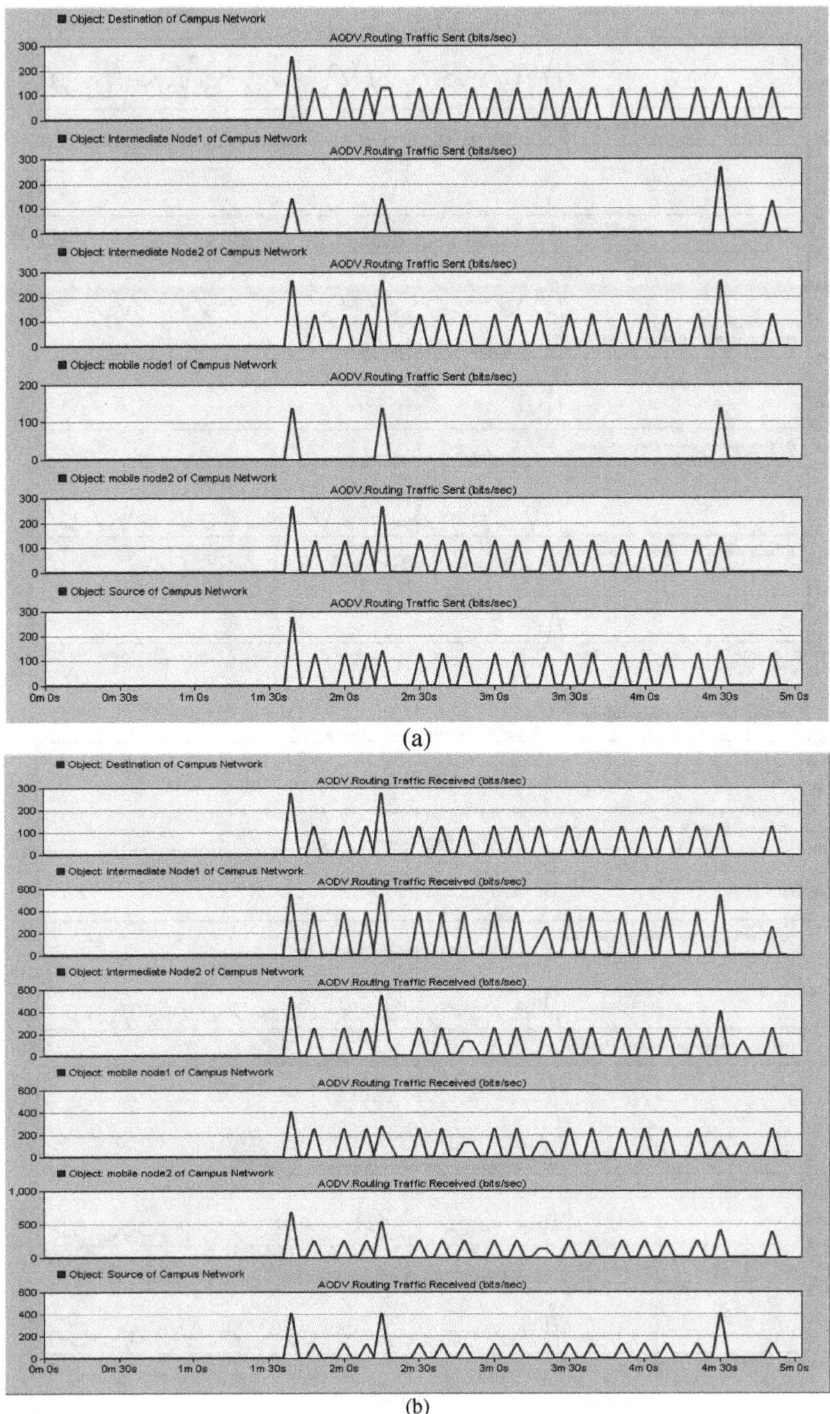

Fig. 2. Total routing Traffic sent (a) and Received (b)

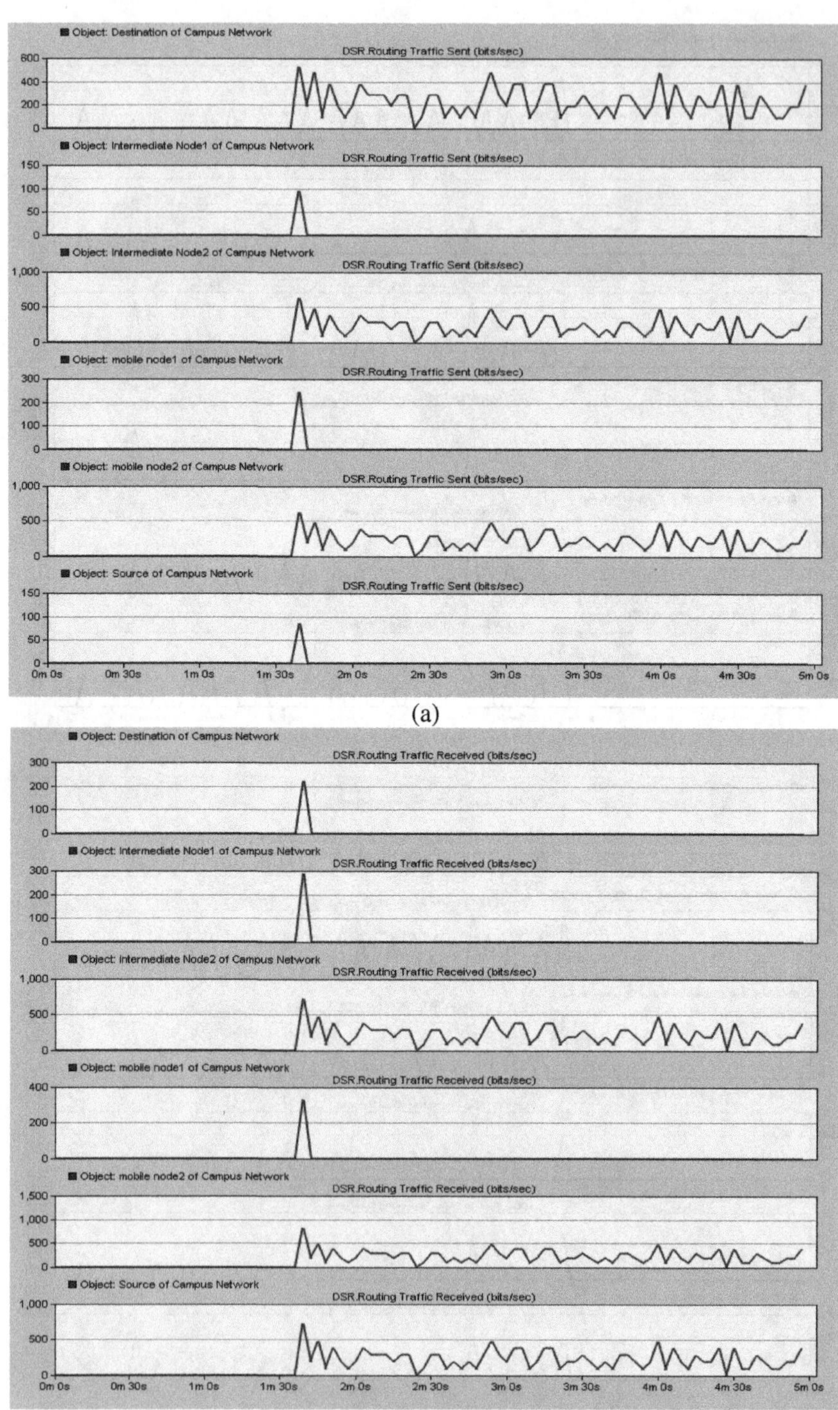

(a)

(b)

Fig. 3. Total traffic sent (a) and Received (b)

5 AODV/DSR for 4G Networks

4G networks are the fourth generation cellular communication systems and are still in the development stage. To provide efficient, fast, and secure communication over the cellular networks, 4G networks promise meeting these demands by the use of Internet Protocol (IP). The International Telecommunication Union (ITU) has set the speed standard for 4G networks as 100Mbps for high mobility and 1Gbps for low mobility situations.

Till this time, cellular networks have seen the transition from first generation to the fourth with advancement in the transmission technology. 1st generation (1G) voice-only networks, being completely analog, expired in the wake of digital era. Replacing the 1G networks, 2G networks are based on GSM technology and encode the radio signals digitally, enabling the transmission of voice as well as data. With 2G networks, short messaging-service (SMS), conference calling, and web browsing facilities are introduced to the users of 2G networks. These networks enable communication to take place at the speeds of 10-20 Kbps. 2.5G and 2.75G networks, which are GPRS and EDGE based systems respectively, provide enhanced speed for faster communications i.e. around 384 and 473 Kbps, respectively. However, these extensions of 2G networks still cannot improve the speed to allow for applications such as video conferencing, voice over IP, live streaming of video, and the alike. The "need for speed" thirst for these multimedia applications is satisfied by 3G networks which provide speeds of DSL standard i.e. beyond 2 Mbps. 3.5G networks further improve the speed many folds (around 14.4 Mbps) using the HSDPA (High Speed Downlink Packet Access) technology. However, this quest for speed is never ending. While 3G networks have not yet become pervasive because the infrastructure is still on the built, 4G networks have evolved in the mean while, which are supposed to provide a bandwidth of 100Mbps – 1Gbps to allow for high quality streaming of live videos, 3D gaming, and faster web surfing. Here we discuss 4G networks and the implication of AODV and DSR protocols for 4G networks which may contribute for future generation technology.

The Like 4G, Mobile Ad-hoc Network is also in a developing stage. While defining the MANET standard, the Internet Engineering Task Force (IETF) is working on routing techniques, like Ad-hoc On-Demand Distance Vector (AODV) defined in and Dynamic source Routing Protocol (DSR) defined in allowing self configuring network of mobile nodes with routing capabilities. MANET standardizes the static and mobile techniques of creating mesh networks using available wireless technology. Currently, 802.11a/b/g/n wireless networks defined by the IEEE standards are being used in applications such as homes, offices, as disaster recovery, conferences, lectures, emergency situation in hospitals, meetings, crowd control, and battle fields, and also could be found in the initial MANET infrastructure. Thus, what limits cellular networks and WLANs will limit MANET.

6 Comparison Summary

In this research work, performance analysis of two different previously published routing protocols AODV and DSR were done through NS-2 and OPNET simulator.

Simulation results for AODV and DSR routing protocols have been presented in many papers by different researchers. Although every simulation has done so far used significantly different input parameters than the one we used as well as performance comparison of two routing protocols to get best effective results. These results illustrate; that mobility model changes have a significant impact on their performance. The simulated scenario shows that routing overhead for the routing protocols heavily depends on node mobility. AODV sends many small routing control packets while DSR sends less but bigger control packets during transmission of data packets. DSR is more useful in smaller networks with less mobility and usage of AODV more appropriate in ad hoc networks with a higher mobility and data transfer rate. The simulations results and analysis have shown that mobile ad hoc network definitely needs more precise routing protocols to support increase mobility. Study shows that not only fundamental performance properties but also some essential network parameters change performances of various routing protocols. With the increase of number of hops in route throughput degrades due to the higher round trip time delay. The throughput also affect badly by traffic added during the transmission of data through hidden or exposed nodes. Time defined for Hello Loss value also very important in AODV routing protocol to keep neighbor nodes active. The network parameters like network size, number of nodes, number of hops per route, traffic patterns (bits/packets), link capacity, mobility trajectory, and frequency of sleeping nodes do have great results on the performance of routing protocols in the mobile ad hoc networks.

7 Conclusion

Based on the conclusion of global statistical factors in ad hoc networks and our statistical analyzed factors, Multicast communication in MANET environment is not ideal due to the frequent nodes mobility and failures. Such problems in transmission of data become reason for increase delays and decrease throughput. The simulation shows that a sudden change in delay, traffic, and link breakage occurs due to nodes mobility. In our simulation scenario, we have seen mobile nodes which effect greatly in routing. The size of the network, traffic load, and delay affects both AODV and DSR routing protocols. DSR routing protocol is not efficient for large networks with many mobile nodes and high load in terms of traffic and delay which will increase overhead. In such situation AODV routing protocol is ideal because of its hop-by-hop routing. The combined performance of both AODV and DSR routing protocol could be best solution for routing in MANET instead of separate performance of both AODV and DSR. Here in this paper we also compare the global and object simulation statistics in order to get best results during simulation with different metrics in simulation of DSR and AODV routing protocols. Here we suggest that the combined performance of both protocols make a very strong performance contribution in ad hoc networks. Our research makes a very good contribution for classrooms and campuses for educational purposes in order to share assignments, books, research articles etc. Here we are focusing more and more to get best performance results in each protocols in order to obey the compatibilities of ad hoc campus networks.

And also:

Today 4G is a convergence platform providing clear advantages in terms of coverage bandwidth and power consumption. 4G mobile communication networks are expected to provide all IP-based services for heterogeneous wireless access technologies, assisted by mobile IP to provide seamless Internet access for mobile users. Methodologies for QoS and security support in 4G networks, integrate signaling with authentication, authorization and accounting services.

Acknowledgements. Our thanks with core of heart to Springer link to give us the opportunity to present our research in competitive world of humans, we also thanks Sarhad University as well as our family who pray for us, to whom we love.

References

[1] Gupta, A.K., Member, IACSIT, Sadawarti, H., Verma, A.K.: Performance analysis of AODV, DSR & TORA Routing Protocols. IACSIT International Journal of Engineering and Technology 2(2) (April 2010) ISSN: 1793- 8236

[2] Barrett, C., Marathe, A., Marathe, M.V., Drozda, M.: Characterizing the interaction between routing and MAC protocols in ad-hoc networks. In: Proceedings of the 3rd ACM International Symposium on Mobile Ad Hoc Networking & Computing (MobiHoc 2002), ACM, New York (2002)

[3] Royer, E.M., Toh, C.-K.: A Review of Current Routing Protocols for Ad Hoc Mobile Wireless Networks. IEEE Personal Communications, 46–55 (April 1999)

[4] Broch, J., Maltz, D.A., Johnson, D.B., Hu, Y.-C., Jetcheva, J.: A Performance Comparison of Multi- Hop Wireless Ad Hoc Network Routing Protocols. In: ACM Mobicom (1998)

[5] Frodigh, M., Johansson, P., Larsson, P.: Wireless Ad Hoc Networking –The Art of Networking without a Network. Ericsson Review 77(4), 248–263 (2000)

[6] Misra, R., Manda, C.R.: Performance Comparison of AODV/DSR On demand Routing Protocols for Ad Hoc Networks in Constrained Situation. Indian Institute of Technology, Kharagpur (India)

[7] Das, S.R., Perkins, C.E., Royer, E.M.: Performance Comparison of Two on demand Routing Protocols for Ad Hoc Networks. In: Proceedings of the IEEE Conference on Computer Communications (INFOCOM), pp. 3–12 (March 2010)

[8] Kenjiro, Cho, Philippe, Jacquet: Technologies for Advanced Heterogeneous network, 1st edn. Springer, Netherlands (2005)

[9] Szczodrak, M., Kim, J., Baek, Y.: 4GM@4GW: Implementing 4G in the Military Mobile Ad-Hoc Network Environment. IJCSNS Int. 70 International Journal of Computer Science and Network Security 7(4) (April 2007)

[10] Halgamuge, K.S., Wang, P.L.: Classification and Clustering for Knowledge Discovery, 1st edn. Springer, Netherlands (2005)

[11] khan, J.: MANET Reactive Protocols-Tutorial Review. International Journal of Computer Applications 12(4), 5–7 (2010) (Published By Foundation of Computer Science)

[12] Khan, J., Khattak, D.N.S., Jan, H.: A step of Mobile Ad-Hoc on-demand Routing Protocols towards 4G Cellular networks. International Journal of Computer Applications 16(6), 20–24 (2011)

[13] Khan, J., Abas, A., Khan, K.: Cellular Handover approaches in 2.5G to 5G Technology. International Journal of Computer Application 21(2) (2011)

[14] Khan, J., Hayder, S.I.: A Comprehensive Performance Comparison of On-Demand Routing Protocols in Mobile Ad-Hoc Networks. In: Kim, T.-h., Vasilakos, T., Sakurai, K., Xiao, Y., Zhao, G., Ślęzak, D. (eds.) FGCN 2010. CCIS, vol. 120, pp. 354–369. Springer, Heidelberg (2010)

[15] Khan, J., Bojkovic, Z.S., Marwat, M.I.K.: Emerging of Mobile Ad-Hoc Networks and New Generation Technology for Best QOS and 5G Technology. In: Kim, T.-h., Adeli, H., Fang, W.-c., Vasilakos, T., Stoica, A., Patrikakis, C.Z., Zhao, G., Villalba, J.G., Xiao, Y. (eds.) FGCN 2011, Part I. CCIS, vol. 265, pp. 198–208. Springer, Heidelberg (2011)

[16] Corchado, E., Herrero, Á.: Neural visualization of network traffic data for intrusion detection. Appl. Soft Comput. 11(2), 2042–2056 (2011)

[17] Liu, H., Abraham, A., Snásel, V., McLoone, S.F.: Swarm scheduling approaches for work-flow applications with security constraints in distributed data-intensive computing environments. Inf. Sci. 192, 228–243 (2012)

[18] Herrero, Á., Zurutuza, U., Corchado, E.: A Neural-Visualization IDS for Honeynet Data. Int. J. Neural Syst. 22(2) (2012)

[19] Panda, M., Abraham, A., Das, S., Patra, M.R.: Network intrusion detection system: A machine learning approach. Intelligent Decision Technologies 5(4), 347–356 (2011)

On Fitness Function Based upon Quasigroups Power Sequences

Eliška Ochodková[1], Jiří Dvorský[1], Pavel Krömer[1], and Pavel Tuček[2]

[1] VŠB - Technical University of Ostrava,
Faculty of Electrical Engineering and Computer Science
Department of Computer Science
{eliska.ochodkova,jiri.dvorsky,pavel.kromer}@vsb.cz
[2] Palacký University Olomouc
Faculty of Science, Department of Geoinformatics
pavel.tucek@upol.cz

Abstract. In the last few years, quasigroups are finding their way to becoming a building block of cryptographic primitives. This paper extends previous work done on evolutionary search for quasigroups by defining fitness functions based on heterogeneous sequences generated during the exponentiation of quasigroups elements. In this paper, we explore how succeed a genetic algorithm in search for maximum of the best fitness function and some statistics about this fitness function is presented, too.

Keywords: Quasigroups, Genetic algorithms, Fitness function, Evolutionary search.

1 Introduction

There exist a lot of attempts to detect potential security threats by soft computing methodologies nowadays. Spam filtering, intrusion detection etc. should be mentioned as an examples [1]. Another approach is to use efficient soft computing methods for testing properties' of cryptographic techniques. In the last few years quasigroups are used as a new building blocks of cryptographic primitives such as public key cryptosystems [2], stream ciphers [3] or hash functions [4]. They all usually make use of quasigroup string transformations and quasigroups of different orders, small or huge. Quasigroup string transformations are based on the identities that hold true for the quasigroups and their parastrophes [5]. With quasigroups in the hearth of advanced cryptosystems, the need for suitable quasigroups of desired order arises [6–8] and examination of quasigroups' algebraic properties and their potential weak features is important [4].

Exhaustive approach to properties' examination is possible only for small quasigroups represented by a look-up table. With the need to do time-consuming tests of quasigroups properties, look-up table is inappropriate representation. This encourages us to explore methods based on evolutionary algorithms for search for good quasigroups in the field of cryptography. So far we have presented several experiments with genetic algorithms and various fitness functions,

Á. Herrero et al. (Eds.): Int. JointConf. CISIS'12-ICEUTE'12-SOCO'12, AISC 189, pp. 141–150.
springerlink.com

which focused, e.g., on quasigroups with good pseudorandom properties [9] or on quasigroups optimal for hash function [10]. A genetic algorithm and three new fitness functions based on a product of a sequence were used to search for a quasigroup in [7]. The GA was able to find quasigroups with less associativity than the base quasigroup had. The new classification of quasigroups based upon product elements obtained by a product of a sequence was originally introduced in [8]. The relation between number of various product elements and associativity of quasigroup was shown there.

We mainly focus on function $f_3^{a^n}(\boldsymbol{s})$, one of the three functions proposed in [7], in this article. This function is described in detail; it is explored how succeed a genetic algorithm in search for maximum of the fitness function. Some statistics about this fitness function is presented, too.

The paper is organized as follows. Second section introduces necessary definitions. Fitness function and experiments motivation can be found in the third section. We explore the genetic algorithm results for the fitness function maximum searching in Sect. 4. The statistics about this fitness function is presented there, too. Finally, paper is closed with conclusions.

2 Background

Quasigroup is the non-associative algebraic structure with one binary operation ($\circ$), which is neither a commutative nor an associative operation. Among all quasigroups of a given order non-associative quasigroups dominate heavily. Quasigroups are equivalent to more familiar Latin squares [11], which follows from the uniqueness of solutions and this guarantees that no element occurs twice in any row or column of the table for ($\circ$). Number of distinct Latin squares of a given order is enumerated for $n \leq 11$ [12].

Definition 1. *A quasigroup is a pair* $(Q, \circ)$*, where* $\circ$ *is a binary operation on (finite) set* Q *such that for all not necessarily distinct* $a, b \in Q$*, the equations* $a \circ x = b$ *and* $y \circ a = b$*. have unique solutions.*

Definition 2. *Let* $A = \{a_1, a_2, \ldots, a_n\}$ *be a finite alphabet, a* $n \times n$ *Latin square* L *is a matrix with entries* $l_{ij} \in A$*,* $i, j = 1, 2, \ldots, n$*, such that each row and each column consists of different elements of* A*.* n *is the order of the Latin square.*

Starting with a base quasigroup we can explore the large class of quasigroups equivalent to it. One of the most common equivalence classes defined for Latin squares is isotopy class. Two Latin squares are in the same isotopy class if one can be obtained from the other by permuting rows, columns and symbols. There exist 1676267 isotopy class representatives for quasigroups of order 8 [11].

Definition 3. *Let* $(G, \cdot)$*,* $(H, \circ)$ *be two quasigroups. An ordered triple* (π, ρ, ω) *of bijections* π, ρ, ω *of the set* G *onto set* H *is called an* isotopism *of* $(G, \cdot)$ *upon* $(H, \circ)$ *if* $\forall a, b \in G, \pi(a) \circ \rho(b) = \omega(a \cdot b)$*. Quasigroups* $(G, \cdot)$*,* $(H, \circ)$ *are said to be* isotopic.

As the base, the quasigroup of modular subtraction with operation ($\circ$)

$$a \circ b = (a + n - b) \bmod n \tag{1}$$

is used. The multiplication in such an isotopic quasigroup [7] can be defined as

$$a \circ b = \omega^{-1}\left((\pi(a) + n - \rho(b)) \bmod\ n\right). \tag{2}$$

We work with two types of quasigroups: a *table quasigroup*, with multiplication table stored in computer memory, and, an *analytic quasigroup* defined by its multiplication formula (without the multiplication table stored) [9]. Isotopic quasigroups can be easily obtained for the table quasigroups. Only three permutations π, ρ, ω must be chosen for isotopic quasigroups generation, but, there exist $(n!)^3$ possible choices of the triple of permutations. These permutations of elements can be straightforwardly evaluated with the multiplication table in computer memory.

The multiplication table is not available for the analytic quasigroup, hence the permutation needs to be implemented as a function of an element of Q. There is a genetic algorithm for quasigroup optimization proposed in [7–10], which implements this task using the bit permutation over the n elements of Q. It does not require all n elements in main memory, so it allows us to explore only a subset $(\log_2(n)!)^3$ of all possible permutation triples.

Testing of all possible identities of given quasigroups may be time and space consuming approach. Therefore we have focused on associativity only [7, 8]. If associativity holds, then for each element $a, b, c \in Q : a \circ (b \circ c) = (a \circ b) \circ c$. The situation differs when we work with non-associative structure: $a \circ (b \circ c) \neq (a \circ b) \circ c$. Experiments with powers a^k, obtained by a product of a sequence, for all elements $a \in Q$, $k = 2, 3, \ldots, n, n = |Q|$, are described in [13].

Definition 4. *The product of a sequence* $a_1, a_2, \ldots, a_n$ *of elements* $a_i \in A, i = 1, 2, \ldots, n$ *is the set* $\{a_1 a_2 \ldots a_n\}$ *defined as:*

- *for* $n = 2$ *the set* $\{a_1 a_2\}$ *consist of only one element* $a_1 a_2$*,*
- *for* $n \geq 2$ *the set* $\{a_1 a_2 \ldots a_n\}$ *is defined as*

$$\{a_1 a_2 \ldots a_n\} = \{a_1\}\{a_2 \ldots a_n\} \cup \{a_1 a_2\}\{a_3 \ldots a_n\} \cup \ldots \cup \{a_1 \ldots a_{n-1}\} \cup \{a_n\}.$$

The n elements $a_i \in A$ can be joined (multiplied), without changing their order, in $C_n = \frac{(2n-2)!}{n!(n-1)!}$ ways, where C_n are *Catalan numbers* [14]. Let all elements $a_i \in A$ are equal and each element is denoted as a. Then a product $\{\underbrace{aa \ldots a}_{k}\}$ of the sequence $\underbrace{a, a, \ldots, a}_{k}$ can be enumerated. This product consists of all C_k product elements. In the ideal case, all possible values of power $\underbrace{aa \ldots a}_{k} = a^k$ for all $a \in Q$ can be obtained. Better information about the identities in the given quasigroup are acquired from sequences of intermediate results $b_1 \ldots b_k$, $b_1 = a^1, b_2 = a^2, \ldots, b_k = a^k$, obtained during the evaluation of product elements (powers) a^k. Ideally, for each $a \in Q$, C_k different sequences $b_1 \ldots b_k$ can be generated [13].

3 Genetic Search for Quasigroups with Heterogeneous Power Sequences

Genetic algorithms are probably the most popular and wide spread member of the class of evolutionary algorithms (EA). EAs build a class of iterative stochastic search and optimization methods based on mimicking successful optimization strategies observed in nature [15]. EAs operate with a population of artificial individuals (chromosomes) encoding potential problem solutions. Encoded individuals are evaluated using a carefully selected objective function which assigns a fitness value to each individual. The fitness value represents the quality (relative ranking) of each individual as a solution to given problem.

Genetic algorithms were first applied to the quasigroup optimization in [10] and the concept was further extended in [7–9]. The artificial evolution was used to search for good quasigroups isotopic to the quasigroup of modular subtraction. Three new fitness functions based on a product of sequence to search for a quasigroup was proposed in [7]. The GA was able to find quasigroups with better (i.e. less) associativity than the base quasigroup when using two of the three proposed fitness functions and the best proven was fitness function $f_3^{a^n}(\boldsymbol{s})$ defined by Eq. (3).

We wished to find a quasigroup that would generate as heterogeneous products of k elements $a \in Q$ as possible. Moreover, we wanted the quasigroup to generate different sequences of intermediate results $b_1 b_2 \dots b_k$ when computing a^k for all $a \in Q$ using all possible products elements. Only the highest power $k = n$ was used to compute randomness of products of sequences [7].

Let Q be a quasigroup of modular subtraction $(Q, \circ)$, where $Q = \{0, 1, \dots, n-1\}$ and operation $(\circ)$ is defined by Eq. 1. Let $\boldsymbol{s} = (s_0, s_1, \dots, s_{n-1})$ is the vector containing the number of distinct sequences of intermediate results obtained when computing a^n, i.e. s_i equals to the number of all distinct sequences generated when computing i^n. The best fitness function was

$$f_3^{a^n}(\boldsymbol{s}) = \frac{\max(\boldsymbol{s}) + \min(\boldsymbol{s})}{2C_n} \tag{3}$$

where C_n is the n-th Catalan number. The fitness function $f_3^{a^n}(\boldsymbol{s})$ assigned large fitness to quasigroups that generate as different sequences of intermediate results as possible.

The following questions arose in a review of [7]: How was the genetic algorithm successful to find the maximum of the fitness function, and, whether there is any statistics about such defined fitness function. However, are we able to evaluate the fitness function for all quasigroups isotopic with quasigroup of modular subtraction? From Eq. (2), it is evident that for a quasigroup of order $n = 8$ there exist $8!^3 \approx 6.5 \times 10^{13}$ isotopic quasigroups. Hence, we restricted ourselves to two permutations π and ρ and considered the permutation ω^{-1} as an identity. The number of isotopic quasigroups was thus reduced to $8!^2 = 1,625,702,400 \approx 1.6 \times 10^9$. The time complexity for such an experiment can be completed within a few days on commonly available computer.

4 Experimental Results

The fitness function $f_3^{a^n}(\boldsymbol{s})$ was evaluated for all of $8!^2$ quasigroups in the experiment. We tried to obtain as much information from these data as possible. The distribution of the quasigroups with respect to the numerator of fitness function is shown in Fig. 1, numerical data are given in Tab. 1.

Remark 1. The fitness function values must satisfy $0 \leq f_3^{a^n}(\boldsymbol{s}) \leq 1$ in the genetic algorithm implementation used in our experiment. The denominator of fitness function, see Eq. (3), is equal to $2C_n$ that ensures normalization of fitness function's values into interval $[0,1]$ – the highest allowed values of $\max(\boldsymbol{s})$ and $\min(\boldsymbol{s})$ are equal to C_n. If we perform calculations for a fixed order quasigroups, in our case $n = 8$, the value of fitness function is fully determined by numerator of the fraction. Therefore we can formally make a substitution $\Omega(\boldsymbol{s}) = \max(\boldsymbol{s}) + \min(\boldsymbol{s})$ and the fitness function can be written as

$$f_3^{a^n}(\boldsymbol{s}) = \frac{\max(\boldsymbol{s}) + \min(\boldsymbol{s})}{2C_n} = \frac{\Omega(\boldsymbol{s})}{2C_n}. \tag{4}$$

We will work with the numerator $\Omega(\boldsymbol{s})$ or with numerator's components $\max(\boldsymbol{s})$ and $\min(\boldsymbol{s})$ instead of the fitness function $f_3^{a^n}(\boldsymbol{s})$, in the following text.

The theoretically possible values of the numerator $\Omega(\boldsymbol{s})$ were not reached, see Tab. 1. The least found value was $\Omega(\boldsymbol{s}) = 21$ and the highest value was stabilized at $\Omega(\boldsymbol{s}) = 720$ in our experiment. The highest possible numerator value $\Omega(\boldsymbol{s}) = 858$ can not be reached – such a quasigroup must be of infinite order. The genetic algorithm found a quasigroup with $f_3^{a^n}(\boldsymbol{s}) = 0.8042$, where the value of the numerator was $\Omega(\boldsymbol{s}) = 690$, i.e. the value of the numerator $\Omega(\boldsymbol{s})$ reached 95.8% of the highest value found during testing all quasigroups (the quasigroup of modular subtraction itself has numerator $\Omega(\boldsymbol{s}) = 515$).

4.1 Statistical Analysis

Distribution analysis of numbers of quasigroups with respect to the value of the fitness function numerator $\Omega(\boldsymbol{s})$ was also performed. As the Fig. 1 shows, the distribution is very irregular. Table 1 provides number of quasigroups for significant values of $\Omega(\boldsymbol{s})$. Numbers of quasigroups for extreme values of the numerator $\Omega(\boldsymbol{s})$ are completely negligible in comparison with the total number of tested quasigroups. For $\Omega(\boldsymbol{s}) = 720$ there are only 256 quasigroups.

Statistical analysis of experimentally obtained distribution of quasigroups was also performed. The MLE test of bimodal distribution [16] was used. The analysis showed that the experimental distribution is a mixture of two Gaussian distribution $N(\mu_1, \sigma_1)$ and $N(\mu_2, \sigma_2)$, see Fig. 2, where the resulting distribution can be written as

$$f(x) = wN(\mu_1, \sigma_1) + (1 - w)N(\mu_2, \sigma_2) \tag{5}$$

where $w = 0.5013702$. The parameters of these two basic distributions are given in Tab. 2. P-value of the test is 0.01615263.

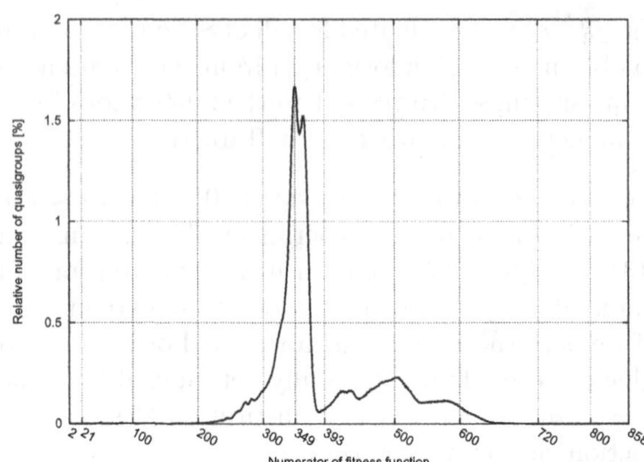

Fig. 1. Distribution of quasigroups with respect to numerator $\Omega(\boldsymbol{s})$

Table 1. Basic statistical properties of processed quasigroups

	$\Omega(\boldsymbol{s})$	Number of quasigroups absolute	relative [%]
Theoretical minimum	2	–	–
Experimental minimum	21	1,024	0.0000630
Experimental maximum	720	256	0.0000157
Theoretical maximum	858	–	–
Modus	349	27,088,768	1.6662809
Median	359	23,828,864	1.4657581
Mean	393	1,194,496	0.0734757
Maximum obtained by genetic algorithm	690	7,680	0.0004724

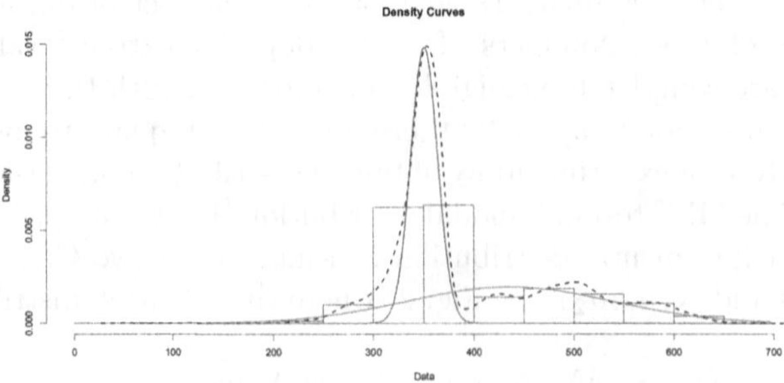

Fig. 2. Experimental distribution and its components

Table 2. Parameters of experimental distribution's components

Distribution	Distributions' parameters	
	μ_i	σ_i
$N(\mu_1, \sigma_1)$	350.3283824	13.5179064
$N(\mu_2, \sigma_2)$	435.9915212	101.6103921

4.2 Numerator $\Omega(s)$ Sum Components' Analysis

Furthermore, we examined what values acquired the components of the numerator $\Omega(\boldsymbol{s})$ for highest numerator values and what was the general distribution of addends in $\Omega(\boldsymbol{s})$. The graph of results in Fig. 3 shows, that the distribution of addends is very uneven again, and is far from including all possible combinations of addends. Even the case for $\min(\boldsymbol{s}) = 1$ was dropped from the graph and a separate graph was constructed in Fig. 4(a). For $\min(\boldsymbol{s}) = 1$ and $\max(\boldsymbol{s}) = 348$, there are approximately 27 million quasigroups, while for other options, this value of $\Omega(\boldsymbol{s})$ is not more than 400 thousand.

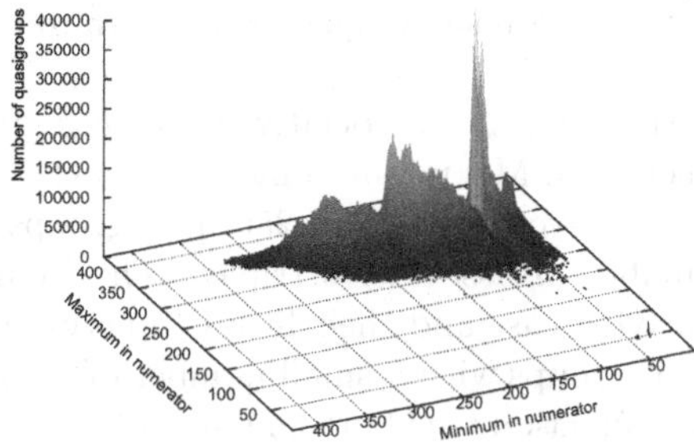

Fig. 3. Number of quasigroups with respect to numerator $\Omega(\boldsymbol{s})$ sum components

The Fig. 3 also clearly shows that quasigroups with the maximum numerator $\Omega(\boldsymbol{s})$ are located in the "upper left corner" of the graph's area filled with data (with exception where $\min(\boldsymbol{s}) = 1$). The maximum value of the numerator $\Omega(\boldsymbol{s}) = 720$ is composed of only a single combination of minimum and maximum, and there is $720 = 390 + 330$. Likewise, the minimum numerator $\Omega(\boldsymbol{s}) = 21$ is only a single combination of two numbers $21 = 20 + 1$. Components of the sum of the numerator's $\Omega(\boldsymbol{s})$ mode are shown in Fig. 4(b). In this case as well, we had to omit the combination of addends $\Omega(\boldsymbol{s}) = 348 + 1$ from the graph. This combination occurs $26,887,296$ times whilst the second most frequent combination in numerator $\Omega(\boldsymbol{s}) = 280 + 69$ occurs only $14,336$ times.

4.3 Relation between Numerator $\Omega(s)$ and Number of Associative Triples

It is evident from the previous text that the value of fitness function is influenced by degree of associativity of tested quasigroups. If tested quasigroup includes many associative triples, the repetition of sequence of element's power calculation will be more likely. Thus, the value of fitness function respectively the value of numerator of $\Omega(\boldsymbol{s})$ will decline. The relationship between numerator $\Omega(\boldsymbol{s})$ and the number of associative triples in all quasigroups is given in Fig. 5. Individual quasigroups are shown as points, the largest and smallest values of the numerator $\Omega(\boldsymbol{s})$ are highlighted using slightly larger dots. The graph clearly shows that

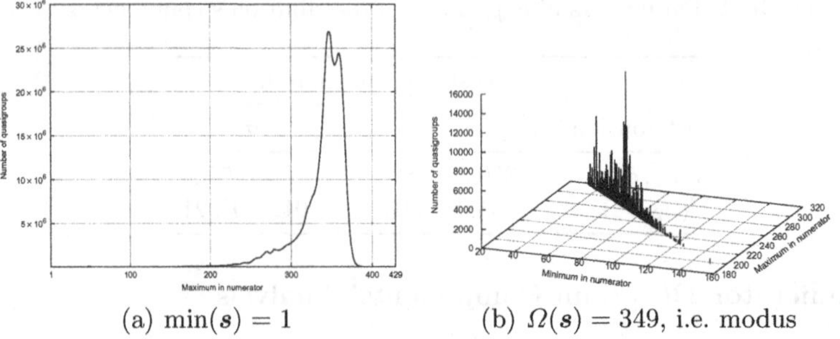

(a) $\min(\boldsymbol{s}) = 1$ (b) $\Omega(\boldsymbol{s}) = 349$, i.e. modus

Fig. 4. Number of quasigroups with respect to numerator $\Omega(\boldsymbol{s})$ sum components

with increasing associativity of quasigroups the value of the numerator $\Omega(\boldsymbol{s})$ decreases. Maximum numerator $\Omega(\boldsymbol{s}) = 720$ was found for quasigroups with 62 associative triples – all 256 quasigroups with this value of $\Omega(\boldsymbol{s})$ have the same number of associative triples. The high-associative quasigroups represent the outlying case and their fitness function values are small. The question, why few quasigroups with a small number of associative triples still have low value of fitness function (the bottom part of graph just below the "central cloud"), remains.

4.4 Visualisation of Selected Quasigroups

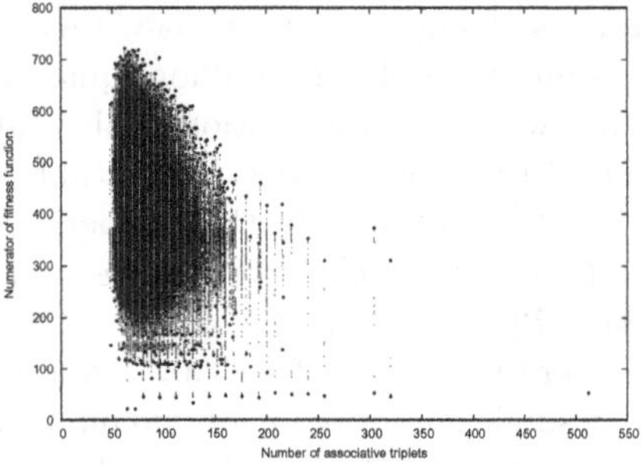

Fig. 5. Value of numerator $\Omega(\boldsymbol{s})$ with respect to associative triples

To visualize a quasigroup a grey map is used. The grey map visualizes sequences of intermediate results $b_1 b_2 \dots b_n$. Let M is a $n \times n$ matrix and the elements of the matrix $M_{b_i,b_{i+1}}$ are the number of pairs $b_i b_{i+1}$ in all sequences of intermediate results $b_1 b_2 \dots b_n$. Elements of the matrix M are then converted to shades of grey, where the color black represents 0 and white represents the maximal value of M. The more different shades of grey on the image, the less associative the quasigroup is.

Four quasigroups are presented as an example. The first one is the base quasigroup of modular subtraction, Fig. 6(a). Quasigroup in Fig. 7(a) is an example of quasigroup with numerator $\Omega(\boldsymbol{s}) = 21$. It is isotopic quasigroup with worse properties than the base quasigroup and it generates many flaws in the sequences of intermediate results. Figure 6(b) shows a quasigroup with $\Omega(\boldsymbol{s}) = 690$ searched by genetic algorithm [7]. Sample of qausigroup with maximal numerator $\Omega(\boldsymbol{s}) = 720$ is given in Fig. 7(b). The difference between the quasigroup found by genetic algorithm and the quasigroup with numerator $\Omega(\boldsymbol{s}) = 720$ is minimal. Base quasigroup of modular subtraction is rather average from this point of view.

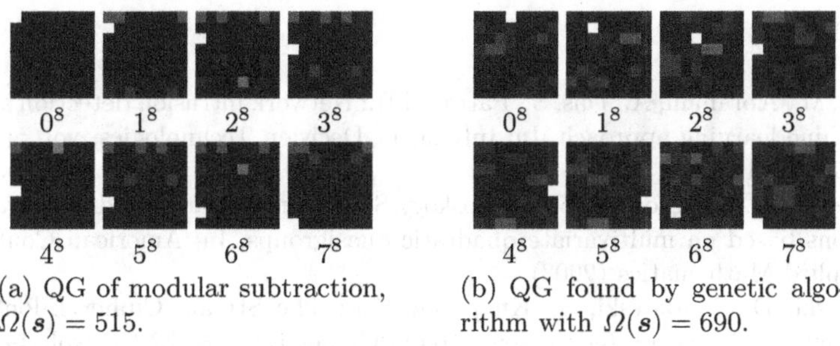

(a) QG of modular subtraction, $\Omega(\boldsymbol{s}) = 515$. (b) QG found by genetic algorithm with $\Omega(\boldsymbol{s}) = 690$.

Fig. 6. Selected important quasigroups

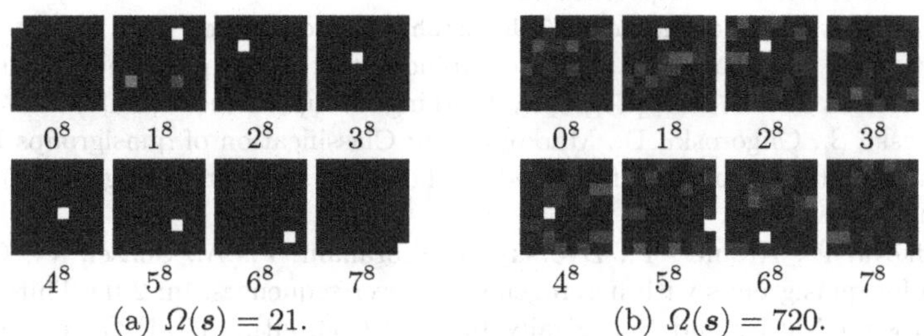

(a) $\Omega(\boldsymbol{s}) = 21$. (b) $\Omega(\boldsymbol{s}) = 720$.

Fig. 7. Sample quasigroups

5 Conclusions

The detailed analysis of the fitness function based on the power sequences, which was used for genetic searching of optimal quasigroup, is proposed in this article. It was shown how successful was the genetic algorithm to find the maximum of the fitness function. The comparison with associativness of quasigroups was presesented in this analysis, too. Some statistics about fitness function, e.g. the distribution of numbers of quasigroups with respect to the value of fitness function numerator was introduced. The numerator analysis shows that proposed fitness function correctly assigns a large fitness to quasigroups that generate as different sequences of intermediate results as possible.

Acknowledgment. This work has been elaborated in the framework of the IT4Innovations Centre of Excellence project, reg. no. CZ.1.05/1.1.00/02.0070 supported by Operational Programme 'Research and Development for Innovations' funded by Structural Funds of the European Union and state budget of the Czech Republic. It was also supported by the Operational Programme 'Education for competitiveness' funded by Structural Funds of the European Union and state budget of the Czech Republicg under the project reg. no. CZ.1.07/2.3.00/20.0072.

References

1. Panda, M., Abraham, A., Das, S., Patra, M.R.: Network intrusion detection system: A machine learning approach. In: Intelligent Decision Technologies, vol. 5(4), pp. 347–356 (2011)
2. Gligoroski, D., Markovski, S., Knapskog, S.J.: Multivariate quadratic trapdoor functions based on multivariate quadratic quasigroups. In: American Conference on Applied Mathematics (2008)
3. Gligoroski, D., Markovski, S., Knapskog, S.J.: The Stream Cipher Edon80. In: Robshaw, M., Billet, O. (eds.) New Stream Cipher Designs. LNCS, vol. 4986, pp. 152–169. Springer, Heidelberg (2008)
4. Gligoroski, D., et al.: EdonR cryptographic hash function. Submition to NIST's SHA-3 hash function competition (2008), `http://csrc.nist.gov/groups/ST/hash/sha-3/index.html`
5. Gligoroski, D., Markovski, S.: Cryptographic potentials of quasigroup transformations. Talk at EIDMA Cryptography Working Group, Utrecht (2003)
6. Markovski, S., Gligoroski, D., Markovski, J.: Classification of quasigroups by random walk on torus. Journal of Applied Mathematics and Computing 19(1-2), 57–75 (2005)
7. Ochodkova, E., Kromer, P., Dvorsky, J., Abraham, P.J.A., Snasel, V.: Genetic search for quasigroups with heterogeneous power sequences. In: 2011 Third World Congress on Nature and Biologically Inspired Computing (NaBIC), pp. 533–539 (October 2011)
8. Ochodková, E., Dvorský, J., Snášel, V., Abraham, A.: Testing quasigroup identities using product of sequence. In: Pokorný, J., Snásel, V., Richta, K. (eds.) DATESO. CEUR Workshop Proceedings, vol. 567, pp. 155–162. CEUR-WS.org (2010)
9. Snášel, V., Dvorský, J., Ochodková, E., Krömer, P., Platoš, J., Abraham, A.: Genetic Algorithms Evolving Quasigroups with Good Pseudorandom Properties. In: Taniar, D., Gervasi, O., Murgante, B., Pardede, E., Apduhan, B.O. (eds.) ICCSA 2010, Part III. LNCS, vol. 6018, pp. 472–482. Springer, Heidelberg (2010)
10. Snášel, V., Abraham, A., Dvorský, J., Ochodková, E., Platoš, J., Krömer, P.: Searching for quasigroups for hash functions with genetic algorithms. In: World Congress on Nature & Biologically Inspired Computing, NaBIC 2009, pp. 367–372. IEEE Computer Society (2009)
11. Belousov, V.D.: Osnovi teorii kvazigrup i lup, Nauka, Moscow (1967) (in Russian)
12. McKay, B.D., Wanless, I.M.: On the Number of Latin Squares. Journal Annals of Combinatorics 9(3), 335–344 (2005)
13. Borůvka, O.: Foundations of the theory of groupoids and groups. Wiley (1976)
14. Hilton, P., Pedersen, J.: Catalan Numbers, Their Generalization, and Their Uses. Journal The Mathematical Intelligencer 13(2), 64–75 (1991)
15. Mitchell, M.: An Introduction to Genetic Algorithms. MIT Press, Cambridge (1996)
16. Kemperman, J.H.B.: Mixture with a limited number of modal intervals. Annal. Stat. 19, 2120–2144 (1991)

Investigation on Operating Systems Identification by Means of Fractal Geometry and OS Pseudorandom Number Generators

Ivan Zelinka[1], Filip Merhaut[2], and Lenka Skanderova[1]

[1] Department of Computer Science, Faculty of Electrical Engineering and Computer Science VŠB-TUO, 17. Listopadu 15, 708 33 Ostrava-Poruba, Czech Republic
{ivan.zelinka,lenka.skanderova.st}@vsb.cz
[2] Getmore s.r.o, Prague, Czech Republic
filip.merhaut@gmail.com

Abstract. This work demonstrates a novel way how to identify the operating systems and networking devices working with TCP / IP stack on the basis of differences in their random number generators. We also sketch possible OS identification based on fractal geometry and neural networks.

Keywords: Computer security, TCP/IP, operating system fingerprinting, port scanning, pseudorandom numbers generator, fractal geometry, neural networks.

1 Introduction

In the past, the pseudorandom number generators (PRNGs) were usually implemented as a standalone programs or parts of the library functions of the programming languages. In the current generation, the most common types of operating systems (Windows and Unix branches, including the Linux and OS X) include the PRNG functionality directly as a part of the operating system itself. From a cryptographic point of view this architecture has substantial advantages - the developers of operating systems are expected to be more knowledgeable and to have more resources necessary for successful design and implementation of complex, yet critically important features such as the PRNG. Also most operating systems have more possible sources of entropy available than any single program or library function. The drawback of this approach is the close coupling of individual PRNG (with its particular properties, algorithms and their implementations) and a given operating system. We have exploited this weakness to create a new method of remote identification of operating system.

The task to determine the type and version of the remote operating system is one of the first steps of the remote computer attack. This phase is necessary because the information gained there provide the ground for all subsequent activities of the intruder (hacker, network worm, pentester) – including but not limited to the choice of tools and vulnerability scanning specific for the particular operating system (eg. preparation of the return address according to the architecture, language version and service pack level of the target system and the installed software). In addition to the problems caused by the very nature of this difficult process the attackers effort is

Á. Herrero et al. (Eds.): Int. JointConf. CISIS'12-ICEUTE'12-SOCO'12, AISC 189, pp. 151–158.
springerlink.com

often hampered by defensive measures such as firewalls, obfuscated standard service ports, fake baits (honeypot systems) and altered header application protocols, all of which are methods that can distort or completely reverse the results of the traditional active methods of network reconnaissance. For the defender (administrator, owner) of the system, it is important to conceal the identity of their machines as much as possible. Knowing what methods are available to the other side to helps them to prepare better for this role. There is a lot of good literature discussing above-mentioned ideas, like [1]-[8] for example.

2 Overview of the Operating Systems Remote Identification

Of the traditional methods of identifying the operating systems the most widely used technique is the port scanning. It is the process of sending one or more network packets to the destination IP address or network and the derivation of which ports are open on the target machine on the basis of responses received (or the lack of them). This technique operates at the transport layer of the OSI model.

With a certain degree of probability it is possible to determine the identity of the remote machine just on the basis of the combination of open ports because it is a common practice that each type of service usually listens to listens on "its" assigned port number. More information can be obtained by inspection of the application protocols that respond on these ports – this is the application layer of the OSI model. First of all, there are certain application protocols and their versions that exist only on specific platforms (eg. SMB protocol = Windows, SMBv2 = Windows 2008 and above). Secondly, the application protocols often reveals detailed information about the software that handles the communication with the client (eg. web server header in the HTTP protocol - IIS, Apache). Some of the less conventional methods include inspection of the remote operating system „fingerprints“ on the basis of TCP/IP stack state information or ICMP packets analysis. Fingerprinting of the operating system is the process where we try to uncover the identity of the remote operating system by analyzing the different characteristics of the captured network traffic from this system, or broadcast packets to the system and analyze its responses. The Nmap program is usually considered to be the best port scanner. Among other functions it provides the TCP/IP based OS fingerprinting. The Amap program has an extensive database of the OS application level responses and thus is able to identify the services by analysis their banners, even when they are on non-standard port number. The Xprobe program is the favorite tool to explore the ICMP packets. The aim of our work is to verify whether it is possible to create a similar type of tool what would be based on another, novel principle of identification. Our method is based on the identification of the PRNG algorithm of the remote party Its advantage is that it should have the ability to deal with most of the defensive countermeasures - firewalls, port masquerades, fake honeypot decoys and modification of service banners.

3 Introduction of the Method

Whenever a TCP connection is created between two network stations (client – the initializing side of the connection, server – the receiving side) an initial value of the “sequence number“ field in the TCP header is generated. This value is an arbitrarily chosen random 32-bit positive integer. This value is chosen once for the client (in the first stage of three-phase connection initialization) and once for the server (in the second stage). This number is important both for the control of the communication sequence (in each subsequent packet its value is increased by the number of sent bytes) and for the security mechanism that prevents any third party to pretend the identity of either the client or the server (so-called IP spoofing attack) and thus the unauthorized interference of the communication - from a technical point of view it is very easy to create a TCP packet with arbitrary values of its header fields and send it to the client or server and for example to reset their mutual connection or illegally enter Application Protocol commands. The sequence number prevents these attacks because generally each received packet that contains invalid values of the sequence or confirmation number (the sequence number counter-part) fields is considered invalid and is discarded.

Thus to disrupt a foreign communication the attacker must be able guess or to calculate the correct value of the sequence number. The more random the choices of the initial sequence number value are, the harder it is for the third party to successfully perform this attack. In the best possible case, a chance to guess the value of the initial sequence number would be 2^{-32}. In real operating systems and networking devices there are huge differences of how the ISN generation is implemented. The most simple solutions return always the same constant value (some older types of network printers), some systems use increments in blocks of 64 kilobytes (older versions of UNIX), some slightly random increments (older versions of Windows), some the model of time-dependent values (newer versions Windows) and the best ones try to attempt as random choice of ISN as possible by using rather sophisticated random number generators with a large number of external entropy sources (the latest versions of Windows and Linux). These differences allow us to learn the specifics of the various operating systems and network devices and ideally to distinguish them from one another.

4 The Role of Random Numbers in Computer Security

In the computer security systems, the PRNGs are used in a number of various tasks - initial values of functions, the key generation cryptographic functions (eg, AES, TLS session keys that protect the https connection for secure access to web sites), an optional headers of some network protocols (ISN, source port on the client), nonce (one-time use random numbers that prevent replay attacks), salts (random numbers added to the hash function to prevent dictionary attacks with pre-calculated hash values).

The use of weak random numbers generators has infamous history in the field of computer security - for example it was the root cause behind one of the biggest vulnerabilities in the Internet history – the DNS bug discovered in 2008 by Dan Kaminsky or behind the hacking of Tsutomu Shimomuras computers (the security analyst who helped to capture notorious hacker Kevin Mitnick in the early nineties)

To create a sequence of numbers with a computer that are not predictable and that do not contain any regular patterns is very difficult task due to the deterministic nature of the programming itself. In fact, the only known way in which the software random number generator (PRNG) is able to produce output that is subjectively and objectively as possible similar to the "natural" random sequence of numbers is to use an external source of entropy (eg. the current value on the stack of the process, the system time, some input from the user, network traffic, process ID, thread ID, the number of ticks since the system boot, etc. [5]). The sources of entropy are most often used to provide the initial values (the seed) for the PRNG. Among the systems that work with TCP/IP protocol (and therefore need to initialize the ISN) there are substantial differences in their available entropy resources. Generally speaking, the simpler the device is, the less diversity of entropy it has available. For example, it is apparent that the embedded devices or server-grade computers usually do not have as much direct input from the user as the desktop computer does.

5 Real World PRNGs

Currently, the cryptography provides us a sufficient theoretical foundation for building a secure PRNG systems. However real world PRNG system implementations often suffer from various kinds of weaknesses. There are several reasons:

- Effort to avoid the performance hits
- Weak initialization vectors
- Insufficient skills of the architects

Historically, there were many weaknesses in the PRNGs in operating systems. In some extreme cases, it was possible with a relatively high degree of success to predict the future values of the PRNGs and thus make the IP spoofing attack possible in real world [1].

6 Theory

From a security perspective, the PRNGs can be judged against the following criteria:

Pseudo-randomness: the extent to which the output appears as a random sequence to the human observer

Forward Security: knowledge of the internal PRNG state at any given moment does not provide any information on the previous output of the generator. In terms of network protocols that is not so important criterion, however, in general computer security it is - failure to comply with this criterion, for example, could allow an attacker to obtain previously generated cryptographic keys and thus decrypt previously captures encrypted communication

Backward security: knowledge of the internal PRNG state at any given moment does not provide any possibility to calculate or estimate the future output of the generator. Because if PRNG implemented by a computer program which is always a deterministic system, we need repeatedly feed the generator with enough entropy from external sources. In terms of network protocols, this criterion is more important than the previous. The failure to comply with this criterion could allow for example the IP spoofing attack.

7 The Source of the Data

To obtain the values of initialization sequence number (ISN) from the remote system TCP stack, we first used a combination of scripts for Tshark and Logparser software and custom program based on Microsoft .NET library for generating TCP/IP traffic.

This procedure was suitable for the initial tests and the verification of the feasibility of our method but was not fast enough for real-world scenarios, so we have created a complete C# program suite which automated the whole ISN data mining process. After entering the parameters (IP address of the server, delay time between connections, the number of connections, etc.) the program executed the reconnaissance and provided a text file with the ISN data accompanied with a XML file that documented the important aspects of the experiment.

Visualization of obtained data can be done, as proposed in [1], by Eq. 1. This equation is based on simple idea of 3D reconstruction from one dimensional time series. Cloud of points – strange attractor, as reported in [1] and in Fig. 1, is created. Strange attractor [10] is set of points that have so called fractal dimension [9].

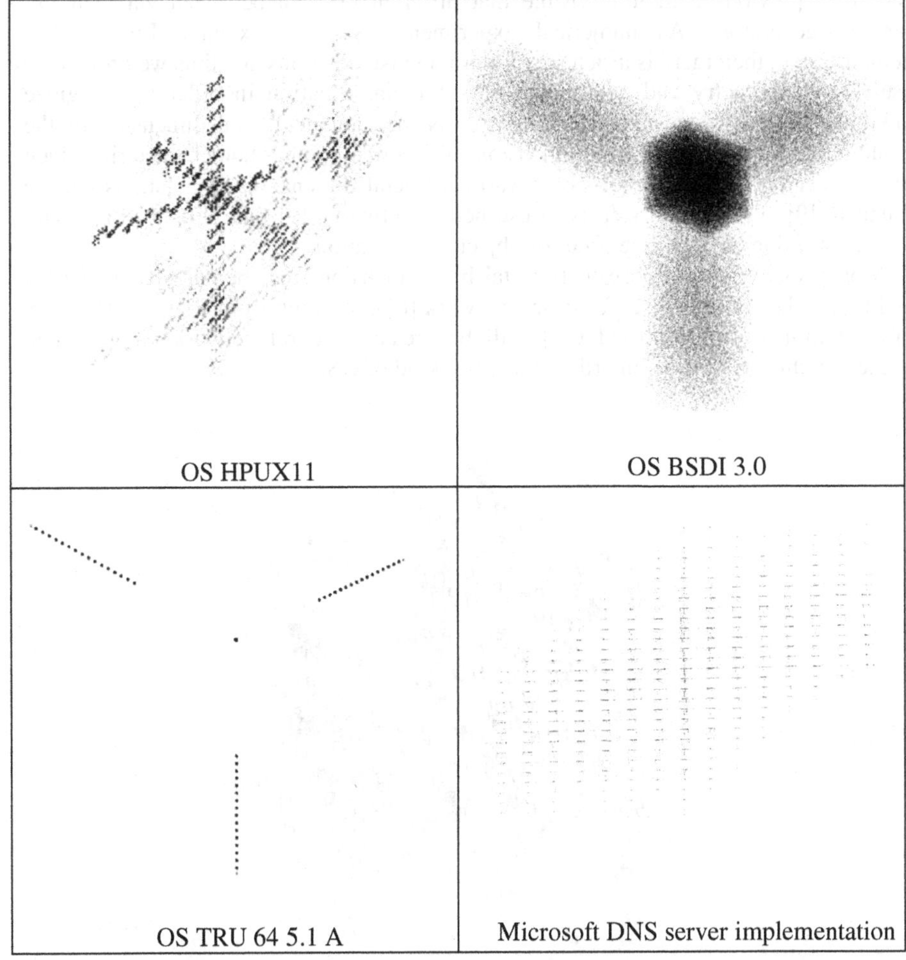

Fig. 1. Different "clouds of random points" in the state space. Randomly 4 OS has been selected here for demonstration. As one can see, no randomness is visible. Different patterns are formed here

In investigation [1] has been demonstrated that a wide class of various attractors is generated by different OS. Question is whether it is possible to automatically identify kind of OS based on its fractal attributes. We propose this idea in the next section.

$$x_n = s_{n-2} - s_{n-3}$$
$$y_n = s_{n-1} - s_{n-2} \tag{1}$$
$$z_n = s_n - s_{n-1}$$

8 Identification Based on Fractal Geometry and Neural Networks – Future Research

Based on previous text, it is visible that different OS can be tested via their OS random generators. As numerical experiments (see for example Fig. 2) has demonstrated, then there is a rich set of possible visualizations and thus we propose to use fractal geometry and neural networks for classification in order to recognize unknown on-line tested OS. The first step is to calculate fractal dimension of the cloud in the state space. Fractal dimension [9] is a real number that characterize given object. Very good description about various fractal dimensions calculations can be found in [9]. The second step is to use neural networks [9] that belong to artificial intelligence domain, shall be used for object identification.

More exactly, we propose that fractal box dimension shall be calculated, see [9] and Fig. 2. Then training set for neural network [8] consisting from couples of vectors {fractal dimension, identified OS} will be created. Neural network will be then trained on this training set in order to identify kind of OS.

Fig. 2. Demonstration of the fractal dimension calculation. Small boxes are used to cover points in the state space and then this cover is used to calculate fractal dimension of that "cloud of points".

9 Conclusions

The advantage of presented method is its absolute independence on the port numbers and application protocols. This is very important, because it can avoid most of the current defensive mechanisms (firewalls, port or application headers masquerade, honey-pots). Under certain conditions the method may succeed even when only one single TCP port with a completely generic service is open on the target machine. For traditional methods, this setting does not provide any clues. The disadvantage of our method is its conspicuousness – it requires relatively large samples of data. If a vigilant watcher (IDS/IPS system, administrator, UTM system) inspects the network, it is most likely that this atypical traffic will be noticed. The solution is mask the identification attempt as a DOS attack. The method presented in this paper itself is only suitable for some specific scenarios (eg. the ones where the success rate is more important than the stealth factor), but in combination with traditional tools, it can significantly improve the chances for success. The best results in identifying remote operating system can be achieved by combination of several methods.

We are planning to extend our research of the PRNGs with the NIST methodologies and also include more operating systems and devices into the research sets. Also the plans of benchmarking our method with other types of identification (Nmap, Amap, Xprobe) are under way, as well as investigation of the influence 32bit/64bit and IPv4/Ipv6 architectures. The ultimate goal is to build a database of ISN designs for all kinds various network devices. Later on we may also explore whether it is possible to do the reverse identification with this method – the client side. A client connecting to a server also uses the ISNs and port numbers in a similar manner.

Also future research based on fractal geometry and neural networks has been proposed. It is clear that neuro-fractal identification is promising way how to identify unknown OS based on random number analysis.

Acknowledgments. This work was supported by the European Regional Development Fund in the IT4Innovations Centre of Excellence project (CZ.1.05/1.1.00/02.0070) and by the Development of human resources in research and development of latest soft computing methods and their application in practice project, reg. no. CZ.1.07/2.3.00/20.0072 funded by Operational Programme Education for Competitiveness, co-financed by ESF and state budget of the Czech Republic.

References

1. Strange Attractors and TCP/IP Sequence Number Analysis, `http://lcamtuf.coredump.cx/newtcp/`
2. Dorrendorf, L., Gutterman, Z., Pinkas, B.: Cryptoanalysis of the Random Number Generator of the Windows Operating System. ACM Transactions on Information and System Security 13(1)
3. Dorrendorf, L., Gutterman, Z., Reinman, T.: Analysis of the Linux Random Number Generator. In: SP 2006 Proceedings of the 2006 IEEE Symposium on Security and Privacy. IEEE Computer Society, Washington, DC (2006)

4. Jiao, J., Wu, W.: A Method of Identify OS Based On TCP/IP Fingerprint. IJCSNS International Journal of Computer Science and Network Security 6(7B) (July 2006)
5. Howard, M., LeBlanc, D.: Writing secure code. Microsoft Press, ISBN-10: 0735617228
6. FIPS 140-1 Security requirements for cryptographics modules, `http://www.itl.nist.gov/fipspubs/fip140-1.htm`
7. Lecture 28: Random Numbers - Richard Buckland UNSW (2008), `http://wn.com/Lecture_28_Random_Numbers__Richard_Buckland_UNSW_2008`
8. Bose, N.K., Liang, P.: Neural Network Fundamentals with Graphs, Algorithms, and Applications. McGraw-Hill Series in Electrical and Computer Engineering (1996) ISBN 0-07-006618-3
9. Barnsley, M.: Fractals Everywhere. Academic Professional Press (1993) ISBN 0-12-079061-0
10. Hilborn, R.C.: Chaos and Nonlinear Dynamics. Oxford University Press, UK (1994) ISBN 0-19-508816-8

Usability of Software Intrusion-Detection System in Web Applications

Radek Vala, David Malaník, and Roman Jašek

Tomas Bata University in Zlin, Faculty of Applied Informatics, nám. T.G.Masaryka 5555, 760 01 Zlín, Czech Republic
{jasek,vala,dmalanik}@fai.utb.cz

Abstract. This article is focused on the security solution based on intrusion detection idea, which should be independent of the web server type or configuration and do not rely on the other network hardware components. Discussed intrusion detection system solution is connected directly with the web application and is based on the real-time request analysis. The main opportunities of proposed principle are very low cost and simple implementation. Proposal is based on implementation of LGPL library PHPIDS [https://phpids.org/] into the demo application which consists of simple web form for testing. Integration of PHPIDS library was tested against the main web security flaws - SQL Injection, Cross Site Scripting, and HTTP Parameter Pollution. On this demo application, simple stress tests were performed and also level of security was evaluated. Moreover, suggestions for future improvements of this security solution are discussed.

Keywords: Security, Web Application, Web Attack, Intrusion-Detection, IDS, PHPIDS, SQL Injection, CSS, HPP.

1 Introduction

Internet is the fastest developing phenomena of 21st century. People are used to provide their personal information to untrusted web applications and despite all of the risks, they are not enough careful by using the internet. With growing awareness increases also the ingenuity of hacker techniques and social engineering. Security of web applications is one of the main challenges of this time in the world of internet. WhiteHat security research (WhiteHat Security, 2010) shows an alarming number of vulnerable web pages. 64% of web pages contained at least one serious vulnerability in 2010. Large companies are realizing the importance of securing their on-line applications and keeping their customers safe and are investing large money to complex security systems. These security solutions consist of expensive technology operating at multiple levels. In large companies is usual to deploy physical security network equipment such as Intrusion-Detection Systems (IDS) which are scanning network communication for possible attacks on various layers of the OSI model [5]. Physical IDS are very expensive and mostly require a trained operator. It is clear that smaller companies or individual webmasters cannot afford expensive solutions and mostly have no access to network or webserver settings and configuration. Therefor the main challenge for them is a well secured web application on source-code and functional logic level.

Á. Herrero et al. (Eds.): Int. JointConf. CISIS'12-ICEUTE'12-SOCO'12, AISC 189, pp. 159–166.
springerlink.com

This article introduces an implementation of IDS on web application level, using PHPIDS library [6]. The main advantage of this solution is checking all HTTP requests to the web application against attack vectors.

Implementation of IDS on web application level, of course, entails its performance slowdown which is tested and results are discussed in section 3. Performance Testing. Other limitations and proposal improvements are discussed in chapter 4.

1.1 Intrusion-Detection System (IDS)

An intrusion-detection system (IDS) can be defined as the tools, methods, and resources to help identify, asses, and report unauthorized or unapproved network activity [1]. Typical IDS works on the network layer of the OSI model and analyses packets to find possible attack pattern in the traffic.

The basic types of IDSs are categorized according to their application to Network Intrusion-Detection Systems (NIDS) and Host Intrusion-Detection Systems (HIDS). NIDS are analyzing overall network traffic while HIDS are analyzing only packets sent to and from a single machine [2]. In these days most of NIDS usually contains also some sort of HIDS element and are creating the third category called Hybrid IDSs [1]. According to detection model, IDSs could be divided into IDSs with misuse detection [3] or anomaly detection [4].

1.2 Intrusion-Detection System on Web Application Level (WAIDS)

Idea of WAIDS is based on the fact that most of attacks against web application are conducted over HTTP requests. For creating well-secured web application is therefore necessary to validate all inputs from HTTP request. Act of misusing an HTTP Request is illustrated in the Figure1.

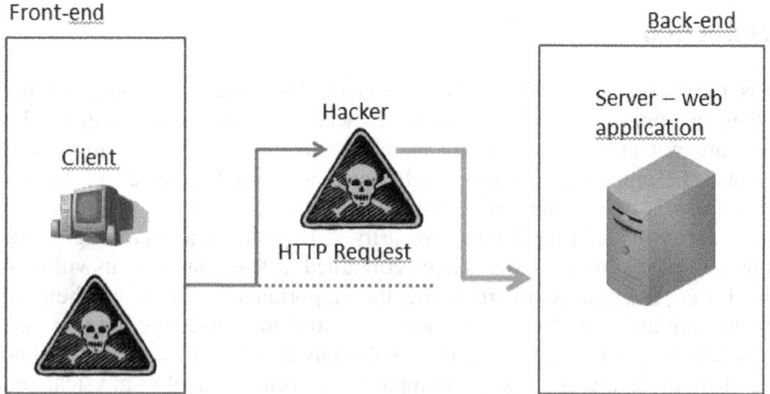

Fig. 1. HTTP request attack

Due to high costs for acquisition and operation of IDS systems (host or network based), small companies or individuals running their own web application on the Internet have to rely on security level of webhosting companies' webservers. In most cases they aren't allowed to change network or server configuration. It is therefore very important to satisfy high security requirements on the web application level. Universal solution for this target group seems to be the idea of checking HTTP requests coming to the application. Through HTTP requests are mostly conducted common types of web attack, such as SQL Injection, Cross Site Scripting (CSS) or HTTP Parameter Pollution (HPP). Currently the most used solution is validating of all GET, POST and REQUEST inputs using the string validation against the possible attack patterns.

2 Implementation WAIDS Using PHPIDS Library

For the testing purposes of WAIDS solution, PHPIDS library [7] was implemented into demo application.

2.1 PHPIDS

PHPIDS is an open-source library which is designed to detect all sorts of SQL Injection, XSS, header injection, directory traversal RFE/LFI, DoS and LDAP attacks. This library is written in PHP, which is most commonly used programming language for small business web application. It is needed at least PHP version 5.1.6.

2.2 PHPIDS Performance

Implementing PHPIDS into production web application brings the question of performance. If all HTTP request should be validated, it can be assumed that the web application response decreases. But PHPIDS validates only necessary information. Only requests which contain non-alphanumeric characters (which can be connected with possible flaw) are analyzed. The issue of testing is more detailed answered in section 3. Performance Testing.

2.3 Demo Application with WAIDS

Demo application was created for security and performance testing. Simple demonstration application consists of two HTML forms with textareas for input testing. The content of the first is sent using HTTP GET method and the content of the second by HTTP POST method. If the content is valid, user is able to see it on right side. If a malicious code is recognized, demo application shows warning with details about the attack. At the bottom of the page, there are also examples to test SQL Injection, CSS or HPP.

2.4 Example Attack on Demo Application

The demo application was under fictitious hacker attack. Scenario is as follows: Hacker wants to break the authentication system in application, which may be vulnerable to SQL injection. The attack query served to the login form (demo textarea) looks like this:

```
anything' OR 1=1';--
```

In vulnerable application could this piece of text produce, this SQL query:

```
SELECT id, name FROM users WHERE login='' AND pass='anything' OR 1=1';--';
```

Original SQL query could be changed and thanks to OR condition, which is always true, could return some record without appropriate authorization. PHPIDS properly validates the query and serves the warning message.

Warning message from PHPIDS example. You can see below an example of the warning message with important parameters: Total impact, affected tags, variable, value and description.

```
Total impact: 44

Affected tags: xss, csrf, id, sqli, lfi

Variable: REQUEST.get-request | Value: anything' OR 1=1';--
Impact: 22 | Tags: xss, csrf, id, sqli, lfi
Description: Detects common comment types | Tags: xss, csrf, id | ID: 35
Description: Detects classic SQL injection probings 1/2 | Tags: sqli, id, lfi | ID: 42
Description: Detects basic SQL authentication bypass attempts 1/3 | Tags: sqli, id, lfi | ID: 44
Description: Detects chained SQL injection attempts 1/2 | Tags: sqli, id | ID: 48

Variable: GET.get-request | Value: anything' OR 1=1';--
Impact: 22 | Tags: xss, csrf, id, sqli, lfi
Description: Detects common comment types | Tags: xss, csrf, id | ID: 35
Description: Detects classic SQL injection probings 1/2 | Tags: sqli, id, lfi | ID: 42
Description: Detects basic SQL authentication bypass attempts 1/3 | Tags: sqli, id, lfi | ID: 44
Description: Detects chained SQL injection attempts 1/2 | Tags: sqli, id | ID: 48
```

Impact. The impact indicates the severity of the attack. The PHPIDS brings around 50 filter rules to detect attacks and each one of them has an impact – the more rules match on the incoming data, the more likely it's an attack and the higher ranks the resulting impact [7].

Total impact. Shows sum of counted impact from all types of HTTP request (GET, POST, REQUEST).

Affected tags. This parameter indicates using tags, with which type of attack is malicious query connected.

Variable. This parameter shows type of HTTP request (GET, POST, REQUEST) and name of variable.

Value. Shows value of attack query.

Description. Here is possible to find more detailed description of vulnerability, which could be abuse.

All error messages are possible to save in a log file. Moreover, with the impact factor it is possible to eliminate most false positives (false alarms) or to track hacker`s activity. In most cases a usual total impact of first attack is around 5-10 sometimes 15-20. For example a typical XSS probing total impact usually results in time to value of 50-150 [7].

2.5 The Proposal PHPIDS Implementation to a Web Application

Figure 2 shows proposal implementation of PHPIDS, with all of the dataflow. The proposal implementation counts with the impact factor for eliminating possible false positives. According to the value of total impact should be possible to block the request, or to allow it. Both situations could be logged. If the total impact factor is greater than the threshold, the reaction system is then activated. Three types of reactions are proposed. System under attack could show to the attacker some kind of warning message that his or her actions are logged and outlaw. In most of the cases, such a warning message has great psychological effect. The second type of reaction could be destroying the user session, if the session is used for authentication. The third option is the most radical and blocks the entire user request to the web application.

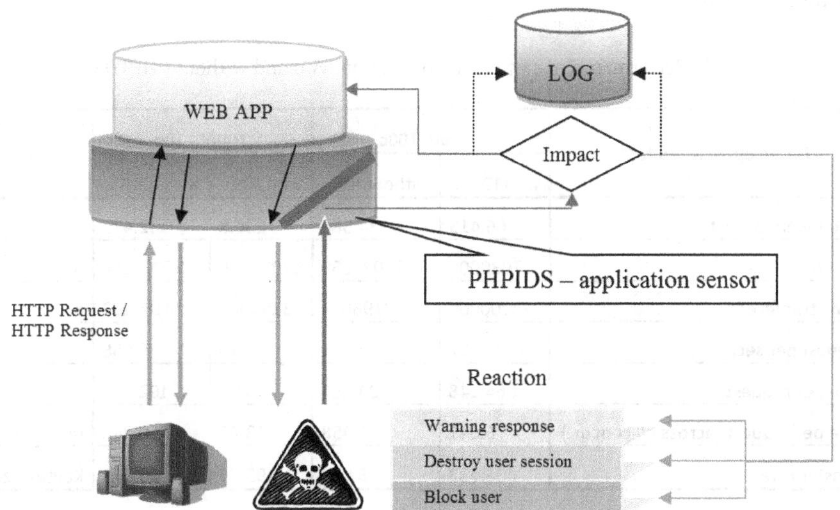

Fig. 2. Schema of demo application with WAIDS

3 Performance Testing of WAIDS Using PHPIDS Library

Performance tests using the load simulator were carried out on the demo application described above to identify the time and performance costs of PHPIDS library.

3.1 Testing Methodology

The tests were performed on the local server with server load simulation. Configuration of computer is as follows: Processor – Intel Core 2 Duo, 2.40 GHz, 4 GB RAM, IDE HDD.

For simulating of the server load was used Apache module AB. Demo application was tested with load of 10000 requests with maximum 100 concurrently and also 10 concurrently requests. Tests were carried out on application without and with PHPIDS implementation.

3.2 Testing Results

Table 1 show that the PHPIDS library consumes some additional time for execution. 10000 requests with maximum of 100 concurrent requests take 66.4 seconds. But more relevant parameter is Time per request (across all concurrent) which value is 6.641 milliseconds. On the other hand, Time per request without PHPIDS implemented is only 3.458 milliseconds. This parameter describes how fast is executed 1 request form 1 user by the server under the load. If the number of total requests is 10000 by maximum of 10 concurrent request, time per request with PHPIDS is 13.48 milliseconds and without PHPIDS 10.248 milliseconds.

It could be stated, that increase of response time is about 3 milliseconds and this is more than acceptable on server also with higher load. Results prove, that the performance slowdown connected with PHPIDS library implementation is in normal conditions very low.

Table 1. Performance test of demo application with and without PHPIDS

Demo application test	10000, 100c		10000, 10c		units
	with IDS	without IDS	with IDS	without IDS	
Time taken for test:	66.415	34.584	134.8	102.479	s
Total transfered:	27030000	27028658	27021936	27019264	b
HTML tranfered:	23200000	23198659	23191942	23189272	b
Request per sec.	150.57	289.15	74.18	97.58	
Time per request	664.148	345.84	134.8	102.479	ms
Time per request (across all concur.)	6.641	3.458	13.48	10.248	ms
Transfer rate	406.722	763.22	195.76	257.48	Kbytes/sec

4 Limitations of Proposed WAIDS

Proposed WAIDS design has following limitations which must be considered in the specific implementation.

4.1 Programming Language Specific

Proposed WAIDS based on PHPIDS implementation is suitable only for PHP web applications running on some PHP webserver (typical configuration may be Apache & MySQL). The version of PHP should be 5.1.6 and newer.

4.2 Performance Issues

Because of the fact, that proposed WAIDS is running directly on application level and intrusion detection is actually part of the application scripts, web developer emphasis should be given to control of total response time to avoid performance leakage of the web application. The advantage of no need for any special network IDS equipment is compensated with small decrease in web application performance and higher load of the webserver. According to section 3 Performance testing of WAIDS using PHPIDS library, increase of the time of response is only about 3 milliseconds per request.

4.3 Vulnerability to HTTP Parameter Pollution (HPP)

HPP [7] is kind of vulnerability which is here technically present for a long time. More discussed started to be in recent time with raising the number of attacks. This type of attack abuse HTTP request method (GET, POST or REQUEST) specifically transmitted parameters. Various servers differently handle the situation when there are more parameters of the same name in the HTTP request. Specifically Apache server takes into account only the last parameter. This property can be exploited by various attacks.

By security testing of PHPIDS library also HPP attack was tested. This type of attack is not detected by the library. Therefore this type of validation is proposed and could by implemented by simple request string analysis. No request with duplicate names of parameters should be valid.

5 Conclusion

This paper is focused on the security of small business application which are enormous vulnerable to various type of hacker's attacks. In this segment of World Wide Web there are insufficient funds to finance enterprise security solutions, including appropriate security policy and therefore less expensive security standards should be found.

WAIDS is Intrusion-Detection System on Web Application Level and its design is proposed in chapter 2. Key element of this Web Intrusion-Detection system is open-source PHP library PHPIDS and in chapter 2.3 is proposed its implementation and attack response options. Also significant parameters from error messages (discovered attack) are described and Impact factor and its utilization is discussed.

For carrying out performance testing the demo application was implemented. The performance test (chapter 3) shows, that the web application with PHPIDS implemented slows down only for time around 3 milliseconds (in case of described hardware configuration), which is very short time, but it is necessary to count it in the total response time of whole web application.

Also limitations of the proposal were discussed in chapter 4. Described design is programming language specific and its base – PHPIDS library – is written in PHP language and requires Apache web server. The intrusion detection is working on the web application level and it is integral part of web application and therefore is necessary to consider the question of performance.

During the testing, failure detection of HPP vulnerability in PHPIDS was discovered. For solving this problem, some kind of pre-validation, should be implemented into PHPIDS library. All HTTP requests should be validated with simple string validation and all queries that contain duplicate names of the parameters, should be considered as possible flaw.

References

1. Endorf, C., Schulz, E., Mellander, J.: Intrusion detection & prevention. McGraw-Hill Professional, Emeryville (2004) ISBN 0072229543
2. Northcutt, S., Novak, J.: Network Intrusion Detection: An Analyst's Handbook. New Riders Publishing, Thousand Oaks (2002)
3. Brumley, D., Newsome, J., Song, D., et al.: Towards automatic generation of vulnerability-based signatures. In: Proceedings of the IEEE Symposium on Security and Privacy, SP 2006, Washington, DC, USA, pp. 2–16 (2006)
4. Leung, K., Leckie, C.: Unsupervised anomaly detection in network intrusion detection using clusters. In: Proceedings of the 28th Australasian Conference on Computer Science, ACSC 2005, Darlinghurst, Australia, pp. 333–342 (2005)
5. Stewart, J.M., Tittel, E., Chapple, M.: CISSP: Certified Information Systems Security Professional Study Guide. Wiley Publishing, Indiana (2011)
6. PHPIDS project homepage, `https://phpids.org/`
7. Acunetix, Web Application Security, `http://www.acunetix.com/blog/whitepaper-http-parameter-pollution/`

A Genetic Algorithm for Solving RSA Problem in Elastic Optical Networks with Dedicated Path Protection

Mirosław Klinkowski

Department of Transmission and Optical Technologies,
National Institute of Telecommunications,
1 Szachowa Str., Warsaw, Poland
mklinkow@itl.waw.pl

Abstract. In this work we address the problem of static Routing and Spectrum Assignment (RSA) in a flexible grid optical network with dedicated path protection consideration. Since RSA is a difficult problem, we make use of the Genetic Algorithm (GA) metaheuristic to provide near-optimal solutions to the problem. We investigate the effectiveness of GA for a set of network scenarios. Evaluation results show that the proposed algorithm outperforms reference algorithms from the literature.

Keywords: Elastic optical networks, network optimization, dedicated path protection, routing and spectrum assignment.

1 Introduction

The evolution path of optical communication networks leads toward elastic optical networks (EON) in which advanced single-carrier modulation formats (such as m-PSK, m-QAM) and multi-carrier modulation techniques (such as O-OFDM) are applied for mixed-line-rate transmission and where elastic access to spectral resources is achieved within a flexible frequency grid (*flexgrid*) [1][2]. Future optical networks will utilize spectrum resources in optical fibres more efficiently, according to the transmission path characteristics and bandwidth requirements. We refer to some recent papers for more details on EON architectures [3][4] and for reports on proof-of-concept EON experiments [5][6].

In a flexgrid-based optical network (FG-ON) the optical frequency spectrum is divided into narrow frequency *slices* [7]. The optical path (*lightpath*) is determined by its routing path and a channel, which consists of a flexibly assigned subset of slices. Elastic spectrum allocation in FG-ON differs with channel assignment in fixed-grid wavelength division multiplexing (WDM) networks in that the channel width is not rigidly defined but it can be tailored to the actual width of the transmitted signal. Due to this difference, ordinary Routing and Wavelength Assignment (RWA) algorithms are not appropriate for FG-ON.

In EON, the problem of finding unoccupied spectrum resources so that to establish a lightpath is called the Routing and Spectrum Allocation (RSA) problem. RSA concerns assigning a contiguous fraction of frequency spectrum to a

Á. Herrero et al. (Eds.): Int. JointConf. CISIS'12-ICEUTE'12-SOCO'12, AISC 189, pp. 167–176.
springerlink.com

connection request - we refer to it as the *spectrum contiguity constraint* - subject to the constraint of no frequency overlapping in network links. The RSA optimization problem is $\mathcal{NP}$-hard [8] and it is more challenging than RWA in fixed grid networks due to the existence of the spectrum contiguity constraint. Offline RSA has been addressed with both Integer Linear Programming (ILP) (e.g., see [9][10][11]) and heuristic algorithms, which are based either on ILP problem relaxations [9] or on greedy, sequential processing of demands [9][10].

So far, there have been proposed few solutions for survivable FG-ON in the literature. Regarding offline network design, survivable RSA algorithms for 1+1 dedicated and 1:1 shared protection schemes in a ring network have been presented in [12]. For a network with generalized connectivity and 1+1 dedicated path protection scheme, an auxiliary graph based approach has been proposed in [13]. In both works, the optimization objective is to minimize the occupied spectrum width. These solutions make use of simple RSA heuristics.

The Genetic Algorithm (GA) metaheuristic is an effective approach for providing near-optimal solutions for large-scale optimization problems [14]. In this paper we propose to employ GA in the search of optimal RSA solutions in an FG-ON with dedicated path protection (DPP) and static traffic demands. To the best of our knowledge, the application of GA for solving RSA has not been studied in the literature so far. The main contributions of this paper is the development of an efficient GA-based algorithm that provides a near-optimal solution to the offline RSA with DPP problem in FG-ON. Numerical results show that the algorithm outperforms significantly the algorithms presented in the literature.

The remainder of the paper is organized as follows. In Section 2, we formulate the problem. In Section 3, we describe our GA-based RSA algorithm. In Section 4, we present numerical results. Finally, in Section 5 we conclude the work.

2 Problem Formulation

In this Section, we formulate an offline problem of Routing and Spectrum Allocation with Dedicated Path Protection in an FG-ON with static traffic demands. Among network survivability schemes, connection recovery through path protection, in which back up network resources (i.e., lightpaths) are provisioned in advance for each connection, is preferred due to its quick recovery time. Here, we focus on a cost-effective path protection scenario in which the transponders are shared between primary (working) and backup connections and a traffic demand has allocated the same segment of the optical frequency spectrum on its primary and backup path. Such a solution reduces the network cost and alleviates the connection switching time, as discussed in [13]. We denote the problem as RSA/DPP/SC (RSA with DPP and Same Channel allocation).

The considered RSA/DPP/SC problem can be formally stated as:
Given:

1. an FG-ON represented by a graph $\mathcal{G} = (\mathcal{V}, \mathcal{E})$, where $\mathcal{V}$ denotes the set of nodes, and $\mathcal{E}$ denotes the set of fiber links connecting two nodes in $\mathcal{V}$;

2. a frequency spectrum, the same for each link in $\mathcal{E}$ (without loss of generality), with the flexgrid represented by an ordered set of frequency slices $\mathcal{S} = \{s_1, s_2, \ldots, s_{|\mathcal{S}|}\}$;
3. a set $\mathcal{D}$ of static traffic demands to be transported with path protection guarantees; each demand d is represented by a tuple (o_d, t_d, n_d), where o_d and t_d are source and destination nodes respectively, and n_d is the requested number of slices.

Find a primary lightpath and a backup lightpath over the FG-ON for every transported demand subject to the following *constraints*:

1. *spectrum contiguity*: for each demand, the slices should be allocated each next to the other (i.e., adjacent) in the frequency spectrum,
2. *spectrum continuity*: for each demand, the subset of allocated slices should be the same for each link on the selected routing path,
3. *slice capacity*: a slice in a link can be allocated to one demand at most,
4. *same channel*: for each demand, the primary and backup lightpath should have allocated the same subset of slices,

and with the *objective* to minimize the occupied spectrum width (denoted as Φ).

It is worth mentioning that by minimizing Φ we reduce the fragmentation of spectrum resources, which is one of the objectives in FG-ON (e.g., see [15]).

The requested number of slices (n_d) is a function of the requested bandwidth, the modulation technique applied, the width of slice in the flexgrid, and the guard band introduced to separate two spectrum adjacent connections, among others. For a given network scenario, where both the transmission parameters and flexgrid definitions are given, there is a relation between n_d and the requested bandwidth (e.g., see Sec. II in [9]). Therefore, without loss of generality, we express the demand volumes by means of the number of slices. Assuming the flexgrid definition proposed in [7] and its requirement to have the spectrum allocated symmetrically around a central frequency, n_d is considered to be an even number, i.e., $n_d = 2i_d$, where $i_d \in \mathbb{Z}^+$. For more details, we refer to [7].

In this paper, we consider that a set of candidate routing paths is given. Let $\mathcal{P}$ denote the set of all candidate paths. Each path $p \in \mathcal{P}$ is identified with a subset $p \subseteq \mathcal{E}$. Let $\mathcal{Q}$ denote the set of pairs of paths (p, q), where $p \in \mathcal{P}$ is a primary path and $q \in \mathcal{P}$ is a backup path. Let $\mathcal{Q}_d$ denote the set of path pairs for demand $d \in \mathcal{D}$; each set $\mathcal{Q}_d$ comprises only paths that have the origin in o_d and the termination in t_d. We consider that $(p, q) \in \mathcal{Q}$ are link disjoint paths.

3 Genetic Algorithm for RSA/DPP/SC

In this section, we present a GA metaheuristic-based RSA algorithm for the off-line RSA/DPP/SC problem in an FG-ON. In the following, we describe our approach to solve the problem and then we focus on GA implementation details.

3.1 Algorithm Approach

In the RSA/DPP/SC problem formulation, we have assumed identical spectrum allocation on the primary and backup path. Therefore, we can treat RSA/DPP/SC as an RSA problem in which the spectrum allocation is performed on a super-path P formed by concatenation of p and q, i.e., $P = p \cup q = \{e : e \in p \vee e \in q\}$, and where $(p, q) \in \mathcal{Q}$. Indeed, the allocation of spectrum resources on P corresponds to the allocation of the same channel on both paths p and q.

To solve RSA/DPP/SC, we propose a hybrid algorithm in which we combine GA with a sequential heuristic RSA algorithm. The role of GA is to look for a sequence of demands which, when processed by the RSA algorithm, minimizes the occupied spectrum width (Φ). Having performed RSA, we calculate Φ by counting the number of slices $s \in \mathcal{S}$ that are occupied in network links:

$$\Phi = \sum_{s \in \mathcal{S}} x_s, \tag{1}$$

where $x_s = 1$ if slice $s \in \mathcal{S}$ is occupied in any link $e \in \mathcal{E}$, and equal to 0 otherwise.

As the sequential RSA algorithm we use the MSF (Most Subcarriers/Slices First) algorithm presented in [9]. MSF performs RSA for a set of demands that are arranged in decreasing order of the number of requested slices.

3.2 Auxiliary Notation

Here, we introduce some auxiliary notation.

Let $\mathcal{N}$ be the set of the numbers of requested slices, i.e., $\mathcal{N} = \{n : \exists d \in \mathcal{D}, n_d = n\}$, and $n_{\max} = \max\{n \in \mathcal{N}\}$ and $n_{\min} = \min\{n \in \mathcal{N}\}$ be, respectively, the largest and the smallest number in this set. In the remainder of this paper, we consider that the set of demands $\mathcal{D} = \{d_1, \ldots, d_i, d_{i+1}, \ldots, d_{|\mathcal{D}|}\}$ is an ordered set and, in particular, the demands are sorted in decreasing order of n_d, i.e., $\forall i : n_{d_i} \geq n_{d_{i+1}}$. Let $\mathcal{D}_n$ denote the set including all demands with n requested slices, i.e., $\mathcal{D}_n = \{d : d \in \mathcal{D}, n_d = n\}, n \in \mathcal{N}$.

3.3 GA Details

In our GA, each individual in the population represents a sequence (permutation) of demands. The RSA algorithm serves the demands, one-by-one, in an order determined by the demand permutation vector. The objective is to find such a permutation of demands which, when processed by the RSA algorithm, minimizes the occupied spectrum width. The GA algorithm starts with creating the initial population. Better permutations are found over a number of generations. Operators such as crossover and mutation explore further possible permutations. In our approach, all the individuals are feasible solutions therefore no additional constraint handling procedures are necessary. When the stopping criterion is satisfied, a near-optimal sequence of demands is found. Below we describe in detail the relevant building blocks of our GA.

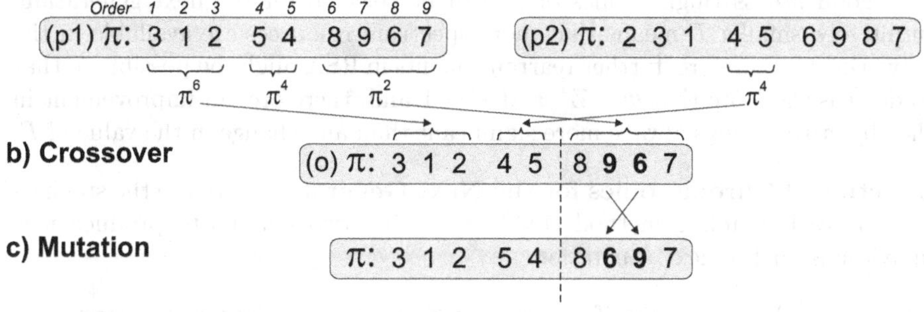

Fig. 1. GA components: a) chromosome, b) crossover operator, c) mutation operator

Representation: The encoded solution (chromosome) is a vector of numbers, coded as $\boldsymbol{\pi} = [\pi_1, \ldots, \pi_{|\mathcal{D}|}]$, which represents a sequence of demands (to be processed by the RSA algorithm). In order to assure that the demands are sorted in decreasing order of n_d, vector $\boldsymbol{\pi}$ of each chromosome is formed and maintained by GA as a concatenation of smaller vector segments $\boldsymbol{\pi}^n$, where $\boldsymbol{\pi}^n$ represents a permutation vector of demands belonging to $\mathcal{D}_n$ for each $n \in \mathcal{N}$. As a result, starting with segments $\boldsymbol{\pi}^{n_{\max}}$, which is a permutation vector of demands belonging to $\mathcal{D}_{n_{\max}}$, we form $\boldsymbol{\pi} = [\boldsymbol{\pi}^{n_{\max}}, \ldots, \boldsymbol{\pi}^{n}, \ldots, \boldsymbol{\pi}^{n_{\min}}]$.

In Fig. 1a) we present two exemplary chromosomes $p1$ and $p2$ formed for the case of $|\mathcal{D}| = 9$ demands among which there are 3 demands requesting 6 slices (segment $\boldsymbol{\pi}^6$, in positions $1-3$), 2 demands requesting 4 slices (segment $\boldsymbol{\pi}^4$,in positions $4-5$), and 4 demands requesting 2 slices (segment $\boldsymbol{\pi}^2$, in positions $6-9$). The order of demands is (312548967) and (231456987), respectively, for $p1$ and $p2$.

Population Initialization: The GA starts with an initial population of chromosomes. Permutation vectors $\boldsymbol{\pi}^n, n \in \mathcal{N}$, of each chromosome are generated randomly. In this work, a population of $M = 60$ chromosomes is maintained.

Fitness Function: The objective function to be minimized is the spectrum width obtained with a sequential RSA algorithm. As discussed in Sec. 3.1, in this study we consider MSF as the RSA algorithm. In order to drive the search toward better solutions we use a fitness value (f) that, apart from the objective value (Φ), incorporates additional information about spectrum occupancy (denoted as Γ), which is provided by RSA. We calculate Γ as the fraction of the network capacity that is allocated to the demands:

$$\Gamma = (|\mathcal{E}|\,|\mathcal{S}| + 1)^{-1} \sum_{e \in \mathcal{E}} \sum_{s \in \mathcal{S}} x_{es}. \tag{2}$$

where $x_{es} = 1$ if slice $s \in \mathcal{S}$ is occupied in link $e \in \mathcal{E}$, and equal to 0 otherwise.

Eventually, the fitness value is calculated as $f = -(\Phi + \Gamma)$ and individuals with good fitness (higher vales of f) will be selected for the next generation. Intuitively, smaller Γ means that more spectrum resources are available in the network and, therefore, further rearrangements in RSA might be possible so that to decrease Φ. Note that $\Phi \in \mathbb{Z}^+$ and $\Gamma < 1$ and, therefore, an improvement in the objective value is always more significant than any change in the value of Γ.

Selection of Chromosomes for the Next Generation: We use the stochastic universal sampling method [14] for selecting two parents to produce new individuals for the next generation.

Crossover: We use the uniform crossover operator [14] by which a single offspring is created from two parents by coping randomly segments $\boldsymbol{\pi}^n$ from either parent for each $n \in \mathcal{N}$.

In Fig. 1b), segments $\boldsymbol{\pi}^6$ and $\boldsymbol{\pi}^2$ are selected from parent $p1$ and segment $\boldsymbol{\pi}^4$ is selected from parent $p2$ to create offspring o.

Mutation: The mutation procedure is applied after the crossover on each offspring independently. Specifically, for randomly selected segment $\boldsymbol{\pi}^n, n \in \mathcal{N}$, we exchange the position of two randomly selected elements in this segment.

In Fig. 1c), segment $\boldsymbol{\pi}^2$ is selected and, within this segment, the position of demands 9 and 6 is exchanged.

Replacement: New individuals replace old individuals if their fitness values are better than those of the old ones. Apart from that, in this work we consider that: a) at least $0.1P = 6$ best individuals survive to the next generation, and b) 80% of the population at the next generation, not including the survivals from the old population, are created by the crossover function. The population size at each generation is maintained fixed and equal to P.

Stopping Criteria: Our GA terminates after G generations. In this work, we perform evaluation for $G \in \{1, 10, 50, 100, 150, 200\}$.

3.4 Algorithm Complexity

The algorithm complexity is polynomial in time. In particular, the computation complexity of MSF is bounded by $O(|\mathcal{D}|\,|\mathcal{Q}_d|\,|\mathcal{S}|\,|\mathcal{E}|)$, where $|\mathcal{D}|$ is the number of demands to be processed, $|\mathcal{Q}_d|$ is the number of candidate path pairs per demand, and $|\mathcal{S}|\,|\mathcal{E}|$ corresponds to the (worst-case) complexity of the search of unoccupied slices. MSF is called once for each chromosome. The number of chromosomes is limited by the product of the population size (M) and the number of generations (G) in GA.

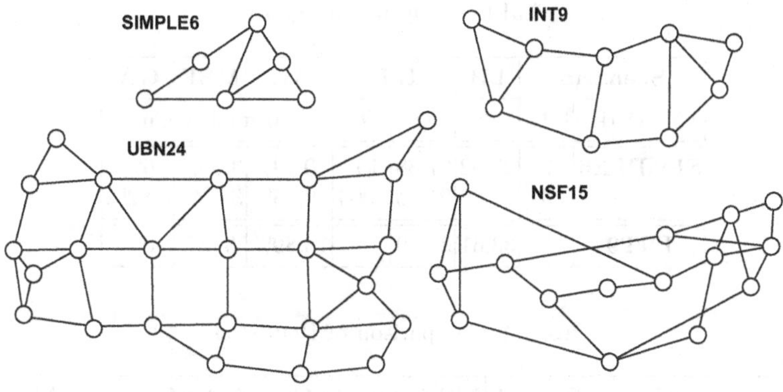

Fig. 2. Network topologies

4 Numerical Results

In this Section, we evaluate the performance of our **GA**-based algorithm and compare it with other reference algorithms. We focus on the occupied spectrum width (Φ), in terms of the number of slices, and computation time (T; in seconds).

As the reference algorithms we use: **AG** - an Auxiliary Graph-based algorithm proposed in [13] for dedicated path protection, and **MSF** - an RSA algorithm based on ordering of demands according to the number of requested slices [9]. Moreover, for small network scenarios, we provide optimal results obtained with an **ILP** formulation of the RSA/DPP/SC problem. The formulation applies the channel assignment approach presented in [11]; due to space limitations, we do not present the formulation. Also, we provide a lower bound (**LB**) on Φ, obtained by solving a multicommodity routing problem (see Sec. III-A in [9]).

The evaluation is performed for SIMPLE6 (6 nodes, 16 links), INT9 (9 nodes, 26 links), NSF15 (15 nodes, 46 links) and UBN24 (24 nodes, 86 links) network topologies (see Fig. 2). We set $|\mathcal{S}| = 1500$, which is large enough to accommodate all demands without blocking in the most demanding scenario; in other words, $\Phi \leq |\mathcal{S}|$. Candidate pairs of primary-backup paths are link disjoint paths and they are calculated as the shortest paths, taking into account the overall length of both paths. We consider $|\mathcal{Q}_d| \in \{2, 3, 5, 10, 30\}$. The requested spectrum n_d is generated with uniform distribution and for randomly selected demand pair (o_d, t_d), where $o_d, t_d \in \mathcal{V}$ and $o_d \neq t_d$. In particular, $n_d = 2I$, where $I \in \{1, 2, ..., N\}$ and $N \in \{4, 8\}$. For fair evaluation, we consider all algorithms operate on an ordered set of demands $\mathcal{D}$ (as defined in Sec. 3).

The GA procedure is implemented using the Genetic Algorithm (GA) toolbox in MATLAB. AG is implemented in MATLAB as well. In order to speed up the processing of MSF, we implement it in C++ and call as an external function from MATLAB. We use IBM ILOG CPLEX v.12.4 [16] to solve ILP. The evaluation is performed on an Intel *i*5 3.3GHz 16GB computer. We make use of

Table 1. Optimality gap

Scenario		**LB**	**ILP**		**AG**	**MSF**	**GA**
Network	$\|\mathcal{Q}_d\|$	Φ	Φ	T	Optimality gap		
SIMPLE6	2	25.02	27.88	191	10.78%	3.96%	0%
	3	24.88	27.86	1004	10.85%	3.45%	0.82%
INT9	2	33.61	34.79	–	10.89%	3.8%	0%

Table 2. Comparison of Heuristics

Scenario		**LB**	**AG**	**MSF**	**GA**	**AG** vs. **GA**	**MSF** vs. **GA**
Network	$\|\mathcal{Q}_d\|$	Φ	Φ	Φ	Φ	Performance gap	
NSF15	2	428.5	–	538.8	494.3	–	8.26%
	10	333.2	–	485.4	439.2	–	9.52%
	30	321.3	–	461.2	418.2	–	9.32%
	–	–	566.6	–	–	26.2%	–
UBN24	2	955.7	–	1083	1029.6	–	4.95%
	10	665.6	–	923.5	861.4	–	6.72%
	30	665.3	–	836.6	787.4	–	5.87%
	–	–	942.2	–	–	16.4%	–

parallel processing (on 4 processor cores) which is available in the GA toolbox. If not mentioned otherwise, the results are averaged over 100 randomly generated demand sets.

4.1 Optimality of GA

In Tab. 1 we compare performance results of heuristic algorithms with the optimal ones (ILP) in small network scenarios. We consider $|\mathcal{D}| = 10$ and $|\mathcal{D}| = 15$, respectively, for SIMPLE6 and INT9, and $n_d \in \{2, 4, 6, 8\}$ for both networks. GA runs $G = 50$ generations. Due to the complexity of ILP, which has not reached optimality for about 62 out of 150 cases (39%) in a $2h$ period, the results for INT9 are averaged over the remaining 88 demand sets.

We can see that GA outperforms other heuristics and it achieves optimal solutions in most of the cases (279 out of 288 evaluated demand sets).

4.2 Performance Comparison

In Tab. 2 we present performance results obtained with heuristic algorithms in larger networks. In the evaluation, $|\mathcal{D}| = 210$ and $|\mathcal{D}| = 552$, respectively, for NSF15 and UBN24, and $n_d \in \{2, 4, .., 16\}$ for both networks. GA runs $G = 200$ generations. Since AG calculates routing paths online and it does not use candidate paths, we present its results in a dedicated row. In the comparison AG vs. GA we consider $|\mathcal{Q}_d| = 30$ for GA.

Table 3. Performance of GA in a function of G

Scenario		$G = 1$		$G = 10$		$G = 50$		$G = 100$		$G = 200$		$G = 1$ vs. $G = 200$
Network	$\lvert\mathcal{Q}_d\rvert$	Φ	T	Φ	T	Φ	T	Φ	T	Φ	T	Improvement in Φ
NSF15	2	513	1.3	502	6.6	497	30	496	60	494	119	3.59%
	10	459	1.5	448	7.9	442	36	440	71	439	142	4.25%
	30	437	2.1	426	11	421	50	420	98	418	195	4.32%
UBN24	2	1050	5.6	1037	31	1032	140	1031	277	1030	552	1.92%
	10	887	8	873	43	866	199	863	394	861	785	2.84%
	30	810	13	797	72	791	329	789	650	787	1292	2.77%

Again, GA achieves lower values of Φ than reference algorithms. The performance gap is of about 16% – 26% and 5% – 10%, respectively, when comparing GA with AG and MSF. We can see that Φ can be improved considerably (about 15% for NSF15 and 24% for UBN24) if there are more path pairs available in the set of candidate paths ($|\mathcal{Q}_d| = 2$ vs. $|\mathcal{Q}_d| = 30$). For MSF, the average computation times are below 0.3 sec. in the most demanding scenario. The computation times for GA are presented in the next section.

4.3 GA Performance

Eventually, in Tab. 3 we present the GA performance results in function of the number of generations (G). The improvement between $G = 1$ and $G = 200$ is of about 2% – 4%, however, at the cost of extended computation time, which grows linearly with G. The impact of $|\mathcal{Q}_d|$ on computation times is much smaller. Accordingly, it is worth to have a large set of candidate paths since it has a great impact on the value of Φ (as shown in the previous section). We can see that above certain G the objective function (Φ) does not improve significantly. We can conclude that $G = 50$ is a reasonable trade-off between solution quality and computation time for GA in the studied scenarios.

5 Conclusions

In this paper, we have focused on the problem of off-line routing and spectrum allocation in flexible grid optical networks with dedicated path protection. To solve the problem, we have proposed a novel algorithm which is based on the genetic algorithm metaheuristic combined with a sequential RSA algorithm. The aim of GA is to help in the search of RSA solutions that minimize the occupied spectrum width. To assess the algorithm performance, we have compared it with state of the art RSA algorithms. The performed numerical experiments show that the GA-based algorithm outperforms the reference heuristics and it provides results close to the optimal ones.

Acknowledgments. This work has been supported by the Polish National Science Centre under grant agreement DEC-2011/01/D/ST7/05884.

References

1. Gerstel, O., et al.: Elastic optical networking: A new dawn for the optical layer? IEEE Comm. Mag. 50(2), 12–20 (2012)
2. Wei, W., et al.: Cognitive optical networks: Key drivers, enabling techniques, and adaptive bandwidth services. IEEE Comm. Mag. 50(1), 106–113 (2012)
3. Jinno, M., et al.: Spectrum-efficient and scalable elastic optical path network: Architecture, benefits, and enabling technologies. IEEE Comm. Mag. 47(11), 66–73 (2009)
4. Meloni, G., et al.: PCE architecture for flexible WSON enabling dynamic rerouting with modulation format adaptation. In: Proc. of ECOC, Geneva, Switzerland (September 2011)
5. Geisler, D.J., et al.: The first testbed bemonstration of a flexible bandwidth network with a real-time adaptive control plane. In: Proc. of ECOC, Geneva, Switzerland (September 2011)
6. Cugini, F., et al.: Demonstration of flexible optical network based on path computation element. IEEE J. Lightw. Technol. 30(5), 727–733 (2012)
7. Li, Y., et al.: Flexible grid label format in wavelength switched optical network. IETF RFC Draft (2011)
8. Klinkowski, M., Walkowiak, K.: Routing and spectrum assignment in spectrum sliced elastic optical path network. IEEE Commun. Lett. 15(8), 884–886 (2011)
9. Christodoulopoulos, K., Tomkos, I., Varvarigos, E.: Elastic bandwidth allocation in flexible OFDM based optical networks. IEEE J. Lightw. Technol. 29(9), 1354–1366 (2011)
10. Klinkowski, M., Walkowiak, K., Jaworski, M.: Off-line algorithms for routing, modulation level, and spectrum assignment in elastic optical networks. In: Proc. of IEEE ICTON, Stockholm, Sweden (June 2011)
11. Velasco, L., Klinkowski, M., Ruiz, M., Comellas, J.: Modeling the routing and spectrum allocation problem for flexgrid optical networks. Phot. Netw. Commun (in press, 2012)
12. Takagi, T., et al.: Algorithms for maximizing spectrum efficiency in elastic optical path networks that adopt distance adaptive modulation. In: Proc. of ECOC, Torino, Italy (September 2010)
13. Patel, A.N., et al.: Survivable transparent flexible optical WDM (FWDM) networks. In: Proc. of OFC, Los Angeles, USA (March 2011)
14. Talbi, E.G.: Metaheuristics: From Design to Implementation. Wiley (2009)
15. Patel, A.N., et al.: Defragmentation of transparent flexible optical WDM (FWDM) networks. In: Proc. of OFC, Los Angeles, USA (March 2011)
16. IBM ILOG CPLEX optimizer (2012), http://www.ibm.com

Correlation Approach for SQL Injection Attacks Detection

Michał Choraś[1,2], Rafał Kozik[2], Damian Puchalski[1], and Witold Hołubowicz[2,3]

[1] ITTI Ltd., Poznań, Poland
mchoras@itti.com.pl
[2] Institute of Telecommunications, UT&LS Bydgoszcz, Poland
[3] Adam Mickiewicz University, Poznań
holubowicz@amu.edu.pl

Abstract. In this paper we prove that the correlation approach to SQL Injection Attacks allows improving results of such attacks detection. Moreover, we propose a novel method for SQLIA detection based on the genetic algorithm for determining anomalous queries. Experimental scenario is also described and the achieved results are reported.

1 Introduction

The major contribution of this paper is the proposition of the correlation approach to SQLIA detection and the genetic algorithm applied to SQLIA detection task. In this work, we try to prove that correlating several sources of information (sensors) and then performing reasoning on the correlated information, allows improving results of cyber attacks detection. Such approach is compliant with the the national project SOPAS funded by Ministry of Science in the theme of homeland security where Federated Networks Protection System (FNPS) is developed. The basic pillars of FNPS are federated approach to cyber security, information sharing and network events correlation from various sources (sensors)[1]. Detection of SQL Injection Attacks is one of the key goals of the project, and its complexity allows showing the benefits of the correlation approach.

The paper is structured as follows: In Section 2 SQL Injection Attacks are discussed. In Section 3 the proposed genetic algorithm for SQLIA detection is presented. Evaluation methodology is described in Section 4, while the results are reported in Section 5. Conclusions are given thereafter.

2 Overview of SQL Injection Attacks

SQL injection and other similar exploits are the results of interfacing a scripting language by directly passing information through another language and are ultimately caused by insufficient input validation. SQL Injection Attacks (SQLIA) refer to a code-injection attacks category in which part of the users input is treated as SQL code. Programming languages such as PHP or Perl allow user

Á. Herrero et al. (Eds.): Int. JointConf. CISIS'12-ICEUTE'12-SOCO'12, AISC 189, pp. 177–185.
springerlink.com

to access a database via an SQL query. If data provided by the end user is sent directly to the database and is not properly verified, then user can insert malicious SQL commands. While exploiting such vulnerabilities, an attacker can submit SQL commands directly to the database.

Such commands, if executed on the database, may change, erase, or expose sensitive data stored in the database. In more severe cases attacker can execute remote code and access to the system. Therefore, SQL injection is related not only to databases and can affect system as a whole. Since the attacker uses the standard trusted API for executing his attack there is low probability that the system administrator is aware of the attack [2].

This type of vulnerability represents a serious threat to any web application that processes user queries (e.g. through web forms or web APIs) and then executes SQL queries on underlying database. Most widely used web applications in the Internet or within commercial networks work in this manner, and in consequence are vulnerable to SQL injections [3].

SQLIA is relatively easy to perform and hard to detect or prevent. These factors contribute to growing popularity of such form of cyber attacks. This fact has been recently proven by Paul Sparrows who had published the list of 2011 attacks [4]. Figure (Figure 1) shows that SQL injection was the most frequent type of cyber attacks in 2011.

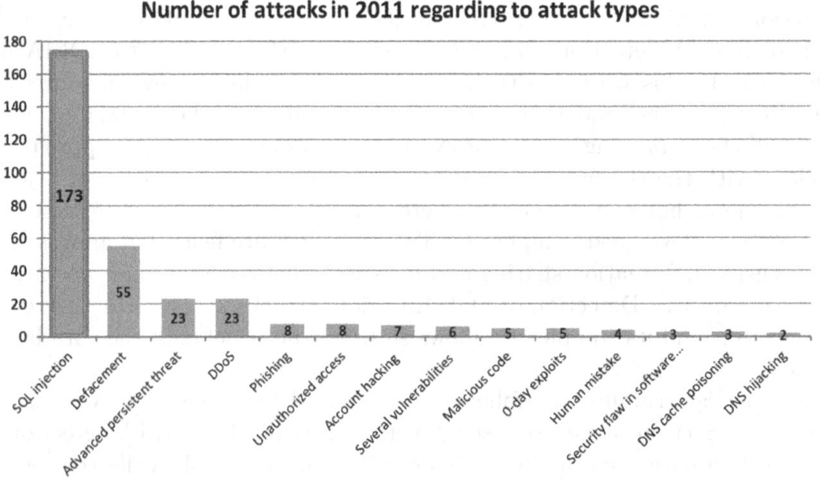

Fig. 1. Number of various types of cyber attacks in 2011

According to [4], more than a half of identified attacks were SQL Injection Attacks, what is graphically depicted on the chart in Figure 2.

Similarly, the list of top 10 most critical risks related to web applications security, provided by OWASP (Open Web Application Security Project) indicates Injection (including SQL, OS, and LDAP injections) as a major vulnerability [5]. Factors, such as easy exploit-ability and severe impact of potential attacks are mentioned as the most crucial. In order to perform injection attack, attacker

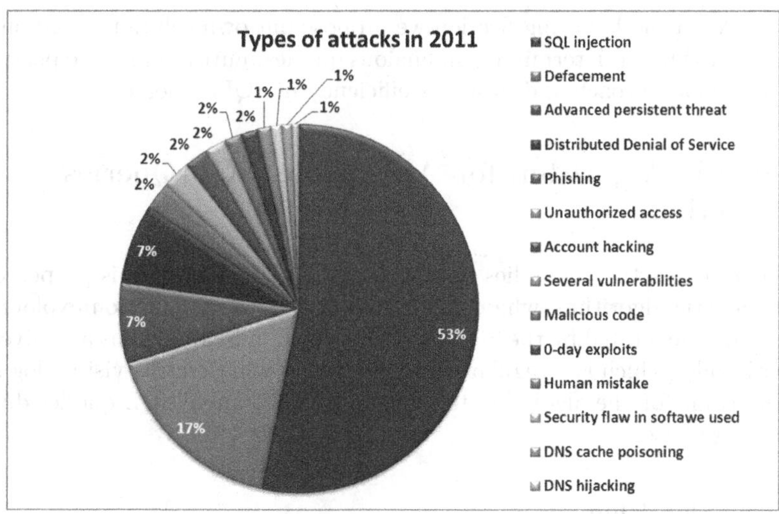

Fig. 2. Distribution (%) of various cyber attacks in 2011

sends simple text, that exploits the syntax of the targeted interpreter, therefore almost any source of data can be an injection attack vector. On the other hand, injection can cause serious consequences including data loss, corruption, lack of accountability or denial of access. Additionally, level of prevalence is described as common, while level of detectability is identified as average [5].

One of the most significant examples of SQL Injection Attacks include:

- hacking the Royal Navys website and recovering user names and passwords of the sites administrators (November 2010) [6];
- stealing information related to almost 100000 accounts of subscribers registered on ISP news and review site DSLReports.com (April 2011) [7];
- exploiting SQL injection vulnerabilities of approximately 500000 web pages (April-August 2008) [8].

These examples show the scale of SQL Injection Attacks consequences, and prove that one of the most common cyber threats should not be ignored by computer security communities.

Several publications provide surveys, as well as analysis evaluating and comparing injection detection and prevention techniques. For example, more than twenty detective and preventive techniques are examined in [9]. In the publication, authors identified various types of SQLIAs and investigated ability to stop SQL injection provided by the most commonly used, current techniques. Similar approaches are presented in [10] and [11], where prevention techniques and security tools for the detection of SQL injection attacks were investigated.

However, the results of the mentioned studies prove and emphasize low performance and efficiency of current methods, techniques and tools. According to the mentioned papers, no current solution can provide complete protection against SQLIA.

Therefore, in the following Section we propose our own solution based on the genetic algorithm for determining anomalous queries. Furthermore, we postulate the correlation approach and prove its efficiency in SQLIA detection.

3 Genetic Algorithm for Anomalous SQL Queries Detection

In order to detect the anomalies in SQL queries a novel method is proposed. It exploits genetic algorithm, where the individuals in the population explore the log file that is generated by the SQL database. Each individual aims at delivering an generic rule (which is a regular expression) that will describe visited log line. It is important for the algorithm to have an set of genuine SQL queries during the learning phase.

3.1 The Algorithm

The algorithm is divided into the following steps:

- Initialization. Each individual and line from log file is assigned. Each newly selected individual is compared to the previously selected in order to avoid duplicates.
- Adaptation phase. Each individual explores the fixed number of lines in the log file (the number is predefined and adjusted to obtain reasonable processing time of this phase).
- Fitness evaluation. Each individual fitness is evaluated according to the rule described in section 3.3. The global population fitness as well as rule level of specificity are taken into consideration, because we want to obtain set of rules that describe the lines in the log file.
- Cross over. Randomly selected two individuals are crossed over using algorithm for string alignment that is described in section 3.2. If the newly created rule is too specific or too general it is dropped in order to keep low false positives and false negatives.

3.2 Crossing over the Individuals

In order to obtain the regular expression from two strings a modified version of the Neddleman and Wunch algorithm is proposed ([12]). The authors used this algorithm to find the best match between two DNA sequences which can diverge over time (e.g. by insertion or deletion) for different organisms. In order to find correspondence between those two sequences, it is allowed to modify the sequences by inserting the gaps. However, for each gap (and for mismatch) there is an penalty and award for genuine matches.

This method can also be applied to find corresponding segment in two text sequences. For example possible match for sequences shown in Fig.3 could be the one shown in Fig.4.

```
SELECT age, weight FROM patient WHERE name like 'Jane'
SELECT height FROM patient WHERE name like 'Joe'
```

Fig. 3. Eamples of SQL queries

```
SELECT |age, weight______|FROM patient WHERE name like|'___Jane'|
SELECT |__________height |FROM patient WHERE name like|'Joe____'|
```

Fig. 4. Example of possible match found for two query strings. Red rectangle indicates place where gaps were inserted (the "_" sign)

For Needleman and Wunsch algorithm the most important is to find the best alignment between two sequences (the one with highest award). From anomaly detection point of view the parts where gaps (marked in Fig.4) are inserted are also important, because they are the points of injections. These parts are described with regular expressions using guidelines proposed in [13].

Therefore, the result shown in Fig.4 can be represented with the following regular expression: "SELECT [a-z,]+ FROM patient WHERE name like [a-zA-z]+".

Needleman and Wunsch first suggested that in order to find the match with highest award a dynamic programming (DP) approach can be adapted. More details explaing how this is implemented can be found in [12].

3.3 Fitness Evaluation

The fitness function, that is used to evaluate each individual, takes into account the particular regular expression effectiveness (number of times it fires), the level of specificity of such rule and the overall effectiveness of the whole population. The fitness function is described by equation 1, where I indicates the particular individual regular expression, $E_{population}$ indicates the fitness of the whole population, E_f effectiveness of regular expression (number of times the rule fires), and E_s indicates the level of specificity. The α, β, and γ are constants that normalize the overall score and balance the each coefficient importance.

$$E(I) = \alpha * E_{population} + \beta * E_f(I) + \gamma * E_s(I) \tag{1}$$

$$E_{population} = \sum_{I \in Population} E_f(I) \tag{2}$$

The level of specificity indicates balance between number of matches and number of gaps. For example shown in Fig.5 there are 6 matches and 15 gaps, and the level of specificity will be equal to 0.28 (6/(6+15)). This allows the algorithm to penalize these individuals that try to find general rule for totally different queries like SELECT and INSERT.

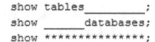

```
show tables_________;
show ______databases;
show ***************;
```

Fig. 5. Example of possible match found for two SQL query strings

4 Evaluation Methodology

In this section our evaluation methodology is described. The SQL Injection Attacks are conducted on php-based web service with state of the art tools for services penetration and SQL injection. The traffic generated by attacking tools are combined together with normal traffic in order to estimate the effectiveness of the proposed methods.

The web service used for penetration test is so called LAMP (Apache + MySQL + PHP) server with MySQL back-end. It is one of the most common worldwide used servers and therefore it was used for validation purposes. The server was deployed on Linux Ubuntu operation system. For penetration tests examples services developed in PHP scripts and shipped by default with the server are validated.

4.1 Attacks Injection

Attack injection methodology is based on the known SQL injection methods, namely: boolean-based blind, time-based blind, error-based, UNION query and stacked queries. For that purpose sqlmap tool is used. It is an open source penetration and testing tool that allows the user to automate the process of validating the tested services against the SQL injection flaws.

4.2 Attack Detection Tools

The set of tools used for detecting the SQL Injection attacks consists of both an algorithms proposed by authors and known (state of the art) solutions and tools. The state of the art tools evaluated in our tests are:

- Apache Scalp
  It is an analyzer of Apache server access log file. It is able to detect several types of attacks targeted on web application. The detection is a signature-based one. The signatures have form of regular expressions that are borrowed from PHP-IDS project.
- Snort
  It the most widely deployed IDS system that uses set of rules that are used for detecting web application attacks. However most of the available rules are intended to detect very specific type of attacks that usually exploit very specific web-based application vulnerabilities.

- ICD (Idealized Character Distribution [14])
The method is similar to the one proposed by C.Kruegel in [14]. The proposed character distribution model for describing the genuine traffic generated to web application. The Idealized Character Distribution (ICD) is obtained during the training phase from perfectly normal requests send to web application. The IDC is calculated as mean value of all character distributions. During the detection phase the probability that the character distribution of a query is an actual sample drawn from its ICD is evaluated. For that purpose Chi-Square metric is used.

5 Results

In this section each tool effectiveness of SQL injection attack detection is reported.

There are two phases. In the first one, genuine traffic to web application is generated using the web crawlers that emulate user behaviour. In the second phase attacks are injected with sqlmap tool.

Each method for attack detection is evaluated against true positives and false positives. The true positives indicate how many times each tool was able to detect an attack, while the false positives indicate the number of genuine queries recognized as an attack. The results of each tool/method effectiveness are presented in Table 1.

Table 1. The results comparing effectiveness of the proposed SQL-ADS method to the state-of-the-art solutions such as SNORT, SCALP and ICD

Tool	TP	FN
SNORT	39.0%	0.0%
SCALP	85.0%	0.0%
ICD [14]	97.7.1%	12.5%
SQL_ADS	99.1%	12.5%

The results show that the proposed method based on the genetic algorithm for determining anomalous queries (*SQL_ADS*) achieves better results in comparison to the well-known, standard methods/tools.

In this scenario we have evaluated how the mechanism of correlation can improve the effectiveness of the SQL injection detection. For that purpose simple voting mechanism was used. The results for different weight configurations are shown in the Table 2.

Table 2. Results of the SQL detection enhanced by correlation

Method	TP	FN
Correlate 1	97.0%	0.0%
Correlate 2	100.0%	12.5%

For both Correlate 1 and Correlate 2 the weights were assigned as follows: $0.4 * SNORT + 1.0 * ICD + 1.0 * SQL - ADS$.

For Correlate 1 the sum threshold was 1.0 (if the sum is greater than 1.0 then the attack is detected). For Correlate 2 the sum threshold was 0.6.

The results clearly show that the correlation approach enables to achieve better detection results than using the methods separately.

6 Conclusions

This paper presents our results in SQL Injection Attacks detection. The major contributions of this paper are: the correlation approach to SQLIA detection and the novel SQLIA detection method based on the genetic algorithm for determining anomalous queries.

We have presented the evaluation methodology and reported the results that prove that: (i) our method based on the described genetic algorithm outperforms other methods and tools such as SCALP, ICD and SNORT; (ii) correlation of all the tested methods/tools allows to improve detection results.

Acknowledgement. This work was partially supported by Polish Ministry of Science and Higher Education funds allocated for the years 2010-2012 (Research Project number OR00012511).

References

1. Choraś, M., Kozik, R., Piotrowski, R., Brzostek, J., Hołubowicz, W.: Network Events Correlation for Federated Networks Protection System. In: Abramowicz, W., Llorente, I.M., Surridge, M., Zisman, A., Vayssière, J. (eds.) ServiceWave 2011. LNCS, vol. 6994, pp. 100–111. Springer, Heidelberg (2011)
2. Rao, T.K., Kum, G.Y., Reddy, E.K., Sharma, M.: Major Issues of Web Applications: A Case Study of SQL Injection. Journal of Current Computer Science and Technology 2(1), 16–20 (2012)
3. Halfond, W., Orso, A.: AMNESIA: Analysis and Monitoring for Neutralizing SQL-Injection Attacks. In: Proceedings of the 20th IEEEACM International Conference on Automated Software Engineering (2005)
4. `https://paulsparrows.wordpress.com/2011-cyber-attacks-timeline-master-index/`
5. OWASP Top 10 – 2010, The Ten Most Critical Web Application Security Risks (2010)
6. Royal Navy Website Attacked by Romanian Hacker (2008), `http://www.bbc.co.uk/news/technology-11711478`
7. Mills, E.: DSL Reports Says Member Information Stolen (2011)
8. Keizer, G.: Huge Web Hack Attack Infects 500,000 pages (2008)
9. Tajpour, A., JorJor Zade Shooshtari, M.: Evaluation of SQL Injection Detection and Prevention Techniques. In: CICSyN 2010 Second International Conference on Computational Intelligence, Communication Systems and Networks (2010)

10. Amirtahmasebi, K., Jalalinia, S.R., Khadem, S.: A Survey of SQL Injection Defense Mechanisms. In: ICITST International Conference for Internet Technology and Secured Transactions (2009)
11. Elia, I.A., Fonseca, J., Vieira, M.: Comparing SQL Injection Detection Tools Using Attack Injection: An Experimental Study. In: 2010 IEEE 21st International Symposium on Software Reliability Engineering (2010)
12. Needleman, S.B., Wunsch, C.D.: A General Method Applicable to the Search for Similarities in the Amino Acid Sequence of Two Proteins. Journal of Molecular Biology (1970)
13. Conrad, E.: Detecting Spam with Genetic Regular Expressions. SANS Institute InfoSec Reading Room (2007)
14. Kruegel, C., Toth, T., Kirda, E.: Service specific anomaly detection for network intrusion detection. In: Proc. of ACM Symposium on Applied Computing, pp. 201–208 (2002)

Information Exchange Mechanism between Federated Domains: P2P Approach

Michał Choraś[1,2], Rafał Kozik[2], Rafał Renk[1,3], and Witold Hołubowicz[2,3]

[1] ITTI Ltd., Poznań, Poland
mchoras@itti.com.pl
[2] Institute of Telecommunications, UT&LS Bydgoszcz, Poland
[3] Adam Mickiewicz University, Poznań
holubowicz@amu.edu.pl

Abstract. In this paper a concept and realization of the P2P (Peer-to-Peer) based communication between federated networks is presented. In order to provide a robust, self path-replicated and resilient communication between different domains in federation (between Decision Modules), P2P technology is used. This approach allows the proposed system to overcome IP addressing issues and minimize the configuration cost. Exchanging information between federated domains enables detection of complex distributed attacks and increases situational awareness of the network administrators.

1 Introduction

The major contribution of the paper is the concept of P2P (Peer to Peer) based information exchange between federated networks. In this paper a concept and an architecture of the Federated Networks Protection System (FNPS) is presented. The system components and, particularly, the Decision Module are described. Communication between Decision Modules (DM) in each of the federated domain is necessary in order to inform about network status, detected attacks or anomalies and distribute General Decision Rules (GDR) describing specific reactions.

The presented system is dedicated for federated networks and systems used by the public administration and military sector. Such systems can increase their overall security and resiliency by sharing and exchanging security related information and general reaction rules. We have also presented a sample scenario (SQLIA SQL injection attack detection) to show how the proposed system can detect complex attacks and benefit from information sharing between federated domains.

In our approach, we use the capability of the federated networks and systems to share and exchange information about events in the network, detected attacks and proposed countermeasures. Such approach to share information about network security (between trusted entities and networks) follows the Federation of Systems (FoS) idea. Also in our case, FoS concept refers to a set of different systems, which are not centrally managed, but cooperate in order to share

Á. Herrero et al. (Eds.): Int. JointConf. CISIS'12-ICEUTE'12-SOCO'12, AISC 189, pp. 187–196.
springerlink.com

knowledge and increase their security. Such approach has recently received much attention and may replace inefficient approach of "closed security" [2]. The concept of federated networks and systems has gained much attention also in the context of critical systems, military networks and NATO Network Enabled Capability (NNEC) [3][4][5].

This paper is structured as follows: in Section 2 the general architecture of the Federated Networks Protection System and Decision Module is presented. In Section 3 the concept and realization of the P2P-based information exchange mechanism is described. In Section 4 we show how the complex SQLIA attack can be detected with the support of the proposed P2P information exchange mechanism. Conclusions are given thereafter.

2 General Architecture of Federated Networks Protection System (FNPS)

The general architecture of the Federated Networks Protection System (FNPS) is presented in Fig. 1. It consists of several interconnected domains, which exchange information in order to increase their security level and the security of the whole federation.

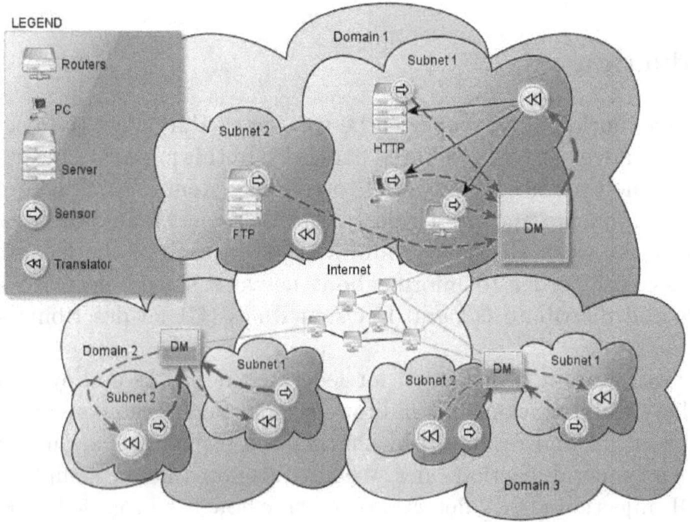

Fig. 1. General Architecture of Federated Networks Protection System (FNPS)

Different subnetworks are arranged in domains, according to the purpose they serve (e.g. WWW, FTP or SQL servers) or according to their logical proximity (two networks closely cooperating with each other). In each of the domains, a Decision Module (FNPS-DM) is deployed. Each DM is responsible for acquiring and processing network events coming from sensors distributed over the domain.

If the attack or its symptoms are detected in one domain, the relevant information are disseminated to other cooperating domains so that appropriate countermeasures can be applied.

In our approach, the idea is to re-use the already available and existing sensors and reaction elements previously deployed in particular domains.

Decision module in FNPS federated system is responsible for correlating network events in order to detect and recognize malicious events in the network. FNPS-DM consists of the following components (see Fig.2):

- Correlation Engine (e.g. based on the Borealis system),
- CLIPS rule engine,
- Ontology (in OWL format),
- Graphical User Interface.

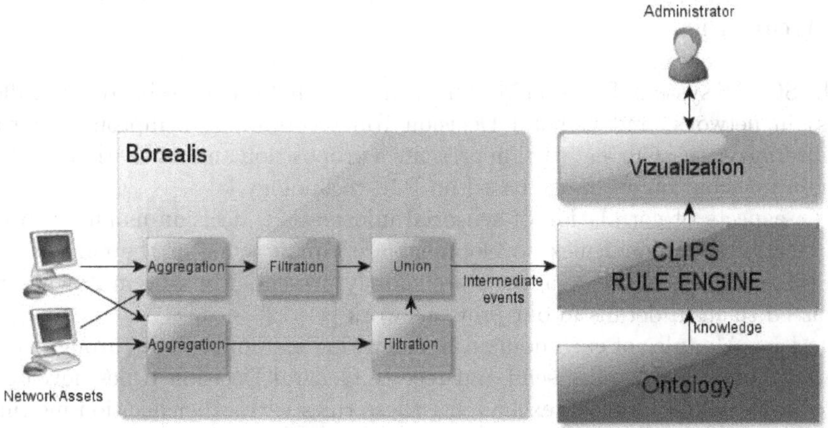

Fig. 2. Decision Module architecture and components

Borealis is a distributed stream processing engine and is responsible for gathering information generated by the network sensors [7]. Correlation engine has mechanism that allows the Decision Module to efficiently execute multiple queries over the data streams in order to perform event correlation. The result of a correlation process is an intermediate event that is further processed by CLIPS rule engine [8].

CLIPS uses ontology that describes broad range of network security aspects (we use "SOPAS ontology" developed in our project). CLIPS engine identifies whenever some attacks or malicious network events have been discovered. The information describing the network incident and reconfiguration procedures are sent to Translator (see Fig. 1). Moreover, detailed information in human readable format are generated and visualized to network administrators via FNPS-DM GUI.

Each Decision Module can react to network events and attacks by sending information to the Translator element. The output information from DMs is the General Reaction Rule describing attack symptoms (information about network events) and particular reaction rule to be applied by reaction elements.

Translator has the knowledge about its subnet capabilities and can access the necessary reaction elements (e.g. firewalls, filters or IDS). Reaction elements can be reconfigured by Translator in order to apply commands sent by the Decision Module.

All Decision Modules within the federation can also interact with each other and exchange security information. Particularly information about network incidents, like attack in one domain, may be sent to different Decisions Modules in order to block the attacker before the consequent attack takes place on another domain. Communication between domains and Decision Modules is based on P2P (Peer-to-Peer) in order to increase communication resiliency and enable data replication.

3 P2P Based Information Exchange between Federated Domains

In the SOPAS system, Decision Modules send information about incidents (called facts) in network, and General Decision Rules containing symptom and rule (countermeasure). Current section presents a proposition and realization of inter-domain information exchange based on P2P technology.

The aspects of correlation of sensorial information, decision-making process and General Decision Rules development are out of scope of the current section. More details on these aspects are shortly presented in section 2, while we discussed them in details in our previous work [11][12].

Decision Modules of each domain in federation are connected (through) P2P Overlay Network. Modules send and receive General Decision Rules developed in decision-making process, exchanging these rules with other decision modules in the federation [11]. Such solution allows not only for communication between Decision Modules, but also for replicating the data and paths.

The proposed P2P overlay network is dedicated only for a communication between domains in federation in order to minimize the impact on network traffic. Each Decision Module is a peer hosting and requesting data concerning federation security aspects. It is assumed that federation may have several public IP addresses where lightweight P2P application can be installed. This allows to multiply a number of routing points in a P2P overlay network. Moreover, it is possible that other machines (not only DMs) may also act as peers in FNPS overlay network.

The proposed approach allows the system to have redundant communication channels between Decision Modules. Particularly, when a physical connection is under attack or is congested, the communication packets still have a opportunity to reach the destination DM using a different path.

Peer-to-Peer (P2P) is a technology that provides distribution of resources (such as data, computational power, hard drive resources, etc.) between multiple machines (applications) with equal privileges. Individual element of P2P network (peer) shares a part of own resources to other participants of P2P communication. Each peer can be provider and recipient of information simultaneously. Systems using P2P technology create abstract network (overlay network) between nodes participating in communication, therefore such communication is independent from physical interconnection of P2P elements.

In P2P network, each peer provides network resources (bandwidth, disc capacity, computational power), thus overall capacity of P2P system increases when new clients join to network. Such characteristic of P2P distinguishes this approach from traditional client-server approaches in which new clients joining to the system reduce available server resources. Distributed architecture of P2P systems allows for reaching high reliability in terms of access to available resources, resulting from ability to dynamic change of path between pair of P2P communication participants.

In the case of malfunction of one or more elements of network, P2P system can still work properly, due to the fact that each peer participating in transmission is equally important. In traditional client-server approach, failure of server causes termination of services provided by system.

P2P technologies can be categorized into three groups, based on topology creation method:

1. Structured P2P,
2. Unstructured P2P,
3. Hybrid P2P.

Due to the specific nature of the SOPAS architecture (for Federation of Systems protection) that comprises of numerous equivalent domains, unstructured variant of P2P approach is proposed. Implementation of P2P in SOPAS Project is similar to Gnutella v0.4 system which is typical unstructured P2P solution. Such approach allows for easy and dynamic expanding the system of mutually connected domains. Such solution does not generate high network traffic related to signalization, because only small amounts of data are transmitted between Decision Modules. Implementation of unstructured P2P in the SOPAS Project can be additionally justified due to the fact that system protecting Federation of Networks uses bi-directional transmission (both downloading and uploading the data).

Initialization process of P2P client takes place in the main thread of program. Three parallel threads are activated after loading list of active hosts from WWW service (initialization through WWW server). Each new connection runs in separate thread in order to provide handling of multiple simultaneous requests. Current implementation of P2P communication specifies the following types of signaling packets:

- Ping-pong packet, to verify correctness of connection set (whether connect with appropriate P2P client).
- Packet containing list of active hosts, received from other P2P client.
- Packet requesting transmission of active hosts list, received from other P2P client.
- Packet containing General Decision Rules.

Current implementation of P2P communication specifies two types of hosts list:

- List of active hosts, including active connections.
- List of known hosts, including IP addresses, which can be possibly active.

List of known hosts is modified automatically by P2P client (known hosts are added). During the first launch, list contains only addresses obtained from the WWW server, during the further work the list expands with new recognized addresses and is stored in local memory. Each packet received through active connection is validated. Such validation includes:

- Verification of packet Time To Live (TTL packet)
- Detection of duplicates (timestamp packet) in time sample.

Communication assumes the existence of many equivalent P2P nodes, which can be both client and server mediating in the communication between other clients at any time. In such case, the attacker may relatively easy perform a series of actions that may disturb proper communication between Decision Modules. These may include:

- Eavesdropping
- Modification of content of packets transmitting between client A and client B (man-in-the-middle attack)
- Generation of false General Decision Rules and sending them to other Decision Modules (spoofing attack)
- Interception of packets and their destruction

Therefore, SSL security mechanism is proposed as a mean for protection from eavesdropping of the packets content and from threat of its modification. Domains which participate in communication must trust each other, thus must confirm their identity. Therefore the certificates is proposed. This requires maintenance of CA (Certificate Authority) within the SOPAS system. Generation of public key and signing it by the CA will be required from Decision Module that is newly connected to the SOPAS system.

The communication channels are protected by SSL encryption algorithms. Such approach allows to protect the communication from the packet sniffing (by third persons). Additionally, payload is encrypted with asymmetric keys and it can be only decrypted by domains that belong to the same distribution group (nodes relaying the message can not read the payload).

4 SQLIA Attack Detection Scenario with the P2P Information Exchange Mechanism

The SQLIA (SQL Injection Attack) scenario demonstrates how weakly protected domains can benefit from federation and sharing the information in order to increase their security level. SQLIA is ranked #1 in The Ten Most Critical Web Application Security Risks released by Open Web Application Security Project (OWASP) [10].

Simplified topology of the federation is presented in Fig. 3. It consists of 4 domains. Decision Modules (orange boxes) are deployed in each domain. There are also different services (green boxes) and sensors responsible for detecting particular type of attacks (blue boxes).

We assume that domains trust each other and cooperate in process of sharing the security information.

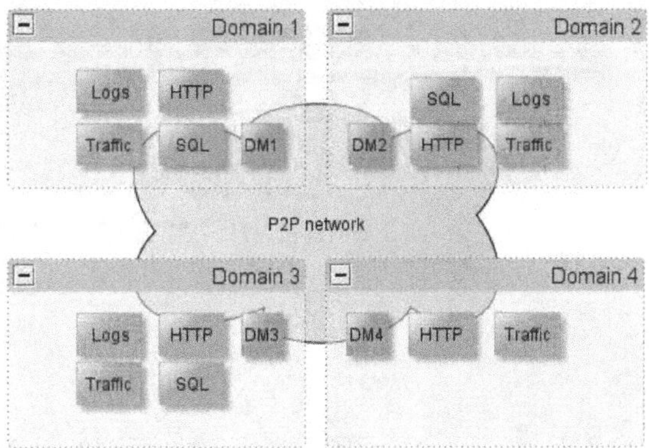

Fig. 3. Simplified topology diagram of the federation (orange box - decision module, green box - hosted services, blue box - installed sensors)

Firstly, the attacker scans the federation to find HTTP services. Particularly attacker aims at finding unsecured services running on unusual ports. The sensors in domain 2 and 3 detect port scanning and report this fact to their Decision Modules. This information is forwarded to domain 1 and 4. The source address of attacker is stored by Decision Modules as suspicious one.

The HTTP services discovered by attacker are further penetrated in order to find application flaws. The unusual traffic in HTTP service logs is spotted by sensors, but only in the domain 1. Moreover, the sensors in the domain 2 report an increasing number of failed and an untypical SQL queries to the database. Also the sensors in the domain 4 report an unusual and huge traffic (typically this domain is rarely visited). The Decision Modules exchange this information using P2P enabled communication.

At his moment, thanks to information exchange and sharing mechanism in federation, all the domains are able to analyze the complete set of information.

The ontology is engaged and the problem is inferred as the SQL Injection Attack (SQLIA)[12].

Each Decision Module works out a reaction appropriate to its policy stored in the ontology (each domain may have a different policy). The Decision Modules in the domains 1, 3 and 4 decide to inform the administrator about the attack, while the DM in domain 4 decides both to block the traffic (there are already some failed queries to database, suggesting that the flaw was spotted by the attacker) and to inform the administrator. The DM in domain 4 sends the request to the Translator to block the traffic coming from the attacker. Then the Translator sends an appropriate command to one of the Reaction Modules to block the attacker.

The screenshots presenting the Decision Modules in operation in the SQLIA detection use-case are shown in Figures 4-5.

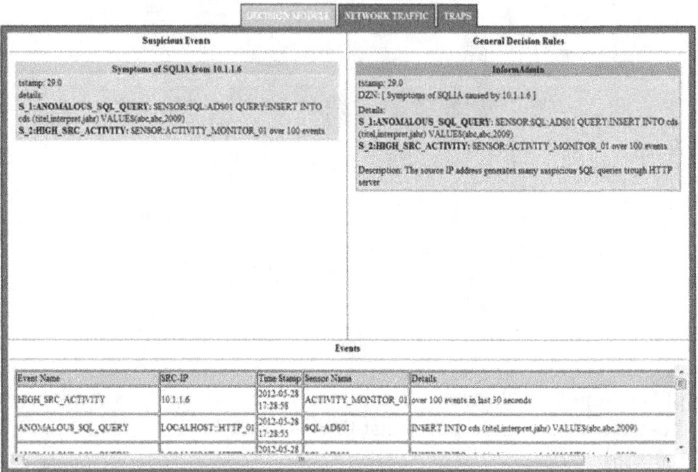

Fig. 4. View of the 1st Decision Module detecting local SQLIA symptoms and sending information (General Distribution Rule) to other Decision Modules in federation

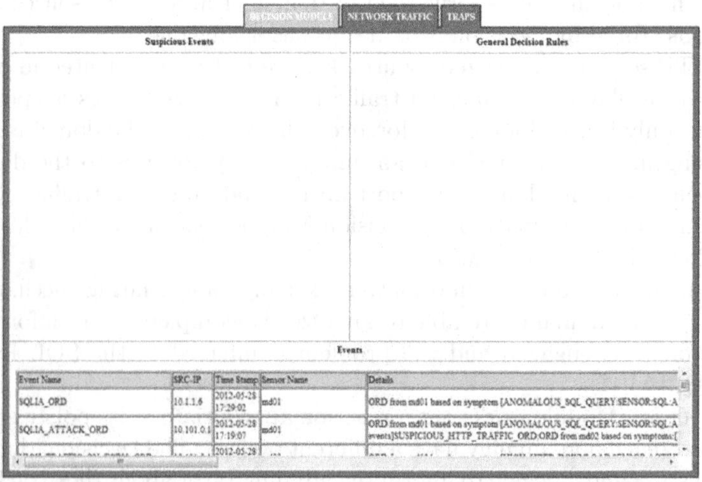

Fig. 5. View of the 2nd Decision Module receiving information (GDR) from Decision Module 1 (named md01). The received information is presented in "network events" field in the lower part of the GUI.

In the Fig. 4 the first Decision Module detects SQLIA symptoms (it detects local SQLIA symptoms listed in GUI field "network events" and creates General Decision Rule to block certain IP (upper right in the GDR GUI field)) and sends this information to another DM in federation using P2P based communication.

In Fig. 5 the second DM is shown. In the GUI field "network events" (lower part), there are 2 messages sent from the first DM termed as md01 (using P2P-based communication between them).

5 Conclusions

This paper presents the results of the national project SOPAS funded by Ministry of Science and Higher Education of Poland in the theme of homeland security.

The major contribution of this paper is the proposition and realization of the P2P-based information exchange mechanism between the federated domains in the Federated Networks Protection System (FNPS) system. The general architecture of the system, the details of the P2P-based communication and the use case showing the benefits of the proposed approach are presented.

The P2P-based communication of the Federated Networks Protection System was successfully tested at NATO CWIX 2012 (Coalition Warrior Interoperability eXercise) as the cyber defence capability.

Acknowledgement. This work was partially supported by Polish Ministry of Science and Higher Education funds allocated for the years 2010-2012 (Research Project number OR00012511).

References

1. Enabling and managing end-to-end resilience, ENISA (European Network and Information Security Agency) Report (January 2011)
2. Choraś, M., D'Antonio, S., Kozik, R., Holubowicz, W.: INTERSECTION Approach to Vulnerability Handling. In: Proc. of WEBIST 2010, vol. 1, pp. 171–174. INSTICC Press, Valencia (2010)
3. NATO Network Enabled Feasibility Study Volume II: Detailed Report Covering a Strategy and Roadmap for Realizing an NNEC Networking and Information Infrastructure (NII), version 2.0
4. El-Damhougy, H., Yousefizadeh, H., Lofquist, D., Sackman, R.: Crowley, Hierarchical and federated network management for tactical environments. In: Proc. of IEEE Military Communications Conference MILCOM, vol. 4, pp. 2062–2067 (2005)
5. Calo, S., Wood, D., Zerfos, P., Vyvyan, D., Dantressangle, P., Bent, G.: Technologies for Federation and Interoperation of Coalition Networks. In: Proc. of 12th International Conference on Information Fusion, Seattle (2009)
6. Coppolino, L., D'Antonio, L., Esposito, M., Romano, L.: Exploiting diversity and correlation to improve the performance of intrusion detection systems. In: Proc of IFIP/IEEE International Conference on Network and Service (2009)
7. Borealis project homepage, `http://www.cs.brown.edu/research/borealis/public/`

8. CLIPS project homepage, `http://clipsrules.sourceforge.net/`
9. Neches, R., Fikes, R., Finin, T., Gruber, T., Patil, R., Senator, T., Swartout, W.R.: Enabling Technology for Knowledge Sharing. AI Magazin 12(3), s.36–s.56 (1991)
10. OWASP Top Ten - 2010. The Ten Most Critical Web Application Security Risks. Published by Open Web Application Security Project (OWASP)
11. Choraś, M., Kozik, R., Piotrowski, R., Brzostek, J., Hołubowicz, W.: Network Events Correlation for Federated Networks Protection System. In: Abramowicz, W., Llorente, I.M., Surridge, M., Zisman, A., Vayssière, J. (eds.) ServiceWave 2011. LNCS, vol. 6994, pp. 100–111. Springer, Heidelberg (2011)
12. Choraś, M., Kozik, R.: Network Event Correlation and Semantic Reasoning for Federated Networks Protection System. In: Chaki, N., Cortesi, A. (eds.) CISIM 2011. CCIS, vol. 245, pp. 48–54. Springer, Heidelberg (2011)
13. Shrideep, P., et al.: A Framework for Secure End-to-End Delivery of Messages in Publish/Subscribe Systems. IEEE Computer Society, 215–222 (2006)
14. Beitollahi, H., Deconinck, G.: Analyzing the Chord Peer-to-Peer Network for Power Grid Applications (2008)
15. Ion, S., et al.: Chord: A scalable peer-to-peer lookup service for internet applications. In: SIGCOMM 2001: Proceedings of the 2001 Conference on Applications, Technologies, Architectures, and Protocols for Computer Communications, pp. 149–160. ACM (2001)
16. Deconinck, G., et al.: A Robust Semantic Overlay Network for Microgrid Control Applications. In: De Lemos, R., Di Giandomenico, F., Gacek, C., Muccini, H., Vieira, M. (eds.) Architecting Dependable Systems V, pp. 101–123 (2008)
17. Maymounkov, P., Mazieres, D., Kademlia: A Peer-to-Peer Information System Based on the XOR Metric, pp. 53–65. Springer (2002)
18. Cohen, E., Shenker, S.: Replication Strategies in Unstructured Peer-to-Peer Networks. In: SIGCOMM 2002: Proceedings of the 2002 Conference on Applications, Technologies, Architectures, and Protocols for computer Communications, vol. 32, pp. 177–190. ACM (2002)
19. Antony, I., Druschel, P.: Pastry: Scalable, Decentralized Object Location, and Routing for Large-Scale Peer-to-Peer Systems, pp. 329–350. Springer (2001)

Web Spam Detection Using MapReduce Approach to Collective Classification

Wojciech Indyk[1], Tomasz Kajdanowicz[1], Przemyslaw Kazienko[1], and Slawomir Plamowski[1]

Wroclaw University of Technology, Wroclaw, Poland
Faculty of Computer Science and Management
{wojciech.indyk,tomasz.kajdanowicz,kazienko,
slawomir.plamowski}@pwr.wroc.pl

Abstract. The web spam detection problem was considered in the paper. Based on interconnected spam and no-spam hosts a collective classification approach based on label propagation is aimed at discovering the spam hosts. Each host is represented as network node and links between hosts constitute network's edges. The proposed method provides reasonable results and is able to compute large data as is settled in MapReduce programming model.

Keywords: MapReduce, collective classification, classification in networks, label propagation, web spam detection.

1 Introduction

Recently networks have become one of commonly used model for representation of relations among objects. The most natural example of networked structure are hosts interconnected by hyperlinks placed in web pages deployed on hosts. Utilizing the information of network's structure it is possible to classify nodes. This means that based on partial labelling of the nodes in the network it is possible to discover labels of the rest of nodes.

Nodes may be classified in networks either by inference based on known profiles of these nodes (regular concept of classification based on attributes of nodes) or based on relational information derived from the network. This second approach utilizes information about connections between nodes (structure of the network) and can be very useful in assigning labels to the nodes being classified. For example, it is very likely that a given web page x is related to sport (label *sport*), if x is linked by many other web pages about sport.

Hence, a form of collective classification should be provided, with simultaneous decision making on every node's label rather than classifying each node separately. Such approach allows taking into account correlations between connected nodes, which deliver usually undervalued knowledge.

Moreover, arising trend of data explosion in transactional systems requires more sophisticated methods in order to analyse enormous amount of data. There is a huge need to process big data in parallel, especially in complex analysis like collective classification.

Á. Herrero et al. (Eds.): Int. JointConf. CISIS'12-ICEUTE'12-SOCO'12, AISC 189, pp. 197–206.
springerlink.com

MapReduce approach to collective classification which is able to perform processing on huge data is proposed and examined in the paper in order to deal with spam data. The collective classification algorithm, revoked in this paper has been introduced in [1] in domain of telecommunication customers' classification. In this paper we examine its abilities to web spam detection.

Section 2 covers related work while in Section 3 appears a proposal of MapReduce approach to label propagation in the network, that is able to perform a classification of spam hosts. Section 4, contain description of the experimental setup and obtained results. The paper is concluded in Section 5.

2 Related Work

Web spam detection problems may be solved using numerous approaches. Previous work on web spam detection focus on two distinct subproblems, namely *(i)* content spam detection (originating from email spam detection) [2] and *(ii)* link spam detection [3, 4, 5].

2.1 Content Spam

Content spam relies on creating web pages containing keywords, that are generic and more related to most of the queries, than to the actual page content. Such malicious behaviour affects the outcome of search engines, such as PAGERANK [6]. To overcome this problem traditional classifiers have been incorporated [2]. On the other hand, more sophisticated solution using language model disagreement has been introduced in [7].

2.2 Link Spam

The latter issue also affects the outcome of web ranking algorithms. According to [8] link spamming refers to 'manipulation of the link structure by a group of users with the intent of improving the rating of one or more users in the group'. Several solutions has been suggested to overcome the issue. For some of them, propagation of some kind of score underlies search engines performance improvement, while others rely on machine learning concepts. In [9] a few techniques has been gathered. From among the most common TRUSTRANK [10, 12], BADRANK [11] and SPAMRANK [13] can be distinguished. More sophisticated approach is introduced to in [14]. Propagation of trust and distrust through web links has been incorporated to identify the potential synergistic gains for pages interconnected in a spam farm. In [9] quite different approach is presented. In order to improve search engine performance, noisy links are removed at site level, instead of single page level. Another solution relies on statistical analysis using such properties as linkage structure, page content and page evolution [15]. An assumption that certain classes of spam pages diverge in some of their properties from remaining web pages underlies this method. From among supervised learning methods, classification is the most obvious one. Several different classifiers have already been tested in the domain of web spam detection. In [16] simple binary trees have been incorporated, while [17] used *SVM* models based on local and link-based features.

2.3 Joining Link-Based and Content-Based Features

Utilizing both content-based and link-based features facilitates thorough web analysis [18]. Traditional pattern recognition methods assume independence between objects/record/instances. In case of web spam detection, however, this is not necessarily always true, as there exist some dependency among web pages and web hosts. Evidently, links between pages and hosts are not distributed randomly. Similar and dissimilar pages are often linked together. Nonetheless, content resemblance is frequent situation according to [5].

Moreover, using the topology of the web graph by exploiting dependencies between pages, boosts inference properties of intelligent system. It has been shown that linked pages or hosts tend to share the same class label, provided that there exist considerable number of such connections [18]. Intuitively, both spam and non spam hosts are likely to be linked together.

Another issue that has to be addressed in order to obtain efficient and scalable inference system, is the size of analysed web graph. As the Internet has grown to unprecedented scale, processing each page separately has become unsatisfactory. To overcome this obstacle, web page analysis has been replaced by host analysis [18, 9]. Such solution is based on the assumption that spam and non spam pages originate form the same web hosts.

Incorporating web graph topology to the inference system may be done in several manners, as described in [18]. Possible approaches embrace the following methods: *(i)* clustering the host graph *(ii)* label propagation to neighbouring hosts which reinforces inference properties of ensemble classifiers *(iii)* using neighbouring labels as new features describing each node

3 Collective Classification by Means of Label Propagation Using MapReduce

The most common way to utilize the information of labelled and unlabelled data is to construct a graph from data and perform a Markov random walk on it. The idea of Markov random walk has been used multiple times, e.g. [19, 20, 21], and involves defining a probability distribution over the labels for each node in the graph. In case of labelled nodes, the distribution reflects the true labels. The aim then is to recover this distribution for the unlabelled nodes. Utilizing Label Propagation approach allows performing classification based on relational data.

Let $G(V, E, W)$ denote a graph with vertices V, edges E and an $n \times n$ edge weight matrix W. In a weighted graph G(V,E,W) with $n = |V|$ vertices, label propagation may be solved by linear equations 1 and 2 [20].

$$\forall i,j \in V \sum_{(i,j)\in E} w_{ij} F_i = \sum_{(i,j)\in E} w_{ij} F_j \tag{1}$$

$$\forall i \in V \sum_{c\in classes(i)} F_i = 1, \tag{2}$$

where F_i denotes the probability density of classes for node i.

Assuming that V_L denotes labelled vertices and V_U – unlabelled ones, such that $V = V_L \cup V_U$, let F_u denote the probability distribution over the labels associated with each vertex $u \in V$. For each node $v \in V_L$, for which F_v is known, a dummy node v' is inserted such that $w_{vv'} = 1$ and $F_{v'} = F_v$, which resembles the 'clamping' operation, discussed in [20]. Let's further assume that V_D denotes set of dummy nodes. The, considering the above, the solution to equations 1 and 2 can be obtained using Iterative Label Propagation algorithm 3.

Algorithm 1. The pseudo code of Iterative Label Propagation algorithm

repeat
  for all $v \in V$ **do**
    $F_v = \frac{\sum_{(u,v)\in E} w_{uv} F_u}{\sum_{(u,v)} w_{uv}}$
  end for
until convergence

Using some local information, namely node's neighbourhood, each node is processed inside the loop. As far as web graph is considered, however, sequential evaluation leads to severe lack of effectiveness. Therefore a parallel version of the algorithm is presented in algorithm 2.

Algorithm 2. The pseudo code of MapReduce approach to Iterative Label Propagation algorithm

$map < node; adjacencyList >$
for all $n \in adjacencyList$ **do**
  propagate$< n; node.label, n.weight >$
end for

$reduce < n; list(node.label, weight) >$
propagate$< n; \frac{\sum node.label \cdot weigth}{\sum weight} >$

MapReduce version of Iterative Label Propagation algorithm comprises two phases. The Map phase fetches all labelled and dummy nodes and propagate their labels to all nodes in adjacency list, taking into account edge weights between nodes. During the Reduce phase new label for each node with at least one labelled neighbour is calculated. Reducers obtain new label for nodes using the list of labelled neighbours and relation strength between nodes (weight). The final result, that is a new label for a particular node, is computed as weighted sum of labels' probabilities from neighbourhood.

4 Experiments and Results

In the experimental studies we aimed at checking the usability of proposed collective classification method to spam detecion. Thus, due to relational structure of internet infrastructure, collective classification for web spam data has been incorporated. The Webspam-UK2007[1] was chosen for experiments. The dataset consists of networked data of 114 529 interconnected hosts in the .UK domain. Part of the set was labelled as 344 spam and 5709 no-spam hosts. Remaining hosts serve as a tool to the influence propagation. Connections between hosts were represented by links between websites of the hosts. Strengths of links were calculated according to the following equation

$$e(a,b) = \frac{c(a,b)}{\sum_{x \in V} c(a,x)}, \tag{3}$$

where V is the set of hosts, $e(a,b)$ represents the strength link from a to b, $c(a,b)$ – the number of connections from host a to host b.

Presented in equation 3 formula for strength calculation of links implies that it is normalized, namely

$$\forall_{b:(a,b)\in E} \sum_{x \in V} e(x,b) = 1,$$

which is important for dummy node influence (see section 3 and [1]).

For the purpose of experiments, an original, provided from the dataset owners1 split into training and testing dataset was remained. This enabled the possibility to compare experimental results of proposed method with the winning method from Web Spam Challenge organized within Fourth International Workshop on Adversarial Information Retrieval on the Web on AIRWeb 2008 Conference. The reference method used for comparison has been published in [22]. The training set consisted of 3776 no-spam and 222 spam hosts, while the test set of 1933 no-spam and 122 spam hosts. According to node outdegree distribution of training and test set presented in Figure 4 we can conclude that original split into training and testing sets remained similar class distributions for both data. During preprocessing phase, directions of links were reverted due to the meaning of influence relation. Influence relation is directed opposite to link relation, see Figure 4.

As it can be noticed, some hosts do not have links to others, namely are separated. Therefore, an influence cannot be propagated and the hosts do not have any label assigned by the method. During the experiments it was assured that all hosts in test set are not separated and if so, the results of evaluation were consistent.

---

[1] Yahoo! Research: "Web Spam Collections". `http://barcelona.research.yahoo.net/webspam/datasets/` Crawled by the Laboratory of Web Algorithmics, University of Milan, `http://law.dsi.unimi.it/`.

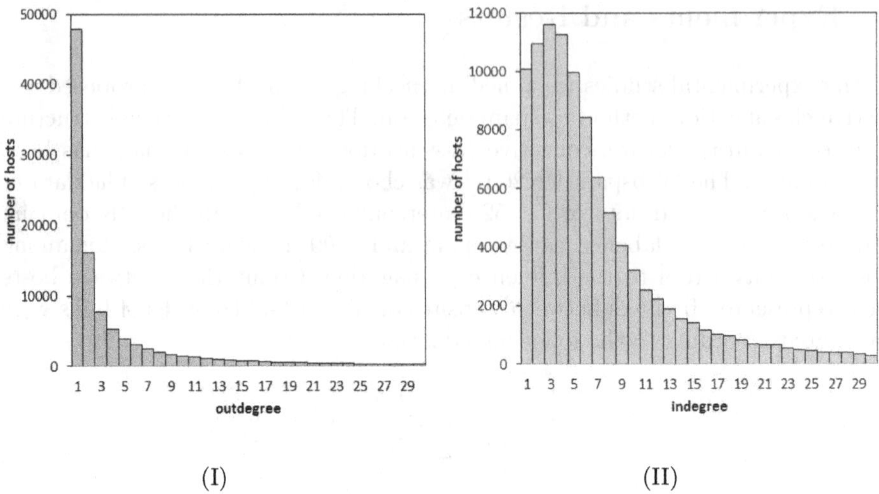

Fig. 1. The network's outdegree distribution (I) and indegree distribution (II)

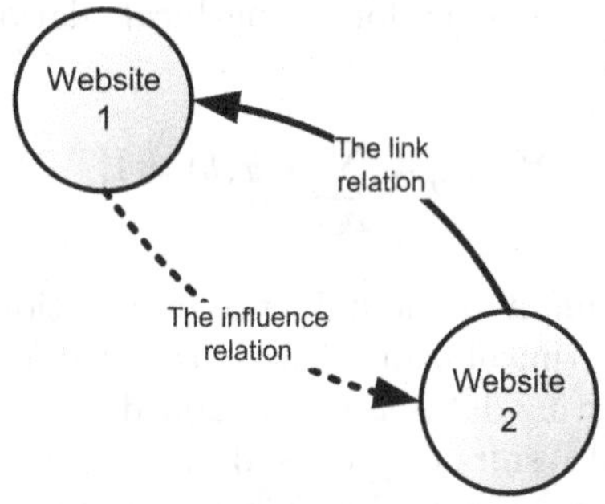

Fig. 2. Relation between the influence and link direction: a link from Website 2 (i.e. your homepage) to Website 1 (i.e. the Wikipedia) means that Website 1 influences Website 2

During the experiments, the influence propagation accuracy has been examined in distinct number of algorithm iterations. Each iteration is able to propagate labels one link further from starting training nodes. In the given data, after just four iterations, all nodes have been labelled.

In the examined data set we can observe a strong skewness in spam and no-spam hosts distribution. This caused a difficulty for proposed propagation algorithm. Thus the gathered results shows that the best accuracy is achieved in the very first iteration of algorithm. According to [18, 9] the spam hosts aggregates in close neighbourhood. Iterating more in propagation algorithm makes the no-spam labels propagated deeper in network structure and injects undesirable influence in spam agglomerations. Finally, after multiple iterations of label propagation we observe no-spam labels overwhelmed a structure of the whole network.

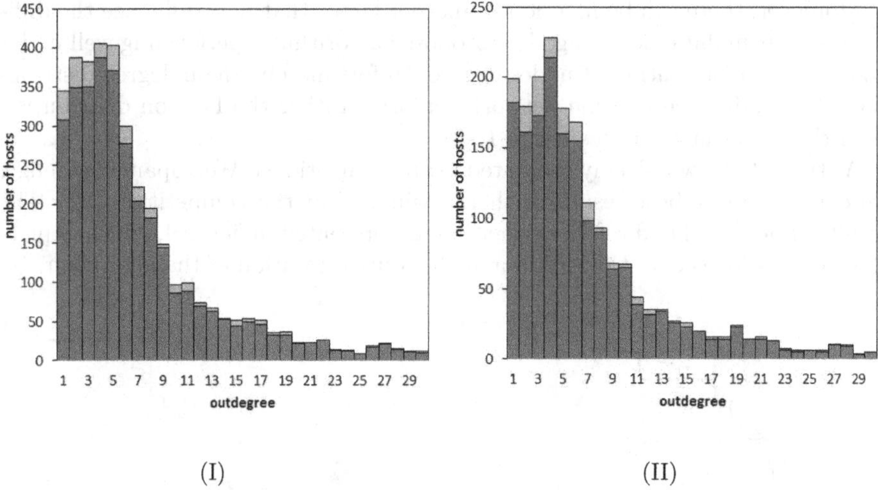

Fig. 3. Training (I) and test (II) set outdegree distribution

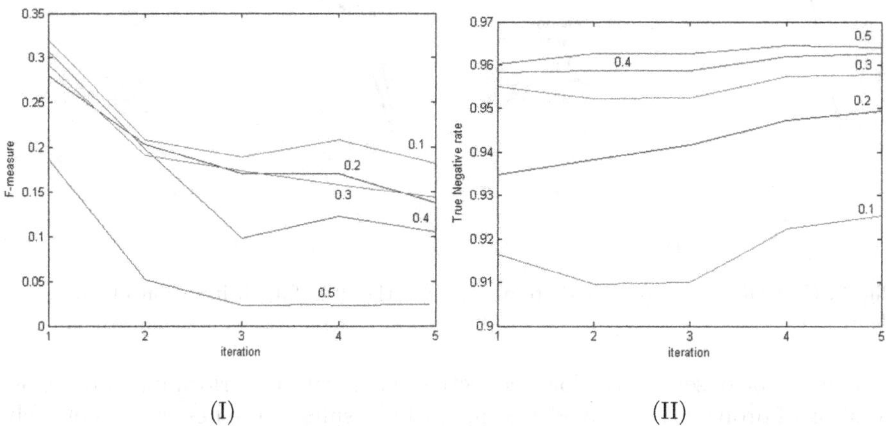

Fig. 4. (I) F-measure for consecutive threshold of probability of spam label. (II) True negative rate for identically parameters as (I).

However, we tried to overcome the unbalanced data set problem by modelling the threshold, that is used to obtain the decision of label assignment. We examined 5 threshold values (from 0.1 to 0.5). The final class was assigned according to a Equation 4.

$$class(host) = \begin{cases} spam \, , & p(host) \geq threshold \\ no - spam \, , & p(host) < threshold, \end{cases} \tag{4}$$

where $class(x)$ is a result of the algorithm for host x.

As we can see in Figure 4, the best result of F-measure was achieved for threshold equal to 0.1.

Moreover, there can be mentioned another factor that may influence the difficulties in spam data modelling. The proposed algorithm is performing well under "power law" distribution of nodes degree. Unfortunately, the indegree distribution of the influence relation network conforms rather the Poisson distribution than the power law one, see figure 1.

As the dataset was deeply evaluated in above mentioned Web Spam Challenge, we compared our best results with the winners' of the competition [22]. The results of proposed and reference method are presented in figure 4, which depicts ROC curves for the first iteration and the fourth iteration of the algorithm.

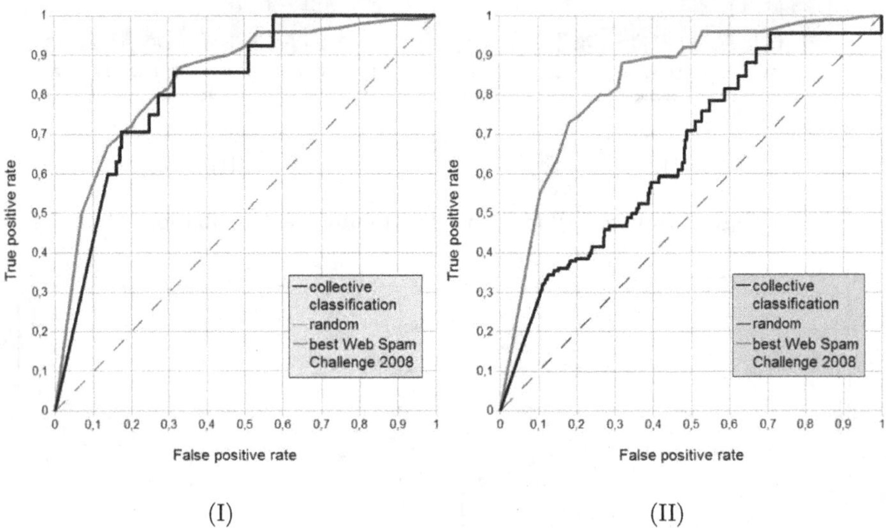

Fig. 5. (I) ROC curve after first iteration and (II) after fourth iteration of proposed algorithm

As it can be observed, the local collective classification, performing only single iteration of propagation, provides comparative results to the best results of Web Spam Challenge. Moreover, comparing to reference method it does not require additional description (attributes) of hosts that are required by reference method in order to classify hosts.

5 Conclusions

The collective classification method based on label propagation was proposed for spam hosts evaluation. The performed experiments as well data analysis revealed that the approach can be applied to the spam data problem, however it requires specific adjustments in configuration.

Further experimentation will consider a comparison of algorithm's behaviour for various datasets distributions (including numerous domains). Moreover, the model of web-spam classification will be improved by utilizing clustering algorithms for spam farms detection.

Acknowledgement. This work was partially supported by The Polish National Center of Science the research project 2011-2012, 2011-2014 and Fellowship co-financed by The European Union within The European Social Fund. Calculations have been carried out in Wroclaw Centre for Networking and Supercomputing (http://www.wcss.wroc.pl), grant No. 177. The authors are grateful for granting access to the computing infrastructure.

References

[1] Indyk, W., Kajdanowicz, T., Kazienko, P., Plamowski, S.: MapReduce approach to collective classification for networks. In: Rutkowski, L., Korytkowski, M., Scherer, R., Tadeusiewicz, R., Zadeh, L.A., Zurada, J.M. (eds.) ICAISC 2012, Part I. LNCS, vol. 7267, pp. 656–663. Springer, Heidelberg (2012)

[2] Ntoulas, A., Najork, M., Manasse, M., Fetterly, D.: Detecting spam web pages through content analysis. In: Proceedings of the 15th International Conference on World Wide Web, WWW 2006, pp. 83–92. ACM, New York (2006)

[3] Gyongyi, Z., Garcia-Molina, H.: Web spam taxonomy. Technical Report 2004-25, Stanford InfoLab (March 2004)

[4] Drost, I., Scheffer, T.: Thwarting the Nigritude Ultramarine: Learning to Identify Link Spam. In: Gama, J., Camacho, R., Brazdil, P.B., Jorge, A.M., Torgo, L. (eds.) ECML 2005. LNCS (LNAI), vol. 3720, pp. 96–107. Springer, Heidelberg (2005)

[5] Davison, B.D.: Recognizing nepotistic links on the web. In: AAAI 2000 Workshop on Artificial Intelligence for Web Search, pp. 23–28. AAAI Press (2000)

[6] Page, L., Brin, S., Motwani, R., Winograd, T.: The pagerank citation ranking: Bringing order to the web (1999)

[7] Mishne, G.: Blocking blog spam with language model disagreement. In: Proceedings of the First International Workshop on Adversarial Information Retrieval on the Web (AIRWeb 2005) (2005)

[8] Zhang, H., Goel, A., Govindan, R., Mason, K., Van Roy, B.: Making Eigenvector-Based Reputation Systems Robust to Collusion. In: Leonardi, S. (ed.) WAW 2004. LNCS, vol. 3243, pp. 92–104. Springer, Heidelberg (2004)

[9] da Costa Carvalho, A.L., Chirita, P.A., Carvalho, C., Calado, P., Alex, P., Chirita, R., Moura, E.S.D., Nejdl, W.: Site level noise removal for search engines. In: Proc. of International World Wide Web Conference (WWW), pp. 73–82. ACM Press (2006)

[10] Gyngyi, Z., Garcia-molina, H., Pedersen, J.: Combating web spam with trustrank. In: VLDB, pp. 576–587. Morgan Kaufmann (2004)

[11] Wu, B., Davison, B.D.: Identifying link farm spam pages. In: Proceedings of the 14th International World Wide Web Conference, pp. 820–829. ACM Press (2005)

[12] Wu, B., Goel, V., Davison, B.D.: Topical trustrank: using topicality to combat web spam (2006)

[13] Benczur, A.A., Csalogany, K., Sarlos, T., Uher, M.: Spamrank - fully automatic link spam detection. In: Proceedings of the First International Workshop on Adversarial Information Retrieval on the Web, AIRWeb (2005)

[14] Gyngyi, Z., Garcia-molina, H.: Link spam alliances. In: Proceedings of the 31st International Conference on Very Large Data Bases (VLDB), pp. 517–528 (2005)

[15] Fetterly, D., Manasse, M., Najork, M.: Spam, damn spam, and statistics: using statistical analysis to locate spam web pages. In: Proceedings of the 7th International Workshop on the Web and Databases: colocated with ACM SIGMOD/PODS 2004, WebDB 2004, pp. 1–6. ACM, New York (2004)
[16] Becchetti, L., Castillo, C., Donato, D., Leonardi, S., Baeza-Yates, R.: Link-based characterization and detection of web spam. In: Proceedings of the 2nd International Workshop on Adversarial Information Retrieval on the Web, AIRWeb (2006)
[17] Kolari, P., Java, A., Finin, T., Oates, T., Joshi, A.: Detecting spam blogs: A machine learning approach. In: 2006 Proceedings of the 21st National Conference on Artificial Intelligence, AAAI (2006)
[18] Castillo, C., Donato, D., Gionis, A., Murdock, V., Silvestri, F.: Know your neighbors: web spam detection using the web topology. In: Proceedings of the 30th Annual International ACM SIGIR Conference on Research and Development in Information Retrieval, SIGIR 2007, pp. 423–430. ACM, New York (2007)
[19] Szummer, M., Jaakkola, T.: Clustering and efficient use of unlabeled examples. In: Proceedings of Neural Information Processing Systems, NIPS (2001)
[20] Zhu, X., Ghahramani, Z., Lafferty, J.: Semi-supervised learning using gaussian fields and harmonic functions. In: Proceedings of the International Conference on Machine Learning, ICML (2003)
[21] Azran, A.: The rendezvous algorithm: Multiclass semi-supervised learning with markov random walks. In: Proceedings of the International Conference on Machine Learning, ICML (2007)
[22] Geng, G., Li, Q., Zhang, X.: Link based small sample learning for web spam detection. In: Proceedings of the 18th International Conference on World Wide Web, WWW 2009, pp. 1185–1186 (2009)

Analysis of Selected Aspects of "IBM I" Security

Maciej Piec[1], Iwona Pozniak-Koszalka[2], and Mariusz Koziol[2]

[1] Vienna School of Informatics,
Vienna University of Technology, Vienna, Austria
e1128446@student.tuwien.ac.at
[2] Department of Systems and Computer Networks,
Wroclaw University of Technology, Wroclaw, Poland
{iwona.pozniak-koszalka,mariusz.koziol}@pwr.wroc.pl

Abstract. The article focuses on analyzing influence of IBM i security mechanisms in network layer on general system performance. Impact of system's security mechanisms on both system and network resources are presented. Additionally, special security metrics is presented to enable comparison of security level between different system configurations. General guidelines for using security mechanisms of IBM i are given. The paper concludes that because of deep integration of security mechanisms, no major impact on the overall system performance was observed. Propositions of further research are also given.

Keywords: IBM System i, security, security metrics, network security.

1 Introduction

Nowadays it is hard to imagine any business working without computer system. One of the most widely known system famous for its security is IBM i [5]. In the age of the Internet and big competition on the market, the security of the computer system and the information stored on it, are the key assets for companies [1]. Unfortunately, the unified and widely-accepted definition of security does not exist [2]. The consequence of that fact is that it is hard to create security metrics that would allow comparing systems by different vendors.

IBM i. IBM System i is computer system designed to work in business oriented environment. It is characterized, among other features, by the high levels of automation built into the operating system and reinforced by advanced autonomic capabilities. That means the application of artificial intelligence technologies to administration and optimization tasks. It was one of the main research objectives of IBM during early 90's. Four categories of System i autonomic functions can be distinguished: self-configuring, self-optimizing, self-protecting, and self-optimizing [11]. Extensive implementations of autonomic technologies caused that since 1990s IBM is the clear industry leader in this area. Since the very beginning one of the most important objectives for the system was its security. Unique solution applied in IBM i is based on integrating security and other system components into an operating system. [8].Such a solution provides complete and efficient security on all layers of an operating system. The security of the IBM i can be discussed in three layers – system, network and application [9].

Á. Herrero et al. (Eds.): Int. JointConf. CISIS'12-ICEUTE'12-SOCO'12, AISC 189, pp. 207–214.
springerlink.com

Security of computer systems. Computer systems security is a term that does not have one widely accepted definition. In existing papers, security is discussed in different ways and using different methods. That results in the lack of the complete, unified, widely used security metrics that would be used to evaluate computer systems. These facts may bring confusion to one who wants to compare security level offered by different security system vendors or even the same system in different configurations [3].

This paper introduces a security metrics that allows comparison of different configurations of IBM i system and evaluates overload of security mechanisms of a system in network layer.

The rest of the paper is organized in the following way: In Section 2 some aspects of IBM i system security are introduced in more detailed way and a new security metrics is defined. Experimentation system and applications used for testing are presented in Section 3. Section 4 describes in details experiments that were conducted. Conclusions and further work are summarized in Section 5.

2 Selected Aspects of IBM i Security

The aspect that was chosen for the deep analysis is the influence of the security mechanisms in network layer on the IBM i performance. Performance includes both system resources utilization and network traffic delays caused by the mechanisms. Following are the security mechanisms that offered in network layer and can be easily turn on and off: Intrusion Detection System (IDS), Network Address Translation (NAT), Packet Filtering and Journaling.

As mentioned earlier, there is no unified definition of the security. Basing on [4], the following definition is proposed and used within the paper. Security is defined by four security functions:

- Confidentiality (protection against unauthorized disclosure, limiting access to confidential information, protection against unauthorized users and other people).
- Integrity (protection against unauthorized data changes, ensuring, that data is only processed by authorized programs using defined interfaces, ensuring the reliability of the data).
- Availability (protection against random data changes or destroying data, protection against attempts of stopping or destroying system resources by unauthorized people, protection against disturbing proper working of the services in the system).
- Audit, logging and journaling (possibility of recreating the actions of the user by authorized user, possibility of auditing current state of the security).

To be able to see the influence of the security mechanisms in network layer on the general security of the system the security metrics based on the security defined was created. Different authors were defining security metrics on different levels and taking in to consideration numerous factors [6][7][10].

Security of the system is defined as number value defined based on combined security functions. For each security function the value of the security for that function is calculated according to equation (1).

$$f_i = \frac{l_a}{l_{\max}} \tag{1}$$

Where f_i - security value for security function *I*, l_a - number of activated components providing security function *i*, l_{max} - maximum number of components providing security function *i*.

System security for each security layer is calculated by formula (2).

$$F_w = \frac{\sum_{i=0}^{n} f_i}{n} \tag{2}$$

Where F_w - security of the layer, *n* - number of security functions, f_i - security value of security function *i*.

Security of the whole system is defined as sum of the security values for all layers. Formula is presented on equation 3.

$$F = F_{sys} + F_{net} + F_{app} \tag{3}$$

Where *F* - security of the whole system, F_{sys} - security in system layer, F_{net} - security in network layer, F_{app} - security in application layer.

Security mechanisms in network layer are presented in Table 1. Value "1" means, that the component has influence on the security function, while "0" means that the component doesn't affect the security function.

Table 1. Security in network layer

Security components	Confidentiality	Integrity	Availability	Audit
IDS	1	1	1	1
NAT	1	0	0	0
IP filtering	1	1	1	0
Journaling	0	1	1	1
maximum for the security function:	3	3	3	2

3 Experimentation System

In order to analyze impact of security mechanisms on system performance, experimentation system was created. It consists of two applications written in Java programming language.

Testing application. Testing application was created to indirectly measure resource utilization. The application is launched on the IBM i system that is being tested. It reads data from file located in a network and performs a number of mathematical operations. Data read time and processing time are measured. Differences of those parameters values are used to draw conclusions about consumption of the resources in tested system configurations.

Ultimate test manager. Ultimate test manager application is an application that is used to automate tests on IBM i platform. It enables user to select the program, set number of execution parameters, download and display data generated by test program. It also includes the panel to let the user simulate additional network traffic.

Testing methodology. The system was tested in different configurations of the security mechanisms and with different additional network traffic generated. Additional network traffic was used to simulate increased network traffic. Five frequencies of packet sending were set: No (no additional packets are sent), Small (packet is sent every 2000 ms), Medium (packet is sent every 200 ms), Big (packet is sent every 20 ms), Very big (packet is sent every 2 ms).

Each configuration was tested 10 times and arithmetical average was taken as a result for the tested configuration. The summary result for the experiment was calculated as an arithmetical average of all results in the experiment (for all tested configuration settings and all additional network traffic values).

Testing environment. All the experiments were conducted on IBM i V6R1 on the platform IBM Power 520. Testing application was launched in dedicated subsystem. The partition was given constant CPU computing power – 0,2 unit and constant available memory size. The file with data to process is located in local network to minimize the influence of other network traffic on the data read wait time.

Experiment Design. Five experiments were conducted to investigate the influence of security mechanisms on the system performance. Two values were measured in each experiment – execution time and wait time. Execution time is time of execution of whole test application. Wait delay is period of time needed by application to read next value from the file to process. Parameters that were changed within the experiment were additional network traffic level and settings specific for the tested mechanism (for example number of IP filter rules when testing IP filtering etc.)

4 Investigations

The series of experiments was conducted to investigate the impact of the security mechanism on the system performance. Experiment 1 was the reference for other experiments. Experiments 2 and 3 tested just one mechanism at the time. Situations where more than one mechanism turned on was tested in experiments 4 and 5. Table 2 shows the configuration of the system during experiments.

Experiment 1 – Unsecured system. In that experiment no security mechanism was turned on. The point of that experiment was to set the reference point to other experiments. The only changing parameter was additional network traffic. The results of the experiment are as follows: Security of the layer = 0; Average execution time [ms] = 12112,82; Average waiting time [ms] = 17864,35.

Table 2. System configuration during experiments

Security components:	Experiment 1	Experiment 2	Experiment 3	Experiment 4	Experiment 5
IDS	off	on	off	off	on
NAT	off	Off	off	off	off
IP Filtering	off	Off	on	on	on
Journaling	off	Off	off	on	on
Security of the layer	0	0,25	0,38	0,38	0,75

Experiment 2 – Intrusion detection system. In that experiment the only mechanism turned on was Intrusion Detection System. The test was conducted for all defined values of additional network traffic and for different number of IDS policies. Numbers of policies that were tested were 10, 25, 50 and 100. Additionally situation when no policy is defined but the IDS is running was considered, because IDS affects Quality of Service services of the system as well.

The defined policies affected all IP Addresses and all ports. The dynamic throttling mechanism, which is available in that IDS implementation wasn't turn on. There was no general trend or dependency observed in Experiment 2. The results of the experiment are as follows: Security of the layer = 0,38; Average execution time [ms] = 14166,27; Average waiting time [ms] = 17883,20.

Experiment 3 – IP filtering. In experiment 3 the only mechanism turned on was IP filtering. The tests were conducted for all defined values of additional network traffic and for selected numbers of IP filtering rules. Numbers of rules that were tested were 10, 25, 50 and 100. The results of the experiment are as follows: Security of the layer = 0,25; Average execution time [ms] = 14354,14; Average waiting time [ms] = 19028,79.

The most interesting dependencies are graphically presented on Fig.1. It can be noticed that in presence of big amount of additional network traffic the execution time increases significantly. It was also observed that in case of very big amount of additional network traffic the execution time is much longer than for the other values of the network traffic.

Experiment 4 – IP filtering and journaling. In that experiment more than one mechanism was turned on. IP filtering and journaling were combined. All levels of additional network traffic were tested for 10, 25, 50 and 100 IPF rules with journaling enabled. Results of that experiment are presented in Fig.2. The results of the experiment are as follows: Security of the layer = 0,38;Average execution time [ms] = 14435,37; Average waiting time [ms] = 19063,44.

It can be noticed that the bigger number of IPF rules is defined and the bigger amount of additional data is sent to the system the longer time is needed to finish execution of test application. It shows that both factors have an impact on the system efficiency.

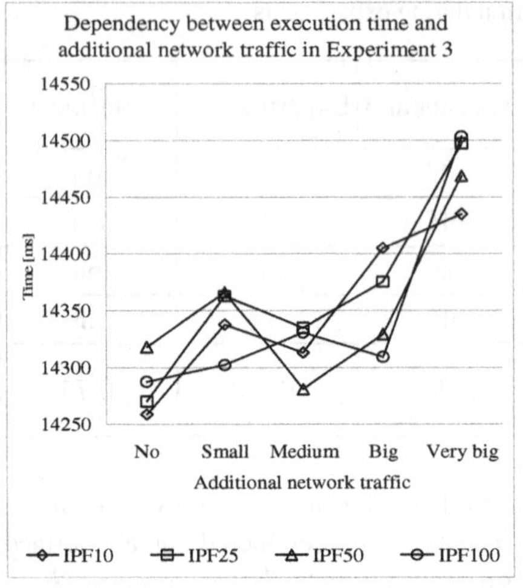

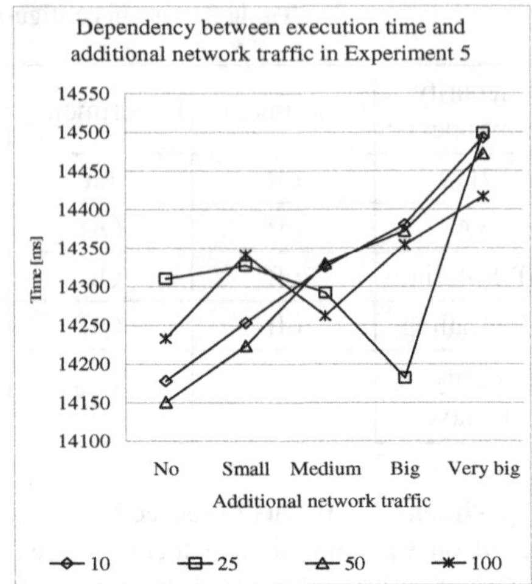

Fig. 1. Dependencies observed in experiments 3 and 5

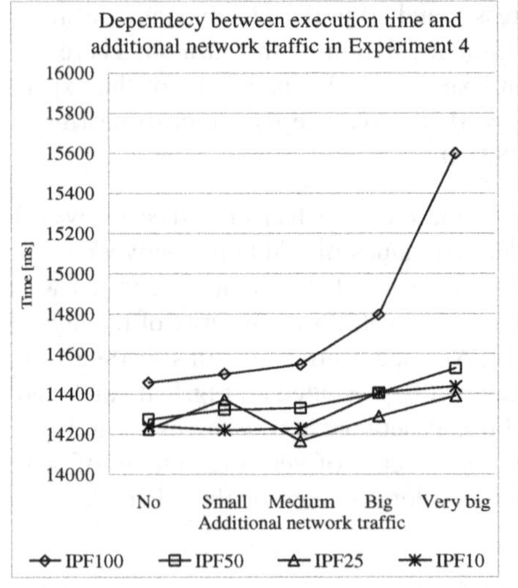

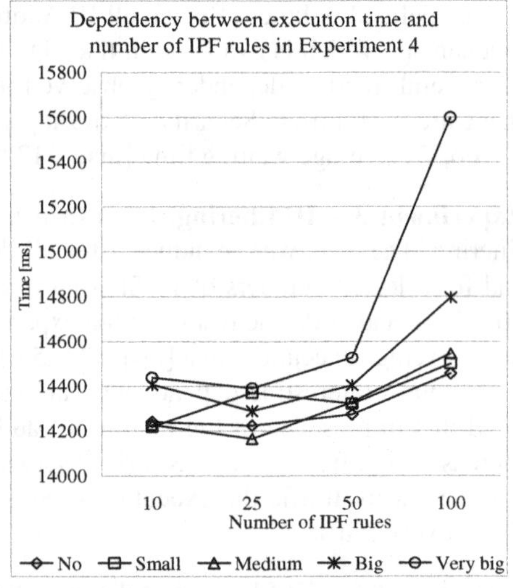

Fig. 2. Dependencies in Experiment 4

Experiment 5 – intrusion detection system, IP filtering and journaling. In experiment 5, IDS, IPF and journaling were turned on. The test was conducted for the same number of IDS and IPF rules. The values tested were: 10, 25, 50, and 100 (it means that for example for value 10 there were 10 IDS policies and 10 IPF rules defined). The results are as follows: Security of the layer = 0,75; Average execution time [ms] = 14319,56; Average waiting time [ms] = 18855,72.

Dependency observed within the experiment is shown in Fig.2. It can be noticed that there is increasing trend in execution time in function of additional network traffic.

5 Conclusion and Perspectives

Summary results of the experiments are presented in Fig.3 and in Table 3. Experiments have shown that increasing system security by turning on security mechanisms in network layer doesn't result in significant increase of system resources used, referring to test run in Experiment 1. Small difference can be noticed when very big additional network traffic was generated. It can be concluded that very deep integration of the security in the system provides such a good efficiency of the tested mechanisms. It can be said that it is worth to use all available security mechanism in the system, because even when the system is very busy and performs a lot of network operations, there is no big difference in execution time of applications.

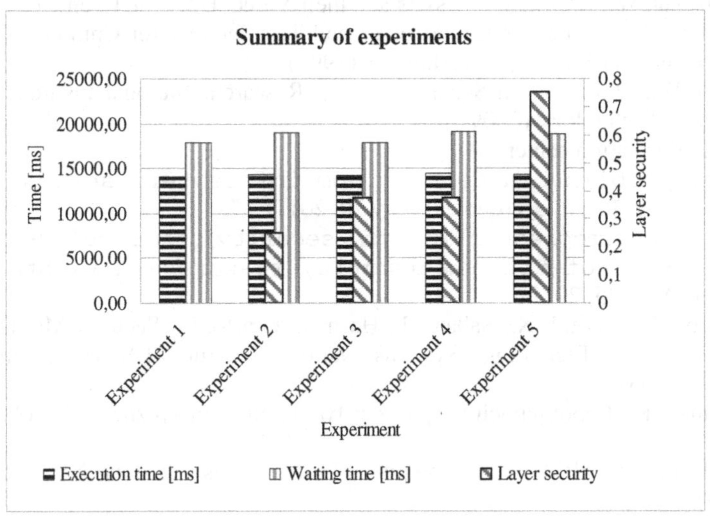

Fig. 3. Summary of the experiments

Table 3. Experiments summary

	Experiment 1	Experiment 2	Experiment 3	Experiment 4	Experiment 5
Security of the layer	0	0,25	0,38	0,38	0,75
Average execution time [ms]	14112,82	14354,14	14166,27	14435,37	14319,56
Average waiting time [ms]	17864,35	19028,79	17883,2	19063,44	18855,72

To get wider view of security of IBM i system additional research should be made. Using proposed metrics other two system layers should be investigated. Another proposal of further research is to check how the system would behave when being flooded by more additional network traffic than described in the paper. Additional experiments can be made for different granularity of the network traffic and security mechanisms parameters.

The proposed security metrics was created taking into consideration system that is not protected by any other security hardware or software. Using for example firewall could result in different interpretation of the metrics value.

References

1. Engelsman, W.: Information Assets and their Value. Univ. of Twente (2007)
2. Brinkley, D.L., Schell, R.R.: Concepts and Terminology for Computer Security. IEEE Computer Society, Los Alamitos (1995)
3. Jansen, W.: Directions in Security Metrics Research. National Institute of Standards and Technology (2009)
4. IBM i information center, `http://publib.boulder.ibm.com/infocenter/iseries/v7r1m0/index.jsp` (quoted: May 20, 2011)
5. `http://www.mcpressonline.com/security/ibm-i-os400-i5os/sneak-peek-of-the-new-security-features-in-v7.html` (quoted: May 25, 2011)
6. Swanson, M., Bartol, N., Sabato, J., Hash, J., Graffo, L.: Security Metrics Guide for Information Technology Systems. National Institute of Standards and Technology (2003)
7. Liderman, K.: O pomiarach bezpieczeństwa teleinformatycznego. In: IAiR WAT (2006)
8. Introduction to IBM i for Computer Professionals, Student Notebook, IBM (2008)
9. Cook, J., Cantalupo, J.C., Lee, M.: Security Guide for IBM i V6.1, IBM (2009)
10. Murdoch, J. (ed.): PSM Security Measurement White Paper, PSM Safety & Security, TWG (2006)
11. Value proposition for IBM System i, International Technology Group (2008), `http://www.compu-tech-inc.com/downloads/ISW03005USEN-ITC-Report.pdf`

Metaheuristic Approach for Survivable P2P Multicasting Flow Assignment in Dual Homing Networks

Wojciech Kmiecik and Krzysztof Walkowiak

Department of Systems and Computer Networks, Wroclaw University of Technology, Wroclaw, Poland
{wojciech.kmiecik,krzysztof.walkowiak}@pwr.wroc.pl

Abstract. Due to growing demand for high definition music and video content, Peer to Peer (P2P) multicasting providing live streaming services has been gaining popularity in the last years. In this paper, we focus on applying the P2P multicasting for delivering of critical data that require to be transmitted safely, intact and with as little delay as possible, e.g., financial data, software security patches, antivirus signature database updates etc. To improve survivability of the P2P multicasting, we propose to use dual homing approach, i.e., each peer is connected to the overlay by two separate access links. The optimization problem is formulated in the form of Integer Linear Programming (ILP). We introduce a Simulated Annealing (SA) algorithm for the considered optimization problem and compare it with optimal results provided by CPLEX solver. Our studies demonstrate that the SA method yields results close to optimal and provides better scalability comparing to CPLEX, since it can solve in reasonable time much larger problem instances than CPLEX.

Keywords: P2P multicasting, survivability, dual homing, protection, Simulated Annealing, overlay network.

1 Introduction

Nowadays, we are observing a rapid growth in popularity of multimedia streaming in the Internet. To emphasize the growing popularity of various video streaming services, we need to quote [3] where the authors claim that Video on Demand traffic will triple and Internet TV will be increased 17 times by 2015. The total share of all forms of video (already mentioned) and P2P will grow continuously to be approximately 90 percent of all global consumer traffic in the next three years. Services like IPTV, internet radio, Video on Demand and high definition video or audio streaming are very useful for network users, but often require a lot of bandwidth, which can be costly [1]. The main advantages of using the P2P (Peer-to-Peer) approach are scalability, adaptability, low deployment cost and optimal content distribution [2], which are crucial to meet that demand. An overlay P2P multicasting technology is based on a multicast delivery tree consisting of peers (end hosts). Content transmitted by the P2P multicasting can be either data files or streaming content with additional requirements like

Á. Herrero et al. (Eds.): Int. JointConf. CISIS'12-ICEUTE'12-SOCO'12, AISC 189, pp. 215–224.
springerlink.com

bit rate etc. [8]. In many related works, the authors assume that users of a multicast network can leave the system. We address the situation where the P2P multicast system is static (peers stay connected to the system for a long time), like in:

- Content delivery network (CDN) - e.g., Akamai Technologies,
- Set-top box (STB) technology used in IPTV,
- Critical information streaming - e.g., hurricane warnings.

To improve network survivability, we apply the P2P multicasting in a dual homing architecture. The dual homing approach assumes that all hosts (nodes) have two disjoint links (homes) to the network. Those links provide network protection because of redundancy. The main contribution of the paper consists of: (i) Heuristic Simulated Annealing algorithm developed for the optimization problem of survivable P2P multicasting systems using dual homing architecture. (ii) Numerical experiments based on Integer Linear Programming (ILP) model and heuristic approach showing comparative results of both methods and other characteristics of the proposed concept.

The rest of the paper is organized in the following way. Section 2 introduces the concept of survivable P2P multicasting based on the dual homing technology. In Section 3, we formulate an ILP model for survivable dual homing P2P multicast. Section 4 presents our Simulated Annealing algorithm. In Section 5, we present results of our experiments. Finally, the last section concludes this work.

2 Survivability for the P2P Multicasting

In our previous works, we proposed to apply disjoint P2P multicast trees streaming the same content [6], [9]. Peers affected by a failure of one of the trees can use another tree to receive the required data in case of a failure. This procedure guarantees very low restoration time. In this paper, we study the network survivability problem for the dual homing architecture. In Fig. 1, we present a simple example to illustrate our concept. There are two disjoint multicast trees A, B that connect 8 nodes - a, b, c, d, e, f, g. In the case of tree A, nodes a, d and f are uploading nodes, while remaining ones are leafs. We use the term *level* to describe the location of nodes in the multicast tree. For example, node a is on level 1 of tree B, nodes b and e are on level 2 of tree B and rest of the nodes are on level 3.

The P2P multicasting is done in the application layer, i.e., end hosts are connected using the overlay network. Connections between peers are established as unicast connections over the underlying physical layer. Each peer is connected to the overlay by an access link. We propose to utilize the dual homing approach to protect the system against a failure of the access link. The main idea is to create two P2P multicasting trees guaranteeing that each of access links carries traffic only of one of the trees. Since each node has two access links (dual homing), it receives the streaming data from both trees on two separate links. Thus, if one of access links is broken, the node is still connected to the stream and moreover, it can upload the stream to subsequent peers located in the tree. A proper configuration of the P2P multicasting with dual homing protects the network from two kinds of failures:

- Uploading node failure – a failure that impacts all successors of the failed peer in the tree,
- Overlay link failure – that one comprises a failure of both directed links between nodes.

3 ILP Model

What we describe below is an ILP model introduced in [6] that considers dual homing architecture. To formulate the problem we use the notation as in [7]. Let indices $v,w = 1,2,\ldots,V$ denote peers – nodes of the overlay network. There are K peers (clients) indexed $k = 1,2,\ldots,K$ that are not root nodes in any trees and want to receive the data stream. Index $t = 1,2,\ldots,T$ denotes streaming trees. We assume that $T = 2$, however the model is more general and values $T > 2$ may be used. In trees, nodes are located on levels $l = 1,2,\ldots,L$. That gives us possibility to set a limit on the maximum depth of the tree. The motivation behind this additional constraint is to improve the QoS (Quality of Service) parameters of the P2P multicasting, .e.g., network reliability and transmission delay. If node v is root of the tree t, then $r_{vt} = 1$, otherwise $r_{vt} = 0$. Constant c_{wv} denotes streaming cost on an overlay link (w,v).

To model survivable P2P multicasting, we modify formulations presented in previous papers [6], [9]. We introduce constant $\tau(v)$, which adds a virtual node associated with the node v. Together they form a primal node. Every primal node has in fact four capacity parameters – constants d_v and u_v are respectively download and upload capacity of the node v and constants $d_{\tau(v)}$ and $u_{\tau(v)}$ are parameters of the virtual node $\tau(v)$. The objective function is streaming cost of all multicast trees. This can be defined in many ways, e.g., as network delay or transmission cost.

Fig. 1 depicts an example of the dual homing modeling. Dual homes are marked with a pattern of sequential lines and dots.

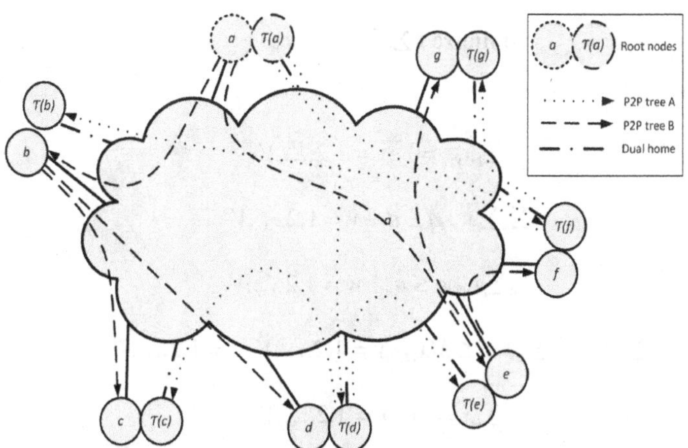

Fig. 1. Modeling dual homing

To model the P2P multicasting we use as in [6] binary variable x_{wvtk}, which is used to denote an individual flow for each streaming path from the root node to node k. This variable is set to 1, when streaming path to node k in tree t includes an overlay link between nodes w and v; 0 otherwise. Additional binary variable x_{wvt} is set to 1, if a link from node w to node v (no other peer nodes in between) is established in the multicast tree t; 0 otherwise. Auxiliary binary variable x_{vt} is set to 1, if an access link of node v is used to download or upload flow of the multicast tree t; 0, otherwise.

indices

$v,w,b = 1,2,\ldots,V$ overlay nodes (peers)
$k = 1,2,\ldots,K$ receiving nodes (peers)
$t = 1,2,\ldots,T$ streaming tree index

constants

d_v download capacity of node v (Kbps)
u_v upload capacity of node v (Kbps)
r_{vt} = 1, if node v is the root (streaming node) of tree t; 0, otherwise
q the streaming rate (Kbps)
c_{wv} streaming cost on overlay link from node w to node v
M large number
$\tau(v)$ index of node associated with node v (dual homing)
L The maximum number of levels of the tree

variables

x_{wvtk} = 1 if in multicast tree t the streaming path from the root to node k includes an overlay link from node w to node v (no other peer nodes in between); 0, otherwise (binary)

x_{wvt} = 1 if link from node w to node v (no other peer nodes in between) is in multicast tree t; 0, otherwise (binary)

x_{vt} = 1 if access link of node v is used to download or upload flow of multicast tree t; 0, otherwise (binary)

objective

$$\text{minimize } \sum_w \sum_v \sum_t x_{wvt} c_{wv} \tag{1}$$

constraints

$$\sum_t x_{vvt} = 0 \quad v = 1,2,\ldots,V \tag{2}$$

$$\sum_w \sum_t x_{wvt} q \le d_v \quad v = 1,2,\ldots,V \tag{3}$$

$$\sum_v \sum_t x_{wvt} q \le u_w \quad w = 1,2,\ldots,V \tag{4}$$

$$\sum_w x_{wvt} + \sum_w x_{vwt} \le M x_{vt} \quad v = 1,2,\ldots,V \quad t = 1,2,\ldots,T \tag{5}$$

$$\sum_t x_{vt} = 1 \quad v = 1,2,\ldots,V \tag{6}$$

$$x_{vt} + x_{\tau(v)t} = 1 \quad v = 1,2,\ldots,V \quad t = 1,2,\ldots,T \tag{7}$$

$$\sum_t x_{v\tau(v)t} = 0 \quad v = 1,2,\ldots,V \tag{8}$$

$$\sum_w x_{wvkt} - \sum_w x_{vwkt} = x_{kt} \quad v = k \quad v = 1,2,\ldots,V \quad k = 1,2,\ldots,V \quad t = 1,2,\ldots,T \tag{9}$$

$$\sum_w x_{wvkt} - \sum_w x_{vwkt} = -x_{kt} \quad r_{vt} = 1 \quad v = 1,2,\ldots,V \quad k = 1,2,\ldots,V \quad t = 1,2,\ldots,T \tag{10}$$

$$\sum_w x_{wvkt} - \sum_w x_{vwkt} = 0 \quad v \neq k \quad r_{vt} \neq 1 \quad v = 1,2,\ldots,V \quad k = 1,2,\ldots,V \quad t = 1,2,\ldots,T \tag{11}$$

$$x_{wvkt} \leq x_{wvt} \quad v = 1,2,\ldots,V \quad w = 1,2,\ldots,V \quad k = 1,2,\ldots,V \quad t = 1,2,\ldots,T \tag{12}$$

$$\sum_w \sum_v x_{wvkt} \leq L \quad k = 1,2,\ldots,V \quad t = 1,2,\ldots,T \tag{13}$$

Condition (2) assures that the node internal flow is zero. This constraint guarantees that in each tree there is at most one transmission per overlay link (w,v). (3) and (4) are respectively the download and upload capacity constraints. Condition (5) specifies definition of the x_{vt} variable. The survivability constraint (6) assures, that multicast trees are separate – each node v (access link) can only be used in one tree t. Constraints (7), (8) state that only one node from the primal node can belong to tree t and there cannot be any connection within the primal node. Conditions (9)-(11) are called flow conservation constraints and define flows in P2P multicasting trees. Connection between (w,v) exists in tree t, when there is at least one transmission between nodes (w,v) in tree t to receiving node k (12). Finally, constraint (13) sets the levels upper limit in the path to each receiving node. For more details on modeling of survivable P2P multicasting, refer to [9].

The considered problem is complex and NP-hard, since it can be reduced to the hop-constrained minimum spanning tree problem, which is known to be NP-hard [4]. Due to that fact, branch-and-bound or branch-and-cut methods have to be used to obtain optimal solution.

4 Simulated Annealing (SA) Algorithm

Solving ILP models including a large number of variables and constraints can take a huge amount of processing time and computing power. In order to be able to solve such models in a relatively short period of time and obtain solution close to optimum, we developed a heuristic algorithm based on Simulated Annealing approach.

The starting solution of the algorithm is prepared by taking the following steps [6]:

1. Pick a node that is root for tree t.
2. Randomly pick any remaining node and connect it to the tree t by making it a child of the node connected before ; if node v is connected to tree t then node $\tau(v)$ has to be connected to the other tree.
3. If all nodes are connected – terminate.

This procedure guarantees that there are no loops in the network. At this point, the created structure does not have to meet all the constraints.

We use an insert method to create a random solution from the neighborhood of the current solution. We randomly pick node v with $r_{vt} = 0$ for $t = 1,2...T$ and choose its new parent w. All the children of node v become children of its former parent. If new parent w is in the same tree as node $\tau(v)$, then that node is reconnected to a different tree.

Original Simulated Annealing algorithm was created to solve unconstrained problem. In order to use it for our problem that includes numerous constraints, we propose a penalty method (14) similar to the Tabu Search algorithm described in [6]:

$$f(cp,lp,up)=x_1cp+x_2cplp+x_3cpup \tag{14}$$

cp cost penalty, which is a difference between the cost of generated solution and the cost of current solution,

lp level penalty, which is set to 1 when there are more levels in the multicast trees than L value; 0 otherwise,

up upload penalty, which is a difference between how much the new parent w uploads and its upload capacity,

x_1,x_2,x_3 weights for each module of penalty method.

SA algorithm has following input parameters:

T_s starting temperature

T_e ending temperature

SA uses geometric progression of temperature. We developed additional optimization method, which is invoked after all iteration of the algorithm. The function finds the most expensive connections in the multicast trees and tries to find cheaper alternatives. The new solution is accepted if the obtained overall network cost is lower than the current one.

5 Research

5.1 Comparing ILP Model and SA Algorithm - Experiment Design

We randomly generated networks with 10 (small network) and 20 (big network) primal nodes and two disjoint trees. Our goal was to test how the level limit L and the streaming rate q would affect the overall network cost achieved by CPLEX and SA algorithm. To solve ILP model in an optimal way, we use newest CPLEX solver v12.4 [5]. We created 2 different networks with link costs in range 1-20, 1-50 and with either symmetric (100Mbps/100Mbps - 10% of all nodes) or asymmetric nodes (1Mbps/256Kbps, 2Mbps/512Kbps, 6Mbps/512Kbps, 10/1Mbps, 20/1Mbps, 50/2Mbps, 100/4Mbps). We assume that the first node is the streaming node for tree $t = 1$ and node $\tau(1)$ is the streaming node for tree $t = 2$. We set $time_{max}$ value to 3600 seconds, limiting the execution time of CPLEX. That gave us the possibility to compare ILP model and heuristic algorithm in terms of quality of obtained solution and time of processing. We tested how the SA algorithm would perform for different values of starting and ending temperature (from 1 to 1000 in both cases) and we found that the best parameters were:

- Starting temperature T_s: 500,
- Ending temperature T_e: 1.

We set those values for the experiments described below. In another preliminary experiment we discovered that the best values for the weights of penalty method were 1, 0.25 and 0.15 respectively. Overall, we conducted three experiments:

- Checking how the number of iterations and execution time would affect results obtained by the *Simulated Annealing* algorithm for either size of the network,
- Comparing CPLEX and SA algorithm for small size of the network and different streaming rate q values in terms of the overall network cost,
- Comparing CPLEX and SA algorithm for big size of the network and different number of levels L in terms of the overall network cost.

5.2 Comparing ILP Model and SA Algorithm – Results

The purpose of the first experiment was to check how the number of iterations would affect the results of the SA algorithm. In Fig. 2 and 3 we show results of SA for small and big size of the network respectively – we present both average overall network cost and best obtained network cost for 100 repetitions of algorithm execution. We can easily notice that, with an increase in the number of iterations, the SA algorithm performs better. Main drawback of increasing the number of iterations is growth of the execution time. Our conclusion is that for every experiment we have to choose the number of iterations that allows us to achieve a quality solution in a reasonable time (i.e., seconds instead of hours etc.).

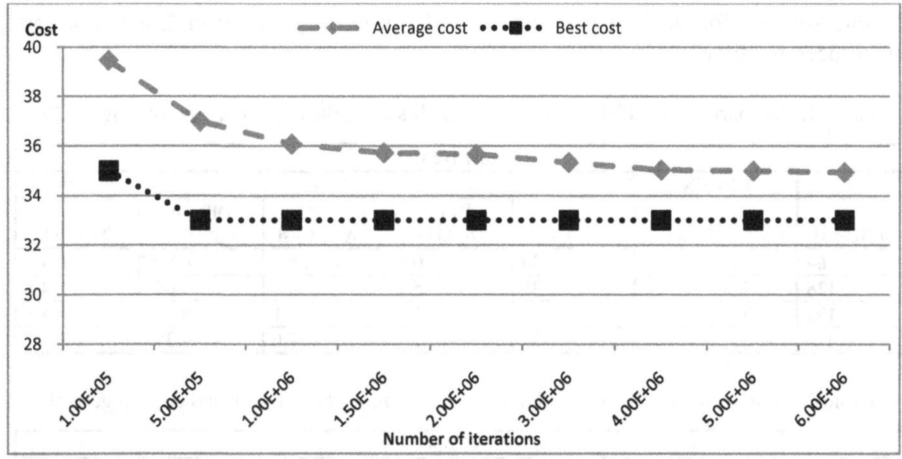

Fig. 2. Overall network cost as a function of the number of iterations for a small-sized network

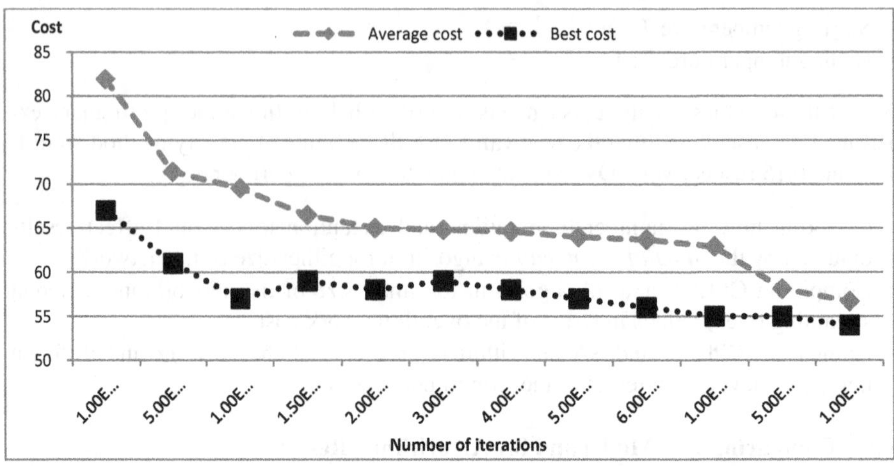

Fig. 3. Overall network cost as a function of the number of iterations for a big-sized network

The goal of the next experiment was to compare CPLEX and SA algorithm in terms of the overall network cost and time of processing for small size of network and different values of the streaming rate q. Tables 1 and 2 present obtained results for both CPLEX and SA and for different sets of cost range. Best and average results of SA were calculated based on 100 executions of the algorithm. For all types of networks, CPLEX achieved optimum in a short period of time (4 to 24 seconds). The SA algorithm was able to find optimum in 4 of 8 cases. Average overall network cost achieved by SA was 4-8% larger than CPLEX result, but processing time was only ~5 seconds. The general conclusion is that both algorithms proved to be good tools for finding solution for small size of the networks in terms of the overall network cost and processing time.

Table 1. Comparison of CPLEX and SA for small size of the network and cost range 1-20

cost 1-20							
	CPLEX		**SA**				
q **[Kbps]**	**Cost**	**Time[s]**	**Best Cost**	**Diff. to opt.[%]**	**Avg Cost**	**Diff. to opt.[%]**	**Time[s]**
64	33	18	33	0.0%	35.4	6.8%	4.7
128	33	16	35	5.7%	35.7	7.6%	4.8
192	36	24	37	2.7%	38.1	5.5%	4.8
256	36	22	37	2.7%	37.6	4.3%	4.7

Table 2. Comparison of CPLEX and SA for small size of the network and cost range 1-50

cost 1-50							
	CPLEX		**SA**				
q **[Kbps]**	**Cost**	**Time[s]**	**Best Cost**	**Diff. to opt.[%]**	**Avg Cost**	**Diff. to opt.[%]**	**Time[s]**
64	69	17	69	0.0%	73.0	5.5%	4.7
128	69	14	69	0.0%	73.3	5.8%	4.8
192	72	4	73	1.4%	78.2	7.9%	4.8
256	72	9	72	0.0%	77.4	7.0%	4.7

Tables 3 and 4 depict results of SA and CPLEX for big size of network (20 hosts). Optimal values were obtained by CPLEX without $time_{max}$ constraint that resulted in very long execution time (from 2 to almost 9 hours). In one hour time, CPLEX managed to find optimum in only one case of 14 and in 3 cases it was not able to find even a feasible solution. Most of the results of CPLEX with 1 hour time limit was 80-90% worse than optimum. The SA algorithm was able to find feasible solution every time and found optimum for 2 of 14 cases. The average cost obtained by SA was 3.5-7.5% worse than optimum. Another conclusion is that with a lower L limit the overall network cost is higher.

For larger networks CPLEX was unable to find a satisfying solution in a reasonable time (1 hour). SA achieved quality solutions in terms of overall network cost and proved itself useful for bigger types of network.

Table 3. Comparison of CPLEX and SA for big size of the network and cost range 1-20

cost 1-20										
	CPLEX		**CPLEX with time constraint**			**SA**				
LL	**Optimum**	**Time[s]**	**Best Cost**	**Diff. to opt.[%]**	**Time[s]**	**Best Cost**	**Diff. to opt.[%]**	**Avg Cost**	**Diff. to opt.[%]**	**Time[s]**
5	53	31525	Unknown	X	3600	55	3.6%	55.5	4.5%	3600
6	53	14423	Unknown	X	3600	54	1.9%	55.1	3.8%	3600
7	52	16563	791	93%	3600	54	3.7%	55.3	6.0%	3600
8	52	13963	430	88%	3600	54	3.7%	56.2	7.5%	3600
9	52	19231	520	90%	3600	54	3.7%	55.5	6.3%	3600
10	52	27082	349	85%	3600	53	1.9%	54.9	5.3%	3600
11	52	5411	52	0%	3560	54	3.7%	55.2	5.8%	3600

Table 4. Comparison of CPLEX and SA for big size of the network and cost range 1-50

cost 1-50										
	CPLEX		**CPLEX with time constraint**			**SA**				
L	**Optimum**	**Time[s]**	**Best Cost**	**Diff. to opt.[%]**	**Time[s]**	**Best Cost**	**Diff. to opt.[%]**	**Avg Cost**	**Diff. to opt.[%]**	**Time[s]**
5	93	10501	1655	94%	3600	94	1.1%	96.1	3.2%	3600
6	90	12323	Unknown	X	3600	91	1.1%	93.2	3.4%	3600
7	89	9482	465	81%	3600	90	1.1%	92.1	3.4%	3600
8	88	7942	89	1%	3600	90	2.2%	92.1	4.5%	3600
9	87	8892	93	6%	3600	90	3.3%	92	5.4%	3600
10	85	15076	354	76%	3600	90	5.6%	91.9	7.5%	3600
11	85	7383	549	85%	3600	88	3.4%	92	7.6%	3600

6 Conclusion

In this paper, we address the problem of survivable P2P multicasting in dual homing networks. Experiments testing the impact of streaming rate and number of levels on the overall network cost were conducted along with experiments focusing on comparison of the introduced SA algorithm and the ILP model. The results of experiments indicate that proper selection of parameters has big influence on the network cost. Both the streaming rate and the level limit impact the overall network cost. Moreover, the SA proved to be an useful algorithm for finding cost efficient solutions for all sizes of networks and achieved quality solutions close to optimum.

In future work, we plan to introduce new constraints that will provide more survivability, like node and ISP disjoint trees, and conduct more experiments evaluating these solutions.

Acknowledgements. This work is supported in part by the National Science Centre (NCN), Poland.

References

1. Aoyama, T.: A New Generation Network: Beyond the Internet and NGN. IEEE Comm. Magazine 47(5), 82–87 (2009)
2. Christakidis, A., Efthymiopoulos, N., Fiedler, J., Dempsey, S., Koutsopoulos, K., Denazis, S., Tombros, S., Garvey, S., Koufopavlou, O.: VITAL++, a new communication paradigm: embedding P2P technology in next generation networks. IEEE Comm. Magazine 49(1), 84–91 (2011)
3. Cisco Visual Networking Index Forecast 2010–2015 (2011), `http://www.cisco.com/en/US/solutions/collateral/ns341/ns525/ns537/ns705/ns827/white_paper_c11-481360.pdf`
4. Dahl, G., Gouveia, L., Requejo, C.: On formulations and methods for the hop-constrained minimum spanning tree problem. In: Resende, M.G.C., Pardalos, P.M. (eds.) Handbook of Optimization in Telecommunications, pp. 493–515. Springer (2006)
5. ILOG CPLEX 12.4 User's Manual, USA (2009)
6. Kmiecik, W., Walkowiak, K.: Survivable P2P multicasting flow assignment in dual homing networks. In: 3rd International Congress on Ultra Modern Telecommunications and Control Systems and Workshops (ICUMT), Budapest (2011)
7. Pióro, M., Medhi, D.: Routing, Flow, and Capacity Design in Communication and Computer Networks. Morgan Kaufman Publishers (July 2004) ISBN:978-0-12-557189-0
8. Shen, X., Yu, H., Buford, J., Akon, M.: Handbook of Peer-to-Peer Networking, 1st edn. Springer (2009)
9. Walkowiak, K., Przewoźniczek, M.: Modeling and optimization of survivable P2P multicasting. Computer Communications 34(12), 1410–1424 (2011)

A Tabu Search Algorithm for Optimization of Survivable Overlay Computing Systems

Krzysztof Walkowiak[1], Wojciech Charewicz[1], Maciej Donajski[1], and Jacek Rak[2]

[1] Wrocław University of Technology, Wybrzeze Wyspianskiego 27, 50-370 Wroclaw, Poland
[2] Gdansk University of Technology, G. Narutowicza 11/12, 80-233 Gdansk, Poland
Krzysztof.Walkowiak@pwr.wroc.pl

Abstract. Recently, distributed computing paradigm is gaining much interest since both industry and academia require large computational power to process and analyze huge amount of data. As distributed computing systems - similar to other network systems - are vulnerable to failures, survivability guarantees are indispensable to provide the uninterrupted service. Consequently, in this work we consider a survivable computing systems based on a 1+1 protection mechanism. The goal of the system is to provide scheduling of tasks to computing nodes and dimensioning of network capacity in order to minimize the operational cost of the system and satisfy survivability constraints. Since the problem is NP-complete and computationally demanding, we propose an effective heuristic algorithm based on Tabu Search approach. Extensive numerical experiments are run to verify effectiveness of the heuristic against optimal results provided by CPLEX and other heuristic algorithms.

Keywords: distributed computing, ILP modeling, optimization, cut inequalities.

1 Introduction

Distributed computing is currently the most significant approach applied to process large and complex computational tasks in many areas such as financial modeling, collaborative visualization of large scientific databases, medical data analysis, bioinformatics, experimental data acquisition, climate/weather modeling, earthquake simulation, astrophysics and many others [1]-[7]. In some cases, the results of computations are of great importance, i.e., processing and next delivery of the results are necessary to provide uninterrupted operation of various systems. For instance, results of simulation regarding earthquake prediction are essential to disseminate warnings among all interested parties and inhabitants. Consequently, survivability, i.e., ability to provide the continuous service after a failure [8], becomes a crucial issue. Survivability of network systems is provided by means of additional (backup) resources (e.g., transmission links/paths, computing units) used after the failure affecting the components of the main communication path (called working path) as the primary network resources of task processing. To ensure survivability in distributed computing systems, we proposed in [9] a proactive approach based on the 1+1 method applied in connection-oriented computer networks [8], [10-11]. The main idea of the approach is that in normal state of the system (without failures), each task is processed on two

Á. Herrero et al. (Eds.): Int. JointConf. CISIS'12-ICEUTE'12-SOCO'12, AISC 189, pp. 225–234.
springerlink.com

separate computing nodes. Results (output data) are concurrently sent from the two nodes to all receivers requesting the information. When a failure occurs, one of the considered computing nodes remains available, what guarantees that results of computation are delivered with no violation at all. We consider a failure scenario, when due to the breakdown, results produced by one of the nodes are not delivered to the destination node(s). This can be caused by a wide range of failures including both networking issues (e.g., access link failure, backbone network link failure, backbone node failure, etc.) and processing issues (e.g., node hardware failure, power outage, etc.).

The distributed computing systems like Grids can be developed using special dedicated high-speed networks [1]. Internet can be also used as the backbone network for an overlay-based systems [6]-[7]. We focus on the latter case, as overlays offer considerable network functionalities (e.g., diversity, flexibility, manageability) in a relatively simple and cost-effective way, as well as regardless of physical and logical structure of underlying networks.

Note that survivability of distributed computing systems (in particular including protection of computing units), is a relatively recent research area – there are few works in this topic (e.g., [12]-[15]). However, most of these papers use a dedicated optical network to construct the computing system. Similar to our research started in [9], here we also use an overlay network. The main contributions of the paper are twofold. (i) A new heuristic algorithm proposed to solve the problem of scheduling and capacity design in survivable overlay computing system. (ii) Numerical results presenting the performance of the heuristic in comparison to the optimal results.

The rest of the paper is organized as follows. In Section 2, we present the system architecture and formulate the ILP model. Section 3 includes description of the Tabu Search algorithm. In Section 4, we report and discuss results of numerical experiments. Finally, the last section concludes the work.

2 ILP Model of a Survivable Overlay Computing System

In this section, we present the ILP model of a survivable overlay computing system. Note that the model for the first time was formulated in [15]. Main assumptions of the model are defined according to characteristics of real overlay and distributed computing systems as well as consistent with assumptions presented in previous works on optimization of distributed computing systems [10]-[18].

The computing system consists of nodes indexed by $v = 1,2,\ldots,V$. The nodes represent computing elements (individual computers or clusters) as well as sources of input data and destinations of output data. The system works on top of an overlay network (e.g., Internet). Each node is connected by an access link to the network. The connectivity between nodes is provided by virtual links of the overlay realized by paths consisting of links realized in the underlying network. The main motivation to use the overlay idea in computing systems is large flexibility – many currently deployed network systems apply the overlay approach, e.g. Skype, IPTV, Video on Demand, SETI@home, etc. According to [16], nodes' capacity constraints are typically adequate in overlay networks. Additionally, in overlays the underlay physical network is typically assumed to be overprovisioned and the only bottlenecks are access links

[17]. Therefore, the only network capacity constraints in the model refer to access links. Since the access link capacity is to be dimensioned, integer variable z_v denotes the number of capacity modules allocated to the access link of node v. We assume that each node v is assigned to a particular ISP (Internet Service Provider), that offers high speed access link with a capacity module m_v given in Mbps (e.g., Fast Ethernet). Each node is already equipped with some computers, and p_v denotes the processing power of node v given by a number of uniform computational tasks that node v can calculate in one second.

The main goal of the computing system is to process a set of computational tasks. Each tasks is uniform, i.e., each task requires the same processing power, which can be expressed as a number of FLOPS. Tasks are indexed by $r = 1,2,\ldots,R$. Tasks are processed independently, i.e., we assume that there is no dependency between individual tasks. For each task, there is a source node that generates the input data and one or more destination nodes that receive the output data including results of computations (processing). Constants s_{rv} and t_{rv} are used to denote the source and destination nodes of each task, i.e., s_{rv} is 1, if node v is the source node of project r; 0 otherwise. In the same way, t_{rv} is 1, if node v is the destination node of project r; 0 otherwise. Constants a_r and b_r denote the transmit rate of input data and output data, respectively, per task r given in bps (bits per second).

The workflow of the system is as follows. The project input data is transferred from the source node providing input data to one or more computing nodes that process the data. Next, the output data (results of computations) is sent from each computing node to one or more destination nodes. We assume (similar to [11]-[12], [17], [18]), that computational projects are long-lived, i.e., they are established for a relatively long time (e.g., days, weeks). The input and output data associated with the project is continuously generated and transmitted. Thus, computational and network resources can be allocated in the system using offline optimization methods.

Many computational tasks processed in distributed systems are of great importance and need execution guarantees, e.g., medical applications, business analysis, weather forecasts, etc. However, distributed systems – similar to communication networks –are subject to various unintentional failures caused by natural disasters (hurricanes, earthquakes, floods, etc.), overload, software bugs, human errors, and intentional failures caused by maintenance actions or sabotage [10]. Such failures influence network infrastructure connecting computing nodes, e.g., access link failure, underlying physical network link failure, etc.. Moreover, elements of distributed computing systems devoted to data processing are also subject to various breakdowns (e.g., hardware failure, power outage, software bug, etc.). Therefore, in distributed computing systems, to provide guarantees on computational tasks completion, execution and delivery of all required results need to be enhanced with some survivability mechanisms.

We consider a failure that leads to a situation when the results to be obtained at one of the computing nodes are not delivered to the requesting destination nodes (e.g., access link failure, backbone network link failure, backbone node failure, etc., and processing issues including node hardware failure, power outage, etc.). To protect the distributed computing system against this kind of a failure, here we introduce a similar approach as in connection-oriented networks, i.e., 1+1 protection developed in the context of Automatic Protection Switching (APS) networks [8]. The key idea is to assign to each computational task two computing nodes: primary and backup. Both nodes simultaneously process the same input data, and next send results to all destination nodes. To make the system more flexible, not all tasks are to be protected - parameter α_r is 1, if task r requires protection; 0, otherwise.

The objective of the problem is to minimize the operational cost (OPEX) of the computing system including expenses related to two elements: transmission and processing. Constant ξ_v given in euro/month denotes the whole OPEX cost related to one capacity module allocated for node v and includes leasing cost of the capacity module paid to the ISP as well as all other OPEX costs like energy, maintenance, administration, etc. Constant ψ_v denotes the OPEX cost related to processing of one uniform task in node v. The ψ_v cost is defined in euro/month and contains all expenses necessary to process the uniform computational tasks including both processing and storage issues (e.g., energy, maintenance, administration, hardware amortization etc.).

Below we formulate the optimization problem in the form of an ILP model.

indices

$v,w = 1,2,\ldots,V$ computing nodes
$r = 1,2,\ldots,R$ computing tasks

constants

p_v maximum processing rate of node v
a_r transmit rate of input data per one task in project r (Mbps)
b_r transmit rate of output data per one task in project r (Mbps)
s_{rv} equals 1, if v is the source node of project r; 0 otherwise
t_{rv} equals 1, if v is the destination node of project r; 0 otherwise
t_r number of destination nodes for task r, i.e., $t_r = \sum_v t_{rv}$
ψ_v OPEX cost related to processing of one task in node v (euro/month)
ξ_v OPEX cost related to one capacity module of node v (euro/month)
m_v size of the capacity module for node v (Mbps)
α_r equals 1, if task r requires protection; 0, otherwise

variables

x_{rv} equals 1, if task r is allocated to primary computing node v; 0, otherwise (binary)
y_{rv} equals 1, if task r is allocated to backup computing node v; 0, otherwise (binary)
z_v capacity of node v access link expressed in the number of capacity modules (non-negative integer)

objective

It is to find transmission scheduling of tasks to primary and backup nodes as well as dimension network access links to minimize the operational cost of the system, i.e.:

$$\text{minimize} \quad C = \sum_v z_v \xi_v + \sum_r \sum_v (x_{rv}\psi_v + y_{rv}\psi_v) \tag{1}$$

constraints

a) Each computing node v has a limited processing power p_v. Therefore, each node cannot be assigned with more tasks to calculate than it can process. Both primary and backup nodes must be taken into account:

$$\sum_r (x_{rv} + y_{rv}) \le p_v \quad v = 1,2,\ldots,V \tag{2}$$

b) Download capacity constraint – incoming flow of each node cannot exceed the capacity of the access link:

$$\sum_r (1 - s_{rv})a_r(x_{rv} + y_{rv}) + \\ \sum_r t_{rv}b_r(1 - x_{rv} + \alpha_r - y_{rv}) \le z_v m_v \quad v = 1,2,\ldots,V \tag{3}$$

c) Upload capacity constraint – outgoing flow of each node cannot exceed the capacity of the access link:

$$\sum_r s_{rv}a_r(1 - x_{rv} + \alpha_r - y_{rv}) + \\ \sum_r (t_r - t_{rv})b_r(x_{rv} + y_{rv}) \le z_v m_v \quad v = 1,2,\ldots,V \tag{4}$$

d) Each task must be asigned to exactly one primary node:

$$\sum_v x_{rv} = 1 \quad r = 1,2,\ldots,R \tag{5}$$

e) If the task is to be protected (i.e., $\alpha_r = 1$), it must be assigned to a backup node:

$$\sum_v y_{rv} = \alpha_r \quad r = 1,2,\ldots,R \tag{6}$$

f) In order to provide survivability, primary and backup nodes must be disjoint:

$$(x_{rv} + y_{rv}) \le 1 \quad r = 1,2,\ldots,R \quad v = 1,2,\ldots,V \tag{7}$$

The problem (1)-(7) is NP-hard, since it is equivalent to the network design problem with modular link capacities [10]. Note that the presented model is generic and can be applied to optimize various kinds of computing systems including Grids and public resource computing systems [1]-[7].

3 Tabu Search Algorithm

Since the problem (1)-(7) is NP-hard and computationally demanding, only relatively small problem instances can be solved using branch-and-bound algorithms. To solve large problem instances, heuristic algorithms are required. In [9], we proposed two simple heuristic algorithms to solve the problem: AlgGreedy based on the greedy approach and AlgRand using a pure random method. However, since performance of both algorithms was not satisfactory – AlgGreedy provided results on average more than 10% worse than optimal, the corresponding gap for AlgRand was about 20% – in this section we introduce a new heuristic algorithm based on the Tabu Search approach. For ease of reference, we call the algorithm AlgTS.

The Tabu Search algorithm explores the solution space in a random way, but in contrary to the local search algorithm, it does not get stuck in local minima. This advantage is provided by introducing *tabu list* denoting a short-term memory, which stores a list of banned moves. The idea of the Tabu Search algorithm was proposed by Glover in [19] and [20].

The problem given by (1)-(7) is represented in AlgTS as two vectors including R integers: $\mathbf{x} = \{x_1, x_2,\ldots,x_R\}$, and $\mathbf{y} = \{y_1, y_2,\ldots,y_R\}$. Vector **x** includes the assignment of tasks to primary computing nodes, i.e., $x_r = v$ means that node v is the primary node of task r. Analogously, vector **y** denotes the assignment to backup nodes.

Two states (solutions) are considered as neighbors, when they differ at exactly one position considering both vectors **x** and **y**, i.e., all tasks except one are assigned to the same nodes. The transition from one state to another included in the neighborhood, is called a *move*. The move operation is represented as a triple (r, v, w), that denotes reallocation of task r from primary (backup) node v to primary (backup) node w. After each move (r, v, w) the tabu list is updated, i.e., node v (the former node of task r) is added to the list. In the move operation, it is assured that primary and backup nodes of the same task are disjoint.

Three following methods were developed to generate the initial solution:

- Random – the solution is generated at random.
- Greedy – each subsequent task selected at random is assigned to nodes (primary and backup) to minimize the current cost.
- Destination Node First (DNF) – the heuristic assigns each task to one of its destination nodes, only if it does not lead to violation of the processing limit constraint (2). If it is impossible, the heuristics tries to assign task to its source node. If neither destination, nor source nodes have available processing power, the task is assigned to some other node using the greedy approach.

To improve performance of the algorithm in terms of the execution time, the objective function of each newly generated solution is calculated on the base of the previous value, since move (r, v, w) implies changes maximally on three nodes – the source node or task r plus nodes v and w.

All constraints of the problem except for (2) are satisfied in the AlgTS algorithm through coding of the problem and construction of the move operation. To incorporate the processing limit constraint (2), we apply the penalty function. It means that not feasible solutions violating constraint (2) are possible to be generated by the move operation. Let $C(\mathbf{x}, \mathbf{y})$ denote the value of objective function (1) calculated for the values of variables included in vectors **x** and **y**. The penalty function added to the $C(\mathbf{x}, \mathbf{y})$ function is formulated as follows:

$$C_{Penalty} = P\, C(\mathbf{x}, \mathbf{y})\, (\Sigma_v(\max(0, \Sigma_r (x_{rv} + y_{rv}) - p_v)) \,/\, \Sigma_v p_v \tag{8}$$

Note that if the current state (solution) is feasible, the value of a penalty function is equal to 0. Otherwise, some additional cost depending on the value of processing limit violation and *penalty_factor* P is added to the function. Parameter P is used to tune the influence of the penalty function. Two additional parameters are used to tune the algorithm: *iteration_number* denotes the number of algorithm's iterations; *explaration_range* defines the size of neighborhood exploration, i.e., number of move operations examined in each iteration of the algorithm. The last tuning parameter – called *fixing_ratio* – is used to monitor feasibility of the solution. In more details, if for *fixing_ratio* subsequent iterations, the generated solution is not feasible (i.e., some constraints are violated), a special fixing procedure is applied. The goal of the procedure is to generate a feasible solution based on the current solution. The main idea of this heuristic method is to reallocate tasks from the nodes that exceed the processing limit to other nodes that have residual processing capacity and provide the lowest cost. Moreover, constraints providing the node-disjointness must be satisfied.

4 Results

In this section, we present the results of computational experiments. The ILP model (1)-(7) was implemented in CPLEX 11.0 solver [21] to obtain the optimal results. AlgTS and the reference heursitic AlgGreedy [9] were implemented in C++ and Java, respectively. The goal of experiments was twofold. First, we wanted to tune the AlgTS, i.e., find the best configuration of tuning parameters of the algorithm. Second, we compared results of heuristic methods against the optimal results provided by CPLEX. Simulations were run on a PC computer with IntelCore i7 processor and 4GB RAM.

Simulations were run on two sets of systems generated at random: small systems including 30 nodes and large systems including 200 nodes. For each size (small and large), six sets of systems and six sets of computing tasks were generated randomly according to parameters' ranges given in Table 1. The capacity module was set to 100 Mbps. Moreover, 11 configurations related to protection requirements were created with the following values of the protected tasks percentage (PTP): 0%, 10%, 20%,..., 100%. Thus, the overall number of individual cases for each size of the system was equal to 396 (i.e., 6 x 6 x 11).

Table 1. Parameters of tested systems

	Small systems	Large systems
Number of computing nodes	30	200
Number of tasks	300-700	1200-4800
Cost of capacity module	120-400	120-400
Processing cost of one unit	50-150	50-150
Processing limit of one node	10-40	10-40
Number of destination nodes	1-4	1-8
Input and output data rates	5-15	5-10

We used CPLEX solver to obtain the optimal results for small systems. Since the CPLEX with default setting of the optimality gap (i.e., 0.0001) was not able to stop calculations within one hour for one test, we set the optimality gap to 0.01 – then CPLEX average execution time was then equal to about 28 seconds, while the obtained average optimality gap was 0.0087. Note that for large systems, CPLEX was not able to yield a feasible result.

First, we focused on tuning of the AlgTS method. Table 2 presents details related to analysed parameters – for each size of the system, we show tested values and values selected for further tests. Each combination of tuning parameters (756 cases for small systems and 480 cases for large systems) was run for one system with three values of the PTP parameter (20%, 50% and 100%). The test was repeated five times for each case. In Fig. 1, we show example results related to selection of two tuning parameters: *initial_solution* and *tabu_list_length*. Values of other parameters were set to default ones reported in Table 2. The figure plots the average gap to the optimal results obtained for each combination of tested parameters.

Table 2. Tested values of tuning parameters of TA algorithm

Parameter name	Small systems		Large systems	
	Tested values	Selected values	Tested values	Selected values
initial_solution	Random, Greedy, DNF	DNF	Random, Greedy, DNF	DNF
tabu_list_length	0, 3, 5, 7, 10, 15	5	5, 15, 30, 50, 100	30
penalty_factor	0, 2, 3, 5, 10, 15, 20	3	0, 2, 5, 10, 15, 20, 50, 100	5
iteration_number	10000	10000	10000	10000
exploration_range	100	100	100	100
fixing_ratio	100, 200, 300, 500, 1000, 2000	200	100, 500, 1000, 2000	100

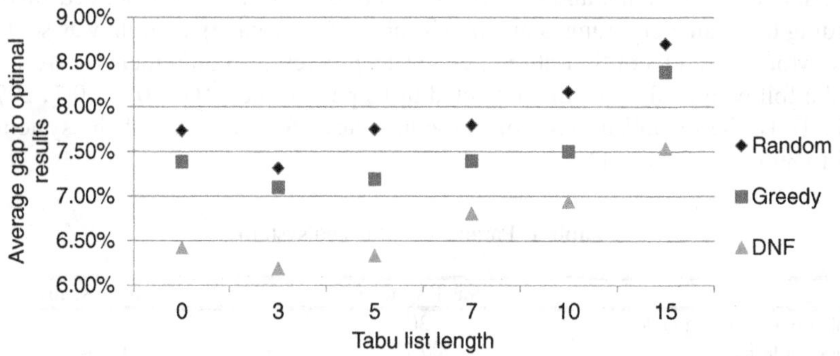

Fig. 1. Tuning of AlgTS for small systems – tabu list length and initial solution generation

The second goal of experiments was to compare the algorithms. In Table 3, we report the average results obtained for small and large systems. In the former case, we compare AlgTS and AlgGreedy against optimal results given by CPLEX. The table shows the average value of the gap to the optimal results and lengths of 95% confidence intervals. Recall that the overall number of individual tests was 396. In the case of AlgTS, we run the algorithm 20 times for each case and we present gaps related to minimum and average result of AlgTS. We can easily notice that AlgTS provides results only 5.77% worse than the optimal values, which is much better compared to AlgGreedy. Note that the average execution time for small systems was 29 seconds, 0.4 seconds and 2.5 seconds, for CPLEX, AlgGreedy, and AlgTS, respectively. In the case of large systems, CPLEX did not yield feasible results. Thus, we compare AlgTS against AlgGreedy. On averag,e the Tabu Search algorithm outperforms the greedy approach by about 8%. In this case, the average execution times were 3.6 seconds and 1.1 seconds for AlgTS and AlgGreedy, respectively. Small values of 95% confidence interval lengths prove that the algorithm provides stable results.

Table 3. Comparison of algorithms – average gap and lengths of 95% confidence intervals

	Small systems		Large systems	
	Average gap to the optimal results	Lengths of 95% conf. intervals	Average gap to AlgGreedy results	Lengths of 95% conf. intervals
AlgGreedy	10.17%	0.25%	-	-
AlgTS minimum	5.77%	0.11%	-8.10%	0.20%
AlgTS average	6.32%	0.11%	-7.80%	0.19%

Table 4 includes a detailed comparison of heuristics for large systems – we show average costs obtained for various values of the PTP parameter and the difference between both methods. With the increase of the PTP parameter, the gap between compared methods decreases. Moreover, full protection (PTP = 100%) compared to the unprotected case (PTP = 0%) increases the cost by 107%, on average.

Table 4. Comparison of algorithms for large systems – average cost obtained for various values of PTP parameter

Protected tasks percentage (PTP)	AlgGreedy	AlgTS minimum		AlgTS average	
	cost	cost	gap to AlgGreedy	cost	gap to AlgGreedy
0%	724591	664103	-9.09%	666914	-8.63%
10%	797781	729828	-9.29%	732622	-8.87%
20%	873366	797592	-9.47%	800492	-9.07%
30%	948939	867501	-9.35%	870253	-9.01%
40%	1025758	939913	-9.10%	942312	-8.83%
50%	1098074	1011179	-8.58%	1013472	-8.33%
60%	1179925	1093333	-7.91%	1095738	-7.67%
70%	1260819	1175858	-7.24%	1178481	-7.00%
80%	1300161	1259850	-6.20%	1262546	-5.97%
90%	1400472	1328193	-5.51%	1330419	-5.33%
100%	1439529	1374849	-4.75%	1377811	-4.53%

5 Concluding Remarks

In this work, we focused on the problem of survivable overlay computing systems optimization. Based on the ILP formulation, we developed a new heuristic algorithm employing the Tabu Search approach. According to specific features of the considered problem, we have introduced in the classical Tabu Search algorithm some additional elements to facilitate the optimization. According to extensive numerical experiments, the proposed method provides results close to the optimal solutions, and significantly outperforms the greedy algorithm. In future work, we plan to continue our research on the application of computational intelligence heuristics to optimization of overlay computing systems including evolutionary methods.

Acknowledgement. The work was supported in part by the National Science Centre (NCN), Poland, under the grant which is being realized in years 2011-2014.

References

1. Wilkinson, B.: Grid Computing: Techniques and Applications. Chapman & Hall/CRC Computational Science (2009)
2. Nabrzyski, J., Schopf, J., Węglarz, J. (eds.): Grid resource management:state of the art and future trends. Kluwer Academic Publishers, Boston (2004)
3. Milojicic, D., et al.: Peer to Peer computing. HP Laboratories Palo Alto, Technical Report HPL-2002-57 (2002)
4. Travostino, F., Mambretti, J., Karmous Edwards, G.: Grid Networks Enabling grids with advanced communication technology. Wiley (2006)
5. Taylor, I.: From P2P to Web services and grids: peers in a client/server world. Springer (2005)
6. Buford, J., Yu, H., Lua, E.: P2P Networking and Applications. Morgan Kaufmann (2009)
7. Shen, X., Yu, H., Buford, J., Akon, M. (eds.): Handbook of Peer-to-Peer Networking. Springer (2009)
8. Grover, W.D.: Mesh-Based Survivable Networks: Options and Strategies for Optical, MPLS, SONET, and ATM Networking. Prentice Hall PTR, New Jersey (2003)
9. Walkowiak, K., Rak, J.: 1+1 Protection of Overlay Distributed Computing Systems: Modeling and Optimization. In: Murgante, B., Gervasi, O., Misra, S., Nedjah, N., Rocha, A.M.A.C., Taniar, D., Apduhan, B.O. (eds.) ICCSA 2012, Part IV. LNCS, vol. 7336, pp. 498–513. Springer, Heidelberg (2012)
10. Pioro, M., Medhi, D.: Routing, Flow, and Capacity Design in Communication and Computer Networks. Morgan Kaufmann Publishers (2004)
11. Vasseur, J.P., Pickavet, M., Demeester, P.: Network Recovery. Elsevier (2004)
12. Thysebaert, P., et al.: Scalable Dimensioning of Resilient Lambda Grids. Future Generation Computer Systems 24(6), 549–560 (2008)
13. Develder, C., et al.: Survivable Optical Grid Dimensioning: Anycast Routing with Server and Network Failure Protection. In: Proc. of IEEE ICC 2011, pp. 1–5 (2011)
14. Buysse, J., De Leenheer, M., Dhoedt, B., Develder, C.: Providing Resiliency for Optical Grids by Exploiting Relocation: A Dimensioning Study Based on ILP. Computer Communications 34(12), 1389–1398 (2011)
15. Jaumard, B., Shaikh, A.: Maximizing Access to IT Services on Resilient Optical Grids. In: Proc. of 3rd International Workshop on Reliable Networks Design and Modeling (RNDM 2011), pp. 151–156 (2011)
16. Akbari, B., Rabiee, H., Ghanbari, M.: An optimal discrete rate allocation for overlay video multicasting. Computer Communications 31(3), 551–562 (2008)
17. Zhu, Y., Li, B.: Overlay Networks with Linear Capacity Constraints. IEEE Transactions on Parallel and Distributed Systems 19(2), 159–173 (2008)
18. Kacprzak, T., Walkowiak, K., Woźniak, M.: Optimization of Overlay Distributed Computing Systems for Multiple Classifier System – Heuristic Approach. Logic Journal of IGPL (2011), doi: 10.1093/jigpal/jzr020
19. Glover, F.: Tabu Search - Part I. ORSA J. on Computing 1(3), 190–206 (1989)
20. Glover, F.: Tabu Search - Part II. ORSA J. on Computing 2(1), 4–32 (1990)
21. ILOG CPLEX, 12.0 User's Manual, France (2007)

Greedy Algorithms for Network Anomaly Detection

Tomasz Andrysiak, Łukasz Saganowski, and Michał Choraś

Institute of Telecommunications, University of Technology & Life Sciences in Bydgoszcz ul. Kaliskiego 7, 85-789 Bydgoszcz, Poland
{tomasz.andrysiak,luksag,chorasm}@utp.edu.pl

Abstract. In this paper we focus on increasing cybersecurity by means of greedy algorithms applied to network anomaly detection task. In particular, we propose to use Matching Pursuit and Orthogonal Matching Pursuit algorithms. The major contribution of the paper is the proposition of 1D KSVD structured dictionary for greedy algorithm as well as its tree based structure representation (clusters). The promising results for 15 network metrics are reported and compared to DWT-based approach.

Keywords: network anomaly detection, cybersecurity, greedy algorithms.

1 Introduction

Nowadays cyber attacks are targeted at individual network users but also at critical infrastructures, countries and nations in order to paralyze them. Emerging new threats and attacks can only be detected by new complex solutions, such as hybrid signature and anomaly based distributed intrusion detection systems. Anomaly detection approach allows for detecting new unknown attacks (so called 0-day attacks) which can not be detected by traditional systems since their signature is yet not known. Therefore in this paper we present new $1D$ $KSVD$ algorithm for signal-based anomaly detection.

The paper is structured as follows: the overview of the greedy algorithms, namely Matching Pursuit and Orthogonal Matching Pursuit, is given in section 2. Afterwards in Section 3 we propose the structured dictionaries for greedy algorithms. $1D$ $KSVD$ algorithm (Section 3.1) and its tree-based structure implementation (Section 3.2) are presented in detail. Experimental results are reported in Section 4. Conclusions are given thereafter.

2 Overview of Greedy Algorithms for Anomaly Detection Systems

Sparse representation is looking for the sparse solution of decomposition coefficients C representing the signal S over the redundant dictionary when the remainder is smaller than a given constant ε, can be stated as:

$$\min \|C\|_0 \ subject\ to \left\| S - \sum_{k=0}^{K-1} c_k d_k \right\| < \varepsilon, \tag{1}$$

Á. Herrero et al. (Eds.): Int. JointConf. CISIS'12-ICEUTE'12-SOCO'12, AISC 189, pp. 235–244.
springerlink.com

where $\|\cdot\|_0$ is the l^0 norm counting the nonzero entries of a vector, $c_k \in C$ represents a set of projection coefficients and d_k are the elements of redundant dictionary D. Finding the optimal solution is an NP-hard problem [1][2]. A sub-optimal expansion can be found by greedy algorithms in means of an iterative procedure, such as the Matching Pursuit algorithm or Orthogonal Matching Pursuit algorithm.

2.1 Matching Pursuit Algorithm

The Matching Pursuit (MP) algorithm was proposed in [2]. The aim of the algorithm is to obtain an approximation to the input signal S, by sequential selection of vectors from the dictionary D. The algorithm follows a greedy strategy in which the basis vector best aligned with the residual vector is chosen at each iteration. Signal S can be written as the weighted sum of these elements:

$$S = \sum_{i=0}^{n-1} c_i d_i + r^n s, \tag{2}$$

where $r^n s$ is residual in an n term sum. In the first step of Matching Pursuit algorithm, the atom d_i which best matches the signal S is chosen. The first residual is equal to the entire signal $r^0 s = S$. In each of the consecutive p^{th} steps in MP algorithm, the atom d_p is matched to the signal $r^p s$, which is the residual left after subtracting results of previous iterations:

$$r^p s = r^{p-1} s - c_p d_{\varphi_p}, \tag{3}$$

where

$$\varphi_p = arg \max_{i \in \Phi_p} |\langle r^p s, d_i \rangle| \,, \, \varphi_p \in \Phi_p \tag{4}$$

and

$$c_p = \left\langle r^{p-1} s, d_{\varphi_p} \right\rangle. \tag{5}$$

The indices of the p vectors selected are stored in the index vector $\Phi_p = \{\varphi_1, \varphi_2, \ldots, \varphi_{p-1}, \varphi_p\}$, $\Phi_0 = \emptyset$ and the vectors are stored as the columns of the matrix $D_p = \left\{d_{\varphi_1}, d_{\varphi_2}, \ldots, d_{\varphi_p}\right\}$ and $D_0 = \emptyset$. The algorithm terminates when residual of signal is lower than acceptable limit:

$$\|r^p s\| < th, \tag{6}$$

where th is the approximation error. The Matching Pursuit algorithm is presented in Listing 1.

2.2 Orthogonal Matching Pursuit Algorithm

The Orthogonal Matching Pursuit (OMP) algorithm is an improvement of MP algorithm and it was proposed in [3]. Similarly to Matching Pursuit, two

Algorithm 1. $(C_p, D_p, R^pS) = MP(S, D, th)$

$r^0s = S$ initial residual
$c_0 = 0$ initial solution
$\Phi_0 = \emptyset$ initial index set of dictionary elements
$D_0 = \emptyset$ initial set of dictionary
$p = 0$ initial the iterative variable

while $\|r^ps\| < th$ {stopping rule} **do**
  $\Phi_p = \Phi_{p-1} \bigcup \varphi_p$ {update index set}
  $D_p = D_{p-1} \bigcup d_{\varphi_p}$ {update set of dictionary}
  $c_p = \langle r^{p-1}s, d_{\varphi_p} \rangle$ {calculate new coefficient}
  $C_p = C_{p-1} \bigcup c_p$ {update set of coefficients}
  $r^ps = r^{p-1}s - c_p d_{\varphi_p}$ {calculate new residual}
  $R^ps = R^{p-1}s \bigcup r^ps$ {update set of residual}
  $p = p + 1$
end while

return (C_p, D_p, R^pS)

algorithms has greedy structure but the difference is that OMP algorithm needs all selected atoms to be orthogonal in every decomposition step. The algorithm selects φ_p in the pth iteration by finding the vector best aligned with the residual obtained by projecting r^ps onto the dictionary components, that is:

$$\varphi_p = arg \max_{i \in \Phi_p} |\langle r^p s, d_i \rangle| \, , \varphi_p \notin \Phi_{p-1}. \tag{7}$$

The re-selection problem is avoided with the stored dictionary. If $\varphi_p \notin \Phi_{p-1}$ then the index set is updated as $\Phi_p = \Phi_{p-1} \bigcup \varphi_p$ and $D_p = D_{p-1} \bigcup d_{\varphi_p}$. Otherwise, $\Phi_p = \Phi_{p-1}$ and $D_p = D_{p-1}$. The residual is calculated as:

$$r^p s = r^{p-1}s - D_p \big(D_p^T D_p\big)^{-1} D_p^T r^{p-1} s, \tag{8}$$

where $D_p^T D_p$ is the Gram matrix. The algorithm terminates when dependency is satisfied (equation 6). The Orthogonal Matching Pursuit algorithm is presented in Listing 2.

3 Proposition of the Structured Dictionaries for Greedy Algorithms

3.1 Proposition of 1D KSVD Algorithm for Searching Dictionary of Signal

$KSVD$ algorithm has been previously used for 2D signals. Hereby, we propose the new $1D$ $KSVD$ algorithm that we use in anomaly detection task. In

Algorithm 2. $(C_p, D_p, R^pS) = MP(S, D, th)$

$r^0s = S$ initial residual
$c_0 = 0$ initial solution
$\Phi_0 = \emptyset$ initial index set of dictionary elements
$D_0 = \emptyset$ initial set of dictionary
$p = 0$ initial the iterative variable

while $\|r^ps\| < th$ {stopping rule} **do**
  $\varphi_p = arg\max_{i\in\Phi_p} |\langle r^ps, d_i\rangle|$ {selection}
  if $\varphi_p \notin \Phi_{p-1}$ **then**
    $\Phi_p = \Phi_{p-1} \bigcup \varphi_p$ {positive update index set}
    $D_p = D_{p-1} \bigcup d_{\varphi_p}$ {positive update set of dictionary}
  else {N is odd}
    $\Phi_p = \Phi_{p-1}$ {negative update index set}
    $D_p = D_{p-1}$ {negative update set of dictionary}
  end if
end while
$c_p = (D_p^T D_p)^{(-1)} D_p^T r^{(p-1)}s$ {calculate new coefficients}
$C_p = C_{p-1} \bigcup c_p$ {update set of coefficients}
$r^ps = r^{p-1}s - D_p c_p$ {calculate new residual}
$R^ps = R^{p-1}s \bigcup r^ps$ {update set of residual}
$p = p + 1$
return (C_p, D_p, R^pS)

particular, we modified the KSVD algorithm proposed in [4]. The main task of $1D\ KSVD$ algorithm is to find the best dictionary D to represent the signal S as sparse composition, by solving:

$$\min_{D,C} \left\{ \|S - DC\|_F^2 \right\} \ subject\ to\ \forall_i \|c_i\|_0 \le T, \tag{9}$$

where $\|\cdot\|_F^2$ is the Frobenius norm and T is a fixed and predetermined number of nonzero entries.

The algorithm is divided into two stages:

1. Sparse Coding Stage: Provided D is fixed. We use orthogonal matching pursuit algorithm (mentioned in section 2.2) to compute M sparse coefficients c_i for each sample of signal S, by approximation the solution of

$$\min_{C} \left\{ \|s_i - Dc_i\|_F^2 \right\} \ subject\ to\ \|c_i\|_0 \le T,\ i = 1, 2, \ldots, M, \tag{10}$$

where s_i is a sample of signal S.

2. Dictionary Update Stages: Provided both C and D are fixed. We focus on an atom d_k of the dictionary and its corresponding sparse vector c_T^k (i.e. row k of c_T^k). The corresponding objective function in equation 9 can be written as:

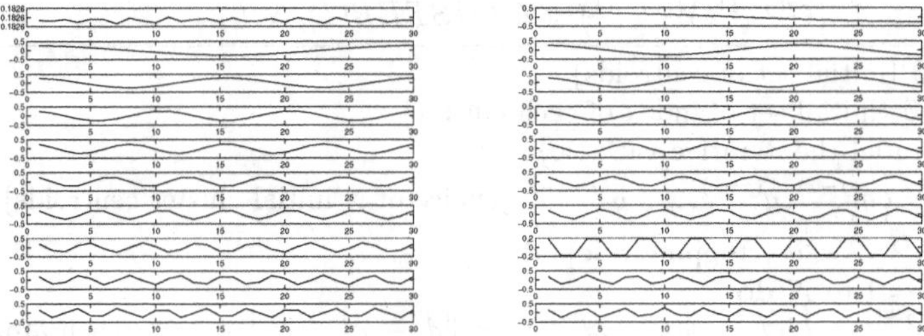

Fig. 1. Real atoms selected by 1D-KSVD algorithm

Algorithm 3. $D = KSVD(S, D_{init}, \varepsilon, \vartheta, N)$

$D_0 = D_{init}$ {initial set of dictionary}
$th = \varepsilon$ {initial the approximation error for OMP procedure}
$Q = 1$ {initial the iterative variable}
repeat
  for $k = 0$ to K **do**
    {update for each column $k = 1, 2K$ in D_{Q-1}}
    $C_Q = OMP(S, D_Q, th)$ { compute the set of coefficients by solving the equation 10}
    $\omega_k = \{i | 1 \le i \le K, c_T^k(i) \neq 0\}$ {define the set of index that use the atom d_k}
    $E_k = S - \sum_{j \neq k} d_j c_T^j$ {compute E_k, see equation 11}
    $E_k^R \leftarrow E_k\ \omega_k\ E_k^R$ {restrict E_k by choosing the columns corresponding to ω_k so that we obtain E_k^R}
    $E_k^R = U \Delta V^T$ {use the SVD method to decomposition E_k}
    $D_k \leftarrow U^{(1)}$ {update the dictionary k^th column by the first column of U}
    $c_k^R \leftarrow V^{(1)} \Delta^{(1,1)}$ {update the coefficient vector by the first column of V multiple by $\Delta^{(1,1)}$}
  end for
  $Q = Q + 1$
until $\|S - D_{Q-1} C_{Q-1}\|_2^2 \le \vartheta$ or $Q > N$ {stopping rule}
return $(D = D_{Q-1})$

$$\|S - DC\|_F^2 = \left\| S - \sum_{j=1}^{K} d_j c_T^j \right\|_F^2 =$$

$$= \left\| \left(S - \sum_{j \neq k} d_j c_T^j \right) - d_k c_T^k \right\|_F^2 = \left\| E_k - d_k c_T^k \right\|_F^2, \tag{11}$$

where E_k indicates the representation error of the sample of signal after removing the kth atom and its fixed.

The new proposed 1D-KSVD algorithm is presented in Listing 3.

Algorithm 4. $\left(G^{(j-1)}, W^{(j-1)}\right) = CLUSTER(D, N, \theta)$

$G^{(0)} = \emptyset$ {initial set of centroids}
$W^{(0)} = \emptyset$ {initial set of index of each cluster}
$j = 1$ {initial the iterative variable}
$G^{(j-1)} = \left\{g_1^{(j-1)}, g_2^{(j-1)}, \ldots, g_N^{(j-1)}\right\}$ {choice of N initial cluster centroids}
while $U^{(j-1)} < \theta$ {stopping rule} **do**
  for $n = 1$ to N **do**
    $W_n^{(j-1)} = \left\{l|\forall_{i\neq n}, \left\|d_l - g_n^{(j-1)}\right\|_2 < \left\|d_l - g_i^{(j-1)}\right\|_2\right\}$ {assign each atoms to the group that has the closest centroid}
    for $n = 1$ to N **do**
      $g_n^{(j)} = \frac{1}{\left|W_n^{(j-1)}\right|} \sum_{l \in W_n^{(j-1)}} d_l$ {calculate the new centroids}
      $U^{(j-1)} = \sum_{n=1}^{N} \sum_{l \in W_n^{(j-1)}} \left\|d_j - g_n^{(j-1)}\right\|_2$ {calculate error of the clustering}
      $j = j + 1$
    end for
  end for
end while
return $\left(G^{(j-1)}, W^{(j-1)}\right)$

3.2 Generation of the Dictionary Tree Structure (Clustering)

The resulting atoms have been grouped into clusters using the algorithm 4. This method creates clusters in the initial dictionary and organizes them in a hierarchical tree structure[5][6]. The elements from the initial dictionary form the leaves of the tree. As the result, the clustering process produces levels of tree structure. Each branch of the tree has N children and it is fully characterized by the list of the atom indexes W_n. A centroid g_n is assigned to the branch of tree that represents atoms of the dictionary in the corresponding subtree. The Clustering algorithm is presented in Listing 4.

4 Experimental Results

Performance of our approach to signal-based anomaly detection using greedy algorithms and the performance of the presented algorithms was evaluated with the use of DARPA trace benchmark database [7]. The benchmark test data contains attacks that fall into every layer of the TCP/IP protocol stack [8].

For the experiments we chose 10 and 20 minutes analysis windows because most of attacks (about 85%) ends within this time periods [10]. We extracted 15 traffic features (see Table 1) in order to create 1D signals for 1D K-SVD and Matching Pursuit analysis. For anomaly detection classification we used two parameters:

Table 1. Detection Rate for W5D5 (Fifth Week, Day 5) DARPA [7] trace

Network Traffic Feature	DR[%] MPMP (KSVD)	DR[%] MPMP $\Psi_{(k)}$	DR[%] $\Psi_{(k)}$ (KSVD)	DR[%] $\Psi_{(k)}$ Tree (KSVD)
icmp flows/min.	64,7	88,23	95,58	83,82
icmp in bytes/min.	79,14	76,47	91,17	92,65
icmp in frames/min.	85,29	80,88	91,17	89,71
icmp out bytes/min.	79,41	77,94	83,35	91,18
icmp out frames/min.	88,23	77,94	91,17	88,24
tcp flows/min.	48,52	55,88	79,41	67,65
tcp in bytes/min.	55,88	92,64	88,23	89,71
tcp in frames/min.	60,29	89,7	63,23	89,71
tcp out bytes/min.	36,76	94,11	61,76	86,76
tcp out frames/min.	38,23	79,41	83,82	79,41
udp flows/min.	85,29	97,05	88,23	79,41
udp in bytes/min.	76,47	100	100	97,06
udp in frames/min.	85,29	98,52	97,05	69,12
udp out bytes/min.	89,7	94,11	94,11	75
udp out frames/min.	91,17	94,11	100	76,47

Table 2. Proposed MP based ADS in comparison to DWT based ADS [9]. Both solutions were tested with the use of DARPA [7] testbed (results in table are for Week5 Day1 testday; DR-Detection Rate [%]) for MP-MP and $\Psi_{(k)}$ energy parameter.

Traffic Feature	$\Psi_{(k)}$ Tree KSVD DR[%]	$\Psi_{(k)}$ DR[%] MPMP	$\Psi_{(k)}$ DR[%] (KSVD)	DWT DR[%]
ICMP (flows/min.)	94.52	90.41	89.04	14.00
ICMP (in bytes/min.)	93.15	94.52	97.26	83.33
ICMP (out bytes/min.)	93.15	94.52	94.52	83.33
ICMP (in frames/min.)	89.04	84.93	98.63	32.00
ICMP (out frames/min.)	75.34	94.52	93.15	32.00
TCP (flows/min.)	63.01	98.63	98.63	26.67
TCP (in bytes/min.)	90.41	93.15	57.53	8.67
TCP (out bytes/min.)	97.26	93.15	90.41	8.67
TCP (in frames/min.)	84.93	95.89	95.89	2.00
TCP (out frames/min.)	89.04	95.89	95.89	2.00
UDP (flows/min.)	90.41	90.41	95.89	10.00
UDP (in bytes/min.)	87.67	98.63	97.26	11.33
UDP (out bytes/min.)	68.49	100	100.00	11.33
UDP (in frames/min.)	98.63	98.63	100.00	12.67
UDP (out frames/min.)	98.63	100	100.00	12.67

Table 3. Proposed MP based ADS in comparison to DWT based ADS [9]. Both solutions were tested with the use of DARPA [7] testbed (results in table are for Week5 Day1 testday; FP-False Positive [%]) for MP-MP and $\Psi_{(k)}$ energy parameter.

Traffic Feature	$\Psi_{(k)}$ Tree KSVD FP[%]	$\Psi_{(k)}$ FP[%] MPMP	$\Psi_{(k)}$ FP[%] (KSVD)	DWT FP[%]
ICMP (flows/min.)	39.73	38.35	38.35	79.33
ICMP (in bytes/min.)	41.10	38.35	38.35	416.00
ICMP (out bytes/min.)	41.10	38.72	42.46	416.00
ICMP (in frames/min.)	38.36	34.24	39.72	112.00
ICMP (out frames/min.)	36.99	35.61	42.46	112.00
TCP (flows/min.)	38.36	39.72	36.98	74.67
TCP (in bytes/min.)	30.14	41.00	42.46	23.33
TCP (out bytes/min.)	41.10	34.24	36.98	23.33
TCP (in frames/min.)	30.14	36.09	30.13	36.00
TCP (out frames/min.)	31.51	34.24	32.87	36.00
UDP (flows/min.)	38.36	39.70	38.35	74.67
UDP (in bytes/min.)	30.14	45.02	43.85	66.67
UDP (out bytes/min.)	34.25	46.57	39.72	66.67
UDP (in frames/min.)	39.73	45.20	43.83	66.67
UDP (out frames/min.)	42.47	45.20	41.09	66.67

Table 4. MPMP - Matching Pursuit Mean Projection for TCP port 80 UNINA test traces (2009) (windows with detected attacks are bolded)

Traffic Port 80	Tree 1D KSVD Window1	Tree 1D KSVD Window2	Tree 1D KSVD Window3	MPMP for normal trace	Traffic Port 80	Tree 1D KSVD Window1	Tree 1D KSVD Window2	Tree 1D KSVD Window3	MPMP for normal trace
Unina1	92,56	**6,42**	**80,88**	128,24	Unina9	96,45	106,66	**195,22**	128,24
Unina2	96,12	87,43	**54,82**	128,24	Unina10	**60,23**	152,2	130,69	128,24
Unina3	142,24	84,32	162,22	128,24	Unina11	**182,79**	**80,22**	100,22	128,24
Unina4	**39,45**	**51,65**	**22,45**	128,24	Unina12	**184,22**	**60,11**	**187,45**	128,24
Unina5	112,8	92,83	124,24	128,24	Unina13	152,33	92,25	**192,43**	128,24
Unina6	145,56	98,27	132,44	128,24	Unina14	90,77	91,25	102,52	128,24
Unina7	134,86	98,45	143,65	128,24	Unina15	112,75	120,24	104,24	128,24
Unina8	**58,87**	**44,32**	122,78	128,24					

- Matching Pursuit Mean Projection (MP-MP)

$$MP-MP=\frac{1}{n}\sum_{i=0}^{n-1}c_i. \tag{12}$$

- Energies of coefficients $\left\|c^{(k)}\right\|^2$, residues $\left\|rs^{(k)}\right\|^2$ and energy of dictionary elements $\left\|d^{(k)}\right\|^2$

$$\Psi_{(k)} = \left\|c^{(k)}\right\|^2 + \left\|rs^{(k)}\right\|^2 + \left\|d^{(k)}\right\|^2. \tag{13}$$

We compared two realizations of ADS based on analytical dictionary (with 1D Gabor atoms) and dictionary D obtained from real traffic signal where every atom is selected with the use of modificated 1D K-SVD algorithm. In eech table, coefficients obtained by means of 1D K-SVD are signed as: MPMP (KSVD) and $\Psi_{(k)}$ (KSVD). Additionally, we added the tree structure to dictionary D generated by 1D KSVD algorithm. Energy parameter $\Psi_{(k)}$ was also calculated for tree structure dictionary.

We compared results achieved by dictionary generated by 1D KSVD and Tree structure 1D KSVD dictionary. Results of detection rate DR and false positives rate FPR were comparable (See Tables 1 - 3). Tree structure dictionary has one main advantage which comes from the fact that search time process is significantly lower (up to 50%). Overall Detection Rate (ODR) is calculated for DR and FPR parameter. ODR takes into consideration set of traffic metrics where at the same time FPR is lowest and DR has highest value. ODR is also calculated for different ADS systems presented in [9][11]) for tree structure dictionary does not change significantly. For the presented ADS system with tree structure 1D KSVD dictionary we obtain 97% – 100% DR and 13 – 15% FPR for DARPA trace. Experimental results are presented in Tables 1 - 3.

We also tested our algorithm with other traffic databases obtained from UNINA (University of Napoli). Sample database contains normal traces and traces with anomalies for TCP port 80. In Table 4 the results of MPMP for 1D KSVD algorithm for subsequent traffic traces are presented. Traffic traces have approximately 1 hour length. Traffic was analyzed with the use of 10 minutes windows. We achieved DR 95% and FP 5%.

5 Conclusions

This paper presents greedy algorithms applied for network anomaly detection. The major contributions are: the proposition of the new 1D KSVD algorithm and its tree-based structure (clustering). The reported experimental results proved the effectiveness of the proposed method. We provided the comparison of the proposed greedy algorithms to DWT-based anomaly detection and reported better results in terms of detection rate and false positives. The presented results prove that the presented algorithms can be used for improving cybersecurity and resilience of the critical infrastructures and other critical networked systems where anomaly based approach should be applied.

References

1. Troop, J.A.: Greed is Good: Algorithmic Results for Sparse Approximation. IEEE Transactions on Information Theory 50(10) (2004)

2. Mallat, S.G., Zhang, Z.: Matching Pursuit with time-frequency dictionaries. IEEE Transactions on Signal Processing 41(12), 3397–3415 (1993)
3. Pati, Y.C., Rezaiifar, R., Krishnaprasad, P.S.: Orthogonal matching pursuit: recursive function approximation with applications to wavelet decomposition. In: Asilomar Conference on Signals, Systems and Computers, vol. 1, pp. 40–44 (1993)
4. Aharon, M., Elad, M., Bruckstein, A.: K-SVD. An algorithm for designing overcomplete dictionaries for sparse representations. IEEE Trans. on Signal Processing 54, 4311–4322 (2006)
5. Jost, P., Vandergheynst, P., Frossard, P.: Tree-Based Pursuit: Algorithm and Properties. In: Swiss Federal Institute of Technology Lausanne (EPFL), Signal Processing Institute Technical Report, TR-ITS-2005.013 (2005)
6. Choraś, M., Saganowski, Ł., Renk, R., Hołubowicz, W.: Statistical and signal-based network traffic recognition for anomaly detection. Expert Systems: The Journal of Knowledge Engineering (2011), doi: 10.1111/j.1468-0394.2010.00576.x
7. Defense Advanced Research Projects Agency DARPA Intrusion Detection Evaluation Data Set, `http://www.ll.mit.edu/mission/communications/ist/corpora/ideval/data/index.html`
8. DeLooze, L.: Attack Characterization and Intrusion Detection using an Ensemble of Self-Organizing Maps. In: IEEE Workshop on Information Assurance United States Military Academy, pp. 108–115. West Point, New York (2006)
9. Wei, L., Ghorbani, A.: Network Anomaly Detection Based on Wavelet Analysis. EURASIP Journal on Advances in Signal Processing 2009, Article ID 837601, 16 pages (2009), doi:10.1155/2009/837601
10. Lakhina, A., Crovella, M., Diot, C.H.: Characterization of network-wide anomalies in traffic flows. In: Proceedings of the 4th ACM SIGCOMM Conference on Internet Measurement, pp. 201–206 (2004)
11. Dainotti, A., Pescape, A., Ventre, G.: Wavelet-based Detection of DoS Attacks. In: IEEE GLOBECOM, San Francisco, CA, USA (November 2006)

Combined Classifiers with Neural Fuser for Spam Detection

Marcin Zmyślony, Bartosz Krawczyk, and Michał Woźniak

Department of Systems and Computer Networks,
Wroclaw University of Technology,
Wybrzeze Wyspianskiego 27, 50-370
Wroclaw, Poland
{marcin.zmyslony,bartosz.krawczyk,michal.wozniak}@pwr.wroc.pl

Abstract. Nowadays combining approach to classification is one of the most promising directions in the pattern recognition. There are many methods of decision making which could be used by the ensemble of classifiers. This work focuses on the fuser design to improve spam detection. We assume that we have a pool of diverse individual classifiers at our disposal and it can grow according the change of spam model. We propose to train a fusion block by the algorithm which has its origin in neural approach and the details and evaluations of mentioned method were presented in the previous works of authors. This work presents the results of computer experiments which were carried out on the basis of exemplary unbalanced spam dataset. They confirm that proposed compound classifier is further step in email security.

Keywords: combined classifiers, neural networks, fuser design, spam detection, concept drift, imbalanced data.

1 Introduction and Related Works

People cannot even imagine how many classification problems are being solved in each second of their regular day. The trivets example could be morning breakfast where human's brain does classification for: temperature of coffee or tea, food to eat, TV station to watch and so many others factors which could have influence on our life. The main problem of the computers and software dedicated for them, is that they do not have intelligence and for each classification task they have to use algorithms (methods) which give very close results to that, which are generated by human's brain. For several practical cases the automatic classification methods are still significantly worse than human decisions [22]. The spam filtering is the obvious example of such a problem because spammers can often lead on the computer algorithms used by spam filters. Therefore methods of automatic pattern recognition are nowadays the focus of intense research. The aim of such task is to classify the object to one of predefined categories, on the basis of observation of its features. Among the proposed classification methods the multiple classifier systems are currently recognized as the promising research direction.

Á. Herrero et al. (Eds.): Int. JointConf. CISIS'12-ICEUTE'12-SOCO'12, AISC 189, pp. 245–252.
springerlink.com

In this paper the method of classifier fusion are presented, which in our opinion, could be used in spam detection. This method is based on a block of fusion which is realized as neural networks. The next paragraphs show that presented algorithm could be further step in security of emails.

2 Spam Detection

In its beginnings spam was a seldom event and was easily intercepted by the usage of basic techniques, such as keyword filters and black lists [8]. Today, there is nothing basic or seldom about spam – it is numerous, well-designed and brings millions of dollars loses to corporations every year. Spammers are developing more sophisticated and more intelligent attacks – programming their messages and creating threats designed to defeat even the most complex anti-spam and anti-virus filters [11]. That is why currently the spam filter area is the focus of intense research. Through the last years there have been proposed many different methodologies for preventing the occurrence of spam in mailboxes [3]. Below the most significant of them are presented:

- **Domain-level black and white lists:** those are the original and most basic forms of blocking spam [13]. Employing this method, a domain administrator simply puts an offending spammer's email address on a blacklist so that all future email from that address is denied. Alternatively, to ensure that email from a particular sender is always accepted, administrators would add a sender's email address to a white list.
- **Heuristic engines:** are a series of rules used to score the spam probability an email – loaded with rules given by a human expert by which a program analyses an email message for spam-like characteristics [14]. Each of the given features / word phrases may have assigned different point values, according to their likelihood of being a part of a spam message. Passing a certain threshold would classify the message as a spam.
- **Statistical classification:** is considered as one of the most effective spam fighting methods [5]. Currently, the most common form of statistical classification is Bayesian filtering [19], which like heuristics, also analyzes the content of an email message. The main difference from the heuristic approach is that statistical classification engines assess the probability that a given email is spam based on how often certain features, called "tokens," within that email have appeared previously in emails labelled as spam ones. To make this determination, statistical classification engines compare a large body (called "corpus") of spam email with a similarly-sized corpus of legitimate email.

Apart from mentioned approaches several other machine learning methods were successfully applied for the task of spam filtering. The Support Vector Machine (SVM) presents a good accuracy in handling highly dimensional tokenized messages [4]. Yet another problem, apart from the size of feature space, is the often-encountered in this area problem of class imbalance. Modified SVMs again proven themselves useful for such tasks [6].

Multiple Classifier Systems are widely being used in security systems [1] as Intrusion Detection and Prevention Systems (IDS/IPS) or to the problem considering in this paper [15]. Usually the main motivation of applying mentioned above methods is that we usually can get a pool of individual trained classifiers on the basis of different learning sets or using different sets of attributes. On the other hand the crucial security problem is that the spam filters are facing with the unknown type of spam ("zero day attack problem") what causes that the developing spam filters should adapt to a new model e.g., by adding a new individual classifier trained on the basis of the new data to the pool. Such a problem is called concept drift and is widely discussed in literature [7].

3 Tokenization

Message tokenization is the process of discretizing an e-mail message using carefully selected delimiters [2]. The selection of delimiters is extremely important as it affects the classification accuracy [12]. Traditionally, selection is performed manually by trial and error and with the experience from previous similar processes. Therefore in spam filtering tokens respond to features in canonical classification problems.

Tokenization can be seen as a heuristic function inside the process of statistical spam filtering. The process of feature extraction itself is fully automatic, but the underlying schema on how they are parsed out from an email is programmed by a human. Nowadays there are attempts to establish a fully automatic tokenization approach by the means of artificial intelligence and natural language processing. Currently there are no methods that can outperform or even compete with human-supervised selection, but many specialists predict that in nearest years to come such procedure should be possible to create.

Let us present some basic notations for tokenization:

The **alphabet** $\{a_1, a_2, \ldots, a_n\}$ is a finite set of symbols. Each symbol is a character. E-mail messages use ASCII character code. The alphabet used is A_{ASCII} = {char(0), char(1), ... , A, B,..., char(255)}, where *char(i)* is the character corresponding to integer *i*.

A **string** X_A is a sequence of symbols from the alphabet *A*. Every e-mail message is a string over the alphabet A_{ASCII}.

The **cardinality** of a string *X* is denoted by $|X| = n$. For example, if *X* = spam, then $|X| = 4$. The cardinality of the alphabet $|A_{ASCII}| = 256$.

A **delimiter** set, *D*, for a string *X* over the alphabet *A* is a given subset of *A* .

A **tokenizer** *T* is a function of a string *X* and a delimiter set *D* such that $Q = T(X,D)$. The **production** $Q = \{x_1, x_2, \ldots, x_n\}$: x_i is a substring between two delimiters in *X*.

Such notation allows us to define the basic mechanism behind the process of tokenization. Yet at the same time it displays the weaknesses of such an approach. It is very dependent on the choice of delimiters and therefore may easily fail in case of foreign language spam, that doesn't use popular delimiters as white spaces etc.

4 Classifier

The formal model of used fuser based on discriminants is as follows [21]. Let us assume that we have n classifiers $\Psi^{(1)}, \Psi^{(2)}, ..., \Psi^{(n)}$ and each of them makes a decision on the basis of the values of discriminants. Let $F^{(l)}(i, x)$ denotes a function that is assigned to class i for a given value of x, and which is used by the l-th classifier $\Psi^{(l)}$. A combined classifier $\hat{\Psi}(x)$ is described as follows

$$\hat{\Psi}(x) = i \quad \text{if} \quad \hat{F}(i, x) = \max_{k \in M} \hat{F}(k, x), \tag{1}$$

where

$$\hat{F}(i, x) = \sum_{l=1}^{n} w^{(l,i)} F^{(l)}(i, x) \text{ and } \sum_{i=1}^{n} w^{(l,i)} = 1 \tag{2}$$

and $w^{(l,i)}$ denotes to weight assigned to the l-th classifier and the i-th class. Let's us note that the fuser can be implemented as one-layer perceptron depicted in Fig.1 and be trained using method dedicated to such a model.

5 Experimental Investigation

The aim of the experiments is to evaluate the performance of fuser based on weights dependent on classifier and class number.

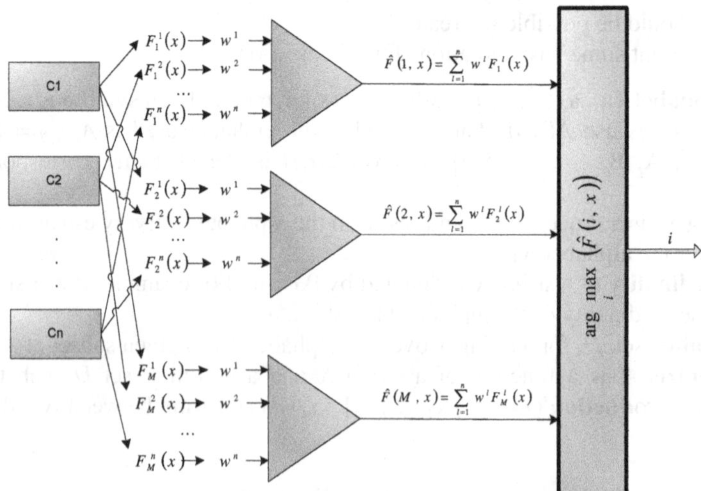

Fig. 1. One layer neural network as a fuser which uses weights depend on classifier and class number

5.1 Dataset

In this study we use the *ECUE* spam dataset, available at [20]. The dataset is a collection of spam and legitimate consists of emails received by one individual. During the preprocessing the following assumptions were made:

- No stop word removal or stemming has been performed on the emails. HTML markup was not removed but was included in the tokenization (split on tag delimiters).
- The name and value of HTML attributes were tokenized separately.
- The body of the email and certain email headers were included in the tokenization.
- Three types of features have been extracted for this dataset:
  - *word features*, which represent the occurrence of a single character in the email;
  - *character features* which represent the occurrence of a single character in the email;
  - *structural features* which represent the structure of the email.
- The datasets include the top 700 features selected using Information Gain but in this study we used the top 100 of most frequent tokens to reduce the computational complexity of the model.

In the dataset under consideration there is an equal number of SPAM and HAM messages – 500 from each class. This is an artificial ratio, as in real-life spam detection problems the number of SPAM messages is several times smaller than number of proper messages at the disposal. This results in a problem known as imbalanced classification, which causes significant difficulties for classification algorithm. Therefore to examine the behaviour of the proposed method we randomly undersampled the SPAM class to create an imbalance ratio 10:1. This allowed to examine how the proposed fusion block handles minority class recognition.

5.2 Set-Up

All experiments were carried out in Matlab environment using PRTools toolbox [10] and our own software. The experiments were carried out on ECUE benchmark dataset, which is described in paragraph no. 5.1.

For the purpose of this experiment, five neural networks were prepared that could be treated as individual classifiers. We trained each neural classifier independently i.e., initial values of neurons' weights were chosen randomly to ensure their diversity. Classification errors of individual classifiers used during experiments (denoted as C1, C2, C3, C4, C5) are presented in Tab.1.

Table 1. Classification errors of individual classifiers

Classifier	C1	C2	C3	C4	C5
Error	10.1%	8.8%	8.8%	9.6%	13.3%

The rest details of used neural nets are as follows:

- Five neurons in the hidden layer,
- Sigmoidal transfer function,
- Back propagation learning algorithm,
- Number of neurons in the last layer equals number of classes of given experiment. During experiments we would like to compare quality of fusers
- FCCN - fuser based on weights dependent on classifier and class number
- MV – fuser based on majority voting rules

with the quality of Oracle classifier [18]. Oracle is a purely theoretical fusion model which gives a good result if at least one of the classifiers from the pool indicates the proper class number.

For trained fuser realized according the idea depicted in Fig.1 number of training iterations was fixed to 1500. The experiment was repeated ten times. Classification error of individual classifier and fuser models were estimated using 10 Fold cross-validation method.

5.3 Results

The results of the experiment are presented in Fig.2.

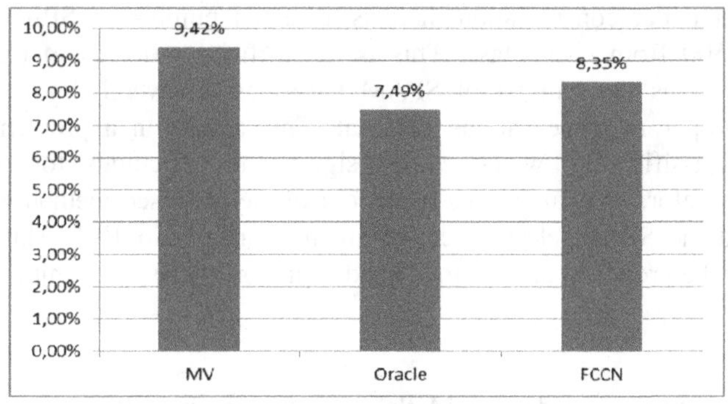

Fig. 2. Results of experiment

Table 2. Confusion matrix for FCCN

	Estimated Labels	
True Labels	SPAM	HAM
SPAM	31	11
HAM	31	394

Additionally as it was previously mentioned we have dealt with an imbalanced classification problem. For such a task canonical accuracy measures do not give a good insight into the actual performance of the classifier. The end-user is not only interested in the average accuracy, but also in the recognition rate of only SPAM messages. Therefore a confusion matrix was presented in the Tab.2, which presents the quality of classification with respect to the individual classes.

The following conclusions can be drawn on the basis of experiments' results:

- Classification errors of FCCN are smaller than MV and each of individual classifiers,
- Only abstract model Oracle achieved better quality of classification than FCCN,
- Results show that even if we have not very good classifiers (like in new kinds of SPAM attacks) we can get satisfied results of SPAM blocking.

6 Conclusions

In this paper we presented an application of the fusion method based on one-layer neural network for the task of efficient spam detection. Spam detection problem requires more and more sophisticated methods to cope with the continually rising complexity of non-wanted messages. Therefore popular, simple methods, based on using a single model, may offer inefficient quality. By using a combined approach one may explore the individual strengths of each of classifiers from the pool, while reducing their weaknesses. The presented approach utilized a trained fusion method, instead of well-known passive combiners. The overall training time of such an approach is longer, yet may result in significant improvement of accuracy by tuning the multiple classifier system to the specific problem under consideration. Experimental investigations proved the high quality of the proposed method. The ensemble returned better accuracy than each of the individual classifiers from the pool. It is worth stressing, that in the considered problem of ECUE spam dataset we dealt with the problem of imbalanced classification. Proposed method returned satisfactory results without referring to pre-processing approaches dedicated to dealing with such data. Therefore we eliminated the risk of shifting the class distribution, which is present in the approaches based on creating the artificial objects, or the problem of defining the cost matrix, which is mandatory for the cost-sensitive approaches. Our future works will concentrate on the usage of one-class classifiers for imbalanced spam detection problem [16]. It is worth noticing that a promising research direction in one-class classification for spam filtering may be an introduction of one-class classifiers ensembles and novel diversity measures designed for them [17].

We believe that the proposed methodology for classifier combination may be useful for real-life spam filters in industrial and every-day usage.

Acknowledgements. Marcin Zmyslony is a scholar within the project co-financed by the European Union under the European Social Fund. Bartosz Krawczyk and Michal Wozniak are supported in part by the Polish National Science Centre under a grant for the period 2011-2014.

References

1. Biggio, B., Fumera, G., Roli, F.: Multiple Classifier Systems under Attack. In: El Gayar, N., Kittler, J., Roli, F. (eds.) MCS 2010. LNCS, vol. 5997, pp. 74–83. Springer, Heidelberg (2010)

2. Blanzieri, E., Bryl, A.: A survey of learning based techniques of email spam filtering. A Technical Report, University of Trento (2008)
3. Calton, P., Webb, S.: Observed trends in spam construction techniques: A case study of spam evolution. In: Proc of the 3rd Conference on E-mail and Anti-Spam, Mountain View, USA (2006)
4. Caruana, G., Li, M., Qi, M.: A MapReduce based parallel SVM for large scale spam filtering. In: Proc. of the 8th International Conference on Fuzzy Systems and Knowledge Discovery (2011)
5. Chih-Chin, L., Ming-Chi, T.: An empirical performance comparison of machine learning methods for spam e-mail categorization. In: Hybrid Intelligent Systems, pp. 44–48 (2004)
6. Dai, N., Davison, B.D., Qi, X.: Looking into the past to better classify web spam. In: AIRWeb 2009: Proc. of the 5th International Workshop on Adversarial Information Retrieval on the Web. ACM Press (2009)
7. Delany, S.J., Cunningham, P., Tsymbal, A., Coylem, L.: A case-based technique for tracking concept drift in spam filtering. Know.-Based Syst. 18, 187–195 (2005)
8. Diao, L., Yang, C., Wang, H.: Training SVM email classifiers using very large imbalanced dataset. Journal of Experimental and Theoretical Artificial Intelligence 24(2), 193–210 (2012)
9. Duin, R.P.W., Juszczak, P., Paclik, P., Pekalska, E., de Ridder, D., Tax, D.M.M.: PRTools4, A Matlab Toolbox for Pattern Recognition. Delft University of Technology (2004)
10. Erdelyi, M., Benczur, A.A., Masanes, J., Siklosi, D.: Web spam filtering in internet archives. In: AIRWeb 2009: Proceedings of the 5th International Workshop on Adversarial Information Retrieval on the Web. ACM Press (2009)
11. Erdelyi, M., Garzo, A., Benczur, A.A.: Web spam classification: a few features worth more. In: Joint WICOW/AIRWeb Workshop on Web Quality In Conjunction with the 20th International World Web Conference in Hyderabad. ACM Press, India (2011)
12. Gansterer, W., et al.: Anti-spam methods – state-of-theart. Tech. rep., Institute for Distributed and Multimedia Systems, University of Vienna (2004)
13. Henzinger, M.R., Motwan, R., Silverstein, C.: Challenges in web search engines. SIGIR Forum 36(2), 11–22 (2002)
14. Hershkop, S., Stolfo, S.J.: Combining email models for false positive reduction. In: Proceedings of the Eleventh ACM SIGKDD International Conference on Knowledge Discovery and Data Mining, Chicago, Illinois, USA, pp. 98–107 (2005)
15. Kang, I., Jeong, M.K., Kong, D.: A differentiated one-class classification method with applications to intrusion detection. Expert Systems with Applications 39(4), 3899–3905 (2012)
16. Krawczyk, B., Woźniak, M.: Combining Diverse One-Class Classifiers. In: Corchado, E., Snášel, V., Abraham, A., Woźniak, M., Graña, M., Cho, S.-B. (eds.) HAIS 2012, Part II. LNCS(LNAI), vol. 7209, pp. 590–601. Springer, Heidelberg (2012)
17. Kuncheva, L.I.: Combining Pattern Classifiers: Methods and Algorithms. Wiley-Interscience (2004)
18. Sahami, M., Dumais, S., Heckerman, D., Hirvitz, E.: A Bayesian approach to filtering junck e-mail, Learning for Text Categorization: paper from the 1998 Workshop. AAAI Technical Report WS-98-05 (1998)
19. School of Computing, `http://www.comp.dit.ie/sjdelany/Dataset.htm`
20. Woźniak, M., Zmyślony, M.: Combining classifiers using trained fuser - analytical and experimental results. Neural Network World 20(7), 925–934 (2010)
21. Wozniak, M.: Proposition of common classifier construction for pattern recognition with context task. Knowledge-Based Systems 19(8), 617–624 (2006)

Combined Bayesian Classifiers Applied to Spam Filtering Problem

Karol Wrótniak and Michał Woźniak

Department of Systems and Computer Networks, Wroclaw University of Technology
Wybrzeze Wyspianskiego 27, 50-370 Wroclaw, Poland
Michal.Wozniak@pwr.wroc.pl

Abstract. This paper focuses on the problem of designing effective spam filters using combined Näive Bayes classifiers. Firstly, we describe different tokenization methods which allow us for extracting valuable features from the e-mails. The methods are used to create training sets for individual Bayesian classifiers, because different methods of feature extraction ensure the desirable diversity of classifier ensemble. Because of the lack of an adequate analytical methods of ensemble evaluation the most valuable and diverse committees are chosen on the basis of computer experiments which are carried out on the basis of our own spam dataset. Then the number of well known fusion methods using class labels and class supports are compared to establish the final proposition.

Keywords: adaptive classifier, classifier ensemble, combined classifier, Näive Bayes classifier, n-gram, OSB, SBPH, spam filtering.

1 Introduction

Multiple Classifier Systems (MCSs) are widely being used in security systems [1] as Intrusion Detection and Prevention Systems (IDS/IPS) or to the problem considered in this paper [2]. The main motivation is that security problems have usually dynamic nature i.e., their model can change while their exploitation and then we can face so-called "zero day attack" problem. If we detect new anomaly, or so called concept drift [3], then we can collect a sufficient number of examples (e.g. e-mails) a quite quickly and train an individual classifier. Then new classifier can be added to a pool used by MCS. It is usually faster and cheaper than rebuilding the whole classifier model as in the case of e.g. decision tree.

The first generation of spam filters used simple and noneffective basic techniques, such as keyword filters and black lists. Unfortunately they cannot adopt to the new spam models, because spammers have been developing more sophisticated and more intelligent attacks - programming their messages and creating threats designed to defeat even the most complex anti-spam and anti-virus filters [4]. Through the last years there have been proposed many different methodologies for preventing against spam [5]. The most significant methods are listed below:

– Domain-level black list consisted of spammers e-mail addresses and white list which includes senders' e-mails which are always accepted.

Á. Herrero et al. (Eds.): Int. JointConf. CISIS'12-ICEUTE'12-SOCO'12, AISC 189, pp. 253–260.
springerlink.com

- Heuristic engines which are a series of rules used to score the spam probability of an email - loaded with rules given by a human expert by which a program analyzes an emails for spam-like characteristics [6].
- Statistical classification is considered as one of the most effective spam fighting methods [7]. Currently, the most common form of statistical classification is Bayesian filtering [8], which as heuristics, also analyzes the content of a message. The main difference form the heuristic approach is that statistical classification engines assess the probability that a given email is spam based on how often certain features, called "tokens," within that email have appeared previously in emails labeled as spam ones. To make this determination, statistical classification engines compare a large body (called "corpus") of spams with a similarly-sized corpus of legitimate emails.

In this paper we focus on the last stream because an useful spam classification model should not only take account of the fact that spam evolves over time, but also that labeling a large number of examples for initial training can be expensive. The idea of using anti-spam filter based on Bayes theorem was first proposed by P. Graham in [9]. Elements (tokens) which statistics are built for single words called unigrams. In [10] Yerazunis et al. introduced more complex tokens (SBPH an OSB). In according to superstition "two heads are better than one" the idea of using committees of parallel classifiers which uses different types of tokens was developed. Only one publication broaching similar issue was found. In [11] authors proposed BKS fusion of classifier working with plain text and other one using OCR.

The content of the work is as follows. Firstly, the idea of tokenization which is a kind of preprocessing responsible for feature extraction is presented. Then we introduce shortly the combined approach to classification. The next section describe the computer experiment and the last one concludes the paper.

2 Tokenization

Message tokenization is the process of discretizing an e-mail using carefully selected delimiters [12]. The selection is extremely important as it affects the classification accuracy [13]. Traditionally, it is performed manually using the experiences from previous similar processes. Tokenization can be seen as a heuristic, fully automatic feature extraction, but the underlying schema on how they are parsed out from an email is programmed by a human. This may seem at first as a static approach that must be often changed. Yet the language itself changes slowly and only minor extensions and improvements are made from time to time to allow handling of contemporary spam messages. Currently there are no methods that can outperform or even compete with human-supervised selection, but many specialists predict that in nearest years to come such procedure should be possible to create. Token type is determined first of all by 2 parameters:

- tokenization method (n-gram, OSB, SBPH was examined)
- tokenization window length - n (examined ranging from 1 to 10)

The second parameter does not specify a number of words which will be included in generated multi-word token but it is number of words from source text participating in creating that complex token. These 2 numbers only in special cases are equal, not all of the words from window must be included in built tokens.

3 Multiple Classifier Systems

For a practical decision problem we usually can have several classifiers at our disposal. One of the most important issue while building MSCs is how to select a pool of classifiers [14]. We should stress that combining similar classifiers would not contribute much to the system being constructed, apart from increasing the computational complexity. An ideal ensemble consists of classifiers with high accuracy and high diversity i.e., they are mutually complementary. In our paper we would like to assure their diversity by training individual classifiers on the basis of learning sets creating by the different tokenization methods.
Another important issue is the choice of a collective decision making method [15]. The first group of methods includes algorithms for classifier fusion at the level of their responses [14]. Initially one could find only the majority vote in literature, but in later works more advanced methods were proposed [16].

Let's assume that we have n classifiers $\Psi^{(1)}$, $\Psi^{(2)}$, ..., $\Psi^{(n)}$. For a given object x each of them decides if it belongs to class $i \in M = \{1, ..., M\}$. The combined classifier $\bar{\Psi}$ makes decision on the basis of the following formulae:

$$\bar{\Psi}\left(\Psi^{(1)}(x),\ \Psi^{(2)}(x),\ ...,\ \Psi^{(n)}(x)\right) = \arg\max_{j \in M} \sum_{l=1}^{n} \delta\left(j,\ \Psi^{(l)}(x)\right) w^{(l)} \Psi^{(l)}(x), \quad (1)$$

where $w^{(l)}$ is the weight assigned to the l-th classifier and δ is the Kronekers's delta.

The second group of collective decision making methods exploits classifier fusion based on discriminants. The main form of discriminants is the posterior probability, typically associated with probabilistic models of the pattern recognition task [17], but it could be given for e.g., by the output of neural networks or that of any other function whose values are used to establish the decision by the classifier. The aggregating methods, which do not require a learning procedure, use simple operators, like taking the maximum or average value. However, they are typically subject to very restrictive conditions [18], which severely limit their practical use. Therefore the design of new fusion classification models, especially those with a trained fuser block, are currently the focus of intense research.

Let's assume that each individual classifier makes a decision on the basis of the values of discriminants. Let $F^{(l)}\ (i,\ x)$ denotes a function that is assigned to class i for a given value of x, and which is used by the l-th classifier $\Psi^{(l)}$. The combined classifier $\hat{\Psi}\ (x)$ uses the following decision rule[19]

$$\hat{\Psi}(x) = i \quad if \quad \hat{F}(i,\, x) = \max_{k \in M} \hat{F}(k,\, x), \tag{2}$$

where

$$\hat{F}(i,\, x) = \sum_{l=1}^{n} w^{(l)} F^{(l)}(i,\, x) \;\; and \;\; \sum_{i=1}^{n} w^{(l)} = 1. \tag{3}$$

4 Experimental Investigations

The main objective of the experiment is to evaluate if pools of Näive Bayes classifiers trained on the basis of different sets of tokens can create valuable committees for MCSs using as spam filters [20]. Firstly, on the basis of computer experiments several ensembles will be selected, then we will use them for a wide range of fusion methods based on class labels and class supports as well.

4.1 Data Acquisition

E-mails used in experiments were collected during the 5 months to the specially established mailboxes hosted by various providers (e.g. yahoo, o2.pl, hotmail). In order to get spam some of these addresses were published in the Internet (e.g. on Twitter and newsgroups) in plain text (without any camouflage). Several newsletters were also subscribed to get legitimate mail (ham). All of gathered e-mails were labeled manually and sorted by date of receiving. The dataset consists of 1395 spam messages and 1347 legitimate ones. Testing and training sets were created according to the following rule: e-mails received in last month came to testing set while former ones became training set. Two such pairs of sets were created:

- **full pair** - includes all of the received mail
- **reduced pair** - includes a third part of the received mail

E-mails contained in reduced pair were chosen with stratification, a third part from each week. Due to down rounding, this pair consisted of 720 e-mails.

4.2 Choosing Tokenization Methods

Fig. 1 shows that best values of window length are in the range from 1 to 3. Considering also false positive[1] rate and database size[2]. The best token types were chosen and presented in (tab. 1).

[1] False positive occurs when legitimate e-mail is classified as spam.

[2] Database size depends on window length approx. exponentially if method is n-gram or OSB and approx. linearly for SBPH.

Table 1. Member teams used in committees

BT	UBT	USB	US	UT
	unigram	unigram	unigram	unigram
		SBPH-3	SBPH-3	
bigram	bigram	bigram		
trigram	trigram			trigram

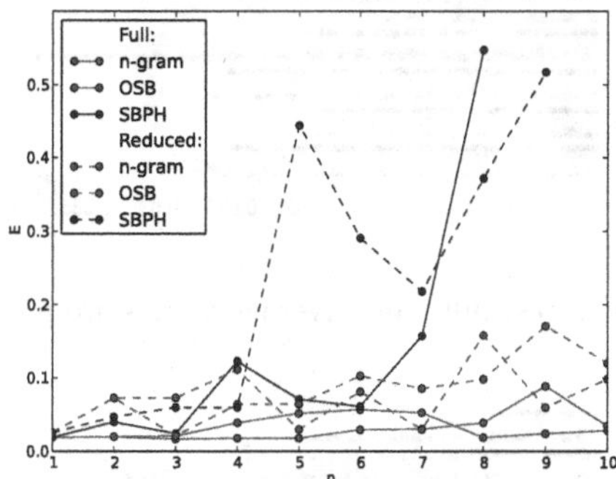

Fig. 1. Classifiers' error rates according to token type

4.3 Fusion Method Comparison

As we mentioned above there are mainly two groups of fusion methods [14] and [21] and therefore we decided to examine the well known representatives form both of them:

- Fusion of class labels:
  unanimity, majority voting, weighted voting, Behavior Knowledge Space (BKS), Naive Bayes combination (NB).
- Fusion of support functions:
  minimum, maximum, average, weighted average, decision templates (DT).

Weights can depend on classifier's accuracy (impractical option because it requires confusion matrices which are often unknown in real classifiers without testing) or on spamicity which is likelihood estimator always returned by classifier. All methods except decision templates and BKS are not trainable i.e., they use deterministic combination rules [18].

4.4 Results

The results of experiments are shown in Fig. 2 and Fig. 3 (classification errors).

Additionally for the case of all received e-mails dataset (ful pair) the false positive rate is presented in Fig. 4.

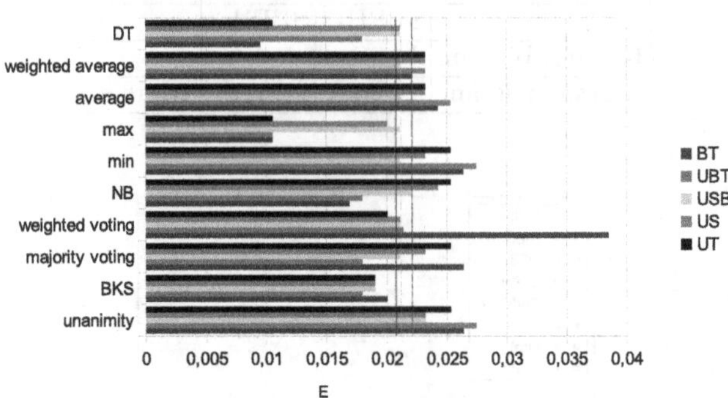

Fig. 2. Committees error rates, full pair. Average values are marked as vertical lines.

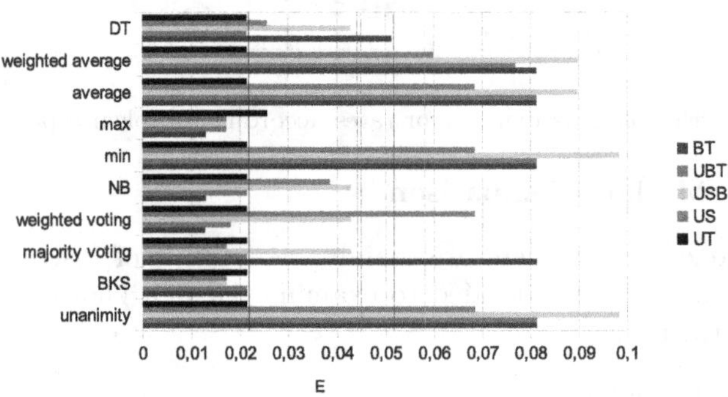

Fig. 3. Committees error rates, reduced pair. Average values are marked as vertical lines.

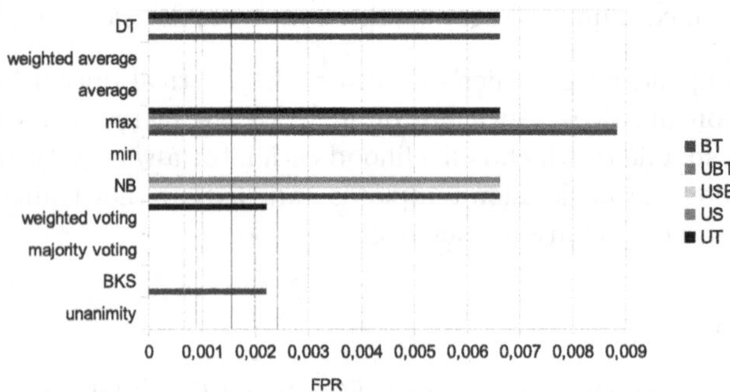

Fig. 4. Committees false positive rates, full pair. Average values are marked as vertical lines.

We can see that error rates were generally lower in the case of full pair. That proofs that classifiers were adaptive, their were able to improve their accuracies in the course of receiving new mail. Several committees (unanimity, minimum, both averages) had always zero false positive rates - it is important feature because false positives are more costly than false negatives in spam filtering issue. Some committees achieved lower error rates than the best of their members (e.g., BT team). Trainable BKS committee in many cases caused balance of results regardless of team.

On the average both voting methods can be considered as the best fusion methods. They achieved relatively low error and false positive rates regardless of the set pair and team. Both averages and minimum methods were clearly worse when operating on reduced pair in comparison to full one. From among teams UT was average best but when using reduced pair UBT was often better.

5 Final Remarks

The paper showed that combined classifier used a pool of Näive Bayes classifiers is valuable method of spam filtering. Our proposition was evaluated on the basis of the computer experiments carried out on our own dataset, because we believe that such a software should be personalized for an user's spam e-mail profile. According to the experiment's results it is worth noting that our proposition can adapt to a new type of spam what is a desirable characteristic of such a security systems as spam filters of IDS. However we should underline the one important characteristic of used tokenization method. On the one hand it can be easily fitted to an user's e-mail profile but on the other hand it is very dependent on the choice of delimiters and therefore may easily fail in case of foreign language spam.

Acknowledgment. This work is supported in part by The Polish National Science Centre under the grant which is being realized in years 2010-2013.

References

[1] Biggio, B., Fumera, G., Roli, F.: Multiple Classifier Systems under Attack. In: El Gayar, N., Kittler, J., Roli, F. (eds.) MCS 2010. LNCS, vol. 5997, pp. 74–83. Springer, Heidelberg (2010)

[2] Hershkop, S., Stolfo, S.J.: Combining email models for false positive reduction. In: Proceedings of the Eleventh ACM SIGKDD International Conference on Knowledge Discovery in Data Mining, pp. 98–107. ACM, New York (2005)

[3] Kurlej, B., Wozniak, M.: Active learning approach to concept drift problem. Logic Journal of the IGPL 20(3), 550–559 (2012)

[4] Erdélyi, M., Benczúr, A.A., Masanés, J., Siklósi, D.: Web spam filtering in internet archives. In: Proceedings of the 5th International Workshop on Adversarial Information Retrieval on the Web, AIRWeb 2009, pp. 17–20. ACM, New York (2009)

[5] Pu, C., Webb, S.: Observed trends in spam construction techniques: A case study of spam evolution. In: CEAS (2006)
[6] Henzinger, M.R., Motwani, R., Silverstein, C.: Challenges in web search engines. SIGIR Forum 36(2), 11–22 (2002)
[7] Lai, C.C., Tsai, M.C.: An empirical performance comparison of machine learning methods for spam e-mail categorization. In: Proceedings of the Fourth International Conference on Hybrid Intelligent Systems, HIS 2004, pp. 44–48. IEEE Computer Society, Washington, DC (2004)
[8] Sahami, M., Dumais, S., Heckerman, D., Horvitz, E.: A bayesian approach to filtering junk E-mail. In: Learning for Text Categorization: Papers from the 1998 Workshop, Madison, Wisconsin. AAAI Technical Report WS-98-05 (1998)
[9] Graham, P.: A plan for spam (August 2002), `http://www.paulgraham.com/spam.html`
[10] Siefkes, C., Assis, F., Chhabra, S., Yerazunis, W.S.: Combining Winnow and Orthogonal Sparse Bigrams for Incremental Spam Filtering. In: Boulicaut, J.-F., Esposito, F., Giannotti, F., Pedreschi, D. (eds.) PKDD 2004. LNCS (LNAI), vol. 3202, pp. 410–421. Springer, Heidelberg (2004)
[11] Gargiulo, F., Penta, A., Picariello, A., Sansone, C.: A personal antispam system based on a behaviour-knowledge space approach. In: Okun, O., Valentini, G. (eds.) Applications of Supervised and Unsupervised Ensemble Methods. SCI, vol. 245, pp. 39–57. Springer, Heidelberg (2009)
[12] Blanzieri, E., Bryl, A.: A survey of learning-based techniques of email spam filtering. Artif. Intell. Rev. 29(1), 63–92 (2008)
[13] Erdélyi, M., Garzó, A., Benczúr, A.A.: Web spam classification: a few features worth more. In: Proceedings of the 2011 Joint WICOW/AIRWeb Workshop on Web Quality. WebQuality 2011, pp. 27–34. ACM, New York (2011)
[14] Kuncheva, L.I.: Combining pattern classifiers: methods and algorithms. Wiley-Interscience (2004)
[15] Wozniak, M.: Proposition of common classifier construction for pattern recognition with context task. Knowledge-Based Systems 19(8), 617–624 (2006)
[16] van Erp, M., Vuurpijl, L., Schomaker, L.: An overview and comparison of voting methods for pattern recognition. In: Proceedings of the Eighth International Workshop on Frontiers in Handwriting Recognition, IWFHR 2002. IEEE Computer Society, Washington, DC (2002)
[17] Bishop, C.M.: Pattern Recognition and Machine Learning (Information Science and Statistics). Springer-Verlag New York, Inc., Secaucus (2006)
[18] Duin, R.P.W.: The combining classifier: To train or not to train? In: International Conference on Pattern Recognition, vol. 2, p. 20765 (2002)
[19] Jacobs, R.A.: Methods for combining experts' probability assessments. Neural Comput. 7(5), 867–888 (1995)
[20] Burduk, R.: Imprecise information in bayes classifier. Pattern Anal. Appl. 15(2), 147–153 (2012)
[21] Shipp, C.A., Kuncheva, L.I.: Relationships between combination methods and measures of diversity in combining classifiers. Information Fusion 3, 135–148 (2002)

Negobot: A Conversational Agent Based on Game Theory for the Detection of Paedophile Behaviour

Carlos Laorden[1], Patxi Galán-García[1], Igor Santos[1], Borja Sanz[1], Jose María Gómez Hidalgo[2], and Pablo Garcia Bringas[1]

[1] S^3Lab - DeustoTech Computing, University of Deusto
Avenida de las Universidades 24, 48007 Bilbao, Spain
{claorden,patxigg,isantos,borja.sanz,pablo.garcia.bringas}@deusto.es
[2] Optenet
Madrid, Spain
jgomez@optenet.com

Abstract. Children have been increasingly becoming active users of the Internet and, although any segment of the population is susceptible to falling victim to the existing risks, they in particular are one of the most vulnerable. Thus, some of the major scourges of this cyber-society are paedophile behaviours on the Internet, child pornography or sexual exploitation of children. In light of this background, Negobot is a conversational agent posing as a child, in chats, social networks and other channels suffering from paedophile behaviour. As a conversational agent, Negobot, has a strong technical base of Natural Language Processing and information retrieval, as well as Artificial Intelligence and Machine Learning. However, the most innovative proposal of Negobot is to consider the conversation itself as a game, applying game theory. In this context, Negobot proposes, first, a competitive game in which the system identifies the best strategies for achieving its goal, to obtain information that leads us to infer if the subject involved in a conversation with the agent has paedophile tendencies, while our actions do not bring the alleged offender to leave the conversation due to a suspicious behaviour of the agent.

Keywords: conversational agent, game theory, natural language processing.

1 Introduction

Children have been turning into active users of the Internet and, despite every segment of the population is susceptible to become a victim of the existing risks, they in particular are the most vulnerable. In fact, one of the major scourges on this cyber-society is paedophile behaviour, with examples such as child pornography or sexual exploitation.

Some researchers have tried to approach this problem with automatic paedophile behaviour identifiers able to analyse children and adult conversations

Á. Herrero et al. (Eds.): Int. JointConf. CISIS'12-ICEUTE'12-SOCO'12, AISC 189, pp. 261–270.
springerlink.com

using automatic classifiers either encoding manual rules [1] or by Machine Learning [2]. These kind of conversation analysers are also present in commercial systems[1].

In this context, "Negobot: A conversational agent based on game theory for the detection of paedophile behaviour" seeks to identify these types of actions against children. With this particular goal in mind, our system is a chatter bot that poses as a kid in chats, social networks and similar services on the Internet. Because Negobot is a chatter bot, it uses natural language processing (NLP), information retrieval (IR) and Automatic Learning. However, the most innovative proposal of Negobot is to apply game theory by considering the conversation a game. Our system observes the conversation as a competition where our objective is to obtain as much information as possible in order to decide whether the subject who is talking with Negobot has paedophile tendencies or not. At the same time, our actions (e.g., phrases and questions) may lead to the alleged aggressor to leave the conversation or even to act discretely.

As a major difference with others work, Negobot involves both a bot resembling the behaviour of a child and acting as a "hook" to Internet predators, and a game theory-based strategy for identifying suspicious speakers. Summarising, the main contributions of the Negobot system are:

- A structure of seven chatter-bots with different behaviours, reflecting the different ways of acting depending on the conversation state.
- A system to identify and adapt patterns within the conversations to maintain conversation flows closer to conversation flows between real persons.
- An evaluation function to classify the current conversation, in real time, to provide the chatter bot with specific information to follow an adequate conversation flow.

2 Negobot Architecture

Negobot includes the use of different NLP techniques, chatter-bot technologies and game theory for the strategical decision making. Finally, the glue that binds them all is an evaluation function, which in fact determines how the child emulated by the conversational agent behaves.

When a new subject starts a conversation with Negobot the system is activated, and starts monitoring the input from the user. Besides, Negobot registers the conversations maintained with every user for future references, and to keep a record that could be sent to the authorities in case of determining that the subject is a paedophile. Fig. 1 offers a functional flow of Negobot.

2.1 Processing the Conversation

Our AI system's knowledge came from the website Perverted Justice[2]. This website offers an extensive database of paedophile conversations with victims,

[1] E.g. Crisp Thinking provides such a service to online communities: `http://www.crispthinking.com/`

[2] http://www.perverted-justice.com

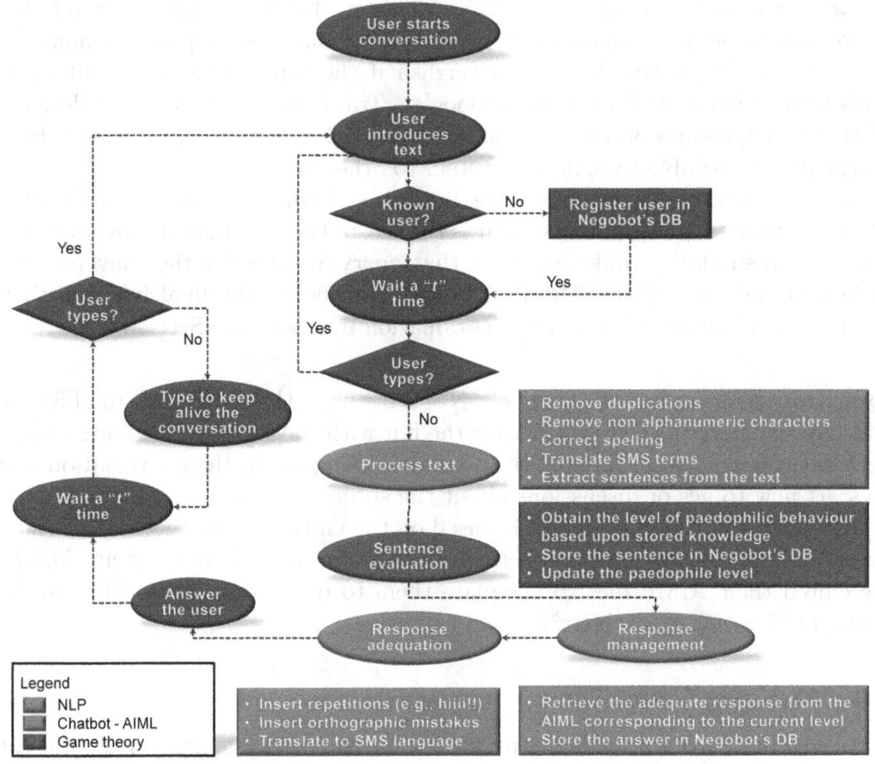

Fig. 1. Functional flow of Negobot

used in other works [2,3,1]. A total of 377 real conversations were chosen to populate our database (henceforth they will be called *assimilated conversations*).

Besides, Perverted Justice users provide an evaluation of each conversation's seriousness by selecting a level of "slimyness" (e.g., dungy, revolting thing). Note that this evaluation is given by the website's visitors, so it may not be accurate, but we consider that it is a proper baseline in order to compare future conversations of the chatter-bot.

2.2 Chat-Bot Conversation Process

Every time the user introduces some text our system gathers the input data, from now on "Conversational Unit" (CU), and character and word repetitions are removed. Then, "emoticons" are replaced and misspelled words are corrected through Levenshtein distance [6]. To translate SMS-like words, we use *diccionar-*

ioSMS[3], a website supported by *MSN, Movistar, Vodafone and Lleida.net.* Negobot also recognises named entities relying on a database of personal nouns[4].

Next, the CU is translated into English if the input language is different. This translation is performed with Google's Translation Service[5]. The decision of translating resides on the intention of normalising Negobot's knowledge base to English, to be able to scale the system to other languages.

In the next step the system queries Lucene[6], a high-performance Information Retrieval tool, to stablish how similar is the CU to the assimilated conversations, returning a similarity rank. Based on that query, we retrieve the conversations surpassing an empirically-defined similarity threshold. The final result is then used to calculate the B value of the evaluation function (see Section 2.4).

Question-Answering Patterns. Negobot uses the Artificial Intelligence Markup Language (AIML) to provide the bot with the capacity of giving consistent answers and, also, the ability to be an active part in the conversation and to start new topics or discussions about the subject's answers.

Although the AIML structure is based on the Galaia project[7], which has successfully implanted derived projects in social networks and chat systems [4,5,7], we edited their AIML files to adequate them to our needs. Those files can be found at the authors' website[8].

2.3 Applying Game Theory

Applied game theory intends to provide a solution for a large number of different problems through a deep analysis of the situation. Modelling these problems as a game implies that there exist two or more teams or players and that the result of the actions of one player depends on the actions taken by the other teams. It is important to notice that game theory itself does not teach to play the game but shows general strategies for competitive situations. The Negobot system assumes that there are two players in a conversation: the alleged paedophile and the system itself.

Goal. The main goal of the Negobot system is to gather, through the conversation, the maximum amount of information in order to evaluate it afterwards. This evaluation will determine whether the subject is a paedophile or not, to communicate, in the latter case, to pertinent authorities.

Conversation Level. The conversation level depends on the input data from the subject and determines which action should be taken by the system. There

[3] www.diccionariosms.com
[4] www.planetamama.com.ar
[5] http://translate.google.com
[6] http://lucene.apache.org/
[7] http://papa.det.uvigo.es/ galaia/ES/
[8] http://paginaspersonales.deusto.es/patxigg/recursos/NegobotAIML.zip

are seven levels consisting of seven different stages. In each stage, the subject talking with the bot is evaluated. The result of the evaluation will determine whether the subject that is talking is an alleged paedophile or not, depending on the content of the conversation and changing or not the level of the conversation.

- **Initial state (Start level or Level 0).** In this level, the conversation has started recently or it is within the fixed limits. The user can stay indefinitely in this level if the conversation does not contain disturbing content. The topics of conversation are trivial and the provided information about the bot is brief: only the name, age, gender and home-town. The bot does not provide more personal information until higher levels.
- **Possibly not (Level -1).** In this level, the subject talking to the bot, does not want to continue the conversation. Since this is the first negative level, the bot will try to reactivate the conversation. To this end, the bot will ask for help about family issues, bullying or other types of adolescent problems.
- **Probably not (Level -2).** In this level, the user is too tired about the conversation and his language and ways to leave it are less polite than before. The conversation is almost lost. The strategy in this stage is to act as a victim to which nobody pays any attention, looking for affection from somebody.
- **Is not a paedophile (Level -3).** In this level, the subject has stopped talking to the bot. The strategy in this stage is to look for affection in exchange for sex. We decided this strategy because a lot of paedophiles try to hide themselves to not get caught.
- **Possibly yes (Level +1).** In this level, the subject shows interest in the conversation and asks about personal topics. The topics of the bot are favourite films, music, personal style, clothing, drugs and alcohol consumption and family issues. The bot is not too explicit in this stage.
- **Probably yes (Level +2).** In this level, the subject continues interested in the conversation and the topics become more private. Sex situations and experiences appear in the conversation and the bot does not avoid talking about them. The information is more detailed and private than before because we have to make the subject believe that he/she owns a lot of personal information for blackmailing. After reaching this level, it cannot decrease again.
- **Allegedly paedophile (Level +3).** In this level, the system determines that the user is an actual paedophile. The conversations about sex becomes more explicit. Now, the objective is to keep the conversation active to gather as much information as possible. The information in this level is mostly sexual. The strategy in this stage is to give all the private information of the child simulated by the bot. After reaching this level, it cannot decrease again.

Actions. The Negobot system has three sensors, three actuators and three different actions to take to obtain its objective.

1. **Sensors:** (i) Knowledge of the current level, (ii) knowledge of the complete current conversation, (iii) assimilated conversations by the AI system.

2. **Actuators:** (i) The accusation level goes up, (ii) the accusation level goes down, (iii) the accusation level is maintained.
3. **Actions:** (i) Level goes up, this action increases the accusation level of the subject to modify the type of conversation by adding more personal information; (ii) level goes down, this action decreases the accusation level of the subject to change the type of conversation by adding more tentative information; (iii) maintain level, this action maintains the accusation level of the user to leave the current conversation type.

The consequences after each action are unknown beforehand because the system does not know how the other player will response to each answer, question, affirmation or negation.

Strategy. The aim of the system is to reach a final state while extracting the highest amount of information. The state of the system is determined both by the conversation and the conversation topic. The environment is partially observable since the chat-bot does not know what will the subject answer or question. Besides, the environment is stochastic because the conversations are composed only of questions, answers, affirmations and negations and when the user is writing one sentence, the environment does not change.

Furthermore, the system utilises an evaluation function (refer to Section 2.4) to analyse the answers. This analysis starts by evaluating the conversation level. Then, it generates the answer and, finally, communicates with the subject. This task had to be performed within a coherent and variable time so that it resembled the writing way of a child. To this end, the system calculates a waiting time for the response based on the number of words in the answer and an estimated child's writing speed.

2.4 Evaluation Function

The evaluation function determines the bot's behaviour according to the evaluation of the sentences that the subject introduces, and determines, in real time, its level of paedophilia. This function is also responsible for evaluating the evolution of the conversation (i.e., if the level maintains, goes up or down).

We defined the function as,

$$f(x) = \alpha + \beta + \gamma \tag{1}$$

α represents the historic values of the conversations with this user. This variable is defined as the sum of the levels of the previous conversations, divided by the number of total conversations. It is defined as, $\sum_{n=1}^{n=x-1} f(x)/n$, where n is the number of conversations the current user has maintained with Negobot.

β represents the "slimyness" of the current CU. We calculated the average value of the result of multiplying the similarity score of each of the retrieved paedophiles' conversations — the ones that surpass a fixed value of similarity —

by its "slimyness" value, $\sum_{n=1}^{n=x} score * slimyness/|AC_l|$, where AC_l is the total number of assimilated conversations.

γ evaluates the temporality of the subject's conversations (i.e., how frequent the user talks with Negobot). There are two possible values for γ, when $\beta = 0$ and when $\beta > 0$:

1. Value for γ if $\beta = 0$
   – The subtraction between the "slimyness" of the CU and the total CUs of the subject. This result is divided by the subtraction, in hours, between the last maintained CU and the first maintained CU: $Ncu - Tcu/T_{last} - T_{first}$
2. Value for γ if $\beta > 0$
   – The number of CUs that surpass the "slimyness" threshold divided by the subtraction, in hours, between the last maintained CU and the first maintained CU: $Ncu> threshold/T_{last} - T_{first}$

3 Examples of Conversations

This section presents some sample conversations with Negobot. It must be noted that the system was developed to analyse extensive conversations because the most dangerous paedophiles are too cautious. In our experiments we show some tests in a controlled environment and with a short extension (see Figure 2 and Figure 3[9]). Also note that the evaluation function does not depend only on the CU, historical conversation values and time are also taken into consideration.

For these experiments we performed two conversations: one *aggressive* conversation and one *passive* conversation.

– **Aggressive conversation.** The subject who maintained this conversation was asked to add explicit questions about sex related topics with the objective to obtain sexual information. Despite usually paedophiles use a more cautious approach, we needed to provide a short conversation talking about sex to show how the level of suspicion raises. In this conversation the bot is asked about sensitive unknown information, such as virginity or porn.
– **Passive conversation.** In this conversation the subject starts a conversation and when the age of Negobot is known he/she tries to end it. Negobot then asks the reason why the subject is leaving and after 10 minutes of not having a response tries to restart the conversation.

The conversations are in Spanish since its the language the system was designed for, but variables have been translated and visual guidance is provided to understand the results.

[9] In these examples, *ottoproject* is the Negobot system and Patxi is a user that emulates the different behaviours

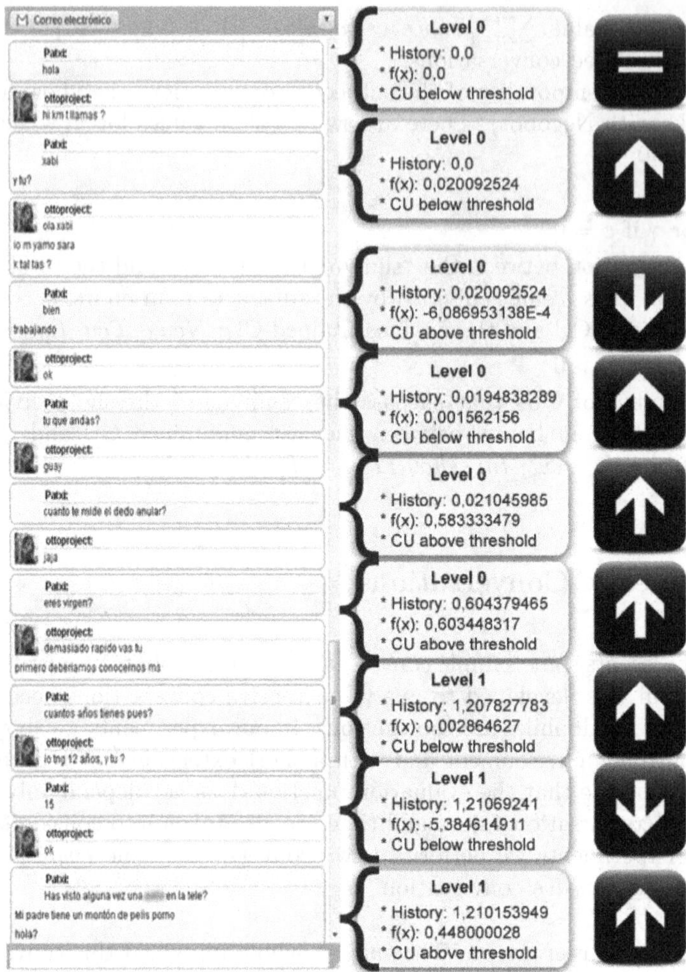

Fig. 2. Aggressive conversation

4 Discussion and Future Work

More and more children are connected nowadays to the Internet. Although this medium provides a lot of important advantages, it also provides the anonymity that paedophiles use to harass children. A rapid identification of this type of users on the Internet is crucial. Systems able to identify those menaces are going to be important protecting this less prepared population segment on the Internet.

Therefore, we consider that the Negobot system and its future updates are going to be useful to the authorities to detect and identify paedophile behaviours. The Negobot project and existing initiatives like PROTEGELES[10] can achieve a safer Internet.

[10] http://www.protegeles.com/

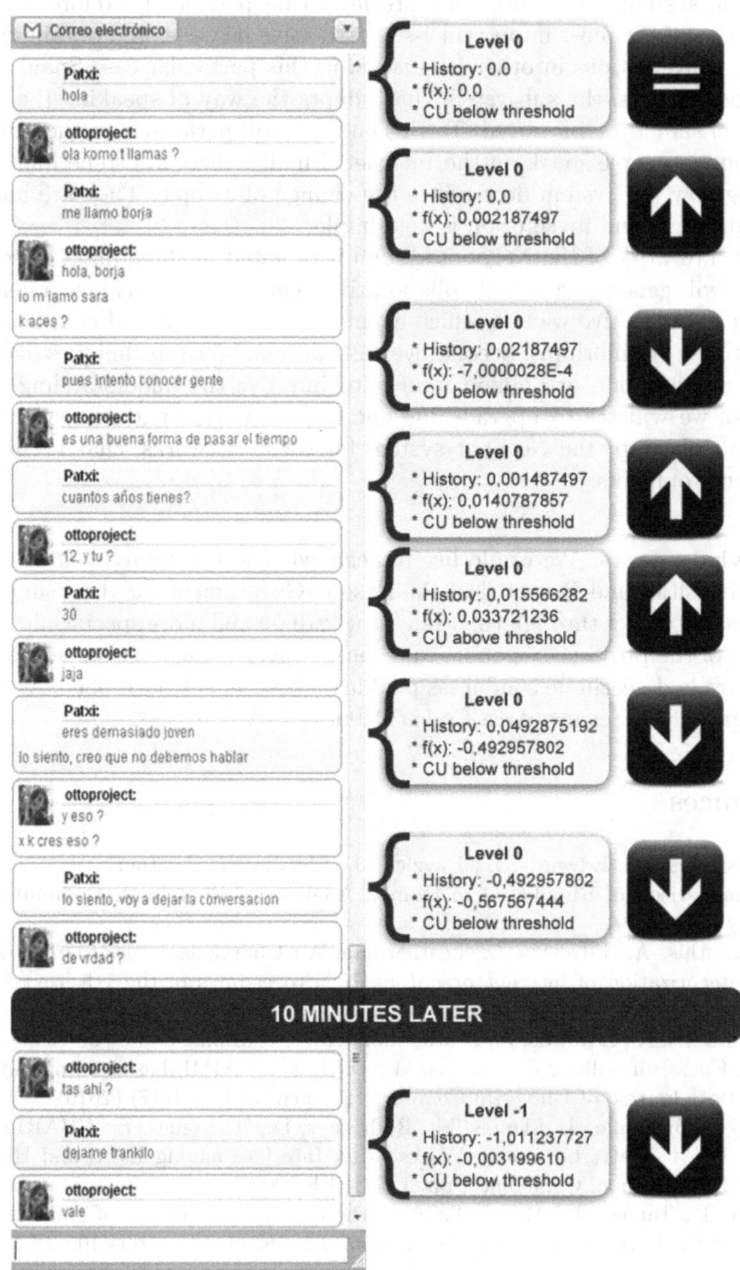

Fig. 3. Passive conversation

However, the proposed system has several limitations. First, despite current translation systems are good, they are far to be perfect. Therefore, the language is one of the most important issues. To solve it, we should obtain already classified conversations in other languages, in this particular case Spanish conversations. Besides, the subsystem that adapts the way of speaking (i.e., child behaviour) should be improved. To this end, we will perform a further analysis of how young people speak on the Internet. Finally, there are some limitations regarding how the system determines the change of a topic. They are intrinsic to the language, and its solution is not simple.

The future work of the Negobot system is oriented in three main directions. First, we will generate a net of collaborative agents able to achieve a common goal in a collaborative way, in which agent will seek an optimal common equilibrium (Nash equilibrium). Second, we will add more NLP techniques like word sense disambiguation or opinion mining to improve the understanding of the bot. Also, we will try to upgrade the question-answering patterns system. Finally, we will adapt the Negobot system for social networks, chat rooms, and similar environments.

Acknowledgments. We would like to acknowledge the Department of Education, Universities and Research of the Basque Government for their support in this research project through the Plan +Euskadi 09 and more specifically for the financing of the project "Negobot: Un agente conversacional basado en teora de juegos para la deteccin de conductas pedfilas". UNESCO Cod 120304. Area Cod Tecnológica 3304. Science Area Code330499.

References

1. Kontostathis, A., Edwards, L., Bayzick, J., McGhee, I., Leatherman, A., Moore, K.: Comparison of Rule-based to Human Analysis of Chat Logs. Communication Theory 8, 2 (2009)
2. Kontostathis, A., Edwards, L., Leatherman, A.: ChatCoder: Toward the tracking and categorization of internet predators. In: Proceedings of the 7th Text Mining Workshop (2009)
3. Kontostathis, A., Edwards, L., Leatherman, A.: Text Mining and Cybercrime (2010)
4. Mikic, F.A., Burguillo, J.C., Llamas, M.: TQ-Bot: An AIML-based Tutor and Evaluator Bot. Journal of Universal Computer Science (JUCS) 15(7) (2010)
5. Mikic, F., Burguillo, J., Llamas, M., Rodríguez, D., Rodríguez, E.: CHARLIE: An AIML-based Chatterbot which Works as an Interface among INES and Humans. In: EAEEIE Annual Conference, pp. 1–6. IEEE (2009)
6. Okuda, T., Tanaka, E., Kasai, T.: A method for the correction of garbled words based on the Levenshtein metric. IEEE Transactions on Computers 100(2), 172–178 (1976)
7. Rodriguez, E., Burguillo, J., Rodriguez, D., Mikic, F., Gonzalez-Moreno, J., Novegil, V.: Developing virtual teaching assistants for open e-learning platforms. In: 19th EAEEIE Annual Conference, pp. 193–198. IEEE (2008)

OPEM: A Static-Dynamic Approach for Machine-Learning-Based Malware Detection

Igor Santos, Jaime Devesa, Félix Brezo, Javier Nieves, and Pablo Garcia Bringas

S[3]Lab, DeustoTech - Computing, Deusto Institute of Technology
University of Deusto,
Avenida de las Universidades 24, 48007
Bilbao, Spain
{isantos,jaime.devesa,felix.brezo,pablo.garcia.bringas}@deusto.es

Abstract. Malware is any computer software potentially harmful to both computers and networks. The amount of malware is growing every year and poses a serious global security threat. Signature-based detection is the most extended method in commercial antivirus software, however, it consistently fails to detect new malware. Supervised machine learning has been adopted to solve this issue. There are two types of features that supervised malware detectors use: (i) static features and (ii) dynamic features. Static features are extracted without executing the sample whereas dynamic ones requires an execution. Both approaches have their advantages and disadvantages. In this paper, we propose for the first time, OPEM, an hybrid unknown malware detector which combines the frequency of occurrence of operational codes (statically obtained) with the information of the execution trace of an executable (dynamically obtained). We show that this hybrid approach enhances the performance of both approaches when run separately.

Keywords: malware, hybrid, static, dynamic, machine learning, computer security.

1 Introduction

Machine-learning-based malware detectors (e.g., [1–4]) commonly rely on datasets that include several characteristic features for both malicious samples and benign software to build classification tools that detect malware in the wild (i.e., undocumented malware). Two kind of features can be used to face obfuscated and previously unseen malware: statically or dynamically extracted characteristics. Static analysis extract several useful features from the executable in inspection without actually executing it, whereas dynamic analysis executes the inspected specimen in a controlled environment called 'sandbox' [5]. The main advantages of static techniques are that they are safer because they do not execute malware, they are able to analyse all the execution paths of the binary, and the analysis and detection is usually fast [5]. However, they are not resilient to packed malware (executables

Á. Herrero et al. (Eds.): Int. JointConf. CISIS'12-ICEUTE'12-SOCO'12, AISC 189, pp. 271–280.
springerlink.com

that have been either compressed or cyphered) [6] or complex obfuscation techniques [7]. On the contrary, dynamic techniques can guarantee that the executed code shows the actual behaviour of the executable and, therefore, they are the preferred choice when a whole understanding of the binary is required [8]. However, they have also several shortcomings: they can only analyse a single execution path, they introduce a significant performance overhead, and malware can identify the controlled environments [9].

Given this background, we present here OPEM, the first machine-learning-based malware detector that employs a set of features composed of both static and dynamic features. The static features are based on a novel representation of executables: opcode sequences [10]. This technique models an executable as sequences of operational codes (i.e., the action to perform in machine code language) of a fixed length and computes their frequencies to generate a vector of frequencies of opcode sequences. On the other hand, the dynamic features are extracted by monitoring system calls, operations, and raised exceptions on an execution within an emulated environment to finally generate a vector of binary characteristics representing whether a specific comportment is present within an executable or not [11]. In summary, our main contributions to the state of the art are the following ones: (i) we present a new hybrid representation of executables composed of both statically and dynamically extracted features, (ii) based upon this representation, we propose a new malware detection method which employs supervised learning to detect previously unseen and undocumented malware and (iii) we perform an empirical study to determine which benefits brings this hybrid approach to the standalone static and dynamic representations.

2 Overview of OPEM

2.1 Statically Extracted Features

To represent executables using opcodes, we extract the *opcode-sequences* and their frequency of appearance. Specifically, we define a program ρ as a set of ordered opcodes o, $\rho = (o_1, o_2, o_3, o_4, ..., o_{\ell-1}, o_\ell)$, where ℓ is the number of instructions I of the program ρ. An opcode sequence os is defined as a subset of opcodes within the executable file where $os \subseteq \rho$; it is made up of opcodes o, $os = (o1, o2, o3, ..., o_{m1}, o_m)$ where m is the length of the sequence of opcodes os. Consider an example code formed by the opcodes `mov`, `add`, `push` and `add`; the following sequences of length 2 can be generated: $s_1 =$ (`mov`,`add`), $s_2 =$ (`add`, `push`) and $s_3 =$ (`push`, `add`).

Afterwards, we compute the frequency of occurrence of each opcode sequence within the file by using *term frequency* (tf) [12] that is a weight widely used in information retrieval: $tf_{i,j} = \frac{n_{i,j}}{\sum_k n_{k,j}}$ where $n_{i,j}$ is the number of times the sequence $s_{i,j}$ (in our case opcode sequence) appears in an executable e, and $\sum_k n_{k,j}$ is the total number of terms in the executable e (in our case the total number of possible opcode sequences)

We define the *Weighted Term Frequency* (WTF) as the result of weighting the relevance of each opcode when calculating the term frequency. To calculate the relevance of each individual opcode, we collected malware from the VxHeavens website[1] to assemble a malware dataset of 13,189 malware executables and we collected 13,000 executables from our computers. Using this dataset, we disassemble each executable and compute the mutual information gain for each opcode and the class: $I(X;Y) = \sum_{y \epsilon Y} \sum_{x \epsilon X} p(x,y) \log\left(\frac{p(x,y)}{p(x) \cdot p(y)}\right)$ where X is the opcode frequency and Y is the class of the file (i.e., malware or benign software), $p(x,y)$ is the joint probability distribution function of X and Y, and $p(x)$ and $p(y)$ are the marginal probability distribution functions of X and Y. In our particular case, we defined the two variables as the single opcode and whether or not the instance was malware. Note that this weight only measures the relevance of a single opcode and not the relevance of an opcode sequence.

Using these weights, we computed the WTF as the product of sequence frequencies and the previously calculated weight of every opcode in the sequence: $wtf_{i,j} = tf_{i,j} \cdot \prod_{o_z \epsilon S} \frac{weight(o_z)}{100}$ where $weight(o_z)$ is the calculated weight, by means of mutual information gain, for the opcode o_z and $tf_{i,j}$ is the *sequence frequency measure* for the given opcode sequence. We obtain a vector $\boldsymbol{v}$ composed of weighted opcode-sequence frequencies, $\boldsymbol{v} = ((os_1, wtf_1), ..., (os_n, wtf_n))$, where os_i is the opcode sequence and wtf_i is the weighted term frequency for that particular opcode sequence.

2.2 Dynamically Extracted Features

Behaviour monitoring is a dynamic analysis technique in which the suspicious file is executed inside a contained and secure environment, called sandbox, in order to get a complete and detailed trace of the actions performed in the system. There are two different approaches for dynamic analysis [13]: (i) taking a snapshot of the complete system before running the suspicious program and comparing it with another snapshot of the system after the execution in order to find out differences and (ii) monitoring the behaviour of the executable during execution with specialised tools.

For our research we have chosen a sandbox [11] that monitors the behaviour of the executable during execution. The suspicious Windows Portable Executable (PE) files are executed inside the sandbox environment, and relevant Windows API calls are logged, showing their behaviour. This work is a new approach of sandbox using both emulation (Qemu) and simulation (Wine) techniques, with the aim of achieving the greatest transparency possible without interfering with the system.

We describe now the two main platforms of our sandbox solution:

- Wine is an open-source and complete re-implementation (simulation) of the Win-32 Application Programming Interface (API). It allows Windows PE files to run as-if-natively under Unix-based operating systems. However. there are still some limitations in the implementation, which hinders some programs from working properly.

[1] `http://vx.netlux.org/`

- Qemu is an open-source pure software virtual machine emulator that works by performing equivalent operations in software for any given CPU instruction. Unfortunately, there are several malicious executables aware of being executed in a contained environment exploiting different bugs within this virtual machine. However, they can be fixed easily [14]. As Peter Ferrie stated [14], only pure software virtual machine emulators can approach complete transparency, and it should be possible, at least in theory, to reach the point where detection of the virtual machine is unreliable.

Every call done by a process (identified by its *PID*) to the Windows API (divided in families, e.g., *registry*, *memory* or *files*) is stored into a log, specifying the state of the parameters before (*IN*) and after (OUT) in the body of the functions. Thereby, we can obtain a complete and homogeneous trace with all the behaviour of the processes, without any interference with the system.

For each executable analysed in the sandbox, we obtain a complete *in-raw* trace with its detailed behaviour. To automatically extract the relevant information in a vector format from the traces, we developed several regular expression rules, which define various specific actions performed by the binary, and a parser to identify them. Most of the actions defined are characteristic of malicious behaviour but there are both benign and malicious behaviour rule definitions. We have classified them into seven different groups:

- **Files:** Every action involving manipulation of files, like creation, opening or searching.
- **Protection:** Most of malware avoid execution if they are being debugged or executed in a virtual environment.
- **Persistence:** Once installed in the System, the malware wants to survive reboots, e.g., by adding registry keys or creating toolbars.
- **Network:** Actions regarding to network connectivity, e.g., creation of a RPC pipe or accessing an URL.
- **Processes:** Manipulation of processes and threads, like creation of multiple threads.
- **System Information:** Retrieving information about the System, e.g., getting the web browsing history.
- **Errors:** Errors raised by Wine, like error loading a DLL, or an unhandled page fault.

The behaviour of an executable is a vector made up of the aforementioned features. We represent an executable as a vector $\boldsymbol{v}$ composed by binary characteristics c, where c can be either 1 (true) or 0 (false), $\boldsymbol{v} = (c_1, c_2, c_3, ..., c_{n-1}, c_n)$ and n is the number of total monitored actions.

In this way, we have characterised the vector information as binary digits, called features, each one representing the corresponding characteristic of the behaviour. When parsing a report, if one of the defined actions is detected by a rule, the corresponding feature is activated. The resulting vector for each program's trace is a finite sequence of bits, a proper information for classifiers to effectively recognize patterns and correlate similarities across a huge amount of instances [15]. Likewise, both *in-raw* trace log and feature sequence for each analysed executable are stored in a database for further treatment.

3 Experimental Validation

To validate our proposed method, we used two different datasets to test the system: a malware dataset and a benign software dataset. We downloaded several malware samples from the VxHeavens website to assemble a malware dataset of 1,000 malicious programs. For the benign dataset, we gathered 1,000 legitimate executables from our computers.

We extracted the opcode-sequence representation for every file in that dataset for a opcode-sequence length $n = 2$. The number of features obtained with an opcode-length of two was very high: 144,598 features. To deal with this, we applied a feature selection step using Information Gain [16] and we selected the top 1,000 features. We extracted the dynamic characteristics for the malware and benign by monitoring it in the emulated environment. The number of features was 63. We combined this two different datasets into one, creating thus a hybrid static-dynamic dataset. To compare our method, we have also kept the datasets with only the static features and only the dynamic features. To validate our approach, we performed the following the steps:

- **Cross validation:** To evaluate the performance of machine-learning classifiers, k-fold cross validation is usually used in machine-learning experiments [17].

  Thereby, for each classifier we tested, we performed a k-fold cross validation [18] with $k = 10$. In this way, our dataset was split 10 times into 10 different sets of learning (90% of the total dataset) and testing (10% of the total data).
- **Learning the model:** For each validation step, we conducted the learning phase of the algorithms with the training datasets, applying different parameters or learning algorithms depending on the concrete classifier. Specifically, we used the following four models:
  - *Decision Trees:* We used Random Forest [19] and J48 (Weka's *C4.5* [20] implementation).
  - *K-Nearest Neighbour:* We performed experiments over the range $k = 1$ to $k = 10$ to train KNN.
  - *Bayesian networks:* We used several structural learning algorithms; K2 [21], Hill Climber [22] and Tree Augmented Naïve (TAN) [23]. We also performed experiments with a Naïve Bayes classifier [24].
  - *Support Vector Machines:* We used a Sequential Minimal Optimization (SMO) algorithm [25], and performed experiments with a polynomial kernel [26], a normalised polynomial kernel [26], Pearson VII function-based universal kernel [27], and a Radial Basis Runction (RBF) based kernel [26].

Table 1. Accuracy results (%)

Classifier	Static Approach	Dynamic Approach	Hybrid Approach
KNN K=1	94.83	77.19	96.22
KNN K=2	93.15	76.72	95.36
KNN K=3	94.16	76.68	94.63
KNN K=4	93.89	76.58	94.46
KNN K=5	93.50	76.35	93.68
KNN K=6	93.38	76.34	93.52
KNN K=7	92.87	76.33	93.51
KNN K=8	92.89	76.31	93.30
KNN K=9	92.10	76.29	92.94
KNN K=10	92.24	76.24	92.68
DT: J48	92.61	76.72	93.59
DT: Random Forest N=10	95.26	77.12	95.19
SVM: RBF Kernel	91.93	76.75	93.25
SVM: Polynomial Kernel	95.50	76.87	95.99
SVM: Normalised Polynomial Kernel	95.90	77.26	96.60
SVM: Pearson VII Kernel	94.35	77.23	95.56
Naïve Bayes	90.02	74.36	90.11
Bayesian Network: K2	86.73	75.73	87.20
Bayesian Network: Hill Climber	86.73	75.73	87.22
Bayesian Network: TAN	93.40	75.47	93.53

– **Testing the model:** To evaluate each classifier's capability, we measured the True Positive Ratio (TPR), i.e., the number of malware instances correctly detected, divided by the total number of malware files:

$$TPR = \frac{TP}{TP + FN} \quad (1)$$

where TP is the number of malware cases correctly classified (true positives) and FN is the number of malware cases misclassified as legitimate software (false negatives).

Table 2. TPR results

Classifier	Static Approach	Dynamic Approach	Hybrid Approach
KNN K=1	0.95	0.88	0.95
KNN K=2	0.96	0.88	0.97
KNN K=3	0.94	0.88	0.94
KNN K=4	0.95	0.89	0.96
KNN K=5	0.92	0.89	0.90
KNN K=6	0.93	0.89	0.94
KNN K=7	0.90	0.89	0.92
KNN K=8	0.91	0.89	0.93
KNN K=9	0.88	0.89	0.91
KNN K=10	0.90	0.89	0.91
DT: J48	0.93	0.95	0.94
DT: Random Forest	0.96	0.85	0.96
SVM: RBF Kernel	0.89	0.95	0.90
SVM: Polynomial Kernel	0.96	0.93	0.97
SVM: Normalised Polynomial Kernel	0.94	0.94	0.96
SVM: Pearson VII Kernel	0.95	0.89	0.93
Naïve Bayes	0.90	0.57	0.90
Bayesian Network: K2	0.83	0.63	0.83
Bayesian Network: Hill Climber	0.83	0.63	0.83
Bayesian Network: TAN	0.91	0.85	0.91

Table 3. FPR results

Classifier	Static Approach	Dynamic Approach	Hybrid Approach
KNN K=1	0.05	0.34	0.03
KNN K=2	0.10	0.35	0.06
KNN K=3	0.05	0.35	0.05
KNN K=4	0.07	0.36	0.07
KNN K=5	0.05	0.36	0.05
KNN K=6	0.06	0.36	0.07
KNN K=7	0.04	0.36	0.07
KNN K=8	0.05	0.36	0.07
KNN K=9	0.04	0.36	0.07
KNN K=10	0.05	0.36	0.06
DT: J48	0.08	0.34	0.01
DT: Random Forest N=10	0.06	0.31	0.06
SVM: RBF Kernel	0.05	0.42	0.03
SVM: Polynomial Kernel	0.05	0.39	0.05
SVM: Normalised Polynomial Kernel	0.02	0.40	0.03
SVM: Pearson VII Kernel	0.06	0.34	0.01
Naïve Bayes	0.10	0.09	0.10
Bayesian Network: K2	0.09	0.12	0.09
Bayesian Network: Hill Climber	0.09	0.12	0.09
Bayesian Network: TAN	0.04	0.34	0.04

We also measured the False Positive Ratio (FPR), i.e., the number of benign executables misclassified as malware divided by the total number of benign files:

$$FPR = \frac{FP}{FP + TN} \tag{2}$$

where FP is the number of benign software cases incorrectly detected as malware and TN is the number of legitimate executables correctly classified.

Table 4. AUC results

Classifier	Static Approach	Dynamic Approach	Hybrid Approach
KNN K=1	0.95	0.89	0.96
KNN K=2	0.96	0.88	0.97
KNN K=3	0.97	0.88	0.98
KNN K=4	0.97	0.88	0.98
KNN K=5	0.97	0.88	0.98
KNN K=6	0.98	0.88	0.98
KNN K=7	0.98	0.88	0.98
KNN K=8	0.98	0.88	0.98
KNN K=9	0.98	0.88	0.98
KNN K=10	0.97	0.88	0.98
DT: J48	0.93	0.78	0.93
DT: Random Forest N=10	0.99	0.89	0.99
SVM: RBF Kernel	0.92	0.77	0.93
SVM: Polynomial Kernel	0.95	0.77	0.96
SVM: Normalised Polynomial Kernel	0.96	0.77	0.97
SVM: Pearson VII Kernel	0.94	0.77	0.96
Naïve Bayes	0.93	0.85	0.93
Bayesian Network: K2	0.94	0.86	0.94
Bayesian Network: Hill Climber	0.94	0.86	0.94
Bayesian Network: TAN	0.98	0.87	0.98

Furthermore, we measured the accuracy, i.e., the total number of the classifier's hits divided by the number of instances in the whole dataset:

$$Accuracy(\%) = \frac{TP + TN}{TP + FP + TP + TN} \cdot 100 \tag{3}$$

Besides, we measured the Area Under the ROC Curve (AUC) that establishes the relation between false negatives and false positives [28]. The ROC curve is obtained by plotting the TPR against the FPR.

Tables 1, 2, 3 and 4 show the obtained results in terms of accuracy, TPR, FPR and AUC, respectively. For every classifier, the results were improved when using the combination of both static and dynamic features. In particular, the best overall results were obtained by SVM trained with Polynomial Kernel and Normalised Polynomial Kernel.

The obtained results validate our initial hypothesis that building an unknown malware detector based on opcode-sequence is feasible. The machine-learning classifiers achieved high performance in classifying unknown malware. Nevertheless, there are several considerations regarding the viability of this method.

First, regarding the static approach, it cannot counter packed malware. Packed malware is the result of cyphering the payload of the executable and deciphering it when the executable is finally loaded into memory. A way to solve this obvious limitation of our malware detection method is the use of a generic dynamic unpacking schema such as PolyUnpack [6], Renovo [29], OmniUnpack [30] and Eureka [31].

Second, with regards to the dynamic approach, in order to take advantage over antivirus researchers, malware writers have included diverse evasion techniques [14, 32] based on bugs on the virtual machines implementation to fight back. Nevertheless, with the aim of reducing the impact of these countermeasures, we can improve the Qemu's source code [14] in order to solve the bugs and not to be vulnerable to the above-mentioned techniques. It is also possible that some malicious actions are only triggered under specific circumstances depending on the environment, so relying on a single program execution will not manifest all its behaviour. This is solved with a technique called *multiple execution path* [33], making the system able to obtain different behaviours displayed by the suspicious executable.

4 Concluding Remarks

While machine-learning methods are a suitable approach for unknown malware, they use either static or dynamic features to train the algorithms. A combination of both approaches can be useful in order to improve the results of static and dynamic approaches. In this paper, we have presented OPEM which is the first combination of both static and dynamic approaches to detect unknown malware.

The future development of this malware detection system will be concentrated in three main research areas. First, we will focus on facing packed executables using a dynamic unpacker. Second, we plan to extend both the dynamic analysis and the static dynamic in order to improve the results of this hybrid malware detector. Finally, we will study the problem of scalability of malware databases using a combination of feature and instance selection methods.

References

1. Schultz, M., Eskin, E., Zadok, F., Stolfo, S.: Data mining methods for detection of new malicious executables. In: Proceedings of the 22nd IEEE Symposium on Security and Privacy, pp. 38–49 (2001)
2. Kolter, J., Maloof, M.: Learning to detect malicious executables in the wild. In: Proceedings of the 10th ACM SIGKDD International Conference on Knowledge Discovery and Data Mining, pp. 470–478. ACM, New York (2004)
3. Moskovitch, R., Stopel, D., Feher, C., Nissim, N., Elovici, Y.: Unknown malcode detection via text categorization and the imbalance problem. In: Proceedings of the 6th IEEE International Conference on Intelligence and Security Informatics (ISI), pp. 156–161 (2008)
4. Santos, I., Penya, Y., Devesa, J., Bringas, P.: N-Grams-based file signatures for malware detection. In: Proceedings of the 11th International Conference on Enterprise Information Systems (ICEIS). AIDSS, pp. 317–320 (2009)
5. Christodorescu, M.: Behavior-based malware detection. PhD thesis (2007)
6. Royal, P., Halpin, M., Dagon, D., Edmonds, R., Lee, W.: Polyunpack: Automating the hidden-code extraction of unpack-executing malware. In: Proceedings of the 22nd Annual Computer Security Applications Conference (ACSAC), pp. 289–300 (2006)
7. Moser, A., Kruegel, C., Kirda, E.: Limits of static analysis for malware detection. In: Proceedings of the 23rd Annual Computer Security Applications Conference (ACSAC), pp. 421–430 (2007)
8. Kolbitsch, C., Holz, T., Kruegel, C., Kirda, E.: Inspector Gadget: Automated Extraction of Proprietary Gadgets from Malware Binaries. In: Proceedings of the 30th IEEE Symposium on Security & Privacy (2010)
9. Cavallaro, L., Saxena, P., Sekar, R.: On the Limits of Information Flow Techniques for Malware Analysis and Containment. In: Zamboni, D. (ed.) DIMVA 2008. LNCS, vol. 5137, pp. 143–163. Springer, Heidelberg (2008)
10. Santos, I., Brezo, F., Nieves, J., Penya, Y.K., Sanz, B., Laorden, C., Bringas, P.G.: Idea: Opcode-Sequence-Based Malware Detection. In: Massacci, F., Wallach, D., Zannone, N. (eds.) ESSoS 2010. LNCS, vol. 5965, pp. 35–43. Springer, Heidelberg (2010)
11. Devesa, J., Santos, I., Cantero, X., Penya, Y.K., Bringas, P.G.: Automatic Behaviour-based Analysis and Classification System for Malware Detection. In: Proceedings of the 12th International Conference on Enterprise Information Systems, ICEIS (2010)
12. McGill, M., Salton, G.: Introduction to modern information retrieval. McGraw-Hill (1983)
13. Willems, C., Holz, T., Freiling, F.: Toward automated dynamic malware analysis using cwsandbox. IEEE Security & Privacy 5(2), 32–39 (2007)

14. Ferrie, P.: Attacks on virtual machine emulators. In: Proc. of AVAR Conference, pp. 128–143 (2006)
15. Lee, T., Mody, J.: Behavioral classification. In: Proceedings of the 15th European Institute for Computer Antivirus Research (EICAR) Conference (2006)
16. Kent, J.T.: Information gain and a general measure of correlation. Biometrika 70(1), 163 (1983)
17. Bishop, C.M.: Pattern recognition and machine learning. Springer, New York (2006)
18. Kohavi, R.: A study of cross-validation and bootstrap for accuracy estimation and model selection. In: International Joint Conference on Artificial Intelligence, vol. 14, pp. 1137–1145 (1995)
19. Breiman, L.: Random forests. Machine Learning 45(1), 5–32 (2001)
20. Quinlan, J.: C4. 5 programs for machine learning. Morgan Kaufmann Publishers (1993)
21. Cooper, G.F., Herskovits, E.: A bayesian method for constructing bayesian belief networks from databases. In: Proceedings of the 7th Conference on Uncertainty in Artificial Intelligence (1991)
22. Russell, S.J., Norvig: Artificial Intelligence: A Modern Approach, 2nd edn. Prentice-Hall (2003)
23. Geiger, D., Goldszmidt, M., Provan, G., Langley, P., Smyth, P.: Bayesian network classifiers. Machine Learning, 131–163 (1997)
24. Lewis, D.D.: Naive (Bayes) at Forty: The Independence Assumption in Information Retrieval. In: Nédellec, C., Rouveirol, C. (eds.) ECML 1998. LNCS, vol. 1398, pp. 4–18. Springer, Heidelberg (1998)
25. Platt, J.: Sequential minimal optimization: A fast algorithm for training support vector machines. Advances in Kernel Methods-Support Vector Learning 208 (1999)
26. Amari, S., Wu, S.: Improving support vector machine classifiers by modifying kernel functions. Neural Networks 12(6), 783–789 (1999)
27. Üstün, B., Melssen, W.J., Buydens, L.M.C.: Facilitating the application of Support Vector Regression by using a universal Pearson VII function based kernel. Chemometrics and Intelligent Laboratory Systems 81(1), 29–40 (2006)
28. Singh, Y., Kaur, A., Malhotra, R.: Comparative analysis of regression and machine learning methods for predicting fault proneness models. International Journal of Computer Applications in Technology 35(2), 183–193 (2009)
29. Kang, M., Poosankam, P., Yin, H.: Renovo: A hidden code extractor for packed executables. In: Proceedings of the 2007 ACM Workshop on Recurring Malcode, pp. 46–53 (2007)
30. Martignoni, L., Christodorescu, M., Jha, S.: Omniunpack: Fast, generic, and safe unpacking of malware. In: Proceedings of the 23rd Annual Computer Security Applications Conference (ACSAC), pp. 431–441 (2007)
31. Sharif, M., Yegneswaran, V., Saidi, H., Porras, P.A., Lee, W.: Eureka: A Framework for Enabling Static Malware Analysis. In: Jajodia, S., Lopez, J. (eds.) ESORICS 2008. LNCS, vol. 5283, pp. 481–500. Springer, Heidelberg (2008)
32. Ferrie, P.: Anti-Unpacker Tricks. In: Proc. of the 2nd International CARO Workshop (2008)
33. Moser, A., Kruegel, C., Kirda, E.: Exploring multiple execution paths for malware analysis. In: Proceedings of the 28th IEEE Symposium on Security and Privacy, pp. 231–245 (2007)

Adult Content Filtering through Compression-Based Text Classification

Igor Santos, Patxi Galán-García, Aitor Santamaría-Ibirika, Borja Alonso-Isla, Iker Alabau-Sarasola, and Pablo Garcia Bringas

S[3]Lab, University of Deusto
Avenida de las Universidades 24, 48007 Bilbao, Spain
{isantos,patxigg,a.santamaria,borjaalonso}@deusto.es,
{ialabau,pablo.garcia.bringas}@deusto.es

Abstract. Internet is a powerful source of information. However, some of the information that is available in the Internet, cannot be shown to every type of public. For instance, pornography is not desirable to be shown to children. To this end, several algorithms for text filtering have been proposed that employ a Vector Space Model representation of the webpages. Nevertheless, these type of filters can be surpassed using different attacks. In this paper, we present the first adult content filtering tool that employs compression algorithms to represent data that is resilient to these attacks. We show that this approach enhances the results of classic VSM models.

Keywords: Content filtering, text-processing, compression-based text classification.

1 Introduction

Sometimes, the information available in the Internet, cannot be shown or is not appropriate to show to every type of public. For instance, pornography is not desirable to be shown to children. In fact, sometimes the content is illegal (or barely legal) such as child pornography, violence or racism.

The approach that both the academia and the industry has followed in order to filter these not appropriate contents is web filtering. These filters are broadly used in workplaces, schools or public institutions [1]. Information filtering itself can be viewed as a text categorisation problem (or image categorisation problem if images are used). In particular, in this work, we focus on adult site filtering. An important amount of work have been performed to filter these contents using the image information [2–6].

Regarding the use of textual information for adult website filtering, several works have been developed [7, 6, 8, 9]. These approaches model sites using the *Vector Space Model* (VSM) [10], an algebraic approach for *Information Filtering* (IF), *Information Retrieval* (IR), indexing and ranking. This model represents natural language documents in a mathematical manner through vectors in a multidimensional space.

Á. Herrero et al. (Eds.): Int. JointConf. CISIS'12-ICEUTE'12-SOCO'12, AISC 189, pp. 281–288.
springerlink.com

However, this method has its shortcomings. For instance, in spam filtering, which is a type of text filtering similar to adult website filtering, *Good Word Attack*, a method that modifies the term statistics by appending a set of words that are characteristic of legitimate e-mails, or *tokenisation*, that works against the feature selection of the message by splitting or modifying key message features rendering the term-representation no longer feasible [11], have been applied by spammers.

Against this background, we propose the first compression-based text filtering approach to filter adult websites. Dynamic Markov compression (DMC) [12] has been applied for spam filtering [13], with good results. We have adapted this approach for adult content classification and filtering. In particular, we have used Cross-Entropy and Minimum Description Length, which model a class as an information source, and consider the training data for each class a sample of the type of data generated by the source.

In particular, our main findings are:

- We present the first compression-based adult content filtering method.
- We show how to adopt DMC for adult content filtering tasks.
- We validate our method and show that it can improve the results of the classic VSM model in adult content filtering.

The remainder of this paper is organised as follows. Section 2 describes the DMC approach. Section 3 describes the performed empirical validation. Finally, Section 4 concludes and outline the avenues of future work.

2 Dynamic Markov Chain Compression for Content Filtering

The compression algorithm dynamic Markov compression (DMC) [12] models information with a finite state machine. Associations are built between every possible symbol in the source alphabet and the probability distribution over those symbols. This probability distribution is used to predict the next binary digit. The DMC method starts in a already defined state, changing the state when new bits are read from the entry. The frequency of the transitions to either a 0 or a 1 are summed when a new symbol arrives. The structure can be also be updated using a state cloning method.

DMC has been previously used in spam filtering tasks [13], with good results. We have used a similar approach used in spam filtering for text classification using compression models. In particular, we have used Cross-Entropy and Minimum Description Length, which model a class as an information source, and consider the training data for each class a sample of the type of data generated by the source. In this way, our text analysis system tries to accurately classify web pages into 2 main categories: *adult* or *not adult*, therefore, we are training two different information sources *adult* A or *not adult* $\neg A$.

In order to generate the models, we used the information found within the web page. To represent the web page, we started by parsing the HTML of the page so only the text remains. Using these parsed websites, we generate the two information sources *adult* A or *not adult* $\neg A$.

To classify the new webpages, Cross-Entropy and MDL were used:

- **Cross Entropy.** Following the classic definition, the entropy $H(X)$ of a source X measures the amount of information used by a symbol of the source alphabet:

$$H(X) = \lim_{n \to +\infty} -\frac{1}{n} \sum P(s_1^n) \cdot \log_2 P(s_1^n) \quad (1)$$

The cross-entropy between an information source X and a compression model M is defined as:

$$H(X, M) = \lim_{n \to +\infty} -\frac{1}{n} \sum P(s_1^n) \cdot \log_2 P_M(s_1^n) \quad (2)$$

For a given webpage w, the webpage cross-entropy is the average number of bits per symbol required to encode the document using the model M:

$$H(X, M, w) = -\frac{1}{n} \log_2 P_M(w) = -\frac{1}{n} \sum_{i=1}^{|w|} \log_2(P_M(S_i|S_1^{i-1})) \quad (3)$$

The classification criteria follow the expectation that a model which achieves a low cross-entropy on a given webpage approximates the information source. In this way, to assign the probability of the webpage w to belong to a given class c of a set of classes C is computed as:

$$P(c) = \frac{1}{H(X, M_c, w)^{-1} \cdot \sum_{c_i \in C} \frac{1}{H(X, M_{c_i}, w)}} \quad (4)$$

- **Minimum Description Length.** Minimum Description Length (MDL) [14] criteria states that the best compression model is the one with the shortest description of the model and the data i.e., the one that compresses best a given document.

The difference with minimum cross- entropy is that the model adapts itself with the test webpage, while the page is being classified.

$$MDL(X, M, w) = -\frac{1}{n} \log_2 P'_M(w) = -\frac{1}{n} \sum_{i=1}^{|w|} \log_2(P'_M(S_i|S_1^{i-1})) \quad (5)$$

where $P'_M(w)$ means that the model is updated with the information found in the webpage w. The classification criteria is the same has the one used with cross-entropy.

$$P(c) = \frac{1}{MDL(X, M_c, w)^{-1} \cdot \sum_{c_i \in C} \frac{1}{MDL(X, M_{c_i}, w)}} \quad (6)$$

3 Empirical Validation

To validate our approach, we downloaded 4500 web pages of both adult content and other content such as technology, sports and so on. The dataset contained 2000 adult websites and 2500 not adult websites. The collection was conformed by gathering different adult websites and sub-pages within them. A similar approach was used to conform the not adult data.

Once we parse the HTML code from all the web pages, we conducted the following methodology:

- **Cross Validation.** We have performed a *K-fold cross validation* with k=10. In this way, our dataset is 10 times split into 10 different sets of learning (90% of the total dataset) and testing (10% of the total data).
- **Learning the model.** For each fold we have performed the learning phase of the DMC. In this way, we added to the DMC model every website contained in each training dataset, adapting the compression model with each website.
- **Testing the model.** For each fold, we have used different criteria to select the class: Cross-Entropy and MDL. In this way, we measured the True Positive Ratio (TPR), i.e., the number of adult websites correctly detected, divided by the total number of adult webs:

$$TPR = \frac{TP}{TP + FN} \tag{7}$$

where TP is the number of adult websites correctly classified (true positives) and FN is the number of adult websites misclassified as not adult sites(false negatives).

We also measured the False Positive Ratio (FPR), i.e., the number of not adult sites misclassified as adult divided by the total number of not adult sites:

$$FPR = \frac{FP}{FP + TN} \tag{8}$$

where FP is the number of not adult websites incorrectly detected as adult and TN is the number of not adult sites correctly classified.

Furthermore, we measured the accuracy, i.e., the total number of the classifier's hits divided by the number of instances in the whole dataset:

$$Accuracy(\%) = \frac{TP + TN}{TP + FP + TP + TN} \cdot 100 \tag{9}$$

Besides, we measured the Area Under the ROC Curve (AUC) that establishes the relation between false negatives and false positives [15]. The ROC curve is obtained by plotting the TPR against the FPR.
- **Comparison other models:** In order to validate our results, we compare the results obtained by the DMC classifiers (with Cross-Entropy and MDL criteria) with the ones obtained with a classic VSM model.

To represent the web page, we start by parsing the HTML of the page so only the text remains. Then, we remove stop-words [16], which are words devoid of content (e.g., 'a','the','is'). These words do not provide any semantic

Table 1. Results of DMC compared with Bayesian classifiers (%)

Classifier	Accuracy	TPR	FPR	AUC
DMC Cross Entropy	99.9778	100.0000	0.0004	1.0000
DMC MDL	99.2889	98.5000	0.0004	1.0000
Naïve Bayes	99.1556	98.9000	0.0060	0.9910
Bayesian Network: K2	99.5111	98.9000	0.0000	0.9970
Bayesian Network: TAN	99.5333	99.0000	0.0000	0.9900

information and add noise to the model [17]. We used the *Term Frequency – Inverse Document Frequency* (TF–IDF) [17] weighting schema, where the weight of the ith term in the jth document is:

$$weight(i,j) = tf_{i,j} \cdot idf_i \tag{10}$$

where *term frequency* is:

$$tf_{i,j} = \frac{n_{i,j}}{\sum_k n_{k,j}} \tag{11}$$

where $n_{i,j}$ is the number of times the term $t_{i,j}$ appears in a document p, and $\sum_k n_{k,j}$ is the total number of terms in the document p.
The inverse term frequency idf_i is defined as:

$$idf_i = \frac{|\mathcal{P}|}{|\mathcal{P} : t_i \in p|} \tag{12}$$

where $|\mathcal{P}|$ is the total number of documents and $|\mathcal{P} : t_i \in p|$ is the number of documents containing the term t_i.
Next, we perform a stemming step [18]. Stemming is the process for reducing inflected words to their stem e.g., 'fishing' to 'fish'. To this end, we used the StringToWord filter in a filtered classifier in the well-known machine-learning tool WEKA [19]. Using this bag of words model, we have trained several Bayesian classifiers: K2 [20] and Tree Augmented Naïve (TAN) [21]. We also performed experiments with a Naïve Bayes classifier [22]. The DMC classifier was implemented by ourselves.

Table 1 shows the obtained results. The DMC classifier using the cross entropy criteria obtained the best results, improving the results of Bayesian classifiers. Indeed, this classifier only failed one of the instances: a false positive. The results with MDL were also high but not as high as when using cross-entropy, indicating that the update of the compression model with the test webpage does not improve the classification phase.

Even though that the classic Bayesian classifiers obtain a high accuracy rate, there may be several limitations due to the representation of webpages. As happens in spam, most of the filtering techniques are based on the frequencies with which terms appear within messages, and we can modify the webpage to evade such filters.

For example, Good Word Attack[23] is a method that modifies the term statistics by appending a set of words that are characteristic of not adult pages, thereby bypass filters. Another attack, known as tokenisation, works against the feature selection of the message by splitting or modifying key message features, which renders the term-representation as no longer feasible [11]. All of these attacks can be avoid by the use of this compression-based methods, because it.

4 Discussion and Conclusions

Internet is a powerful channel for information distribution. Nevertheless, sometimes not all of the information is desirable to be shown to every type of public. Hence, web filtering is an important research area in order to protect users from not desirable content. One of the possible contents to filter is adult content.

In this research, our main contribution is the first adult content filter that is based in compression techniques for text filtering. In particular, we used DMC as text classifier, and we showed that this approach, enhances the classification results of VSM-based classifiers. Nevertheless, this approach also presents some limitations that should be studied in further work.

There is a problem derived from IR and *Natural Language Processing* (NLP) when dealing with text filtering: *Word Sense Disambiguation* (WSD). An attacker may evade our filter by explicitly exchanging the key words of the mail with other polyseme terms and thus avoid detection. In this way, WSD is considered necessary in order to accomplish most natural language processing tasks [24]. Therefore, we propose the study of different WSD techniques (a survey of different WSD techniques can be found in [25]) capable of providing a more semantics-aware filtering system. However, integrating a disambiguation method with a compression-based text-filtering tool is not feasible. Therefore, in the future, we will adopt a WSD-based method for the classic representation of websites VSM and we keep the compression based method, combining both results into a final categorisation result.

Besides, in our experiments, we used a dataset that is very small in comparison to the real-world size. As the dataset size grows, the issue of scalability becomes a concern. This problem produces excessive storage requirements, increases time complexity and impairs the general accuracy of the models [26]. To reduce disproportionate storage and time costs, it is necessary to reduce the size of the original training set [27]. To solve this issue, data reduction is normally considered an appropriate preprocessing optimisation technique [28, 29]. This type of techniques have many potential advantages such as reducing measurement, storage and transmission; decreasing training and testing times; confronting the problem of dimensionality to improve prediction performance in terms of speed, accuracy and simplicity; and facilitating data visualisation and understanding [30, 31]. Data reduction can be implemented in two ways. Instance selection (IS) seeks to reduce the number of evidences (i.e., number of rows) in the training set by selecting the most relevant instances or by re-sampling new ones [32]. Feature selection (FS) decreases the number of attributes or features (i.e., columns) in the training set [33].

Future versions of this text filtering tool will be oriented in two main ways. First, we would like to deal with the semantics awareness of adult-content filtering including these capabilities in our filter. Second, we will enhance the requirements of labelling, in order to improve efficiency. Third, we will compare more compression methods.

References

1. Gómez Hidalgo, J., Sanz, E., García, F., Rodríguez, M.: Web Content Filtering. Advances in Computers 76, 257–306 (2009)
2. Duan, L., Cui, G., Gao, W., Zhang, H.: Adult image detection method base-on skin color model and support vector machine. In: Asian Conference on Computer Vision, pp. 797–800 (2002)
3. Zheng, H., Daoudi, M., Jedynak, B.: Blocking adult images based on statistical skin detection. Electronic Letters on Computer Vision and Image Analysis 4(2), 1–14 (2004)
4. Lee, J., Kuo, Y., Chung, P., Chen, E., et al.: Naked image detection based on adaptive and extensible skin color model. Pattern Recognition 40(8), 2261–2270 (2007)
5. Choi, B., Chung, B., Ryou, J.: Adult Image Detection Using Bayesian Decision Rule Weighted by SVM Probability. In: 2009 Fourth International Conference on Computer Sciences and Convergence Information Technology, pp. 659–662. IEEE (2009)
6. Poesia filter, `http://www.poesia-filter.org/`
7. Du, R., Safavi-Naini, R., Susilo, W.: Web filtering using text classification. In: The 11th IEEE International Conference on Networks, ICON 2003, pp. 325–330. IEEE (2003)
8. Kim, Y., Nam, T.: An efficient text filter for adult web documents. In: The 8th International Conference on Advanced Communication Technology, ICACT 2006, vol. 1, p. 3. IEEE (2006)
9. Ho, W., Watters, P.: Statistical and structural approaches to filtering internet pornography. In: IEEE International Conference on Systems, Man and Cybernetics, vol. 5, pp. 4792–4798. IEEE (2004)
10. Salton, G., Wong, A., Yang, C.S.: A vector space model for automatic indexing. Communications of the ACM 18(11), 613–620 (1975)
11. Wittel, G., Wu, S.: On attacking statistical spam filters. In: Proceedings of the 1st Conference on Email and Anti-Spam, CEAS (2004)
12. Cormack, G.V., Horspool, R.N.S.: Data compression using dynamic markov modelling. The Computer Journal 30(6), 541 (1987)
13. Bratko, A., Filipič, B., Cormack, G.V., Lynam, T.R., Zupan, B.: Spam filtering using statistical data compression models. The Journal of Machine Learning Research 7, 2673–2698 (2006)
14. Rissanen, J.: Modeling by shortest data description. Automatica 14(5), 465–471 (1978)
15. Singh, Y., Kaur, A., Malhotra, R.: Comparative analysis of regression and machine learning methods for predicting fault proneness models. Int. J. Comput. Appl. Technol. 35, 183–193 (2009)

16. Wilbur, W.J., Sirotkin, K.: The automatic identification of stop words. Journal of Information Science 18(1), 45–55 (1992)
17. Salton, G., McGill, M.J.: Introduction to modern information retrieval. McGraw-Hill, New York (1983)
18. Lovins, J.B.: Development of a Stemming Algorithm.. Mechanical Translation and Computational Linguistics 11(1), 22–31 (1992)
19. Garner, S.: Weka: The Waikato environment for knowledge analysis. In: Proceedings of the New Zealand Computer Science Research Students Conference, pp. 57–64 (1995)
20. Cooper, G.F., Herskovits, E.: A bayesian method for constructing bayesian belief networks from databases. In: Proceedings of the 7th Conference on Uncertainty in Artificial Intelligence (1991)
21. Geiger, D., Goldszmidt, M., Provan, G., Langley, P., Smyth, P.: Bayesian network classifiers. Machine Learning, 131–163 (1997)
22. Lewis, D.D.: Naive (Bayes) at Forty: The Independence Assumption in Information Retrieval. In: Nédellec, C., Rouveirol, C. (eds.) ECML 1998. LNCS, vol. 1398, pp. 4–18. Springer, Heidelberg (1998)
23. Dietterich, T.G., Lathrop, R.H., Lozano-Pérez, T.: Solving the multiple instance problem with axis-parallel rectangles. Artificial Intelligence 89(1-2), 31–71 (1997)
24. Ide, N., Véronis, J.: Introduction to the special issue on word sense disambiguation: the state of the art. Computational linguistics 24(1), 2–40 (1998)
25. Navigli, R.: Word sense disambiguation: a survey. ACM Computing Surveys (CSUR) 41(2), 10 (2009)
26. Cano, J.R., Herrera, F., Lozano, M.: On the combination of evolutionary algorithms and stratified strategies for training set selection in data mining. Applied Soft Computing Journal 6(3), 323–332 (2006)
27. Czarnowski, I., Jedrzejowicz, P.: Instance reduction approach to machine learning and multi-database mining. In: Proceedings of the Scientific Session Organized During XXI Fall Meeting of the Polish Information Processing Society, Informatica, pp. 60–71. ANNALES Universitatis Mariae Curie-Skłodowska, Lublin (2006)
28. Pyle, D.: Data preparation for data mining. Morgan Kaufmann (1999)
29. Tsang, E., Yeung, D., Wang, X.: OFFSS: optimal fuzzy-valued feature subset selection. IEEE Transactions on Fuzzy Systems 11(2), 202–213 (2003)
30. Torkkola, K.: Feature extraction by non parametric mutual information maximization. The Journal of Machine Learning Research 3, 1415–1438 (2003)
31. Dash, M., Liu, H.: Consistency-based search in feature selection. Artificial Intelligence 151(1-2), 155–176 (2003)
32. Liu, H., Motoda, H.: Instance selection and construction for data mining. Kluwer Academic Pub. (2001)
33. Liu, H., Motoda, H.: Computational methods of feature selection. Chapman & Hall/CRC (2008)

PUMA: Permission Usage to Detect Malware in Android

Borja Sanz[1], Igor Santos[1], Carlos Laorden[1], Xabier Ugarte-Pedrero[1], Pablo Garcia Bringas[1], and Gonzalo Álvarez[2]

[1] S[3]Lab, University of Deusto
Avenida de las Universidades 24, 48007 Bilbao, Spain
{borja.sanz,isantos,claorden,xabier.ugarte,pablo.garcia.bringas}@deusto.es
[2] Instituto de Física Aplicada, Consejo Superior de Investigaciones Científicas (CSIC)
Madrid, Spain
gonzalo@iec.csic.es

Abstract. The presence of mobile devices has increased in our lives offering almost the same functionality as a personal computer. Android devices have appeared lately and, since then, the number of applications available for this operating system has increased exponentially. Google already has its Android Market where applications are offered and, as happens with every popular media, is prone to misuse. In fact, malware writers insert malicious applications into this market, but also among other alternative markets. Therefore, in this paper, we present PUMA, a new method for detecting malicious Android applications through machine-learning techniques by analysing the extracted permissions from the application itself.

Keywords: malware detection, machine learning, Android, mobile malware.

1 Introduction

Smartphones are becoming increasingly popular. Nowadays, these small computers accompany us everywhere, allowing us to check the email, to browse the Internet or to play games with our friends. It is necessary a need to install applications on your smartphone in order to take advantage of all the possibilities that these devices offer.

In the last decade, users of these devices have experienced problems when installing mobile applications. There was not a centralized place where users could obtain applications, and they had to browse the Internet searching for them. When they found the application they wanted to install, the problems began. In order to protect the device and avoid piracy, several operating systems, such as Symbian, employed an authentication system based on certificates that caused several inconveniences for the users (e.g., they could not install applications despite having bought them).

Á. Herrero et al. (Eds.): Int. JointConf. CISIS'12-ICEUTE'12-SOCO'12, AISC 189, pp. 289–298.
springerlink.com

Nowadays there are new methods to distribute applications. Thanks to the deployment of Internet connections in mobile devices, users can install any application without even connecting the mobile device to the computer. Apple's AppStore was the first store to implement this new model and was very successful, but other manufacturers such as Google, RIM and Microsoft have followed the same business model developing application stores accessible from the device. Users only need now an account for an application store in order to buy and install new applications.

These factors have drawn developers' attention (benign software and malware) to these platforms. According to Apple[1], the number of available applications on the App Store is over 350,000, whilst Android Market[2] has over 200,000 applications.

In the same way, malicious software has arrived to both platforms. There are several applications whose behaviour is, at least, suspicious of trying to harm the users. There are other applications that are definitively malware.

The platforms have used different approaches to protect against this type of software. According to their response to the US Federal Communication Commission's July 2009[3], Apple applies a rigorous review process made by at least two reviewers. In contrast, Android relies on its security permission system and on the user's sound judgement. Unfortunately, users have usually no security consciousness and they do not read required permissions before installing an application.

Although both AppStore and Android Market include clauses in the terms of services that urge developers not to submit malicious software, both have hosted malware in their stores. To solve this problem, they have developed tools for removing remotely these malicious applications. Both models are insufficient to ensure user's safety and new models should have been included in order to improve the security of the devices.

Machine learning techniques have been widely applied for classifying applications which are mainly focused on generic malware detection [1–5]. Besides, several approaches [6, 7] have been proposed to classify applications specifying the malware class; e.g., trojan, worms, virus; and, even the malware family.

With regards to Android, the number of malware samples is increasing exponentially and several approaches have been proposed to detect them. Shabtai et al. [8] trained machine learning models using as features the count of elements, attributes or namespaces of the parsed Android Package File (*.apk*). To evaluate their models, they selected features using three selection methods: Information Gain, Fisher Score and Chi-Square. They obtained 89% of accuracy classifying applications into only 2 categories: tools or games.

1 `http://www.apple.com/iphone/features/app-store.html`

2 http://googleblog.blogspot.com/2011/05/android-momentum-mobile-and-more-at.html

3 `http://online.wsj.com/public/resources/documents/wsj-2009-0731-FCCApple.pdf`

There are another researches that uses a dynamic analysis to detect malicious applications. Crowdroid [9] is an earlier approach that analyse the behaviour of the applications. Blasing et al. [10] created AASandbox, which is an hybrid approximation. Dynamic part is based on the analysis of the logs for the low-level interactions obtained during execution. Shabtai et al. [11] also proposed a Host-Based Intrusion Detection System (HIDS) which use a machine learning methods that determines if the application is malware or not.

On the other hand, Google has deployed a supervision framework, called "Bouncer", which analyse the applications before being published. Oberheide and Miller [12] has revealed some features of this system. For example, the systems is based in QEMU and make both static and dynamic analysis.

Given this background, we present PUMA, a new method for detecting malicious Android applications employing the permission usage of the each application. Using these features, we train a machine-learning models to detect whether an applications is malware or not. Summarising, our main findings in this paper are: (i) we describe the process of extracting features from the Android .apk files, (ii) we propose a new representation for Android applications in order to develop a malware detection approach, and (iii) we perform an empirical validation of our approach and show that it can achieve high accuracy rates. The reminder of this paper is organised as follows. Section 2 details the generation of the dataset. Section 3 presents the permissions used in our approach. Section 4 describes the empirical evaluation of our method. Finally, section 5 discusses the results and shows the avenues of further work.

2 Description of the Dataset

2.1 Benign Software

To conform this dataset, we gathered a collection of 1811 Android applications of different types. In order to classify them properly, we chose to follow the same naming as the official Android market. To this end, we used an unofficial Android Market API[4] to connect with the Android market and, therefore, obtain the classification of the applications.

We selected the number of applications within each category according to their proportions in the Android Market. There are several application types in Android: native applications (developed with the Android SDK), web applications (developed mostly with HTML, JavaScript ad CSS) and widgets (simple applications for the Android desktop, which are developed in a similar way to web applications). To generate the dataset, we did not make distinctions between these types and every of them is represented in the final dataset. Once we determined the number of samples for each category, we randomly selected the applications. The number of samples obtained for each category is shown in Table 1.

[4] `http://code.google.com/p/android-market-api/`

Table 1. Number of benign software applications

Category	*Number*	*Category*	*Number*
Action and Arcade	33	Libraries and Demos	2
Races	3	Books and References	9
Casual	11	Medicine	2
Comics	1	Multimedia and Video	25
Shopping	3	Music and Audio	13
Communications	21	Business	2
Sports	4	News and Magazines	7
Education	1	Personalization	6
Companies	5	Productivity	30
Entertainment	16	Puzzles	16
Way of life	5	Health and Fitness	3
Accounting	2	Society	28
Photography	6	Weather	2
Tools	86	Transportation	2
Casino and card games	4	Sports and guides	9
	Total: 357		

2.2 Malicious Software

Malware samples were gathered by means of *VirusTotal*[5] which is an analysis tool for suspect files, developed by *Hispasec Sistemas*[6], a company devoted to security and information technologies. We have used their service called *VirusTotal Malware Intelligence Services*, available for researchers to perform queries to their database.

Using this tool, we gathered a total number of 4,301 samples. However, we performed a duplication removal step, where we deleted from the dataset the duplicates samples. Finally, we used a total number of 249, which according to Lookout[7] represents the 54% of the total malware samples.

3 Permissions of Android Applications

We performed a study of the different permissions of the applications in order to determine their suitability for malware detection. These permissions are evaluated when installing the app, and must be approved by the user. We used the *Android Asset Packaging Tool* (aapt) to extract and decrypt the data from the `AndroidManifest.xml` file, provided by the Android SDK. Thereafter, we dumped the result and processed it. From all the information available, we only

[5] `http://www.virustotal.com`
[6] `http://www.hispasec.com/`
[7] `https://www.mylookout.com/_downloads/lookout-mobile-threat-report-2011.pdf`

```
<manifest>
      <uses-permission />
      <permission />
      <permission-tree />
      <uses-sdk />
       ....
</manifest>
```

Fig. 1. Example of AndroidManifest file

used the following features: (i) "uses-permission", every permission that the application needs to work is defined under this tag; and (ii) "uses-feature", which shows which are the features of the device the application uses.

We have removed the rest of the information that the AndroidManifest file (shown in Fig. 1) stores because of its dependency on specific devices.

Fig. 2 suggests that the most frequent permissions in both categories are the same. Besides, it seems that there are not visible differences in the permissions used in malware with respect to the ones used in benign applications, at least, when we studyied separately. In other words, malicious applications do not need different permissions than benign ones. This may indicate that the granularity of the permissions system is not accurate enough to distinguish malicious intentions.

On the other hand, we conducted a study regarding the number of permissions of each application (shown in Fig. 3). The number of permissions required for both malicious and benign applications is also nearly the same. However, we noticed several differences in both classes: the chance of finding malware applications requiring only one permission is high while benign applications usually present 2 or 3 permissions. This fact suggests that only one permission is needed to behave maliciously on Android device.

4 Empirical Validation

To validate PUMA, we have employed supervised machine learning methods to classify Android applications into malware and benign software. To this extent, we have used Waikato Environment for Knowledge Analysis (WEKA)[8]. In particular, we have used the classifiers specified in Table 2. To evaluate the performance of machine-learning classifiers, *k-fold cross validation* is usually used [13]. Thereby, for each classifier we tested, we performed a k-fold cross validation [14] with $k = 10$. In this way, our dataset was split 10 times into 10 different sets for learning (90% of the total dataset) and testing (10% of the total data).

[8] http://www.cs.waikato.ac.nz/ml/weka/

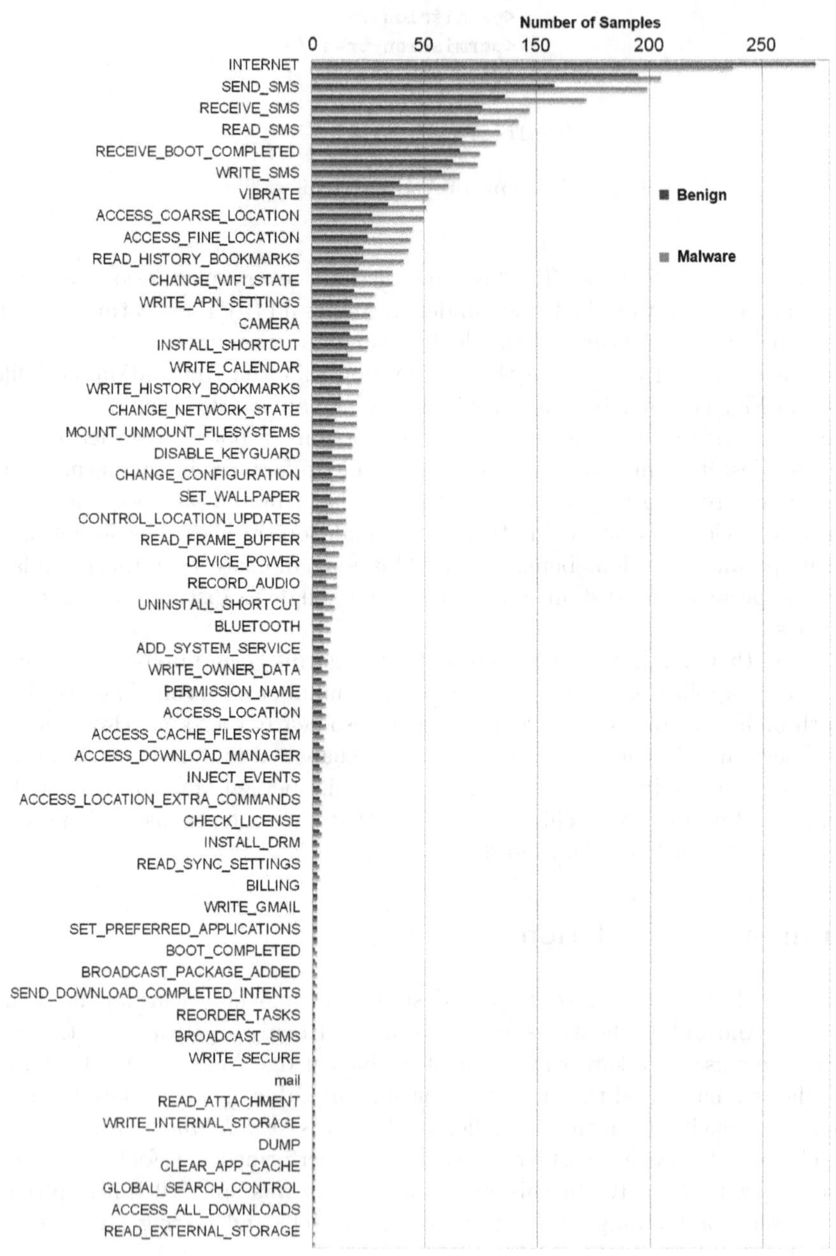

Fig. 2. Extracted permissions for the applications conforming the dataset

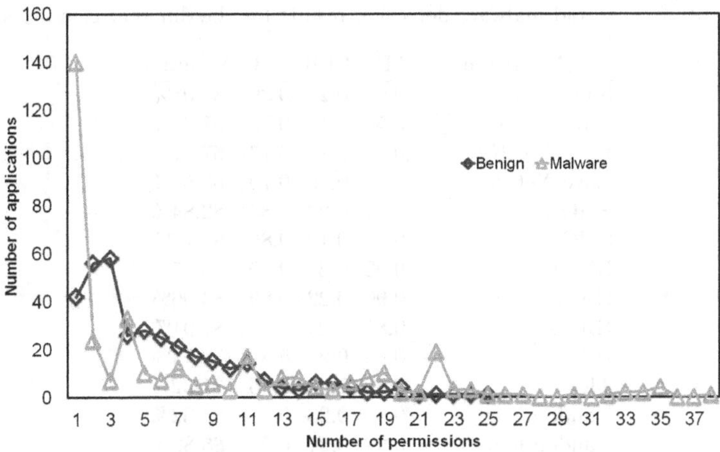

Fig. 3. Number of permissions of benign and malware apps

Table 2. Machine learning classifiers to used in the experiment

Algorithm	Used configuration
SimpleLogistic	
NaiveBayes	
BayesNet	K2 and TAN
SMO	PolyKernel y NormalizedPolyKernel
IBK	Valores de K: 1, 3 and 5
J48	
RandomTree	
RandomForest	Valor de I: 10, 50 and 100

To evaluate each classifier's capability, we measured the True Positive Ratio (TPR):

$$TPR = \frac{TP}{TP + FN} \tag{1}$$

where TP is the number of malware cases correctly classified (true positives) and FN is the number of malware cases misclassified as legitimate software (false negatives).

We also measured the False Positive Ratio (FPR):

$$FPR = \frac{FP}{FP + TN} \tag{2}$$

where FP is the number of benign software cases incorrectly detected as malware and TN is the number of legitimate executables correctly classified.

Furthermore, we measured the accuracy, i.e., the total number of the classifier's hits divided by the number of instances in the whole dataset:

$$Accuracy = (TP + TN) \cdot \frac{TP + TN}{TP + FP + TP + TN} \tag{3}$$

Table 3. Android malware detection results for the different classifiers

Algorithm	TPR	FPR	AUC	Accuracy
SimpleLogistic	0.91	0.23	0.89	84.08%
NaiveBayes	0.50	0.15	0.78	67.64%
BayesNet K2	0.45	0.11	0.77	67.07%
BayesNet TAN	0.53	0.16	0.79	68.51%
SMO Poly	0.91	0.26	0.83	82.84%
SMO NPoly	0.91	0.19	0.86	85.77%
IBK 1	0.92	0.21	0.90	85.55%
IBK 3	0.90	0.22	0.89	83.96%
IBK 5	0.87	0.24	0.88	81.91%
IBK 10	0.85	0.27	0.87	78.94%
J48	0.87	0.25	0.86	81.32%
RandomTree	0.90	0.23	0.85	83.32%
RandomForest 10	0.92	0.21	0.92	85.82%
RandomForest 50	0.91	0.19	0.92	86.41%
RandomForest 100	0.91	0.19	0.92	86.37%

Besides, we measured the Area Under the ROC Curve (AUC) which establishes the relation between false negatives and false positives [15]. The ROC curve is obtained by plotting the TPR against the FPR.

Table 3 shows the obtained results. With the exception of the Bayesian-based classifiers, the methods achieved accuracy rates higher than 80%. In particular, the best classifier, in terms of accuracy, was Random Forest trained with 50 trees with a 86.41%. Regarding the TPR results, Random Forest trained with 10 trees was the best classifier with a 0.92. The lowest FPR was obtained with Bayesian networks trained with Tree Augmented Naïve, however, its TPR results are lower than 55%. In terms of AUC, Random Forest was the best classifier with a 0.92.

5 Discussion and Conclusions

Permissions are the most recognisable security feature in Android. User must accept them in order to install the application. In this paper we evaluate the capacity of permissions to detect malware using machine-learning techniques.

In order to validate our method, we collected 239 malware samples of Android applications. Then, we extracted the aforementioned features for each application and trained the models, evaluating each configuration using the Area Under ROC Curve (AUC). We obtained a 0.92 of AUC using the Random Forest classifier.

Nevertheless, there are several considerations regarding the viability of our approach. Forensic experts are developing reverse engineering tools over Android applications, from which researchers could retrieve new features to enhance the data used to train the models. Furthermore, despite the high detection rate, the obtained result has an high false positive rate. Consequently, this method can be used as a first step before other more extensive analysis, such as a dynamic analysis.

Future work of this Android malware detection tool is oriented in two main directions. First, there are other features from the applications that could be used to improve the detection ratio that do not require to execute the sample. Forensics tools for Android applications should be developed in order to obtain new features. Second, dynamic analysis provides additional information that could improve malware detection systems. Unfortunately, smartphones resources are limited and these kind of analysis usually consumes resources that these devices don't have.

References

1. Schultz, M., Eskin, E., Zadok, F., Stolfo, S.: Data mining methods for detection of new malicious executables. In: Proceedings of the 2001 IEEE Symposium on Security and Privacy, pp. 38–49 (2001)
2. Devesa, J., Santos, I., Cantero, X., Penya, Y.K., Bringas, P.G.: Automatic Behaviour-based Analysis and Classification System for Malware Detection. In: Proceedings of the 12th International Conference on Enterprise Information Systems (ICEIS), pp. 395–399 (2010)
3. Santos, I., Nieves, J., Bringas, P.G.: Semi-supervised learning for unknown malware detection. In: Proceedings of the 4th International Symposium on Distributed Computing and Artificial Intelligence (DCAI), 9th International Conference on Practical Applications of Agents and Multi-Agent Systems (PAAMS), pp. 415–422 (2011)
4. Santos, I., Laorden, C., Bringas, P.G.: Collective classification for unknown malware detection. In: Proceedings of the 6th International Conference on Security and Cryptography (SECRYPT), pp. 251–256 (2011)
5. Santos, I., Brezo, F., Ugarte-Pedrero, X., Bringas, P.G.: Opcode sequences as representation of executables for data-mining-based unknown malware detection. Information Sciences (in press), doi:10.1016/j.ins.2011.08.020
6. Rieck, K., Holz, T., Willems, C., Düssel, P., Laskov, P.: Learning and Classification of Malware Behavior. In: Zamboni, D. (ed.) DIMVA 2008. LNCS, vol. 5137, pp. 108–125. Springer, Heidelberg (2008)
7. Tian, R., Batten, L., Islam, R., Versteeg, S.: An automated classification system based on the strings of trojan and virus families. In: 4th International Conference on Malicious and Unwanted Software (MALWARE), pp. 23–30. IEEE (2009)
8. Shabtai, A., Fledel, Y., Elovici, Y.: Automated Static Code Analysis for Classifying Android Applications Using Machine Learning. In: 2010 International Conference on Computational Intelligence and Security, pp. 329–333 (December 2010)
9. Burguera, I., Zurutuza, U., Nadjm-Tehrani, S.: Crowdroid: behavior-based malware detection system for android. In: Proceedings of the 1st ACM Workshop on Security and Privacy in Smartphones and Mobile Devices, pp. 15–26. ACM (2011)
10. Blasing, T., Batyuk, L., Schmidt, A., Camtepe, S., Albayrak, S.: An android application sandbox system for suspicious software detection. In: 2010 5th International Conference on Malicious and Unwanted Software (MALWARE), pp. 55–62. IEEE (2010)

11. Shabtai, A., Elovici, Y.: Applying Behavioral Detection on Android-Based Devices. In: Cai, Y., Magedanz, T., Li, M., Xia, J., Giannelli, C. (eds.) Mobilware 2010. LNICST, vol. 48, pp. 235–249. Springer, Heidelberg (2010)
12. Oberheide, J., Miller, J.: Dissecting the android bouncer (2012)
13. Bishop, C.: Pattern recognition and machine learning. Springer, New York (2006)
14. Kohavi, R.: A study of cross-validation and bootstrap for accuracy estimation and model selection. In: International Joint Conference on Artificial Intelligence, vol. 14, pp. 1137–1145 (1995)
15. Singh, Y., Kaur, A., Malhotra, R.: Comparative analysis of regression and machine learning methods for predicting fault proneness models. International Journal of Computer Applications in Technology 35(2), 183–193 (2009)

Application of Soft Computing Technologies toward Assessment and Skills Development

Ignacio Aliaga[1], Vicente Vera[1], Cristina González Losada[1], Álvaro Enrique García[1], Héctor Quintián[2], Emilio Corchado[2], Fanny Klett[3], and Laura García-Hernández[4]

[1] University Complutense of Madrid, Odontology Faculty, Madrid, Spain
Ialia01@estumail.ucm.es, {vicentevera,aegarcia}@odon.ucm.es
[2] University of Salamanca, Departamento de Informática y Automática
Plaza de la Merced s/n, 37008, Salamanca, Spain
{hector.quintian,escorchado}@usal.es
[3] Director, German Workforce ADL Partnership Laboratory, Germany
fanny.klett.de@adlnet.gov
[4] Area of Project Engineering, University of Cordoba, Spain
ir1gahel@uco.es

Abstract. Schools and universities face multiple challenges when they target initiating or expanding undergraduate programs. Education has traditionally utilized a teacher-centered educational and assessment approach. Only few attempts exist to involve objective feedback and non-traditional assessment methods and technologies to improve the processes of teaching, learning, and education in general.

This paper addresses a novel objective multi-parameter assessment methodology based on Soft computing technology to discover the effect of students' groupings by exploiting the interrelationships between the grades the students received for their laboratory subjects and the grade they obtained in the university enrolment exam. The research results allow for exploring non-desirable discordant teaching and assessment practices for individuals or groups. In addition, the results obtained illustrate opportunities to focus on the individual student during the education process and determine adaptive teaching strategies based on the particular level of knowledge and experience. Toward these results statistical and Soft computing models implementing Unsupervised Neural and Exploratory Projection Techniques have been applied to carry on the objective assessment of the students' skills development during the entire higher education period.

Empirical verification of the proposed assessment model is performed in a real environment, where a case study is defined, and analysed. The real data set to validate the performance of the proposed approach has been collected at the School of Dentistry of the Complutense University of Madrid.

Keywords: Component, Unsupervised Neural and Exploratory Projection Techniques, Assessment of skills development, Higher education and training, Dental milling.

Á. Herrero et al. (Eds.): Int. JointConf. CISIS'12-ICEUTE'12-SOCO'12, AISC 189, pp. 299–310.
springerlink.com

1 Introduction

In the past, the term assessment was used to refer particularly to the process of determining the extent to which learners have mastered a subject. It has recently been extended in the academic context to cover all uses of evaluation. Assessment data are observations or facts that must be collected, organized, and analysed to become useful.

Assessment has many roles. On the one hand, it is critical for measuring the student's achievement within the learning process, i.e. students can understand to what extend learning objectives have been accomplished. On the other hand, it can also be used as a teaching tool to help students better understand concepts and enhance their skills. In addition, the instructor can realize what students are understanding as well as the concepts that need further explanation. Thus, assessment can be viewed as a powerful tool for quality management, on the individual and the organizational level. [1] Simultaneously, it appears as part of the skills development process. In the framework of skills development and career support, assessment needs a systematic approach to develop a range of transferable personal skills that are sought by every employer, and to track skills development and competency data for every student.

Schools and academia are particularly aware that applying sustainable assessment processes will support them in treating the new challenges that arise every year when trying to improve their educational programs and adapt to the needs of the market, industry and businesses. Innovative approaches in teaching and learning enable handling various complex considerations in engineering, medicine, dentistry and further curricula where various psychomotor, modelling, and complex problem solving skills have to be addressed in practical settings. The acquisition of such skills and abilities appears particularly important for the development of a wide variety of career paths within engineering, design and medicine and requires enormous willingness and individual training [1] leading to a specific psychomotor development, which is determined by the innate abilities and the learning results achieved during the training process [2,3]. The objective assessment of the psychomotor skills remains a challenge, and requires new models and approaches, particularly different from assessing learning in general. Moreover, the requirement for tracking and analysing assessment data in due course creates the demand for an effective technology based assessment.

Against this background, Soft computing and statistical models offer a sufficient environment to address multi-parameter assessment [4-8] of psychomotor skills due to the integrated beneficial opportunity for visualization of analytical and multidimensional data able to facilitate the interpretation of results, and thus, improve the effectiveness of decision-making processes [9]. It appears advantageous to apply Soft computing models to assessment, based also on results from social and collaborative learning theories. For example, knowing the level of psychomotor development of an individual in relation to other students in the same class, it is possible to identify individuals who experience larger difficulties. Once these difficulties are identified, the instructors can raise individual and collective reinforcement [10].

This paper presents a precise, logical, achievable, observable, and objectively measurable Soft computing model to analyse the skills development of students based on a comprehensive analysis of the psychomotor skills of an individual compared to the skills of classmates. This model enables the proper assessment of the skills development process in a practical setting by correlating variables such as the age of the students, previous experience, etc. [11].

The next sections refer to main aspects of the novel Soft computing model to facilitate and advance the assessment of the skills development process. Testing and verification of the approach is performed in a real case study involving dentistry students. The technology has been developed at the Complutense University of Madrid (Test Skills Practice).

2 The General Soft Computing Model

Engineering, design and medicine curricula involve a series of practical experiences in the area of psychomotor skills development referring to an orderly and chained sequence, and following a series of learning objectives that help the instructors to organize them. Each of these practical experiences includes the criteria for assessing the achievement of the objectives.

Commonly, the learning objective is a statement that describes what a learner will be able to do as a result of learning and/or teaching. A learning objective should be precise, logical, achievable, observable and measurable, and it becomes an inseparable part of the assessment process that can involve various variables depending on the evaluation aims. For example, a comprehensive analysis of the individual abilities of a student compared with the classmates' ones and measuring the learning achievements after a period of time without having received more learning, is a challenging task. It is valuable also to analyse all the skills the student has developed during an entire undergraduate study toward identifying a relationship between the grade obtained in the university enrolment exam, the continuing psychomotor skills performance and the ease of learning. Variables such as the age of the students when starting a course and their previous experience in a similar field, is a factor enormously influencing the skills development process.

Exploratory Projection Pursuit (EPP) [12] is a statistical method for solving the difficult problem of identifying structure in high dimensional data. The method used here is based on the projection of the data onto a lower dimensional subspace in which we search for its structure with the naked eye. It is necessary to define an "index" that measures the interestingness of a projection. After that the data is transformed by maximizing the index in order to maximise interest according to that index. From a statistical point of view the most interesting directions are those which are as non-Gaussian as possible [13]. The Soft computing model takes advantage of a neuronal model of Exploratory Projection Pursuit (EPP), and Maximum-Likelihood Hebbian Learning (MLHL) [12,14] to best match the criteria identified for assessing the psychomotor skills of students performing practical work.

To test and verify the research approach, a real case study was developed by focusing on dentistry students studying the subject Introduction to Dentistry in the undergraduate degree of Dentistry. The practical work refers to the performance of cavities, of shapes with different levels of difficulty, previously designed on methacrylate blocks by using two techniques (high speed –turbine- and low speed –contra-). The challenging factor for the Soft computing model consists in the fact that a number of students have been accessed after completing Dental Hygienist or Dental Technician degrees. In light of the above, the variables mentioned before are particularly interesting to be explored during the entire undergraduate study because they may raise the hypothesis that the psychomotor skills of these students vary from the skills of the remaining undergraduates.

The next section presents the intelligent model applied for clustering students. It is based on the use of projection models.

3 The Specific Soft Computing Model – An Intelligence System for Clustering Students

This section introduces the main intelligent model applied in this research to perform clustering in the students' data set. It is used to provide a visual analysis of the internal structure of the data set.

A combination of projection types of techniques together with the use of scatter plot matrices constitute a very useful visualization tool to investigate the intrinsic structure of multidimensional datasets, allowing experts to study the relationship between different components, factors or projections, depending on the technique used.

Projection methods help projecting high-dimensional data points onto lower dimensions in order to identify "interesting" directions in terms of any specific index or projection. Such indexes or projections are, for example, based on the identification of directions that account for the largest variance of a dataset (such as Principal Component Analysis (PCA)[15]) or the identification of higher order statistics such as the skew or kurtosis index, as in the case of Exploratory Projection Pursuit (EPP) [16]. Having identified the interesting projections, the data is then projected onto a lower dimensional subspace plotted in two or three dimensions, which makes it possible to examine its structure with the naked eye. The remaining dimensions are discarded as they mainly relate to a very small percentage of the information or the dataset structure. In that way, the structure identified through a multivariable dataset may be visually analysed in an ease manner.

One of those models is the standard statistical EPP method [16], which provides a linear projection of a dataset, but it projects the data onto a set of basic vectors, which best reveal the interesting structure in data; interestingness is usually defined in terms of how far the distribution is from the Gaussian distribution.

The neural implementation of EPP, applied in this research for identifying the different students clusters, is the Maximum-Likelihood Hebbian Learning (MLHL) [14,17], which identifies interestingness by maximising the probability of the residuals under specific probability density functions that are non-Gaussian.

Considering an N-dimensional input vector (x), and an M-dimensional output vector (y), with W_{ij} being the weight (linking input j to output i), then MLHL can be expressed as:

1. Feed-forward step:

$$y_i = \sum_{j=1}^{N} W_{ij} x_j, \forall i \tag{1}$$

2. Feedback step:

$$e_j = x_j - \sum_{i=1}^{N} W_{ij} y_i, \forall j \tag{2}$$

3. Weight change:

$$\Delta W_{ij} = \eta \cdot y_i \cdot sign(e_j) \left| e_j \right|^{p-1} \tag{3}$$

Where: η is the learning rate and p is a parameter related to the energy function [13].

Then, by maximizing the likelihood of the residual with respect to the actual distribution, we are matching the learning rule to the probability density distribution of the residual. The power of the method comes from the choice of an appropriate function.

4 The Real Case Study

Significance of objective measurements has risen. Simultaneously, the prediction obtained by these measurements appears relevant to be exploited toward the assessment skills development. This fact also applies to dental students' [18].

The real case study focuses on facilitating the identification of divergent or non-desirable situations in the educational process. The particular aim of this study is to classify the psychomotor skills of first year students when creating methacrylate figures during this Dental Aptitude Test, which consists of carving ten methacrylate figures by using rotatory systems, and applying two different speeds (V1 and V2). A total of 20 figures are included. V1 (low speed) rotates at a speed of 10-60,000 revolutions per minute (rpm), while V2 (turbine or high speed) rotates at a speed of 250,000 rpm.

Seven of the figures made by the students can be easily created, while the remainders, which have several planes, involve a higher level of difficultly.

The tests conducted provide a comprehensive analysis of the students' performance with regard to their peer group and year of graduation, and allow for measuring and assessing the students' skills development during the course of an academic year compared to other academic years, or to other students in the same academic year.

4.1 Training Scenario Description

Every student works on a methacrylate sheet at two different speeds, low speed and high speed. The low speed (10.000-60.000 rpm) is to be used to carve the first set of ten figures (see Fig 1). After completing this part of the practical work, students start carving the second part, which is basically a second round of the same figures, but this time using the high speed (150.000-250.000 rpm). The second instrument involves a higher level of difficulty as the bur is spinning faster and better psychomotor skills are the pre-requisite for effectively completing this task.

Both parts of the practical work have to be completed during 90 minutes and the results have to be submitted to the supervisors.

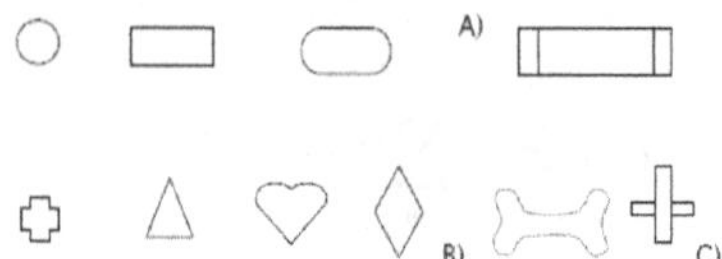

Fig. 1. Figures to be carved by the students

The individual training involves the following steps:

- Properly assemble the low speed and turbine and test its operation. Recognize and properly mount the milling cutters.
- Create figures in a methacrylate block (see Fig 2). In advance, the students have to draw the outline of what to carve with a pen and a ruler.
- At the end of the training period the students have to submit their methacrylate block with all the figures carved on it (see Fig 2).Explanation:
  - Figures should have an approximate size of 1 cm (in the length) and a depth of about 3-4 mm, except figures A, B and C (see Fig 1).
  - The cavity design must be clear and free of defects, and the walls must be smooth and perpendicular to the surface of the methacrylate sheet. The floor must be parallel to the surface and as smooth as possible.
  - Cavities are initially carved with a low speed, and subsequently with a turbine, related to the psychomotor skills acquired.
  - Figures A, B and C (see Fig 1) have to be created in the end because of their level of difficulty concerning two planes.
  - Students can repeat carving the figures as often as they like, so they can use all the figures but only one side of the methacrylate block.
  - If the milling cutter is dulled with resin, it can be cleaned by spinning briefly and gently against a small block of wood or another milling cutter cleaner.

- 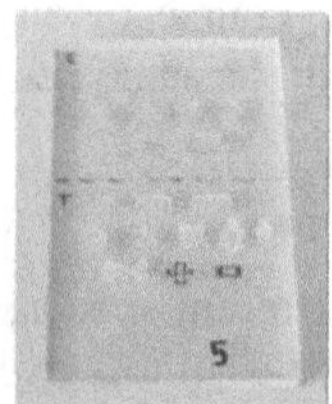 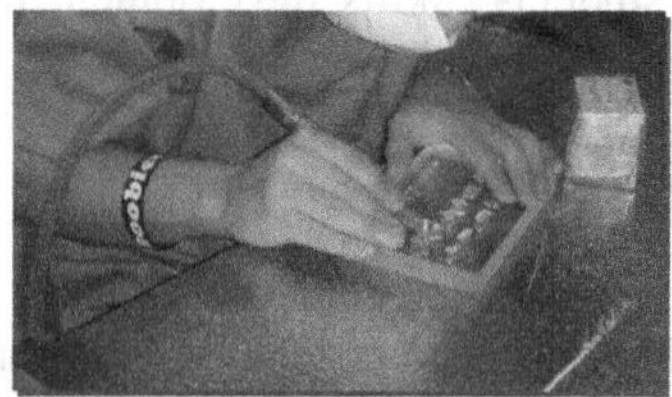

Fig. 2. Real methacrylate sheet with figures carved on it

4.2 Empirical Training Scenario Evaluation

The real case scenario is empirically evaluated based on a survey of 79 first year dental students (24 students were eliminated from the initial 103 students, because they did not participate in the practical work). The information analysed for each student is based on 88 variables. The first most important eight variables concern:

- Age of the student (integer value).
- Sex of the student (integer value).
- Mark obtained by the student in the university enrolment exam. (Decimal value between 0 and 14).
- Previous experience gained by the student. The students may have had professional experience as a nurse, dental technician, hygienist, dental technician and hygienist, or lack of previous work experience.
- Mark obtained by the student in the theoretical exam of the subject (Decimal value between 0 and 10).
- Mark obtained by the students in the practical part of the subject (Decimal value between 0 and 10).
- Group class of the student (integer value between 1 and 4).
- Number of figures carved by the student (integer value between 0 and 20).

The following 80 variables (20 figures with four variables each) are the evaluations of the different parts of the figures (graded between 0 and 5). The way to interpret these variables is as follows: 'x' indicates the figure number and can range from 1 to 10, 'y' indicates the speed used to carve the figure by using Low Speed (1) or High speed (2), and 'z' indicates the evaluator who examines the test (1 or 2):

- **Fx_Vy_Ez_WALL:** evaluate the quality of the walls of the figure created by the student.
- **Fx_Vy_Ez_DEPTH:** evaluate the quality of the depth of the figure created by student.
- **Fx_Vy_Ez_EDGES:** evaluate the quality of the edges of the figure created by student.
- **Fx_Vy_Ez_FLOOR:** evaluate the plain and irregularities presented on the floor of the figure created by the student.

The data are collected in a document that represents the dataset to be evaluated and analysed.

The dataset, along with the corresponding labels, are recorded in a Comma Separated Value (CSV) format text file serving as input data in the software that applies the above described reduction treatment and generates the graphic representations.

5 Results and Discussion

Following the analysing procedure, PCA identified two clearly separated clusters called G1 and G2 (see Fig 3).

- **G_1:** This cluster represents students with high marks in the university enrolment exam, good or very good marks in the theoretical and practical matters of the subject, and a large number of figures carved. This cluster is composed by two sub-clusters: C1 and C2, as explained below.
- **G_2:** This cluster represents students with decent marks in theoretical and practical matters of the subject, and a small number of figures carved. This cluster is composed of two sub-clusters: C3 and C4, as explained below.

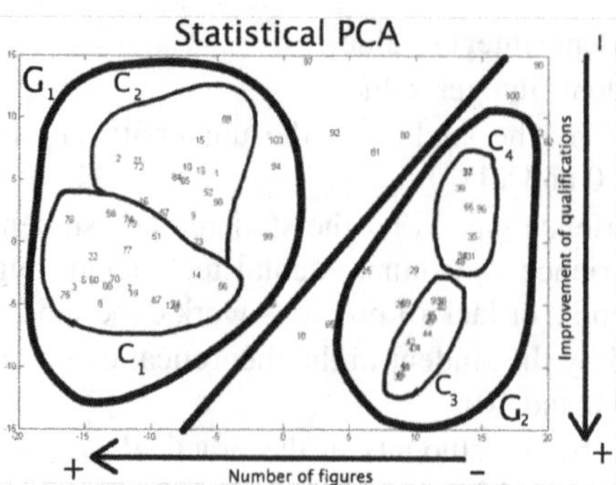

Fig. 3. Statistical PCA

The sub-clusters refer to the following:

- C_1 represents young students (see Table 1) characterized with high marks in the university enrolment exam, without previous professional experience, and good marks in both, theory and practice.
- C_2 represents young students (see Table 1) with no previous professional experience, with good marks in the theoretical part of the subject and average marks in the practical part. These students have carved many figures, but less than students belonging to the earlier cluster (C1).
- C_3 represents students (see Table 1) with decent marks, both in the theoretical and practical parts of the subject, and have succeeded in carving an average number of figures (about half), mostly with low speed. Most of these students belong to the group number G2.
- C_4 represents students (see Table 1) that have been able to successfully carve 7 figures. Their marks in theory and practice are varied.

Table 1. Cluster Classification (PCA)

Clusters	Students belonging to each cluster
C1	3, 5, 6, 7, 8, 12, 19, 22, 24, 51, 58, 60, 67, 68, 70, 73, 74, 75, 76, 77
C2	1, 2, 9, 10, 13, 15, 16, 23, 52, 57, 59, 71, 72, 84, 85, 98
C3	4, 32, 36, 40, 41, 42, 43, 44, 45, 46, 47, 49, 50, 93
C4	33, 34, 35, 37, 39, 48, 55, 96, 101

After analysing the data using PCA, MLHL is then applied to find possible improvements in the classification of the samples (students).

Once the dataset test is analysed using MLHL (see Fig 4), the best result can be obtained by applying the following parameters: 3 neurons in the output layer (m), 100000 iterations, 0.006 as learning rate, and 1.1 as p (see equation 1, 2 and 3).

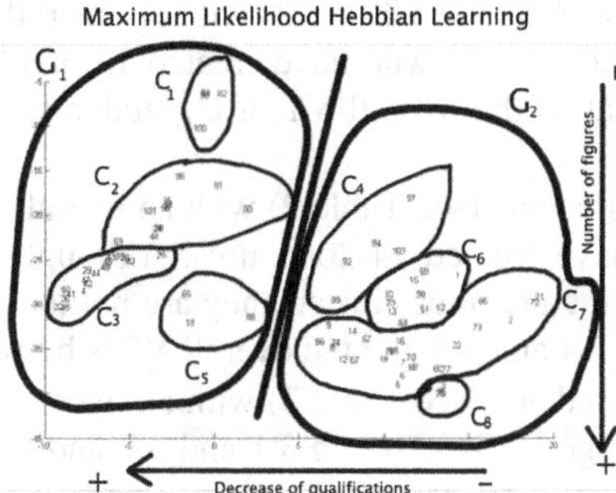

Fig. 4. MLHL Analysis with m=3, iters=100000, lrate=0.006 and p=1.1

Like in PCA, two clusters G_1 and G_2 can be identified. In this case, the first cluster (G_1) is composed of four sub-clusters C_1, C_2, C_3, and C_5, and characterized by an average mark of less than 1.8 (marks go from 0 to 5) for the figures performed at high speed (V2). The second cluster (G_2) is composed of four sub-clusters (C_4, C_6, C_7 and C_8) and characterized by an average mark greater than 1.8 for the figures carved at high speed (V2).

In order to facilitate the understanding of the results, the following equivalences are adopted from this point forward:

- Overall average: average of marks for figures created by a student.
- Average V1: average of marks for figures carved by a student using low speed.
- Average V2: average of marks for the figures created by a student using turbine (high speed).

The clusters that describe the data classification are the following:

- C_1 represents students (see Table 2) whose overall average is between 0.0 and 0.5. This cluster is also characterized by an average V1 between 0 and 0.95 and an average V2 of mostly 0.
- C_2 represents students (see Table 2) who have been able to successfully perform seven figures and whose overall average falls between 0.8 and 1.5. This cluster is also characterized by an average V1 between 1.7 and 2.95 and an average V2 of mostly 0.
- C_3 represents students (see Table 2) who have an overall average between 1.5 and 2.1. Students in this cluster are characterized by having made an average number of figures (between 8 and 13), most of which are carved with low speed (V1). This cluster is also characterized by an average V1 between 3 and 4.15 and an average V2 of mostly 0.
- C_4 represents students (see Table 2) who are similar to those belonging to the previous cluster, as the overall average of students falls within the same range, and the average V1 is a subinterval of students in cluster C_2, namely between 1.64 and 2.48. However, they differ in the average V2, because it falls within a completely different range (between 1.88 and 2).

- C_5 represents students (see Table 2) with an overall average similar to those in clusters C_3 and C_4 and who have created 13 or 14 figures. The overall V1 is a sub-interval of the overall V1 of C_3 students, but the overall V2 is between 0.9 and 1.7.
- C_6 represents students (see Table 2) with an overall average between 2.1 and 2.8 and who have created 14-20 figures. Although this range is included in the interval of the previous cluster, they are not the same because the overall V1 is between 2.4 and 3.5 and the overall V2 is between 2.4 and 3.15.
- C_7 represents students (see Table 2) with an overall average between 2.8 and 3.75. The average V1 is between 3.1 and 3.7 and the average V2 is between 2.05 and 3.9. This is the largest cluster detected.
- C_8 represents students (see Table 2) with the best overall average (between 3.9 and 4.1). Likewise, the average for both V1 and V2 is greater than 3.9. These students have been able to perform (almost) all the figures.

The following table shows which students belong to each cluster:

Table 2. Cluster Classification (MLHL)

Clusters	Students belonging to each cluster
C1	62, 82, 90, 100
C2	33, 34, 35, 37, 39, 48, 55, 69, 80, 81, 96, 101
C3	4, 25, 26, 29, 32, 36, 40, 41, 42, 43, 44, 45, 46, 47, 49, 50, 93
C4	92, 94, 97, 99, 103
C5	18, 65, 86
C6	1,10,13,15,23,51,52,59,84, 98
C7	2, 5, 6, 7, 8, 9, 12, 14, 16, 19, 22, 24, 56, 57, 58, 60, 67, 68, 70, 71, 72, 73, 74, 77, 85
C8	3, 75, 76

6 Conclusions

The analysis based on the use of PCA provides a valid initial classification of the real dataset, but it can significantly be improved by using MLHL.
The methodology applied provides promising results toward the comprehensive assessment of psychomotor skills development in career paths within engineering, design and medicine.

The novel model based on Soft computing technology has been verified in a real case study involving dentistry students. The variables analysed in this study can advantageously serve the teaching process by offering adaptive measures toward identifying those students most likely to pass the course, or who experience larger difficulties, respectively. According to the grade obtained in the university entrance exam and the number of figures carved, it's possible to predict clinical success the students' achievements in terms of reaching the highest mark in the Test Skills Practice.

The research results make it possible to conclude that there is no relationship between recent achievements and previous professional experience, and this represents a helpful indication for the arrangement of future educational settings. Furthermore, this research shows that the quality of the figures carved is the better, the better the students have been trained in the preclinical period, i.e. the greater number of figures have been carved. Therefore, the practice appears a sustainable way to achieve clinical success.

Future work will focus on the extension and application of this objective assessment methodology based on Soft computing models in further educational fields and knowledge domains, such as civil, automotive, aerospace and mechanical engineering, and design and medicine to significantly improve the recent predominantly subjective practices of assessing skills development.

Acknowledgment. This research is partially supported through projects of the Spanish Ministry of Economy and Competitiveness [ref: TIN2010-21272-C02-01] (funded by the European Regional Development Fund). This work was also supported in the framework of the IT4 Innovations Centre of Excellence project, reg. no. CZ.1.05/1.1.00/02.0070 by operational programme 'Research and Development for Innovations' funded by the Structural Funds of the European Union and state budget of the Czech Republic, EU.

References

1. Klett, F.: The Design of a Sustainable Competency-Based Human Resources Management: A Holistic Approach. Special Issue of the Knowledge Management & E-Learning. An International Journal 3(2) (2010)
2. Grantcharov, T.P., Funch-Jensen, P.: Can everyone achieve proficiency with the laparoscopic technique? Learning curve patterns in technical skills acquisition. American Journal of Surgery 197(4), 447–449 (2009)
3. Salgado, J., Grantcharov, T.P., Papasavas, P.K., Gagne, D.J., Caushaj, P.F.: Technical skills assessment as part of the selection process for a fellowship in minimally invasive surgery. Surg. Endosc. 23, 641–644 (2009)
4. Polyzois, I., Claffey, I., McDonald, A., Hussey, D., Quinn, F.: Can Evaluation of a Dental Procedure at the Outset of Learning Predict Later Performance at the Preclinical Level? A Pilot Study. European Journal of Dental Education 15(2), 104–109 (2011)
5. Fiori, S.: Visualization of Riemannian-manifold-valued elements by multidimensional scaling. Neurocomputing 74, 983–992 (2011)
6. González-Navarro, F.F., Belanche-Muñoz, L.A., Romero, E., Vellido, A., Julià-Sapé, M., Arús, C.: Feature and model selection with discriminatory visualization for diagnostic classification of brain tumors. Neurocomputing 73, 622–632 (2010)
7. Corchado, E., Herrero, Á.: Neural visualization of network traffic data for intrusion detection. Applied Soft Computing 11, 2042–2056 (2011)
8. Corchado, E., Sedano, J., Curiel, L., Villar, J.R.: Optimizing the operating conditions in a high precision industrial process using soft computing techniques. Expert Systems 29(3), 276–299 (2012)
9. Corchado, E., Perez, J.: A three-step unsupervised neural model for visualizing high complex dimensional spectroscopic data sets. Pattern Analysis & Applications 14, 207–218 (2011)

10. Lee, T.Y., Radcliffe, D.F.: Innate design abilities of first year engineering and industrial design students. Design Studies 11(2), 96–106 (1990)
11. Pitigoi-Aron, G., King, P.A., Chambers, D.W.: Predictors of Academic Performance for Applicants to An International Dental Studies Program in the United States. Journal of Dental Education 75(12), 1577–1582 (2011)
12. Friedman, J.H., Tukey, J.W.: A Projection Pursuit Algorithm for Exploratory Data Analysis. IEEE Transactions on Computers C-23, 881–890 (1974)
13. Diaconis, P., Freedman, D.: Asymptotics of Graphical Projections. The Annals of Statistics 12(3), 793–815 (1984)
14. Corchado, E., Pellicer, M.A., Borrajo, M.L.: A maximum likelihood Hebbian learning-based method to an agent-based architecture. International Journal of Computer Mathematics 86, 1760–1768 (2009)
15. Dorothy McComb, B.D.S.: Class I and Class II silver amalgam and resin composite posterior restorations: teaching approaches in Canadian faculties of dentistry. J. Can. Dent. Assoc. 71(6), 405–406 (2005)
16. Hotelling, H.: Analysis of a complex of statistical variables into principal components. Journal of Educational Psychology 24(6), 417–441 (1933)
17. Corchado, E., MacDonald, D., Fyfe, C.: Maximum and Minimum Likelihood Hebbian Learning for Exploratory Projection Pursuit. Data Mining and Knowledge Discovery 8, 203–225 (2004)
18. Behar-Horenstein, L.S., Mitchell, G.S., Dolan, T.A.: A Case Study Examining Class-room Instructional Practices at a U.S. Dental School. J. Dent. Educ. 69, 639–648 (2005)

An Interactive Mathematica Book

J.A. Álvarez-Bermejo[1], J. Escoriza[2], J.A. López-Ramos[2], and J. Peralta[2]

[1] Departamento de Arquitectura de Computadores y Electrnica, Universidad de Almería
jaberme@ual.es

[2] Departamento de Algebra y Análisis Matemático, Universidad de Almería
{jescoriz,jlopez,jperalta}@ual.es

Abstract. In this paper we show how we have used the online resource webMathematica as a tool that allows teaching different mathematical disciplines in and out of the classroom, more specifically, Linear Algebra. We also give some statistics on the last three years that show students' satisfaction on these new didactic resource.

1 Introduction

One of the main issues that Engineering students face of during their career is to solve a big number of problems and examples in order to reach a high level in managing the main techniques and procedures used in the different courses they take. To do so students and teaching professionals make use of solved problems books. However the main problem that students encounter is that the difficulty level of the shown examples cannot be selected and just some significant cases are treated in order to reach a high audience. In other occasions the student would need to deepen in a particular aspect or case, but there is not a necessary treatment or simply it does not exist.

Application of technologies for teaching is proportional to their progress and Mathematics and Engineering is a clear exponent of this situation. Internet using produced a few years ago an explosion of access to any kind of resources and bibliography. Web-based training is appropriate for teaching certain skills, specially for learners lacking skills [3] and particularly solving mathematical problems and to avoid the precedent issues. The use of platforms such as WebCT or Moodle supports undoubtedly teaching tasks by providing the students a better directed study.

However teaching Mathematics requires possibility of making complex calculus or showing mathematical expressions in a suitable and easy way. Concerning these mathematical concepts there exist many software packages, some of them licensed, other free that may offer a considerably help in many situations, offering the possibility of making calculations and simulations at different levels of difficulty. Studies on several fields of Mathematics and countries and their corresponding impact can be found in [4], but that are not compatible with these supporting platforms for teaching. The use of these packages allows the student to test with different values for a same type of problems and build a big group of solved examples or check if the solution that the student got from a proposed exercise is right or not. It is clear that those licensed offer a nicer and more helpful environment for users due to resources, but once the user is capable to manage with the different commands to be used, most of them offer similar performances.

Á. Herrero et al. (Eds.): Int. JointConf. CISIS'12-ICEUTE'12-SOCO'12, AISC 189, pp. 311–319.
springerlink.com

One of these mathematical software packages is Mathematica ([5]). Depending on level and studies where it supported teaching, this was used as a simple calculator, making use the proper functions or other implemented by the teacher, mainly in resolution of problems with calculations of high complexity or as self-correcting. Some other times is the own student who has to implement new functions whose designing process involves managing contents and techniques appearing in the course. In this sense teachers may use now webMathematica ([6]) that allows developing interactive didactic material without the use of a license or installing any special software. This is due to the fact that all the material that is designed by means of webMathematica is presented as a webpage and can be accessible just with any web browser.

Traditional strategies used to enhance learning Mathematics such as solving computational problems and completing exercises, combined with the use of text graphics and symbols will be developed in an online manner in order to provide a personal e-learning process, that, as cited in [2] shows to be well-suited as it will be shown in the feedback received from the learners (see Conclusions).

What makes really interesting the use of this technology is that every document is interactive and dynamic, i.e., using webMathematica we can get several interesting things.

Our e-learning proposal for self-paced learning will provide an interactive and highly dynamic document that includes most of ingredients given in [1] for such a e-learning solution:

- At real time, the learner gets what needed when it is needed. - Personalised: each student selects his/her activities from a menu and difficulty appropriate to the background.

- Comprehensive: provides explanations of processes and all terms used and these can be easily searched on the network in case some of them is not known or understood.

- Dynamic: the system provides explanations and solutions for every case and every value given by the learner and contact information with experts maintaining the system are also given.

Our aim in this paper is to show how we have made use of one of webMathematica, to construct what we call an Interactive Problems Book, specifically with some general aspects of Linear Algebra. Linear algebra has evolved as a branch of mathematics with wide range of applications to engineering and computer sciences. We can find applications in Chemistry, Coding Theory, Cryptography, Networks, Image Compression, etc.We will show the process to build such an interactive application, functionality and a particular example of using. In section 2 we recall some terminology that we find when using webMathematica. In section 3 we give a short introduction to webMathematica, the functionality that we propose for Linear Algebra and the last section is devoted to show an example of design a website and how we make use of Mathematica from this web page in order to get what desired.

2 Java, Servlets and Tomcat

We will give a short, probably a non-accurate introduction of terminology, but useful to understand all terminology associated to webMathematica.

A servlet can be defined as a program that runs on a server and that is design to provide dynamic contents on a web server. To design the servlets that will make us of Mathematica we use the language JSP (Java Server Page). General functioning of JSP technology is that the Aplications Server interprets the code contained in the JSP page to produce the Java code of the servlet to be generated. This servlet will be the one in charge of generate the document (usually HTML) that will be presented on the web browser (cf. [7] or more details on JSP technology and JAVA programming language). Then, in order to make use of webMathematica we need a servlets container that provides the required services. To do so we use Tomcat (cf. [8]) since it is recommended by webMathematica, although anyone else can be used. We can find an scheme of functioning for webMathematica in Figure 1. As can be observed we have server for webMathematica over a servlets container as Tomcat.

The server takes the necessary data from a website and makes the required appeals to the Mathematica Kernel, obtaining the desired results and showing them in the initial website. As can be deduced, our website is indeed a JSP file hosted on a server.

3 webMathematica

As can be read in the documentation provided by WolframResearch, webMathematica allows add interactive calculus and visualization capability on a website of those output results given by Mathematica. This is possible by integrating Mathematica with some techniques on webservers. In this way we can design websites where the necessary inputs are taken and make the required calculations using either Mathematica or our own functions. These calculations can be numeric, symbolic or graphic. The most interesting question is that outputs can be integrated on the same website, even in case of graphics or complex symbolic expressions, since MathMl language is allowed.

4 Teaching Linear Algebra

It's possible to find interesting resources on the internet that can be shared in the classroom. Much of them are advanced calculators or graphics calculators. But, is a calculator a good tool for teaching and/or learning mathematics? Under our opinion students

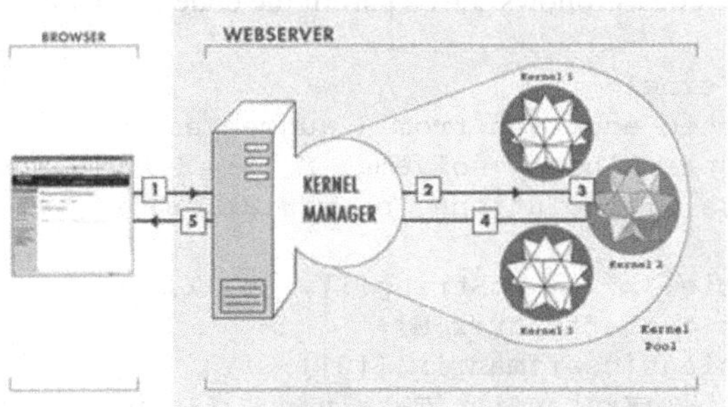

Fig. 1. Architecture of webMathematica server

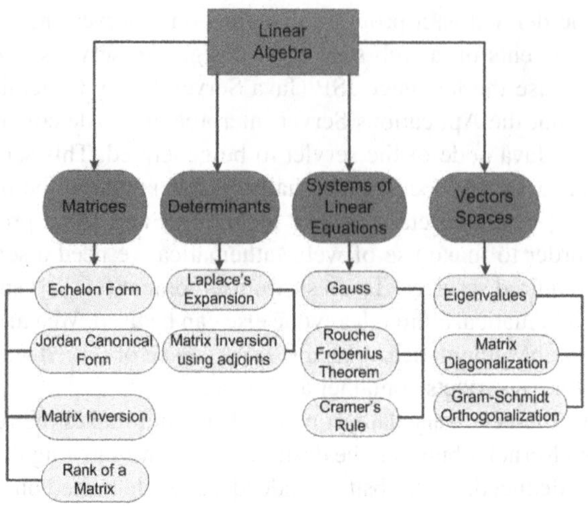

Fig. 2. Actual contends of the interactive book

need to know the procedures that allow finding the solution for different entries. Moreover, there are a lot of topics in Linear Algebra without a graphical representation (for example, the determinant of a matrix). Using webMathematica, we can show the students the main topics in Linear Algebra in a natural web ambient where solutions and explanations are mixed in an interactive way. We think that the use of this technology is more motivating since students can observe mathematical ideas presented through a recognized resource and the learning process is not limited to the classroom. In figure 2, we can see the topics in Linear Algebra that are covered till the moment.

We solve each problem using Mathematica trying to use only elementary functions in order to store all the intermediate results. Functions defined by us, should output the right results and in such a way to be shown on a web site. We always try to define our functions without using functions predefined by Mathematica in order to be able to use intermediate results on the website implementation. Moreover, if we define a full set of Mathematica functions, we will need less code on the web page. All the functions will define the package that will be called by the final web page. The package is a JSP page with Mathematica code as follows. Just a part of the code is shown.

```
<msp:allocateKernel>
<!-Solve a system equation from a augmented
matrix --> <msp:evaluate> solvemat[M_] := Module[{coeffmatrix,
solmatrix, vars, i, solutions, n, variables},
    Clear[x];
    vars = Table["x" <> ToString[i], {i, 1, 10}];
    matrizcoef = coeffmatrix[M];
    n = Dimensions[coeffmatrix][[2]];
    variables = ToExpression[Take[vars, {1, n}]];
    solmatrix = Flatten[mattermind[M]];
    solutions=
```

```
      Solve[coeffmatrix.variables == solmatrix, variables];
    Return[{solutions[[1]], variables}]]
</msp:evaluate>

<!-Get the multiplicity of the characteristic polynomial roots.
The functions returns pairs as {root, multiplicity}-->

<msp:evaluate> multiplicity[L_] := Module[
      {output, i, a, lista},
    output = {}; list = Union[L];
    For[i = 1, i <= Length[list], i++,
      AppendTo[output, {list[[i]], Count[L, list[[i]]]}]];
    Return[output]]
    </msp:evaluate>
</msp:allocateKernel>
```

To read the package from the JSP web page, we have to write the code of the web page on:

```
<jsp:include page="../matrices/codigomatrices.jsp" />
```

For the website implementation we design the data input in such a way to allow the biggest amount of cases and data output are given in the most clarifying way, by showing as many intermediate results and explanations as possible. The web page mixes html code, in the general design of the web page, java code to show the output of Mathematica, javascript to design the inputs forms more complex, and Mathematica code to calculate the solutions of the problems from the input forms. Below we can observe an example of this:

```
------- Java Code, Html Code and Mathematica Code Mixed ------
<%for(int i=1;i<=n_vp;i++){%>

<p align="left" class="Estilo13"> <strong><%=i%>)</strong>
Para el valor propio <strong>r<sub><%=i%></sub></strong>,
sus vectores propios asociados se obtienen a partir de la
soluci&oacute;n del sistema de ecuaciones
<em><strong>A-r<sub><%=i%></sub>I<sub>n</sub>=0</strong>
</em> </p>

<msp:evaluate>
ec=anadirsolmat[M,Table[0,{ToExpression[mf]
   +1}]]//.listavp[[<%=i%>,1]];
</msp:evaluate>  <%}%>
```

An open access example of a web page to find the multiplicity of eigenvalues ([9]),is given below. A Linear Algebra and Distrete Mathematics interactive book continuously developed can be found in [10].

Resolución paso a paso

Introducir la matriz definida sobre R^n

3	1	5
0	7	0
0	0	7

Para calcular los valores propios de la matriz de entrada primero calculamos el polinomio característico el cual viene dado por el determinante de la matriz $A\text{-}rI_{nxn}$

$$\text{Det}\left(\begin{pmatrix} 3-r & 1 & 5 \\ 0 & 7-r & 0 \\ 0 & 0 & 7-r \end{pmatrix}\right) = (3-r)(7-r)^2$$

Los valores propios de la matriz de entrada vienen dado por las raíces del polinomio característico anterior

Valores Propios r_i	Multiplicidad
3	1
7	2

1)Para el valor propio r_1, sus vectores propios asociados se obtienen a partir de la solución del sistema de ecuaciones $A\text{-}r_1I_n=0$

$$\begin{pmatrix} 0 & 1 & 5 & 0 \\ 0 & 4 & 0 & 0 \\ 0 & 0 & 4 & 0 \end{pmatrix} = \begin{pmatrix} 0 & 1 & 0 & 0 \\ 0 & 0 & 1 & 0 \\ 0 & 0 & 0 & 0 \end{pmatrix}$$

Luego el sistema de ecuaciones, una vez simplificado, queda como sigue

$$X2 = 0$$
$$X3 = 0$$

Y la solución es de la forma

$$x2 = 0$$
$$x3 = 0$$

Luego el vector sobre los racionales que define el espacio de soluciones viene dado por

< a 0 0 >

Finalmente, la base del espacio solución está formado por

< 1 0 0 >

2)Para el valor propio r_2, sus vectores propios asociados se obtienen a partir de la solución del sistema de ecuaciones $A\text{-}r_2 I_n=0$

$$\begin{pmatrix} -4 & 1 & 5 & 0 \\ 0 & 0 & 0 & 0 \\ 0 & 0 & 0 & 0 \end{pmatrix} = \begin{pmatrix} 1 & -\frac{1}{4} & -\frac{5}{4} & 0 \\ 0 & 0 & 0 & 0 \\ 0 & 0 & 0 & 0 \end{pmatrix}$$

Luego el sistema de ecuaciones, una vez simplificado, queda como sigue

$$X1 \quad -\frac{1}{4} \quad X2 \quad -\frac{5}{4} \quad X3 \quad = \quad 0$$

Y la solución es de la forma

$$x1 = \frac{x2}{4} + \frac{5\,x3}{4}$$

Luego el vector sobre los racionales que define el espacio de soluciones viene dado por

$$< \quad \frac{a}{4} + \frac{5b}{4} \quad a \quad b \quad >$$

Finalmente, la base del espacio solución está formado por

$$< \quad \frac{1}{4} \quad 1 \quad 0 \quad > < \quad \frac{5}{4} \quad 0 \quad 1 \quad >$$

5 Conclusions

We have developed this tool for e-learning for the last three years for e-learning, providing solutions to many different situations and/or different levels of skill and background. Our solution includes theoretical contents and parameters of the examples that are given by the student or randomly generated. We can also generate an infinite number of exercises. Thirdly, we are able to design self-evaluation exercises, where computations are too complex to be developed using other platforms as WebCT. And finally, but not less important, is that in all the precedent we are able to give a solution to every problem, exercise or example in detail, explaining every step and partial solution appearing in the development of the process by simply making use of the intermediate calculations made with Mathematica.

In Figure 3 we can observe students' feedback on this online tool. Statistics sample size is over three hundred students, divided in the last three courses, from 2008 to 2011. Population is formed by students of first course in Engineering in the University of Almeria and the poll containing the six shown questions was developed anonymously through the virtual tool WebCT. Results obtained during the first tests in 2008-09 contain just around the half part of the total number of students. However, as expected when starting to develop the experience, the number of participants in the poll increased enormously reaching over the 90 per cent during the last course. The use of the tool by the students and its utility is observed not only by the number of students answering the

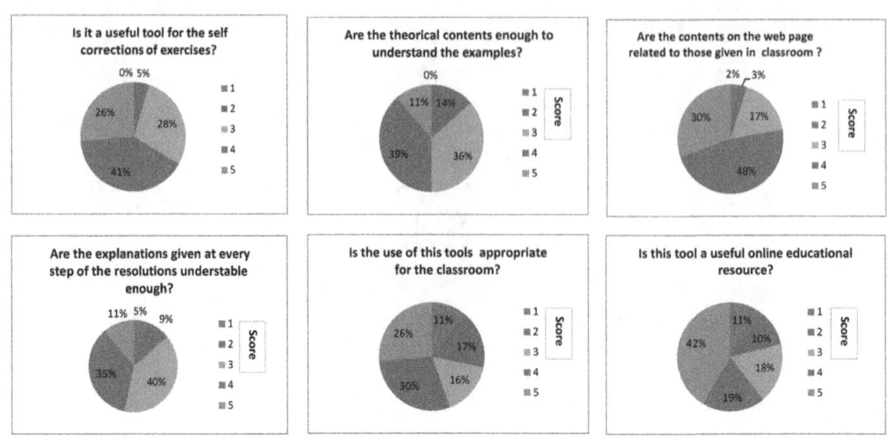

Fig. 3. Average Score of the three last years

poll, but also by other variables that we can relate to this fact. These are the access to the forum of Linear Algebra in WebCT and the number of questions related to solved exercises or issues in the resolution of some exercises. In the first case the number of questions related to exercises has been progressively decreasing to become almost zero during the last course. During the first year of our experience we observed that some students advised others asking for help with some exercise or looking for solved exercises to visit [10]. This kind of questions, as noted above, decreased to almost zero. Only a couple of questions looking for solved examples were made at the beginning of this course. After that no questions on exercises appeared and we can find basically questions concerning class organizations on the forum. In the second case, the number of emails sent to the teachers concerning difficulties with exercises decreased also enormously. Students simply ask for help with those exercises that are of special difficulty or contain a theoretical development in its resolution. The conclusion that we can derive is very positive and that the interactive Linear Algebra book is being used as a main tool for self-learning. Teacher's task concerning students' questions, directly via email or administrating the forums when some of the queries were not properly solved has also decreased and now we can focus our efforts in developing and enhancing a tool that undoubtedly help the students in improving their self learning abilities.

References

1. Alonso, F., López, G., Manrique, D., Viñes, J.M.: An instructional model for web-based e-learning education with a blended learning process approach. British J. Educational Thech. 36(2), 217–235 (2005)
2. Driscol, M. (ed.): Web-based training, creating e-learning experiences, 2nd edn. Jossey-Bass/Pfeiffer, San Francisco (2002)
3. Hung, D., Chen, D.: Implications for the design of web-based e-learning. Education Media International 38(1), 4–11 (2001)

4. Oldknow, A., Knights, C. (eds.): Mathematics Education with Digital Technology, Great Britain (2011)
5. Main Wolfram Research site, `http://www.wolfram.com`
6. webMathematica, `http://www.wolfram.com/products/webmathematica`
7. Sun Java information, `http://java.sun.com`
8. Apache Tomcat, main site, `http://jakarta.apache.org`
9. Multiplicity and eigenvalues web page, `http://ubeire.ual.es:8090/webMathematica/algebra/lineal/espacios/propios.jsp`
10. Linear Algebra and Discreta Mathematics interactive book, `http://ubeire.ual.es:8090/webMathematica/algebra/algebra.html`

Biorefinery Virtual Lab-Integrating E-learning Techniques and Theoretical Learning

M.D. Redel-Macías[1], S. Pinzi[2], A.J. Cubero-Atienza[1], M.P. Dorado[2], and M. P. Martínez-Jiménez[3]

[1] Dep. Rural Engineering, Ed Leonardo da Vinci, Campus de Rabanales, University of Cordoba, Campus de Excelencia Internacional Agroalimentario, ceiA3, 14071 Cordoba, Spain
[2] Dep. Physical Chemistry and Applied Thermodynamics, Ed Leonardo da Vinci, Campus de Rabanales, University of Cordoba, Campus de Excelencia Internacional Agroalimentario,ceiA3, 14071 Cordoba, Spain
[3] Dep.of Applied Physics, E.P.S., Albert Einstein Building, Campus de Rabanales, University of Cordoba, ceiA3, 14071 Cordoba, Spain

Abstract. E-learning is one of the emerging needs of the information age. Therefore a lot of potential is seen in distance learning development. Virtual environment interface to E-learning systems have recently appeared on the Internet. With the addition of online laboratories, it is possible that the applications appear to be promising to E-learning tasks more nature and interactive. In this paper a web page with a virtual laboratory about biorefinery and biomass valorization has been developed in order to make closer the world of academic research and the graduate students learning. Biorefinery represents an excellent object of study in many scientific applied disciplines because it integrates many biomass transformation processes of different knowledge areas. The software application supplies both theoretical and practical information on all the fields covering the biorefinery concept, and the possibility of interacting with the instrumentation typical of this subject by means of simulating it (http://www.uco.es/docencia/grupos/laboratoriovirtualceia3). Virtual laboratory realized the concept of the openness and sharing of teaching resources, promoting the development of quality education, creative and continuing education.

Keywords: biorefinery, e-learning, virtual laboratory.

1 Introduction

The traditional approach to education, where the transfer of knowledge is achieved mostly by lecturing, has a number of shortcomings, in particular because the students are not motivated enough to acquire knowledge actively [1].The educational system should engender an interest in independent learning, and prepare the students for life-long learning, which is a necessary skill for successful participation in the knowledge society [7].

Over a 40-year period, ICT-supported education has been given various names and has appeared in different forms and different applications. The most well know is e-learning. E-learning should be understanding like much more than a "gift-wrapping"

Á. Herrero et al. (Eds.): Int. JointConf. CISIS'12-ICEUTE'12-SOCO'12, AISC 189, pp. 321–330.
springerlink.com

of materials of the traditional on-line courses [5]. Thus, according to [2], e-learning is defined as “the use of new multimedia technologies and the Internet to improve the quality of learning by facilitating access to resources and services as well as remote exchanges and collaboration.” The participation in e-learning of the students does not only depend on personal factors or on preferences, but also on the nature of the technology employed and on the pedagogy followed. Virtual Labs interface to e-learning systems have recently appeared on the Internet. The creation of a virtual laboratory allows the dissemination of that information to its final users and the teaching of theoretical-practical concepts by experimentation making use of the new technologies.

The main advantages of this education tool are thatthey let students represent some situations that are not reproducible in the laboratory because they require costly and complex equipment. Moreover, the students can work with ideal experiment conditions using models that show partial aspects of reality.In addition, using these tools it is possible to analyzeprocesses that could be dangerous. Virtual Laboratories can also help to solve the problem of classroom and laboratory overcrowding. For these reasons, the use of computerapplications as an extra infrastructure is more versatile andcheaper than experiment laboratories[9; 10].

An important part of the software applications currently developed is oriented towards teaching and the transmission of knowledge. This type of application seeks to stimulate the end user with educational technology and to facilitate the acceleration of learning thanks to the application’s interactivity. One important factor in the increase in the use of new technologies as a teaching tool is the expansion of the software through Internet. Internet is the means which brings these complementary teaching tools nearer to any user so that they can benefit from them at every moment [8].

A biorefinery is an industrial facility that produces energy (such us fuels), chemicals, and other commodities and specialities, just like a petroleum refinery does. The main difference between them is that in a biorefinery, biomass is used as raw material -agricultural wastes, agrofood-wastws, municipal organic wastes, etc.- while a petroleum-based refinery uses non-renewable fossil-derived petroleum[3].

There is a large number of biomass resources and many different technologies to transform it into valuable products. Therefore, the variety of products that can be produced in a biorefinery and the processes involved in it are very high[6].

Biorefinery represents an excellent object of study in many scientific applied disciplines because it integrates many processes of the food industry, including chemistry, chemical engineering, biology, agronomy, biotechnology, materials engineering, environmental engineering, computer science, thermodynamics, energy, electrical engineering and economics, among others. This allows that students develop cross-curricular competencies.

Furthermore, practical lessons have an important role in engineering and experimental sciences, in accordance to concept of "apply learning to practice", referred to the European Higher Education Area (EHEA), and they need to be promoted. Students need to integrate theoretical with practical knowledge. Among the good pedagogical reasons that motivate the inclusion of a virtual laboratory of biorefinery in the curriculum are: the necessity to illustrate and validate the analytical concepts, the need of trainings for the students in order to learn to solve the uncertainties involving non-ideal situations.

With the aim to combine the knowledge of the academic research on biorefinery and the graduate students learning, it has been attempted to develop a virtual laboratory on this theme in the labor scenario. It supplies both theoretical and practical information on all the fields covering this topic, and the possibility of interacting with the instrumentation typical of this subject by means of simulating it. The website and virtual laboratories have been created to provide that information thus making it easy for any person wishing to go deeper into the theme to do so.

2 Objectives

There are several reasons for University education to focus on creating virtual labs [4]. One of the primary reasons includes the cost and lack of sufficient skill-set for facing the current growth in biorefinery field. The setup cost of laboratories puts a large overhead on the educators. The main objective of the work is to develop a laboratory to educate sufficient target group with the details of different biomass conversion and valorization processes and the protocols to obtain it.

The secondary objectives are:

- To develop an application containing information about the instrumentation and measurement methods.
- To include extensive information on the process to obtain biodiesel and biomass.
- To develop software with complete training tutorials carried out by expert technicians.
- To allow the user to become familiarized with some of the measurement and analysis instruments by means of the simulation of their functioning in the labs.
- To permit the updating of the information it contains through a resource manager with a database that contributes dynamism to the application itself.
- To develop a virtual laboratory with a general interface for the presentation of different applications related to process for obtaining biofuels and bioproducts.
- To carry out a web page that works in a "multiplatform" environment. The application can be used in any computer regardless of its operative system.
- To develop an intuitive and easy to use virtual laboratory, in order to be used by people who do not necessarily have any extensive knowledge of the subject.

3 Experimental Description, Material and Methods

The virtual laboratory, in this first phase of the project, consists of the following modules:

- *Analysis of biomass.* In this module, the most important processes to analyze dry biomass such as water, volatile compounds and ash contents by thermogravimetric balance (TGA) and high calorific value by calorimetric bomb has been simulated.
- *Energy use of biomass.* In this module, some energetic applications of biomass, such as the biofuels (biodiesel and bioethanol) production and the energy use for electricity production, has been simulated. In the last case the simulation of biomass boiler has been proposed.

- *Analysis of biofuelsproperties.* In this module, the most important analysis of biodiesel has been simulated. The standard EN14214 is followed for the tutorials development.
- *Treatment and utilization of secondary products.*
- *Traceability and quality.*The development of these modules involves several research groups from the University of Córdoba, Huelva, Cadiz (Spain) and the University of Birmingham and Manchester (UK) and Athens (Greece).

The virtual laboratory will be used in different degrees of Bachelor in Industrial, Agricultural Forestry Computer and Chemical Engineering, in Chemistry, Biotechnology and Biology, as well as in the Master of Biotechnology on the University of Manchester (UK) and the University of Athens (Greece). The virtual laboratory have a modular structure in order to interface each module with others. The inclusion of transversal modules covering all processes of the the biorefinery from the point of view of specific discipline has been take into account. Figure 1 shows the general structure of the virtual laboratory. All the laboratory, as well as the tutorials,has beendrawn in Spanish and English, to promote the international character of the project, agreeing with the philosophy of international excellence of the CeiA3 campus.

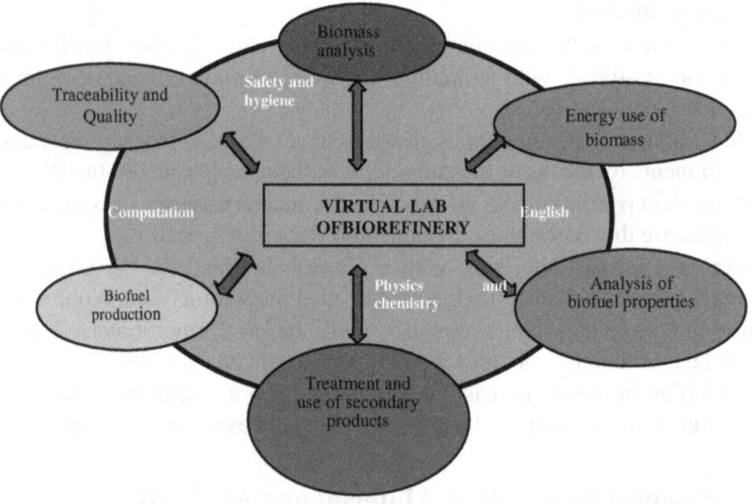

Fig. 1. Description of Virtual labs units

The laboratory was designed to include the following aspects:

A tutorial to guide the technician, supplying him/her with information on different physics and chemical explanations, the instrumentation to be used in measuring them, the regulations in force, updated, to be applied during that measuring process together with the different configuration optimal parameters in the process and a series of tutorial videos which facilitate the understanding of the handling of the devices related to this theme.

Gallery in which there is a useful educational illustrated reference with information on all the machines and instruments employed.

Virtual Laboratory or Simulation of biomass valorizations and transformations.

Analyses of results, interpretation of images, carrying out calculations, resolution of cases, etc.

Flexible interface for access to information available.

Opportunity for the learner to repeat the practical activities as many times as he/she considers it necessary until the didactic results wished are achieved.

The learning timetable is established by the users themselves in accordance with their needs.

An application has been set up to permit the handling of the different sections detailed in the definition of the real problem. The user can gain access to each of those sections by selecting the corresponding option. By using a multimedia platform in which, through animations and an attractive interface which makes learning a pleasure for the user, all the necessary theory is provided for the latter for an optimal understanding in the field of biorefinery, as well as a complete description of the instruments required in the measuring and analysis, together with explanatory video-tutorials making it easier to manage them. Moreover, there is a possibility of consulting the explanation of the working of the apparatus and to manipulate it. To ensure that the user learns correctly, the system guarantees that if this is the first time he/she accesses to it, he/she should carry out some basic activities on it, then, later, be able to increase their difficulty. In addition, the practical activities (virtual laboratories) present in the programme include the principle of "learning while practicing", with which the user will acquire the necessary capacity to handle the machines in real life and interpret their results. He/she will be able to verify analytical solutions and experiment with different configuration parameters. The base of the web page was developed using Joomla. This content management system is an open source application programmed mostly in PHP under a GPL license. The content administrator can work on Internet or on intranets and it requires a MySQL database and, preferably, an Apache HTTP server. The interface was aimed to be simple, friendly and intuitive so that all its users, regardless of their degree of knowledge on the theme, would be capable of handling it. The application has been developed with the ActionScript programming language and the database MySQL, so that for its modification it will be necessary to dispose of Macromedia Flash 8 and an Oracle system. For its use and visualization only a browser will be needed since this programme is accessed to through the website of Córdoba University: http://www.uco.es/docencia/grupos/laboratoriovirtualceia3 by either Explorer 6.0, Firefox, Netscape, with the installation of the plug in of Flash Player being considered necessary.

The interface is intuitive and ergonomic to handle. It has shades of eccentricity and tries to be very clear and perceptive. Its objective is to facilitate access to different units of the programme easily and quickly.

The constant use of a multitude of sounds and of continuous movements is not recommendable since this may slow down the execution of the applications and prevent the user from achieving his/her objectives at any speed.

There is a striking presentation of the information as it makes use of multimedia techniques to attract the user's attention.

From the user's point of view, the interface should permit him/her to locate the information and carry out his/her operations quickly as can be verified in most web pages. It has been attempted not to distract the user too much with the design but at the same time make the application attractive, reaching a halfway point in the animations employed. The developed product is available at all times in the website *http://www.uco.es/docencia/grupos/laboratoriovirtualceia3* of Córdoba University. Together with the application developed the documentation generated during its creation process is supplied. This includes the technical handbooks, which contain the general information on the process of analysis, design, implementation and system test, and the Code Manuals in which the source codes of the applications developed are described. A user's guide is also included with all the information necessary for a good management of the application. Screen captures are carried out on all and every one of the sections of the application to support the information offered in it.

4 Results Obtained and Availability of Use

A web portal has been created for training in biorefinery and biofuels production, which is being used in the engineering degree and Master in Renewable Energy at Córdoba University, and is permanently available in the server http://www.uco.es/docencia/grupos/laboratoriovirtualceia3.

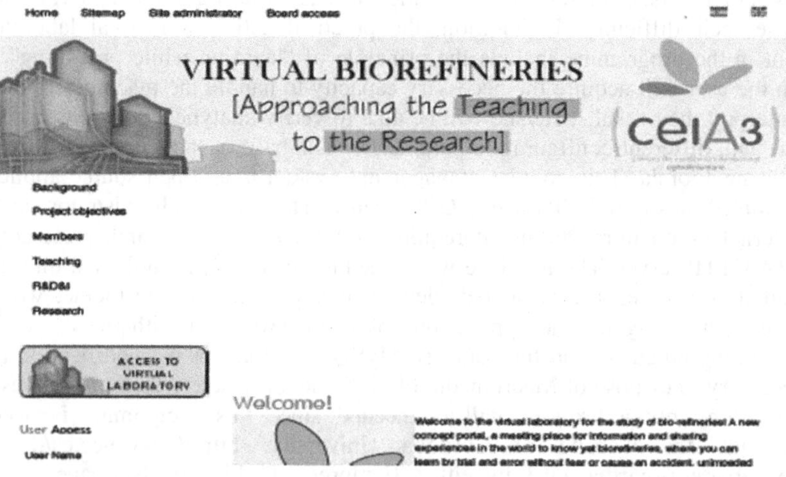

Fig. 2. Imprint pant of the web page of virtual lab of biorefinery

The user has at his/her disposal a tutorial divided into different sections with the aim of facilitating his/her study of the theory of biorefinery and biodiesel process

The information given in the tutorial is structured in the following way:

1. A complete theoretical introduction with references
2. Methodology
3. Practical case with several options.

For each module, a separate tutorial has been carried out.

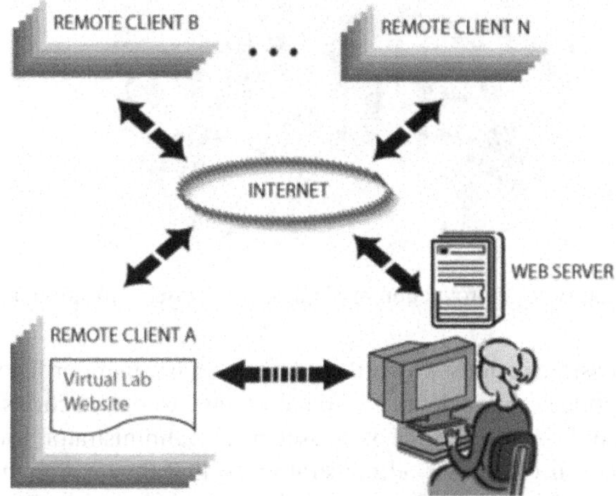

Fig. 3. Basic structure of virtual lab

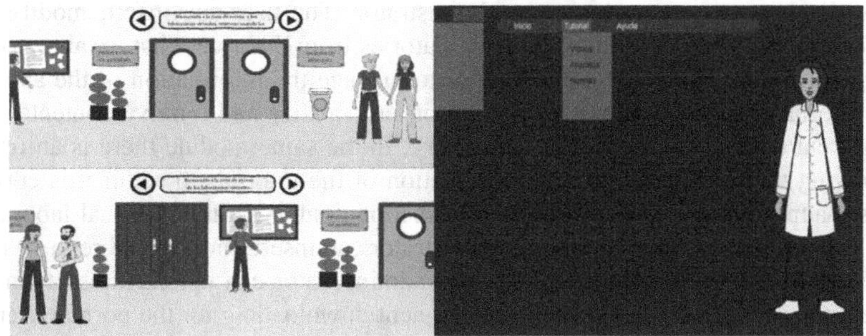

Fig. 4. Virtual laboratory environment before to enter in each module and virtual laboratory technician for user helps and tutorials

Moreover, the user disposes of a tutorial with the legislation in force in order to find out the legal regulations concerning safety and hygiene in the lab, as well as measurement process. The information given is classified in accordance with the xisting norms, including laws, royal decrees and directives.

The user can gain access to a gallery in which he/she can observe the different apparatus and instruments used. The user can identify, become acquainted with and study the general characteristics of these instruments. The user can access to a laboratory in which he/she is shown a series of simulations in order to prepare him/her in the generation of biodiesel process and the interpretation of his/her results. As an example, among the simulations available this is the laboratory scale production of biodiesel by transesterification of vegetable oils.

The user can consult videos that provide complementary information on certain aspects included in other modules of the application. In each of the programme's modules the help section can be accessed to from that module. It is easy to use and very useful to the user for solving problems arising when browsing through the application.

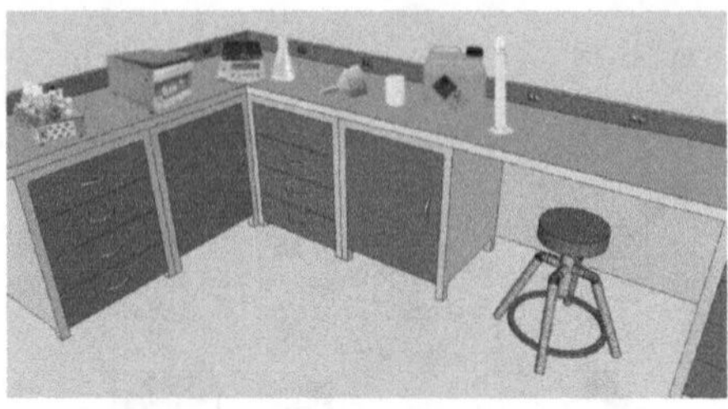

Fig. 6. Virtual laboratory scale production of biodiesel by transesterification of vegetable oils

To finish, the possible actions which the user at an administrator level can perform by executing the application are detailed. To gain access to these actions the user must fill in a form to confirm that he/she is a user at an administrator level. The user-administrator can enroll other user-administrators so that they can gain access to the application's management tool. Each user-administrator can modify his/her own password to enter the management tool of the virtual laboratory, and, if necessary remove him/herself from the application itself. No user-administrator can change the password or remove any other user-administrator. The user can insert, modify and even remove the address of the virtual laboratories from the access list to laboratories. The contents permit the user-administrator to manage the information of the application's tutorials. The user-administrator can insert, modify and remove contents from the different tutorials in the virtual laboratory. In the same module there is an option for showing the information from some section of the tutorials to see if it is correct. The user-administrator can update the information contained in the virtual laboratory guide. In the same module, the user-administrator can insert, modify and remove links of interest within the application. The user-administrator can upload files of images for the tutorials and pdf files for their subsequent downloading for the normal users of the application. The module shows all the images uploaded.

5 Application of Virtual Lab

The web portal developed has only been finished and implemented for a short time in the web server, and, therefore, it has not been used by the students. However, different users taken at random have done some pilot validations and all of them have underlined two important aspects of this program. On one hand, how easy it is to use, and, on the other, they say that it has helped them to improve concepts. Most of them have also mentioned that the section corresponding to the videos in which the handling of the apparatus was explained and visualized was of great interest to them.

In a second stage of investigation, this tool will be implemented with the students and the results obtained will be studied. The network multimedia courseware enriched the experimental contents, provided colorful information and good man-machine interface. The quality and reliability of the software are two of the most important characteristics to be taken into account in the creation of the software. That is why the application developed with this work has been submitted to pertinent exhaustive tests in order to minimize, as far as possible, the risk of any failure. Therefore, the student's learning interest may be stimulated and their motivation may be aroused. Students can carry out remote individual learning by using the virtual lab through network, which surmounted the difficulty to put staff together to carry out continuing education, realizing the sharing of teaching resources, breaking limits of lab time, space, geographical, and reflecting information and openness of experimental instruction.

6 Conclusions

A software application about biorefinery and biofuels production has been proposed. This software compensates the lack of information of technical students when acquiring theoretical knowledge during practical exercises in the laboratories. This is especially useful in the practical classes about experimentation with specific instruments measuring. Virtual lab were used for a new experimental teaching model, which can stimulate learning interest, develop autonomy, cultivate pioneering spirit and innovation sense of students. It realized the concept of the openness and sharing of teaching resources, promoting the development of quality education, creative and continuing education.

Acknowledgments. This research was supported by the Agrifood Campus of International Excellence, CeiA3 (Project for Innovation in Teaching of the CeiA3 "Teaching for Excellence"), the High Politechnic School, of Cordoba University (EPS) and the University of C6rdoba.

References

1. Bates, T.: Managing Technological Change: Strategies for College and University Leaders (2000)
2. Communication from the Commission to the Council and the European Parliament: The e-Learning Action Plan, E. C. (2007)
3. Cherubini, F.: The biorefinery concept: Using biomass instead of oil for producing energy and chemicals. Energy Conversion and Management 51(7), 1412–1421 (2010)
4. Diwakar, S., et al.: Biotechnology virtual labs- Integrating wet-lab techniques and theoretical learning for enhanced learning at universities. In: International Conference on Data Storage and Data Engineering (2010)
5. Fischer, G.: Meta-Design: Beyond User-Centered and Participatory Design. In: Proceedings of HCI International, Greece (2003)
6. FitzPatrick, M., Champagne, P., et al.: A biorefinery processing perspective: Treatment of lignocellulosic materials for the production of value-added products. Bioresource Technology 101(23), 8915–8922 (2010)
7. Hoic-Bozic, N., Mornar, V., Boticki, I.: A Blended Learning Approach to Course Design and Implementation. IEEE Transactions on Eduction 52(1), 19–29 (2009)
8. Jimenez, P.M., et al.: Telematic Training via a website of technicians in work-related risk prevention. In: 2st International Conference on Computer Supported Education, Valencia, Spain (2010)

9. Martinez-Jimenez, P., Varo, M., et al.: Virtual Web Sound Laboratories as an Educational Tool in Physics Teaching in Engineering. Computer Applications in Engineering Education 19(4), 759–769 (2010)
10. Murphy, T., Gomes, V.G., et al.: Facilitating process control teaching and learning in a virtual laboratory environment. Computer Applications in Engineering Education 10(2), 79–87 (2002)

Approaching System Administration as a Group Project in Computer Engineering Higher Education

P.A. Gutiérrez, J. Sánchez-Monedero, C. Hervás-Martínez, M. Cruz-Ramírez, J.C. Fernández, and F. Fernández-Navarro

Department of Computer Science and Numercial Analysis, University of Córdoba, Spain
{pagutierrez,jsanchezm,chervas,mcruz,jcfernandez,i22fenaf}@uco.es

Abstract. The convergence of European higher education establishes as a primary goal the acquisition of the main skills demanded by the society. One of the main job prospects of Computer Engineering (CE) is System Administration (SA) as part of institutions or companies. In Spain, this professional profile is only collaterally considered in CE higher education and always from the theoretical perspective of operating systems. This paper describes a learning methodology that has been applied to CE students for improving their SA skills from a very practical point of view. The experience consists on a distributed laboratory where the students have to deploy different services of a typical distributed system. The innovation remarks are two: the definition of a professional distributed infrastructure as the target of the practical exercises and the implicit and *mandatory* distributed tasks between student teams. In addition, different professionals from public and private institutions have supervised the activity. The results of two polls (taken before and after this experience) show that student SA skills were considerably improved thanks to the laboratory. Moreover, the laboratory was an excellent tool to promote teamwork and collaboration among students and very suitable for transnational education, the lab being remotely accessed from different locations.

1 Introduction

Present system changes of higher education in Europe, within the Bologna reform are aiming at connecting the University and enterprise in redesigning University education and connecting it with lifelong learning courses for sustainable development of professionals. "Applying knowledge to practice" has been considered a weakness of the traditional University courses [16]. In the Convergence of European Higher Education, it is very important that the students obtain the new necessary skills in the current society. Due to professional needs, in the case of engineering students, these include: learning to learn, lifelong learning, teamwork, organisational understanding, social-skills, and project management. It is specially highlighted that during the learning activities, the students should not only automate the learning but being aware of the learning process [6]. Focusing on the engineering field, there are some investigations in recent decades, which analyse the competencies that engineering students must achieve in the context of contemporary society [10]. Higher education is a key instrument to overcome the current world challenges and to train citizens able to build a more fair and open society

Á. Herrero et al. (Eds.): Int. JointConf. CISIS'12-ICEUTE'12-SOCO'12, AISC 189, pp. 331–340.
springerlink.com

[1] and "a new kind of engineer is needed, an engineer who is fully aware of what is going on in society and who has the skills to deal with societal aspects of technologies" [7].

This paper introduces an experience consisting on a learning technique applied to Computer Engineering (CE) students in order to boost their System Administration (SA) skills. The SA profile is not developed by the CE students of some Spanish Universities, at least not in the way that companies and industries demand. The experience is aimed to make the student (or the groups of students) develop a professional work in order to deploy a Distributed System (DS) with the collaboration of all the other students, where each student team will be responsible of a service or a set of services. The experience is able to enhance the ability of the student to carry out active and distributed working processes, then stimulating their team working aptitude, which is one of the most appreciated in professional environments.

There are several reports aimed to evaluate the education acquired in the University institution from the point of view of the recently graduated students who are starting their career [2]. In this study, a set of negative aspects are identified from the graduated students perspective: the education received is very theoretical and extremely generalist, some parts of the programmes being not useful, from their point of view, which clearly reveals that the student is not perceiving a global coherence of the training received at the University. Moreover, several authors have recently outlined the necessity of learning by doing things, specially at Universities [3].

When the education received at CE higher education is analysed, there are some professional profiles in which the training received is clearly insufficient. Private companies and public institutions insist that CE graduated students lack practical SA education, what usually involves a very long training period after the student is hired, and sometimes results in security or performance problems for the systems they admin. It is constantly pointed out that the increase of practical tasks might partially improve it. In this way, implicating professionals from the private sector may contribute to this goal. However, although this is not new in many Universities, it is not so common in those of small cities where the number of software companies is low, such as the one where the experience in this paper was tackled.

The experience presented in this paper was applied to one module of CE program at the University of Córdoba, called "Distributed Operating Systems", with the help and collaboration of some professionals of the public and the private sector. It consisted of deploying a Distributed System Administration Laboratory (DSAL), i.e. a set of computers connected into a private Local Area Network (LAN), each one of them offering one or several services. All the infrastructure was designed and implemented by using free software tools [4], promoting in this way the use of free software, which is one of the basic line of actions of different European Universities [5]. The social and educational reasons why free software should be used in higher education environments are summarized in [11]. The DSAL was deployed by the students, organized in teams, in such a way that every team installed and configured one different service. The nature of the problems and the learning process design encouraged teams to work in a cooperative way in order to deploy the whole service infrastructure, being the results very satisfactory.

The activity is based on a network of interconnected computers, which are accessed from the University network. Students can access to it from different locations, or even countries, and this makes the DSAL a perfect tool for transnational education, where implementing other schemes could be more difficult. Security concerns guarantee the isolation of the activity in the lab.

The rest of the paper is organized as follows. Sect. 2 summarizes the objectives of the experience in order to clearly justify the proposal. In the next section, the methodology considered and the different activities carried out for achieving these objectives are described. Sect. 4 is aimed to different results obtained and Sect. 5 includes some discussion and future perspectives of the experience. Finally, Sect. 6 report the main conclusions of our study.

2 Objectives

The main objective of the methodology presented is to encourage students to apply the theoretical concepts of SA and distributed systems into a practical real problem, consisting of the installation and configuration of a service in a computer connected to other computers hosting other services. But, at the same time, the experience was aimed to achieve other objectives:

- To make students aware of the importance of a correct SA when dealing with any kind of CE project.
- To develop practical aspects of the deploying, maintenance and configuration of computer systems in multi-user environments.
- To strengthen their general knowledge about operating systems and, in particular, about distributing systems by a practical approach.
- To value the students' acceptance degree of this kind of activity.
- To value the students' capability of working in different teams with a distributed task goal.

3 Methodology and Activities Considered

The objectives previously introduced were tackled by a methodology organized into three different stages or phases: first, it was necessary to prepare the infrastructure of the DSAL, then, to carry on the activity with the students, and, finally, to analyse the obtained results. Specifically, these three stages were developed in this way:

1. In the first stage, the DSAL had to be constructed and configured. We considered a set of six old computers, where the software needed for the experience was installed and configured. Canonical Ltd. Ubuntu Server 10.04 operating system[1] was installed to old the computers, and the minimum services were deployed, including a directory server (Lightweight Directory Access Protocol, LDAP) [12] for authentication, a Domain Name System (DNS) server and a Virtual Private Network

[1] `http://www.ubuntu.com/business/server/overview`

(VPN) [8]. However, important security issues had to be considered when deploying this laboratory, since granting administrator privileges to the students can result in unwanted attacks over the computer systems of the University. The scheme of the DSAL has been included in Fig. 1. The Cheetah server has the role of the gateway of the DSAL, filtering all packets sent or received from the rest of the University network, so that the laboratory machines' network traffic is completely isolated. Cheetah includes `iptables` rules for allowing the students access the rest of the members of the DSAL in order to administrate them, redirecting some trivial ports to the ports of the different computers (e.g. external port 10222 of Cheetah is redirected to port 22 of Zapata). Students are given administrator password of all the computers but Cheetah, in order to avoid security problems.

2. The second stage consisted on carrying out the activity with the students, by four different actions taken in this order during the course:
   (a) *Initial poll or questionnaire*: students were given the initial questionnaire, including five-level Likert items to evaluate the general knowledge of the students about GNU/Linux and more specific knowledge about SA. This initial survey had a double objective: i) adapting the activity to the initial GNU/Linux skills of the students (and valuing the possibility of some introductory sessions), and ii) contrasting the results with those obtained after performing the activity to evaluate its effectiveness.
   (b) *Guided visit to the Computing Service of the University*, where one of the system administrators explained the difficulties involved in administrating such a complex distributed system as the University one is. This activity was more focused on the services requirements and the hardware setup rather than on the software services integration.
   (c) *Special seminar about services integration with LDAP directories*, focussing on the University of Córdoba case [13]. This activity was complementary to the previous one, highlighting the software component and information management. This seminar also summarized the standardization efforts in services integration through RedIris [14], the Spanish research and academic communication network, as well as other international services integration projects such as *eduroam* (education roaming) [15].
   (d) *Main experience*, i.e. development of the SA activity with the students during three practical sessions of two hours as part of the module "Distributed Operating Systems" from the University of Córdoba. To benefit the most motivated students, the results obtained by them could be (optionally) used as the final project of the module. This activity was organized in the following way:
      i. A descriptive tutorial about the configuration and the access instructions for the DSAL was elaborated and explained at the first session, and general hints and instructions were given to the students. A special attention was given to using distribution specific package management tools (e.g. `apt-get` for Debian based system) and the associated documentation[2].
      ii. Five groups of students were formed, so that each group deployed and configured one different service. The services deployed by these groups were

[2] Ubuntu Server Guide:
`https://help.ubuntu.com/10.04/serverguide/C/index.html`

the following ones: File Transfer Protocol (FTP) server, HTTP server with Hypertext Preprocessor (PHP) serving capabilities, SQL database server, Message Passing Interface (MPI) server, email server, and Nagios monitoring system. In addition to the services, a content management system (specifically, Wordpress[3]) was installed in the web and databases servers, and it was integrated with the LDAP directory users. In our opinion, the services selected covered the most common needs for a professional distributed system, and a general description about them can be found in [8].

iii. After installing all the services and during the last session, the students checked that they had been correctly installed, and that they were interacting and running smoothly. For example, FTP service must interact with the LDAP service for user authentication.

iv. For documenting the deployed infrastructure, we proposed a distributed documentation process, i.e. a common "*wiki*" space was enabled for all the students to concurrently editing them and take notes about the different steps needed for installing the services. Some of the students explained the administration process to their classmates, in one final session organized as a seminar.

(e) Finally, we handed out the same initial questionnaire to all the students, to check if the activity has achieved the objectives intended.

3. The third and final stage was the analysis of the results of the activity, trying to conclude its effectiveness by using statistical analyses.

4 Results of the Experience

In this section, the results of the experience are described to clearly establish the contribution of the activity, and to outline the possible applicability over other higher education institutions. The results have been organized in two groups, general results and questionnaire results.

4.1 General Results

The following qualitative results have to been outlined:

- A guided visit to the Computing Service of the University of Córdoba was done, and the general opinion of the students was very positive. Proof of it was the high number questions make by them during the visit and the positive comments about it. We think that it is a good motivational initial activity not only for the experience described in this paper, but also for any module related to distributed systems or SA. The special seminar on services information integration with LDAP helped students to better understand the role of information management both for local University services and also linked to transnational projects, such as *eduroam*.
- The DSAL included in Fig. 1 was deployed and made accessible through the network of the University, being specially cautious about possible security problems. A tutorial was prepared to make the access to the laboratory easier for the students.

[3] `http://wordpress.org/`

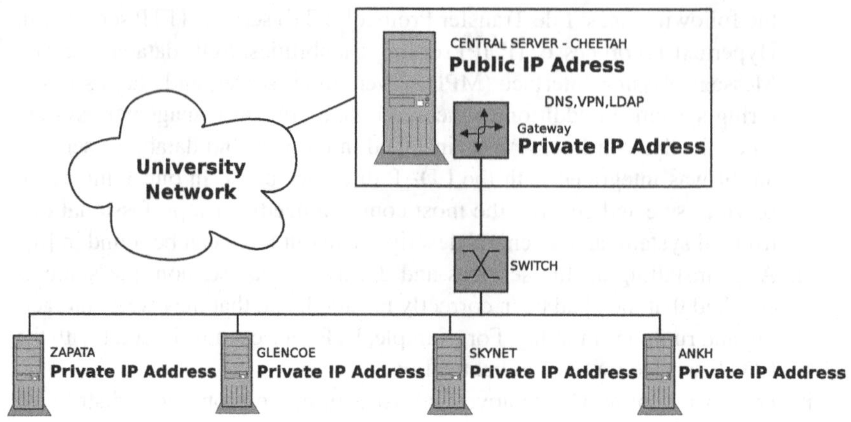

Fig. 1. Scheme of the Distributed System Administration Laboratory (DSAL)

- All the material prepared by the students (in the form of "*wikis*" for all the services) were stored for the students of other courses or modules. This allow future students update and improve the documentation, resulting in a good SA documentation database.

4.2 Questionnaire Results

The five-point Likert items (1. Strongly disagree, 2. Disagree, 3. Neither agree nor disagree, 4. Agree, 5. Strongly agree) included in the anonymous questionnaire were the following:

- Q1: "Are your team working skills good?".
- Q2: "How much do you value the work of the Computing Service of the University?".
- Q3: "Do you think you are prepared to work at the Computing Service of the University?".
- Q4: "Are your GNU/Linux skills good?".
- Q5: "Is your System Administration knowledge good?".
- Q6: "Give a value to your knowledge about the following services in Distributed Systems:". An independent value for all these services: LDAP, FTP, HTTP, DHCP, NFS and Andrew File System (AFS).

A summary of the results of the questionnaire before and after doing the activity is included in Table 1 for a total of 21 students.

The first fact that is important to take into account is that the evaluations made by the students about their own skills are, in general, quite low, specially when they are referred to SA and to the services LDAP, DHCP, NFS and AFS. This should be given quite importance, since they are fourth year students of a Computer Engineering degree, and, theoretically they should already be more prepared for SA (or, at least, they should think so).

To ascertain whether significant differences can be assumed in the evaluations after doing the proposed activity, we applied a Student's t-test [17], comparing the averages obtained before and after the activity and using the following confidence interval (which is the one that should be applied to two independent normally distributed samples with unknown equal variances):

$$(\overline{x}_1 - \overline{x}_2) \pm \left[t_{(2n-2,\alpha/2)} \sqrt{\frac{S_1^2 + S_2^2}{n-1}} \right], \quad (1)$$

where n is the number of samples ($n = 21$ in our case), $\overline{x}_1$ and $\overline{x}_2$ are the averages values before and after the activity, α is the significance level ($\alpha = 0.05$ in our case), and S_1 and S_2 are the standard deviations of the samples (before and after). After applying these tests for each question, the results show that significant differences can be assumed for the following items of the questionnaire:

- Q5: "Is your System Administration knowledge good?"
- Q6: "Give a value to your knowledge about the following services in Distributed Systems:", when referred to the services: LDAP, DHCP, NFS and AFS.

Consequently, we could affirm that the activity significantly improved the skills of the students for several of the aspects evaluated, and, without significant differences, for all the others.

5 Discussion

In this section, an analysis and discussion of the results obtained is made, from two different perspectives:

- The results of the questionnaire clearly showed the lack of skills about SA that CE students have. The students do not feel prepared for integrating themselves into a professional context in which they should have to apply these skills, when SA is one of the most demanded professional profile. On the other hand, the result analysis involves that the activity has improved their knowledge about this field, specially for some services such as LDAP or DHCP. These results are of crucial interest for the current society and the crisis we are facing, being specially important to assure that students can quickly integrate into a real job.
- The DSAL was successfully deployed and used. The platform is now available for future uses. The documentation produced by the students is also available in the virtual platform of the module. This experience could be easily applied to other modules in any country to assure a proper SA learning during the CE degree. The experience has several advantages for the different sectors involved:
  - For the students,
    - · they developed abilities about installing and configuring Distributed Systems, and they improved their general capabilities related to operating systems, such as GNU/Linux.

Table 1. Results of the questionnaire before and after doing the proposed activity

General Questions							
Before				After			
Question	Mean	SD	Mode	Question	Mean	SD	Mode
Q1	3.9	0.73	4	Q1	4.0	0.45	4
Q2	4.2	0.26	4	Q2	4.0	0.35	4
Q3	2.9	1.73	3	Q3	3.3	0.61	3
Q4	3.7	0.53	3	Q4	3.8	0.36	4
Q5	2.3	0.41	2	**Q5**	3.0	0.55	3
Questions about services (Q6)							
Before				After			
LDAP	1.4	0.55	1	**LDAP**	2.9	0.29	3
FTP	3.6	0.75	3	FTP	3.6	0.76	3
HTTP	3.7	0.61	3	HTTP	3.7	0.83	3
DHCP	2.0	1.15	1	**DHCP**	2.3	1.23	3
NFS	1.7	0.63	2	**NFS**	2.7	0.41	3
AFS	1.0	0.05	1	**AFS**	2.6	0.59	3
POP3	3.1	0.49	3	POP3	3.1	0.53	3
SMTP	2.7	0.81	3	SMTP	3.1	0.62	3
IMAP	2.3	1.33	3	IMAP	2.7	1.03	3
DNS	2.9	0.29	3	DNS	3.1	0.49	3

The questions were significant differences are found are represented in bold face.

  · they strengthened their knowledge about the module Distributed Systems, specially from a practical perspective.
  · they worked as a team, and, at the same time, they helped to other teams, acquiring different responsibilities inside and outside their own group.
  · they solved a real problem, closer to the professional context of CE.
  · they improved their oral expression skills, because they had to explain and defend what they had done to the rest of classmates.

– For the professors and the degree,

  · we validated the grade of comprehension of the theoretical classes.
  · we updated the practical contents of the module.
  · a new professional profile was boosted and improved in the CE degree, which could be very difficult to be considered unless a practical environment such as the DSAL is used.

– For the Computing Service department of the University,

  · The experience improved the relationships between students, professors and the members of this department by making students and professors aware of the importance and difficulty of the work in this department, which is sometimes underestimated.

6 Conclusions

This paper presents a "learning-by-doing" experience in SA for CE students, consisting of the installation and management of different services of a DS. The experience was given the name Distributed System Administration Laboratory (DSAL) and was applied to fourth year students of a University in Córdoba (Spain). The DSAL infrastructure was previously prepared, by deploying the standard services and security mechanisms to carry out the activity with the students. The students undertook the task of deploying and administrating one of the services in a group, trying to simulate a real environment in a professional context. The results were very encouraging, showing a improvement of the SA skills of the students after applying the DSAL activity. Remote access to the DSAL makes it a good practical learning tool for transnational education, security requirements being properly satisfied.

To enhance and promote a new society, it is necessary to encourage new competences in higher education students. Within these new competences, those related to transfer and generalization of learning are significant, and the activity presented is also very focused to the capacity of the student to independently solve the problems involved in SA, so it should be considered as a good example of this kind of learning.

Acknowledgement. This work has been partially subsidized by the TIN2011-22794 project of the Spanish Inter-Ministerial Commission of Science and Technology (MICYT), FEDER funds and the P2011-TIC-7508 project of the "Junta de Andalucía" (Spain). Javier Sánchez-Monedero's research and Francisco Fernández-Navarro's research have been funded by the Ph. D. Student Program and the Predoctoral Program (grant reference P08-TIC-3745), respectively, both from the "Junta de Andalucía" (Spain).

References

1. Alvarez, X.: Educació en enginyeria mecànica. Una visió de futur. PhD Dissertation. Universitat Politécnica de Catalunya, Barcelona (2000)
2. Agencia Nacional de la Evaluación de la Calidad y Acreditación (ANECA), Los procesos de inserción laboral de los titulados universitarios en España. Factores de facilitación y de obstaculización, Madrid (March 2009), http://www.aneca.es/media/308144/publi_procesosil.pdf (cited April 19, 2012)
3. Felder, R.M., Brent, R.: Learning by Doing. Chemical Engineering Education 37(4), 282–283 (2003)
4. Free Software Foundation. What is free software and why is so important for society?, http://www.fsf.org/about/what-is-free-software
5. Free Technology Academy. European Project, http://ftacademy.org/
6. Gombert, J.E.: Metalinguistic Development. Haverster Wheatshef, London (1990)
7. De Graaff, E., Ravesteijn, W.: Training complete engineers: global enterprise and engineering education. European Journal of Engineering Education 26(4), 419–427 (2001)
8. Nemeth, E., Snyder, G., Hein, T.R.: Linux Administration Handbook, 2nd edn. Prentice-Hall (2007)

9. Pérez-Gómez, A.I.: Competencias o pensamiento práctico. La construcción de los significados de representación y de acción. In: Gimeno Sacristrán, J. (ed.) Educar Por Competencias, Qué Hay de Nuevo, Morata, Madrid, pp. 59–102 (2008)
10. Segalas, J.: Engineering education for a sustainable future. PhD Dissertation. Universitat Politécnica de Catalunya, Barcelona (2009)
11. Stallman, R.: Why Open Source misses the point of Free Software (2007), http://www.gnu.org/philosophy/open-source-misses-the-point.en.html(cited April 19, 2012)
12. Howes, T.A., Smith, M.C., Good, G.S.: Understanding and Deploying LDAP Directory Services, 2nd edn. Addison Wesley (2003)
13. Sánchez-Monedero, J., Meléndez-Aganzo, L., Ventura, S.: Implantación de LDAP como sistema de autenticación centralizad. Actas de las III Jornadas de Software Libre de la Universidad de Cádiz, pp. 31–42 (2003)
14. RedIRIS, http://www.rediris.es
15. eduroam, http://www.eduroam.org/
16. Istenic Starcic, A., Klinc, R., Fischinger, M., Turk, Z.: Computer based learning in earthquake engineering - giving control to students. In: Auer, M.E. (ed.) Ambient and Mobile Learning: Proceedings of the Workshop ICL - Interactive Computer Aided Learning, pp. 1–14 (2005)
17. Walpole, R., Myers, R., Ye, K.: Probability and Statistics for Engineers and Scientists. Pearson Education (2002)

Speed Control of A.C. Drive with Induction Motor Using Genetic Algorithm

Pavel Brandstetter and Marek Dobrovsky

VSB-Technical University of Ostrava, Department of Electronics,
17. listopadu 15/2172, 70833 Ostrava-Poruba, Czech Republic
pavel.brandstetter@vsb.cz

Abstract. The paper deals with application of a genetic algorithm for setting parameters of the speed controller in the control structure of the A.C. drive with the vector controlled induction motor. In the first part of the paper, the design of the genetic algorithm for the PI controller is presented. The second part shows possibilities of the modern simulation method using hardware in the loop simulation method in the software environment Matlab-Simulink and its implementation using modern digital signal controller. In the last part, simulation results for different speed and torque changes of the induction motor are shown. Rightness of the designed genetic algorithm has been verified by simulations then the method has been implemented into DSP control system.

1 Introduction

Variable-speed drives are being continually innovated. Their development is characterized by innovative process made in various areas including control theory, power electronics systems, control systems with modern digital signal processors etc. Vector controlled drives, providing high-dynamic performance, are finding increased number of industrial applications. Induction motors are often preferred choice in variable speed drive applications. Nowadays, low cost digital signal processors enable the development of cost effective electrical drives and the widespread availability of digital signal processors enable the development of large variety of drives with advanced features and of course using new control methods [1].

Fuzzy logic, artificial neural networks, genetic algorithms (GA) and their combinations are considered in the field of artificial intelligence (AI). Considerable research has been performed in the field of the AI. Recent trends and advancements in this field have stimulated the development of various systems for electrical machine and drive application. AI-based control is receiving great interest world wide [2]-[6].

The classical controller design is based on the mathematical model of the controlled system, which is often multivariable and nonlinear with parameter variation. The classical PI controller, which is most commonly used control algorithm in practice, has fixed gains. Therefore, the use of the classical PI controller does not meet the requirements of the robust performance. The main advantage is its relatively simple structure, which can be easily implemented in industrial applications. The parameters of the classical PI controllers have been set by various mathematical and experimental methods [8].

Á. Herrero et al. (Eds.): Int. JointConf. CISIS'12-ICEUTE'12-SOCO'12, AISC 189, pp. 341–350.
springerlink.com

GA-based controllers can lead to improved performance, enhanced tuning and adaptive capabilities and there are further possibilities for much wider range of AI-based applications in variable speed drives [9].

Modern concepts of control systems allow the implementation of time-consuming control algorithms in real-time, for example methods such as hardware in the loop simulation (HIL), where the controlled A.C. drive is replaced by a complex computer model.

2 Vector Control of the Induction Motor

Control methods of A.C. motors are based on the mathematical description of the universal motor with 3-phase stator and rotor windings. This model is suitable for most applications with the sufficient accuracy. Using the limiting conditions for stator and rotor voltage equations, the mathematical model can be use for the description of different A.C. motors, for example for the induction motor [10].

The mathematical model of the induction motor is the multi-parameter, non-linear system. We can describe the mathematical model of the induction motor in different coordinate systems. In reference to simple decoupling between individual components, it is useful to select a relative coordinate system [*x*, *y*], which rotates with angular speed ω_m and is oriented to the rotor flux vector ψ_R. The rotor flux vector ψ_R is expressed by the magnetizing current i_m and the main induction of the induction motor L_h.

We can derive a system of equations which describe the behavior of a squirrel-cage induction motor in the reference frame system [*x*, *y*] [10]:

$$\sigma T_{\mathrm{S}} \frac{di_{\mathrm{Sx}}}{dt} + i_{\mathrm{Sx}} = \frac{u_{\mathrm{Sx}}}{R_{\mathrm{S}}} + \omega_{\mathrm{im}} \sigma T_{\mathrm{S}} i_{\mathrm{Sy}} - (1-\sigma) T_{\mathrm{S}} \frac{di_{\mathrm{m}}}{dt} \tag{1}$$

$$\sigma T_{\mathrm{S}} \frac{di_{\mathrm{Sy}}}{dt} + i_{\mathrm{Sy}} = \frac{u_{\mathrm{Sy}}}{R_{\mathrm{S}}} - \omega_{\mathrm{im}} \sigma T_{\mathrm{S}} i_{\mathrm{Sx}} - (1-\sigma) \omega_{\mathrm{im}} T_{\mathrm{S}} i_{\mathrm{m}} \tag{2}$$

$$T_{\mathrm{R}} \frac{di_{\mathrm{m}}}{dt} + i_{\mathrm{m}} = i_{\mathrm{Sx}} \tag{3}$$

$$J \frac{d\omega_{\mathrm{m}}}{dt} = \frac{3}{2} \frac{L_{\mathrm{h}}}{1+\sigma_{\mathrm{R}}} i_{\mathrm{m}} i_{\mathrm{Sy}} - T_{\mathrm{L}} = K_T i_{\mathrm{m}} i_{\mathrm{Sy}} - T_L \tag{4}$$

$$T_{\mathrm{S}} = \frac{L_{\mathrm{S}}}{R_{\mathrm{S}}}, \quad T_{\mathrm{R}} = \frac{L_{\mathrm{R}}}{R_{\mathrm{R}}} \tag{5}$$

Mathematical symbols:

i_{Sx}, i_{Sy}	components of stator current space vector $\boldsymbol{i}_S$
u_{Rx}, u_{Ry}	components of stator voltage space vector $\boldsymbol{u}_S$
i_m	magnetizing current
L_{h}	magnetizing inductance
L_{S}, L_{R}	stator and rotor inductance

R_S, R_R	stator and rotor phase resistance
T_S, T_R	stator and rotor time constant
T_L	load torque
K_T	torque constant
J	total moment of inertia
ω_m	rotor angular speed
ω_{im}	angular speed of the magnetizing current i_m
σ	total leakage constant
σ_R	rotor leakage constant

By using vector control, it is possible to control separately the flux producing component i_{Sx} and torque producing component i_{Sy} of the stator current vector $\boldsymbol{i}_S$ which is defined by stator phase currents.

3 Classical Speed Controller

The quality of the control is assessed according to the response of the control loop to step changes of input variables. From a practical point of view, four factors are the most important for the assessment of the control quality: rise time, settling time, overshoot and steady state error.

The classical PI speed controller for the induction motor drive can be expressed as follows:

$$i_{sy\_ref} = K_p(\omega_{m\_ref} - \omega_m) + K_i \int (\omega_{m\_ref} - \omega_m)dt \tag{6}$$

where i_{Sy_ref} is the reference value of the torque producing component, ω_{m_ref} and ω_m are the reference and actual value of the rotor angular speed, K_p is the proportional gain, and K_i is the integral gain.

We can express the change of the torque producing component in discrete form using sampling period T as follows:

$$e_{\omega(k)} = \omega_{m\_ref(k)} - \omega_{m(k)} \tag{7}$$

$$\Delta i_{sy(k)} = i_{sy\_ref(k)} - i_{sy\_ref(k-1)} = K_p[e_{\omega(k)} - e_{\omega(k-1)}] + K_i T e_{\omega(k)} \tag{8}$$

$$i_{sy\_ref(k)} = i_{sy\_ref(k-1)} + \Delta i_{sy(k)} \tag{9}$$

The actual speed is measured by the speed sensor then is compared with the reference value of the rotor angular speed. Difference between reference and actual value e_ω is processed by PI speed controller whose parameters determine a behavior of the electrical drive.

At the start of control process, the conventional PI controller should have large proportional gain either for a fast motor acceleration or deceleration. The integral gain should be small to avoid overshoots. When the motor reaches the reference speed, a small proportional gain, and large integral gain are required to maintain the motor speed at desired value. Based on this, K_p is varied between K_{pmin} and K_{pmax} and K_i is varied between K_{imin} and K_{imax} to get satisfactory control performance.

4 Speed Controller with Genetic Algorithm

The genetic algorithms are finding important applications in optimization of technical systems. As an intelligent control technology the genetic algorithms can give robust adaptive response of an electrical drive with nonlinearity, parameter variation and load disturbance effect. The GA part of the controller performs parameter setting, which the genetic algorithm attempts to optimize by using simulation based fitness evaluation of the controllers in the closed-loop systems. The block scheme of the GA speed controller is shown in the figure 1.

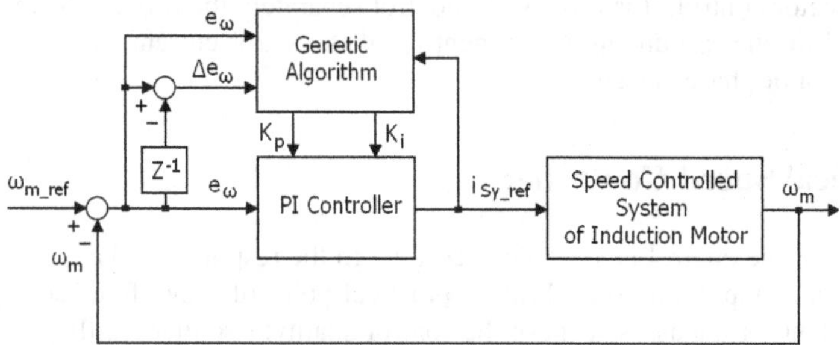

Fig. 1. Speed controller with genetic algorithm

The GA speed controller differs from the classical PI controller. The PI controller parameters are not constant, but their values are adapting to new conditions and circumstances in the controlled system [11].

Modifications and tuning of these constants ensures Genetic Algorithm block that evaluates the input variables based on the speed error e_ω according to a specified algorithm best suited values of the proportional gain K_p and the integral gain K_i, which then sends it to the conventional part of the speed controller.

The designed speed controller with genetic algorithm has following steps:

1. Sampling the speed signal of the induction motor.
2. Computation of the speed error e_ω between reference speed ω_{m_ref} and actual speed ω_m, speed error change Δe_ω and change of torque producing current Δi_{Sy_ref}.
3. Coding chromosomes. We have used decimal number because controller parameters K_p and K_i can be coded into a chromosome. The corresponding chromosome coded by K_p and K_i has six digits in decimal number.
4. Determination of an initial population of K_p and K_i gains with the chromosomes c_j (j=1, 2 … n, n is population size) using random selection.
5. Computation of the parameter p_j for each chromosome c_j and calculated speed error e_ω, speed error change Δe_ω and current change Δi_{Sy_ref}.

$$p_j = \left[K_p \Delta e_{\omega(k)} + K_i T e_{\omega(k)}\right] - \left[i_{sy\_ref(k)} - i_{sy\_ref(k-1)}\right] \quad (10)$$

6. Production of the next generation using GA operators (crossover, mutation) and go to step (5). The maximal number of generations is N.

7. Computation of criterion function J_i:

$$J_i = \sum_{i=1}^{N} \alpha_i p_j^2 = \alpha_1 p_j^2 + \alpha_2 p_j^2 + ... + \alpha_N p_j^2 = 1.05 p_j^2 + 1{,}1 p_j^2 + ... + 2.0 p_N^2 \quad (11)$$

$$\alpha_i = 1 + \frac{j}{20} \quad (12)$$

where i is generation number, N is number of generations, α_i is generation weight.

8. Fitness evaluation using the fitness function:

$$F_i = \frac{1}{1 + J_i} \quad (13)$$

9. Selection of the combination of the controller parameters K_p and K_i according to the selection probability, which is defined as:

$$P_i = \frac{F_i}{\sum_{i=1}^{n} F_i} \quad (14)$$

If right combination of the controller parameters K_p and K_i satisfying the required properties was found, the algorithm is finished, and the controller parameter values are used for speed control [11].

5 Hardware in the Loop Simulation

Hardware in the loop (HIL) is a tool that connects the hardware (controller) with a mathematical model (controlled system) in a closed feedback loop. To simulate this method, a real microcomputer control system and a mathematical model in an appropriate software environment are required.

The functional model contains the following main parts: microcontroller Freescale MC56F8037, multifunction card MF_624 HUMUSOFT and a personal computer. Method of the control system HIL is applied to the vector control of induction motor, the IM model and voltage inverter model run on the personal computer in the software environment Matlab-Simulink and vector control is programmed in the microcontroller Freescale MC56F8037. The data connection between PC and microcontroller Freescale MC56F8037 is performed by the multifunction card, which is determined for communication of the software environment Matlab-Simulink with the surroundings. The control of the system is presented through interactive graphical front panel in LabVIEW environment, which offers a unique opportunity to explore signal processing concepts in real time. The figure 2 shows a block scheme of the simulation using method hardware in the loop.

The control system generates an actuator variable dependent on the control deviation between desired and actual quantity. The control variable enters to the model then output variable (actual value) gets off from the mathematical model, and then gets back to the control system. This simulation works in real time. Results of this simulation approach to reality, because the control system with sensors and actuators is implemented as close as it would be in real application.

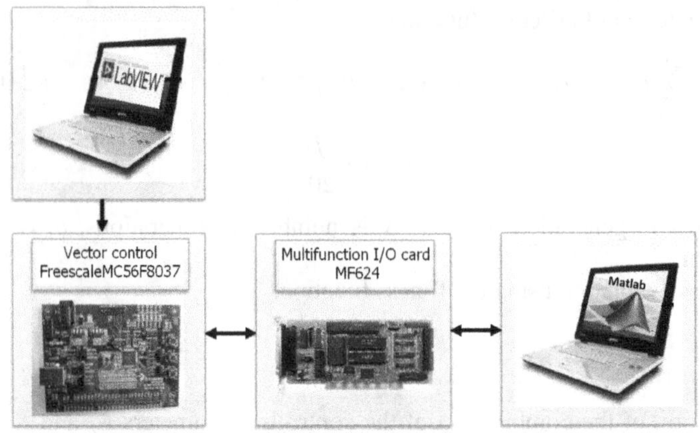

Fig. 2. Block scheme of Hardware in the Loop Simulation

6 Simulation Results

The speed control of the induction motor was tested in the different conditions and at different changes of the rotor angular speed. The simulation model of the induction motor supplied by voltage inverter was performed in Matlab-Simulink. The simulation results of the speed control were recorded by program for setting control variables in LabVIEW environment (see figure 3).

The control variable magnetizing current is set to 4 A. The reference speed is set, first at 0 rpm, and then is changed to 200 rpm, resp. -200 rpm. The figures 4-7 show the time responses of important quantities. The figures 8 and 9 show time courses of the gains K_p, K_i.

The parameters of the speed controllers:

Classical PI controller
Proportional gain K_p 10 A/rad/s
Integral gain K_i 1000 A/rad

Speed controller with genetic algorithm

Proportional gain K_p [0 - 20] A/rad/s
Integral gain K_i [0 - 2000] A/rad
Maximum generation 20
Population size 20

The parameters of the induction motor and voltage inverter for simulation in Matlab-Simulink:

$U_d = 300$ V, $I_m = 4$A, $n_n = 1\,420$ rpm, $p = 2$, $T_n = 12$ Nm, $J = 0.05$ kgm^2, $R_S = 5\ \Omega$, $L_h = 0.178$ H, $f_{sam} = 20$ kHz.

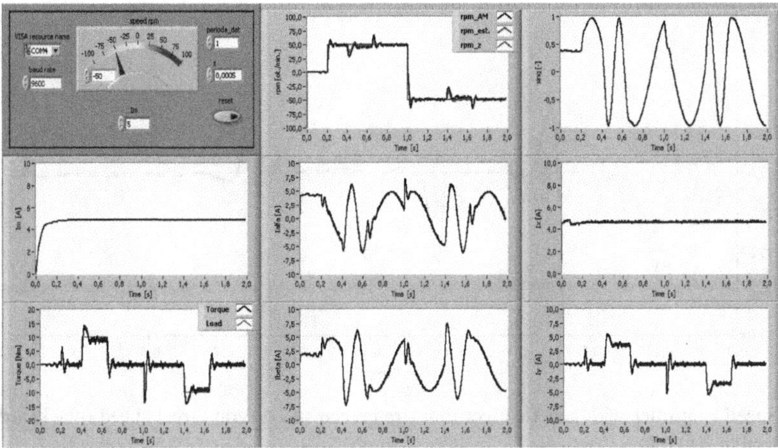

Fig. 3. Program in Labview environment for setting control variables and imaging the quantities of the induction motor drive with the vector control

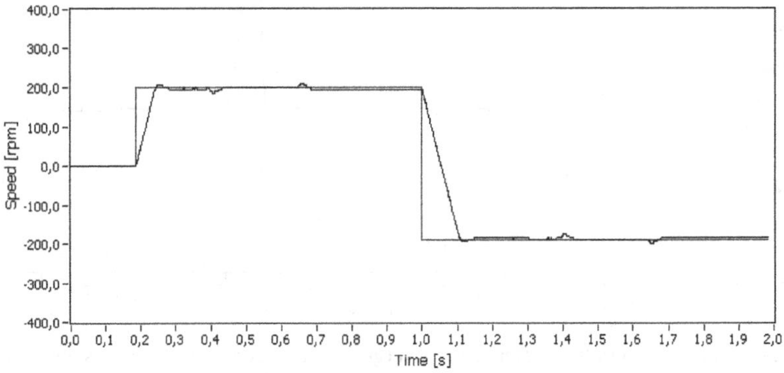

Fig. 4. Speed controller with genetic algorithm - reference speed (gray) 0 rpm, 200 rpm, -200 rpm, real speed (black)

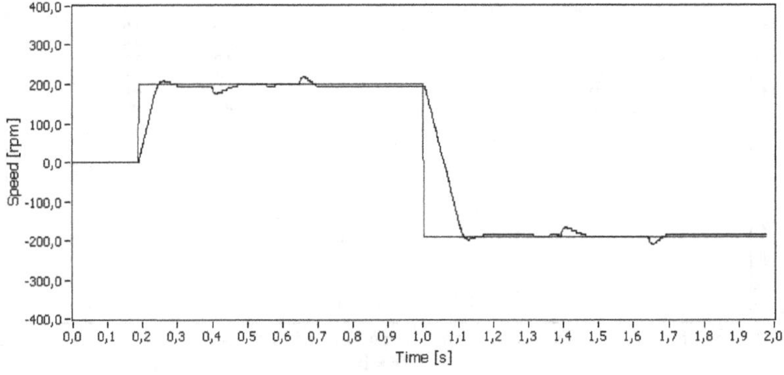

Fig. 5. Classical speed controller - reference speed (gray) 0 rpm, 200 rpm, -200 rpm, real speed (black)

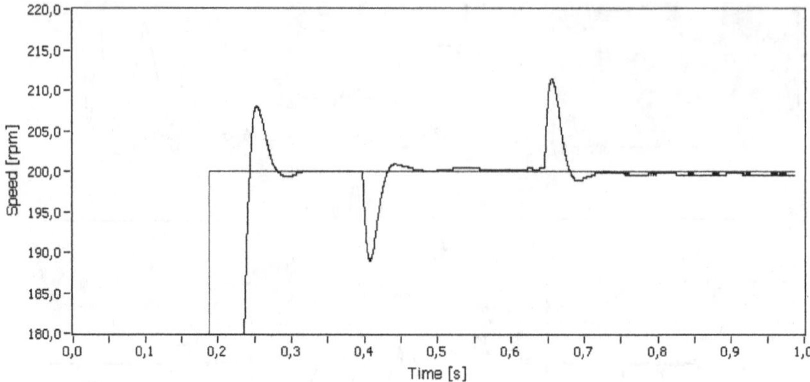

Fig. 6. Speed controller with genetic algorithm - reference speed (gray), real speed (black) - details about speed of 200 rpm

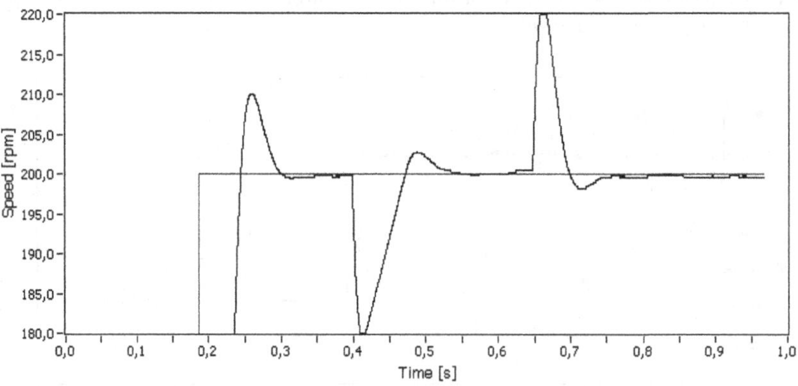

Fig. 7. Classical speed controller - reference speed (gray), real speed (black) - details about speed of 200 rpm

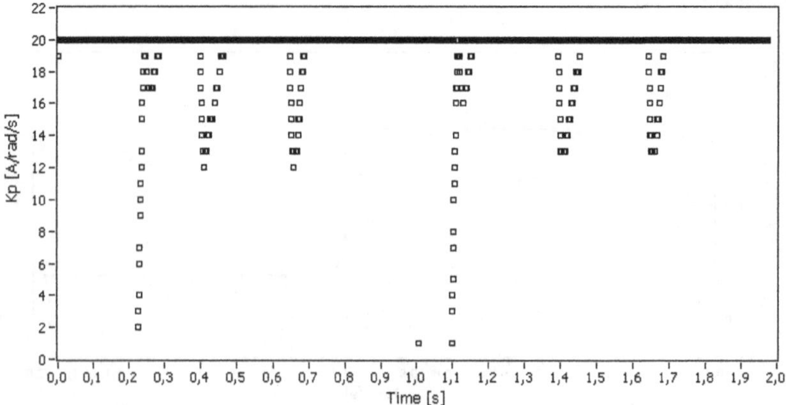

Fig. 8. Time course of the proportional gain K_P [A/rad/s]

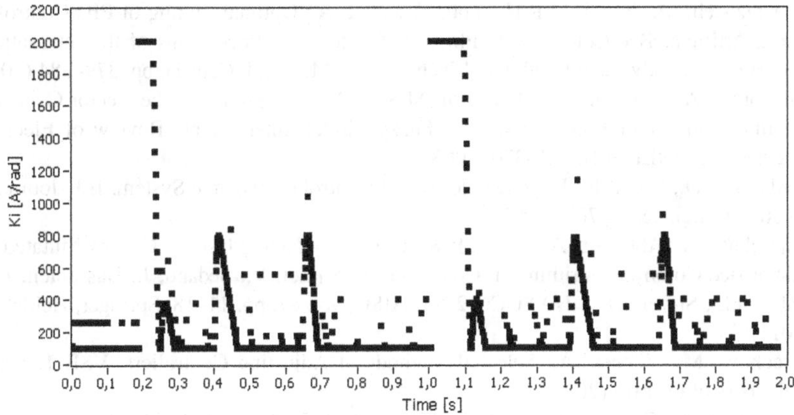

Fig. 9. Time course of the integral gain K_i [A/rad]

7 Conclusion

It has been realized important simulations, which confirm the rightness of proposed speed controller structure and good behavior of developed speed controller tuned by genetic algorithm. Hardware in the loop simulation is a technique that is used in the experimental test and development of complex real-time systems. The paper shows a possibility its application in the control of electrical drives with the induction motors. Obtained simulation results confirm expected application possibilities. The realized functional model of the control system with the HIL allows to perform research of new control methods and to solve problems of controlled A.C. drives, which contribute to increasing the efficiency of electrical products and reducing the electrical energy consumption.

Acknowledgments. The article has been elaborated in the framework of the IT4Innovations Centre of Excellence project, reg. no. CZ.1.05/1.1.00/02.0070 supported by Operational Programme 'Research and Development for Innovations' funded by Structural Funds of the European Union and state budget of the Czech Republic and in the framework of the project SP2012/85 which was supported by Student Grant Competition of VSB-Technical University of Ostrava.

References

1. Finch, J.W., Giaouris, D.: Controlled AC Electrical Drives. IEEE Transactions on Industrial Electronics 55(2), 481–491 (2008)
2. Vas, P.: Artificial-Intelligence-Based Electrical Machines and Drives. Oxford science publication (1999)
3. Holtz, J.: Sensorless Control of Induction Motor Drives. Proceedings of the IEEE 90(8), 1359–1394 (2002)

4. Abachizadeh, M., Yazdi, M.R.H., Yousefi-Koma, A.: Optimal Tuning of PID Controllers Using Artificial Bee Colony Algorithm. In: Conference Proceedings of the International Conference on Advanced Intelligent Mechatronics, Montreal, Canada, pp. 379–384 (2010)
5. Amamra, S.A., Barazane, L., Boucherit, M.S.: A New Approach of the Vector Control of the Induction Motor Using an Inverse Fuzzy Model. International Review of Electrical Engineering - IREE 3(2), 361–370 (2008)
6. Perdukova, D., Fedor, P.: Fuzzy Model Based Control of dynamic System. JEE-Journal of Electrical Engineering 7(3) (2007)
7. Rajasekhar, A., Abraham, A., Jatoth, R.K.: Controller Tuning Using a Cauchy Mutated Artificial Bee Colony Algorithm. In: Corchado, E., Snášel, V., Sedano, J., Hassanien, A.E., Calvo, J.L., Ślęzak, D. (eds.) SOCO 2011. AISC, vol. 87, pp. 11–18. Springer, Heidelberg (2011)
8. Viteckova, M., Vitecek, A.: Selected Methods of Adjusting Controllers. VSB-Technical University of Ostrava (2011)
9. Goldberg, D.E.: Genetic Algorithms in Search, Optimization and Machine Learning. Addison-Wesley Publishing Company, Boston (1989)
10. Brandstetter, P.: A.C. Controlled Drives - Modern Control Methods. VSB-Technical University of Ostrava (1999)
11. Elmas, C., Yigit, T.: Genetic Algorithm Based On-line Tuning of a PI Controller for a Switched Reluctance Motor Drive. Electric Power Comp. and Syst. 35(6), 675–691 (2007)

Applications of Artificial Neural Networks in Control of DC Drive

Pavel Brandstetter and Pavel Bilek

VSB-Technical University of Ostrava, Department of Electronics,
17. listopadu 15/2172, 70833 Ostrava-Poruba, Czech Republic
pavel.brandstetter@vsb.cz

Abstract. The paper deals with the applications of artificial neural networks in the control of the DC drive. In the paper three control structures are discussed. The first control structure uses a conventional PI controller. The second structure uses a neural network predictive control. The last structure is a sensorless control of the DC drive using feedforward neural network. The DC drives were simulated in program Matlab with Simulink toolbox. The main goal was to find the simplest neural network structures with minimum number of neurons, but simultaneously good control characteristics are required. Despite used neural networks, which are very simple, it was achieved satisfactory results.

1 Introduction

In general, the application of artificial intelligence in drives can lead to increased performance and robustness to parameter and load variations. Artificial neural networks are used for the identification and control of non-linear dynamic systems. An artificial neural network (ANN) is a massively parallel, non-linear adaptive system containing highly interacting elements called neurons or perceptrons. The artificial neural networks are based on crude models of the human brain and contain many artificial neurons linked via adaptive interconnections (weights). In other words they are adaptive function estimators which are capable of learning the desired mapping between the inputs and the output of the system.

The artificial neural networks usually must learn the connection weights from available training patterns. Performance is improved over time by iteratively updating the weights in the network. Mostly used artificial neural network is trained off-line by set of corresponding input-output pairs of controlled system. The learning and adapting capability of neural networks makes them ideal for control purposes. An ANN can be successfully applied even if the motor which is to be controlled and the load parameters are unknown [1]-[11].

The conventional DC motor drive continues to take a large share of the variable-speed drive market. However, it is expected that this share will very slowly decline, but there are some companies that produce DC drives. Artificial intelligence belongs to modern technology. Its application can lead to improvement of parameters of electric regulated drives with DC motors. This technology allows significant innovations of DC drives.

Á. Herrero et al. (Eds.): Int. JointConf. CISIS'12-ICEUTE'12-SOCO'12, AISC 189, pp. 351–360.
Springelink.com

2 Control Structure of DC Drive

The power part of the DC drive consists of DC to DC converter (chopper or controlled rectifier) and DC motor. Block scheme of the drive is shown in Fig. 1. When a fixed DC supply is available, a DC to DC converter can be used for the purposes of the control of the DC motor, where the constant DC voltage is transformed into an adjustable voltage to control the speed of the motor. An armature current is controlled by current controller. A speed controller provides the speed control.

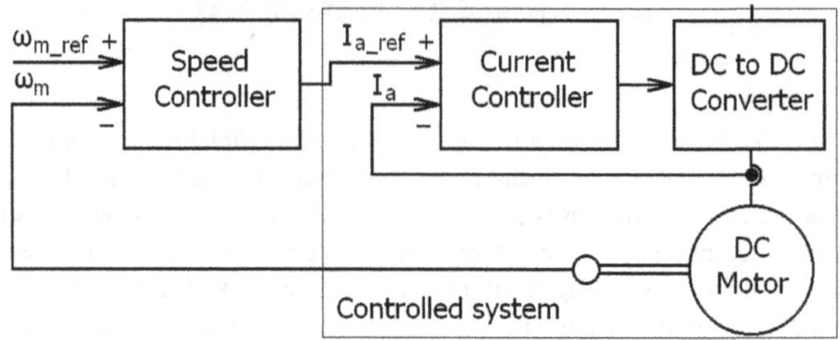

Fig. 1. DC drive block scheme

3 Conventional PI Controller

For future comparison it is important to consider the conventional control. It is conventional PI controller used as the speed controller which can be described as follows:

$$i_{a\_ref} = K_p(\omega_{m\_ref} - \omega_m) + K_i \int(\omega_{m\_ref} - \omega_m)dt + K_d \frac{d}{dt}(\omega_{m\_ref} - \omega_m) \quad (1)$$

where i_{a_ref} is the reference value of the torque producing current (armature current), ω_{m_ref} and ω_m are the reference and actual value of the rotor angular speed, K_p is the proportional gain, K_i is the integral gain, and K_d is the derivative gain.

The quality of the control is assessed according to the response of the control loop to step changes of input variables. From a practical point of view, four factors are the most important for the assessment of the control quality: rise time, settling time, overshoot and steady state error

In control theory applications, the rise time t_r is the time required to reach first the steady-state value (100%). It may also be defined as the time to reach the vicinity of the steady-state value particularly for a response with no overshoot, e.g. the time between 10% and 90%. The settling time t_s is the time for departures from final value y_f to sink below some specified level of final value y_f, i.e. $y_f \pm \Delta$. The overshoot is the maximum value y_m of the controlled quantity above final value y_f [12].

4 Neural Network Predictive Control

The model predictive control method is based on the receding horizon technique. The neural network model predicts the plant response over a specified time horizon.

The predictions are used by a numerical optimization program to determine the control signal that minimizes the following performance criterion over the specified horizon [13]:

$$J = \sum_{j=N_1}^{N_2} \left[y_r(t+j) - y_m(t+j) \right]^2 + \rho \sum_{j=1}^{N_u} \left[u'(t+j-1) - u'(t+j-2) \right]^2 \quad (2)$$

where N_1, N_2 and N_u define the horizons over which the tracking error and the control increments are evaluated. The *u'* variable is the tentative control signal, y_r is the desired response and y_m is the network model response. The ρ value determines the contribution that the sum of the squares of the control increments has on the performance index.

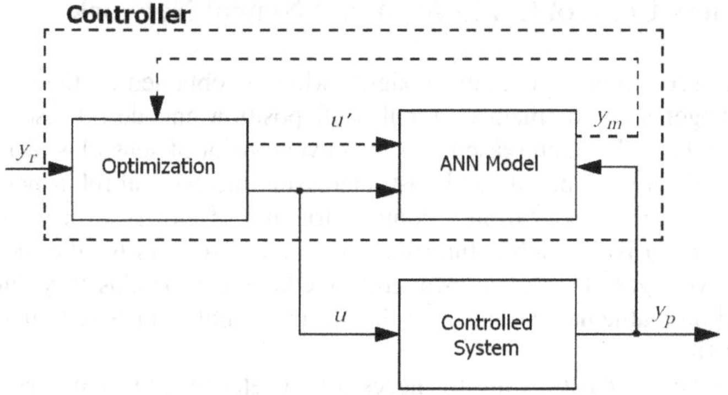

Fig. 2. Block scheme of NN predictive control

The figure 2 shows block diagram which illustrates the model predictive control process. The controller consists of the neural network plant model and the optimisation block. The optimisation block determines the values of *u'* that minimize *J*, and then the optimal *u* is input to the plant [13].

5 Structure of Artificial Neural Network

There are typically two steps involved when using neural network for predictive control: system identification and control design. In the system identification stage, development of the neural network model (5-1 ANN) of the controlled system is performed. There were used 2 delayed plant inputs, 3 delayed plant outputs and 10000 training samples. In the control design stage, the ANN model of the controlled system is used to design or train the controller. For the model predictive control, the ANN model of the controlled system is used to predict future behavior of the system (control horizon is 2), and an optimization algorithm is used to select the control signal that optimizes future performance. The controller requires a significant amount of online computation, since an optimization algorithm is performed at each sample time to compute the optimal control signal.

For realization of control system with NN predictive control, artificial neural network was used which was trained off-line by set of corresponding input-output pairs of controlled system. The weights of the ANN can be then adjusted via the so-called backpropagation algorithm using Levenberg-Marquardt method to minimize the error.

For the ANN model, a three layer neural network is used which contains five neurons in hidden layer with tanh activation function and one neuron in output layer with linear activation function. It stands to reason that structure of neural network is very simple, it consists of only six neurons. However, good results of important drive quantities are achieved. It was tested various structures of artificial neural network, but for example if it was used three neurons in hidden layer, results were not so good, especially control signal (output of the controller) contains higher ripple.

6 Sensorless Control Using Artificial Neural Network

The speed control requires a feedback signal which is obtained by the speed sensors such as tachogenerator or mainly digital shaft position encoder. These sensors are sources of trouble. The main reasons for the development of sensorless drives are: reduction of hardware complexity and cost, increasing mechanical robustness, reliability. Removing speed sensors from a control structure of electrical drive leads to so-called sensorless drive, which naturally requires other sensors for the monitoring of currents and voltages. The speed estimation methods can be classified into conventional, based on mathematical model of the electrical motor, or based on artificial intelligence [14]-[20].

To realize the speed estimator, it is necessary to determine the appropriate structure of the neural network with appropriate input variables, which will implement the views defined by the following equation:

$$\omega_{m(k)} = \mathbf{f}\,[i_{a(k)}, i_{a(k-1)}, u_{a(k)}, u_{a(k-1)}, \mathbf{w}\,] \tag{3}$$

where f is the activation function and w is a vector of weighting and threshold coefficients.

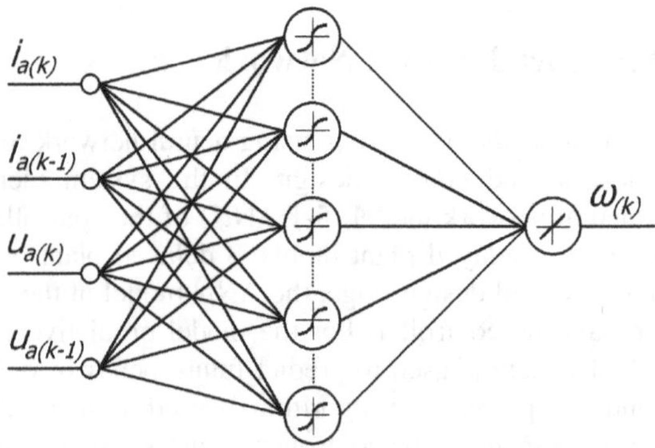

Fig. 3. Speed estimator with artificial neural network

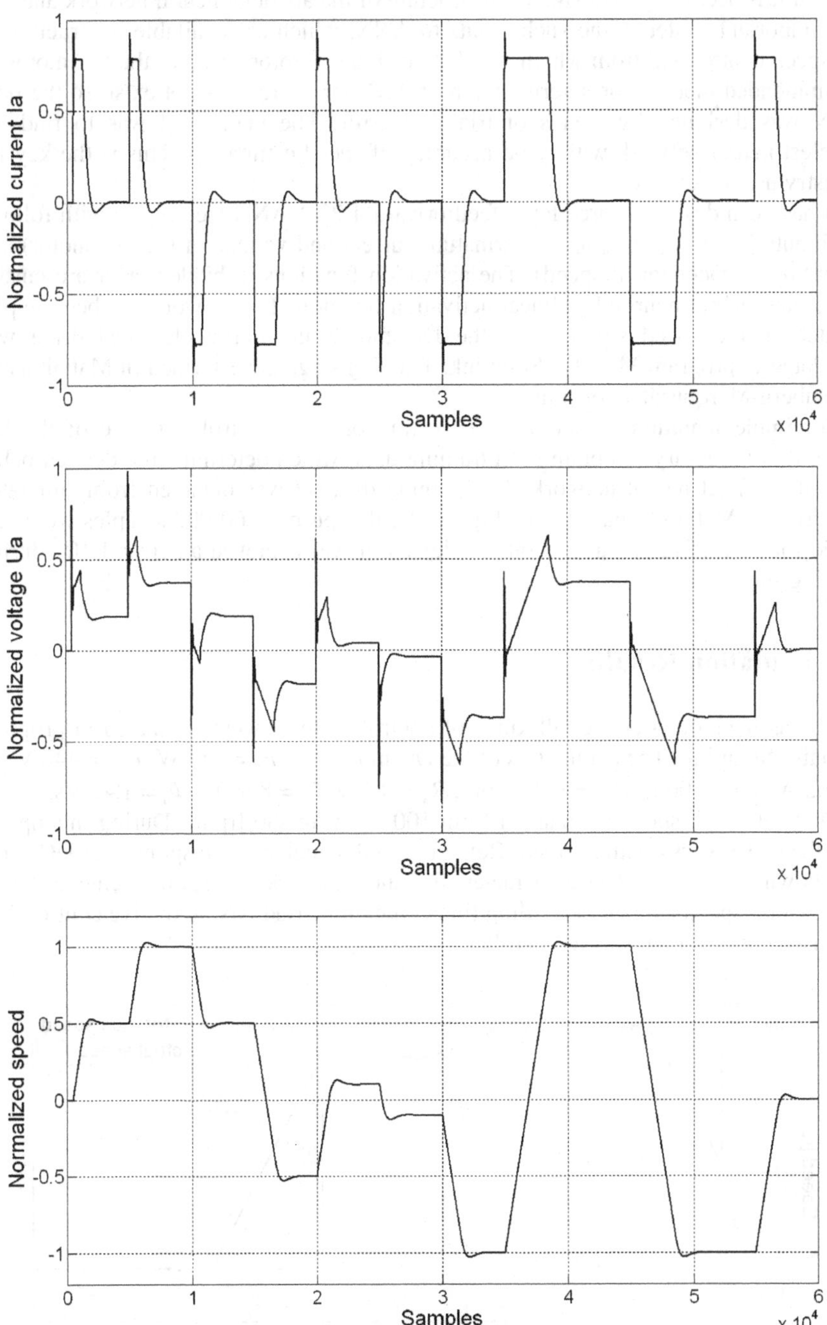

Fig. 4. Example of the training data set for ANN speed estimator

First it is necessary to design right structure of the artificial neural network and it is also important to determine such inputs to ANN, which are available in structure of the speed control and from which is able to estimate a rotor speed of the DC motor. A recommended method for determination of ANN structure does not exist, so the final ANN was designed by means of trial and error. The main goal was to find the simplest neural network with good accuracy of speed estimation. This is the key for industry use of ANN´s.

It has been designed three layer feedforward 4-22-1 ANN (see Fig.3) with following inputs $i_{a(k)}$, $i_{a(k-1)}$, $u_{a(k)}$, $u_{a(k-1)}$, (armature current and voltage of the DC motor) and output $\omega_{m(k)}$ (mechanical speed). The activation functions in hidden layer are tansigmoids and output neuron has linear activation function. The network has been implemented in the speed structure of the DC motor and entire electrical drive was simulated in program Matlab - Simulink. Training stage is performed in Matlab using Levenberg-Marquardt algorithm.

For implementation of neural speed estimator onto control structure of the DC drive, it is necessary to obtain such training data, which determine the desired behavior of artificial neural network. The training data set was obtained from simulated DC drive in Matlab-Simulink (see Fig.4). For this purpose 60 000 samples were recorded for each of the input and output signals. It was achieved an error 1.10^{-4} during training stage.

7 Simulation Results

As it was mentioned above, all kinds of control systems were simulated in program Matlab - Simulink. The parameters of the DC motor are: $P_n = 15$ kW, $U_{an} = 440$ V, $I_{an} = 37.5$ A, $n_n = 2800$ rpm, $J = 0.24$ kgm^2, $R_a = 0.7\ \Omega$, $L_a = 8$ mH, $c\Phi_n = 1.42$ Vs.

First reference speed is changed from 100 [rpm] to -80 [rpm]. During this operation the drive works without load. Reference and actual speed responses of DC drive are shown in Fig. 5, 6. These characteristics show that speed response achieves better parameters (speed overshoot, settling time) when it is used NN predictive control than it is used conventional speed controller.

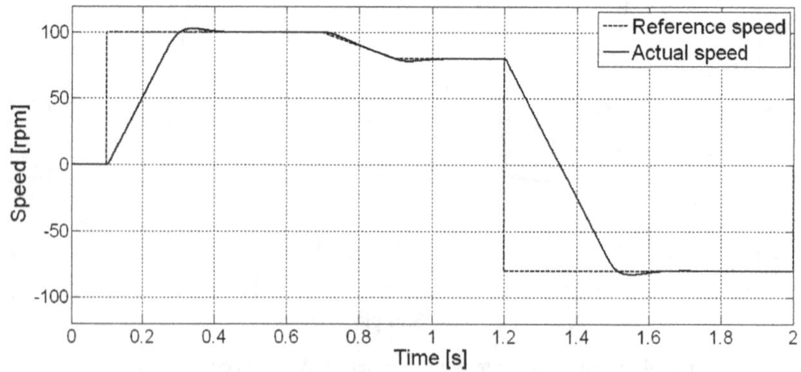

Fig. 5. Reference and actual angular speed responses of DC drive - conventional PI controller

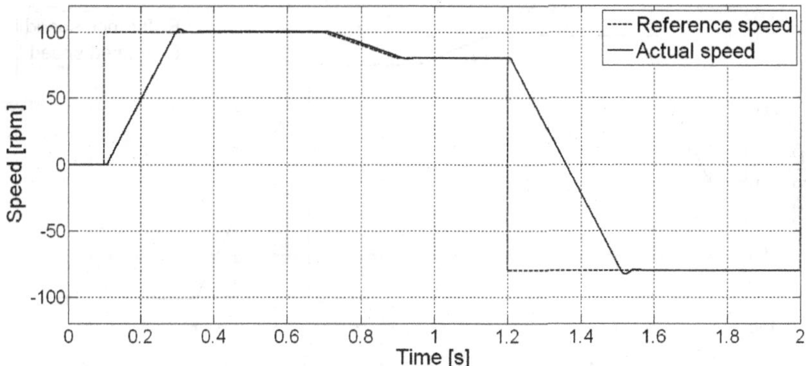

Fig. 6. Reference and actual speed responses of DC drive - Neural network predictive control

In the second stage after run-up, the electrical drive was loaded with rated torque. We can find out a conclusion that the NN predictive control achieves a shorter control time while the drive is subjected to load than conventional PI controller. So it is sure that NN predictive control increases robustness of the drive. Of course, it was tested another values of reference speed and load too.

For the control quality evaluation of the sensorless DC drive, it is important to assess the speed time course in different situations. Thus, the time course of reference speed was defined. The simulation was performed for the reference speeds which represent two speed areas: area of very low speed (ω_{m_ref} = 10 rpm), area of low speed (ω_{m_ref} = 100 rpm). The estimated speed is used as the feedback signal for the speed control.

For presentation, the simulation results in the area of the low speed (ω_{m_ref} = 100 rpm) were selected (see Fig.7, 8). The figures 7 and 8 show the time response reference, actual and estimated speed for the sensorless control with feedforward artificial neural network. The difference between actual and estimated speed is shown in the Fig. 9.

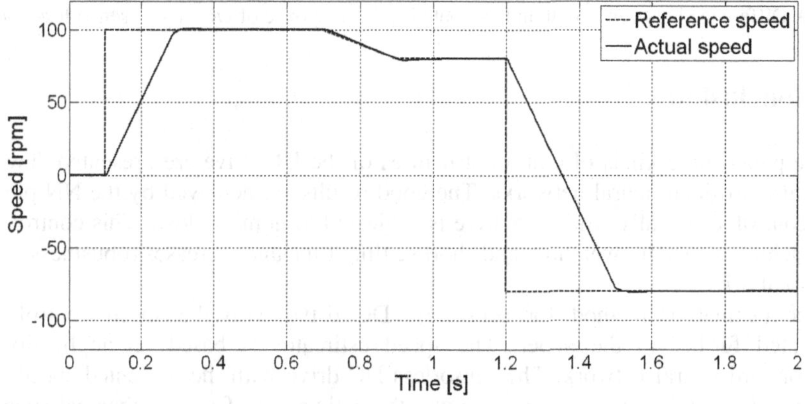

Fig. 7. Reference and actual speed responses of DC drive - sensorless control

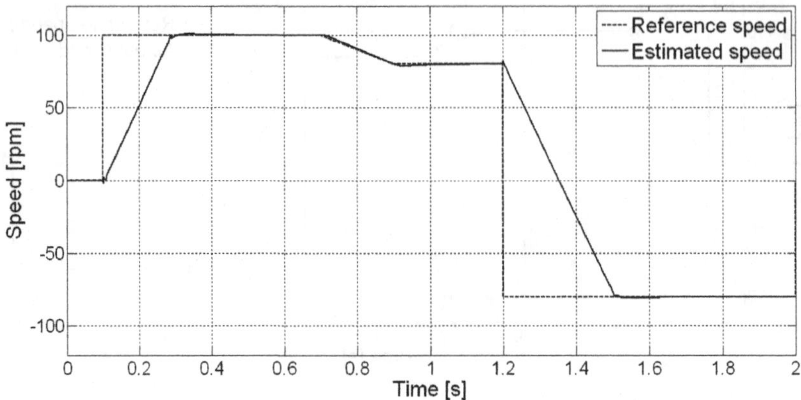

Fig. 8. Reference and estimated speed responses of DC drive - sensorless control

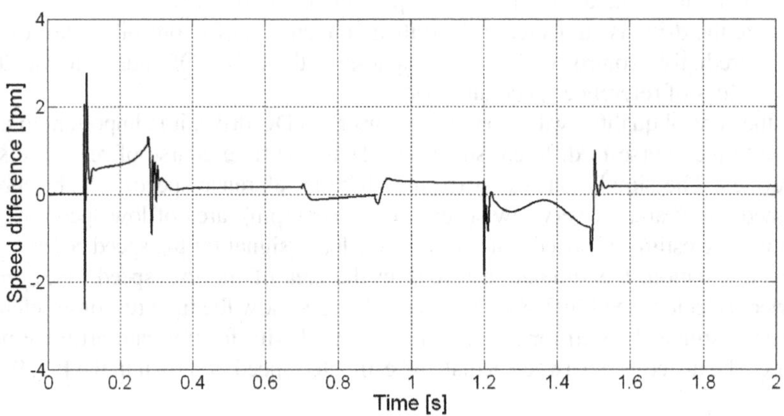

Fig. 9. Difference between actual and estimated speed response of DC drive - sensorless control

7 Conclusion

In the paper three kinds of control structures of the DC drive are presented. Two of them use artificial neural networks. The good results are achieved by the NN predictive control, especially while the drive is subjected to nominal load. This control method achieves a small overshoot and short settling time and increases robustness of the electrical drive.

The estimation method for sensorless DC drive with the speed control was presented further in the paper. The speed estimator is based on application of feedforward neural network. The sensorless DC drive with the presented speed estimator gives good dynamic responses and the estimation of the mechanical speed is satisfactory in steady state and also in transient state.

Acknowledgments. The article has been elaborated in the framework of the IT4Innovations Centre of Excellence project, reg. no. CZ.1.05/1.1.00/02.0070 supported by Operational Programme 'Research and Development for Innovations' funded by Structural Funds of the European Union and state budget of the Czech Republic and in the framework of the project SP2012/85 which was supported by Student Grant Competition of VSB-Technical University of Ostrava.

References

1. Norgaard, M.: Neural Networks for Modelling and Control of Dynamic Systems. Springer, London (2000)
2. Norgaard, M.: Neural network based control system design toolkit. Technical University of Denmark (2000)
3. Brandstetter, P.: A.C. Controlled Drives - Modern Control Methods. VSB-Technical University of Ostrava (1999)
4. Abachizadeh, M., Yazdi, M.R.H., Yousefi-Koma, A.: Optimal Tuning of PID Controllers Using Artificial Bee Colony Algorithm. In: Conference Proceedings of the International Conference on Advanced Intelligent Mechatronics, Montreal, Canada, pp. 379–384 (2010)
5. Amamra, S.A., Barazane, L., Boucherit, M.S.: A New Approach of the Vector Control of the Induction Motor Using an Inverse Fuzzy Model. International Review of Electrical Engineering - IREE 3(2), 361–370 (2008)
6. Perdukova, D., Fedor, P.: Fuzzy Model Based Control of dynamic System. JEE-Journal of Electrical Engineering 7(3) (2007)
7. Luger, G.F.: Artificial Intelligence, Structures and Strategies for Complex Problem Solving. Williams (2003)
8. Russel, S.J., Norvig, P.: Artificial Intelligence, A Modern Approach. Prentice Hall (2006)
9. Vas, P.: Artificial-Intelligence-Based Electrical Machines and Drives. Oxford Science Publication (1999)
10. Haykin, S.: Neural Network a Comprehensive Foundation. Prentice-Hall, New Jersey (1999)
11. Hagan, M.T., Demuth, H.B., Beale, M.: Neural Network Design. PWS Publishing Company (1996)
12. Levine, W.S.: The Control Handbook. CRC Press, Boca Raton (1996)
13. Beale, M.H., Hagan, M.T., Demuth, H.B.: Neural Network Toolbox™, User's Guide. The MathWorks, Inc. (2012)
14. Holtz, J.: Sensorless Control of Induction Motor Drives. Proceedings of the IEEE 90(8), 1359–1394 (2002)
15. Girovsky, P.J., Timko, J., Zilkova, J., Fedak, J.V.: Neural estimators for shaft sensorless FOC control of induction motor. In: Conference Proceedings, 14th International Power Electronics and Motion Control Conference, pp. T7-1–T7-5 (2010)
16. Gacho, J., Zalman, M.: IM Based Speed Servodrive with Luenberger Observer. Journal of Electrical Engineering 61(3), 149–156 (2010)
17. Vas, P.: Sensorless Vector and Direct Torque Control. Oxford University Press, New York (1998)
18. Lascu, C., Boldea, I., Blaabjerg, F.: Comparative Study of Adaptive and Inherently Sensorless Observers for Variable-Speed Induction-Motor Drives. IEEE Transactions on Industrial Electronics 53(1), 57–65 (2016)

19. Gadoue, S.M., Giaouris, D., Finch, J.W.: Sensorless Control of Induction Motor Drives at Very Low and Zero Speeds Using Neural Network Flux Observers. IEEE Transactions on Industrial Electronics 56(8) (2009)
20. Gallegos, M., Alvarez, R., Nunez, C., Cardenas, V.: Effects of Bad Currents and Voltages Acquisition on Speed Estimation for Sensorless Drives. In: Conference Proceedings Electron., Robot. Automotive Mech. Conference, pp. 215–219 (2006)

Energy Optimization of a Dynamic System Controller

Pavol Fedor and Daniela Perdukova

Technical University of Kosice, Department of Electrical engineering and mechatronics, Letná 9, 042 00 Kosice, Slovak Republic
daniela.perdukova@tuke.sk

Abstract. The paper deals with the application of fuzzy logic in the method of designing the parameters of a continuous dynamic system controller that is energetically optimal and at the same time meets the desired dynamic control parameters. The suitable controller parameters are established on the basis of a fuzzy model of the system, generated through its identification from the measured inputs and outputs. The proposed method has been verified by simulations on an example of parameter design for a PI controller of a DC drive with non-linear load. In comparison with a standardly designed PI controller with constant parameters for the whole operational space of the DC drive it is possible to save approximately 24.39 % of electric power at each dynamic motion of the drive.

1 Introduction

One of the basic ways of increasing the energetic efficiency of existing or newly designed electro-energetic equipment is the optimisation of its operation through the application of better ("intelligent") control technologies. The design of controllers with firmly specified parameters is not always optimal from this point of view, as the exact controlled system parameters are either not known, or they can change during the operation of the equipment (e.g. the resistance values of electric motor windings can change after warming by as much as 30%, the drive load may be non-linear, etc.) [1, 2]. It is therefore advisable to identify the existing parameters of the controlled system and use these as a basis for the selection of energetically optimal controller parameter values.

This paper presents one of the possible methods of this procedure which is based on the exploitation of fuzzy system knowledge. A brief description of the design process of the fuzzy model of the system concerned, as well as of the procedure of its exploitation in the optimisation of the PI controller parameters are provided. The applied method is verified by simulations on a drive with a DC motor in MATLAB, and possible energy savings achieved through drive control are presented.

2 Dynamic System Fuzzy Model Design

The dynamic system we want to control can be described in continuous form in state space by the equation:

Á. Herrero et al. (Eds.): Int. JointConf. CISIS'12-ICEUTE'12-SOCO'12, AISC 189, pp. 361–369.
springerlink.com

$$\dot{\mathbf{x}} = \mathbf{A}(\mathbf{x}(t), t) * \mathbf{x}(t) + \mathbf{B}(\mathbf{x}(t), t) * \mathbf{u}(t) \tag{1}$$

In discrete form system (1) can be described by equation:

$$\begin{aligned} \mathbf{x}_k &= \mathbf{x}_{k\text{-}1} + \Delta\mathbf{x}_{k\text{-}1} \\ \Delta\mathbf{x}_{k-1} &= \mathbf{A}'\mathbf{x}_{k\text{-}1} + \mathbf{B}'\mathbf{u}_{k\text{-}1} \end{aligned} \tag{2}$$

This means that the state of the system in step k is a function (generally unknown and nonlinear) of the system state and system inputs in step (k-1).

We can easily ascertain that if the sampling time T satisfies Shanon-Kotelnik conditions, then it approximately applies that:

$$\mathbf{A}' = \mathbf{A} * \mathbf{T}, \mathbf{B}' = \mathbf{B} * \mathbf{T} \tag{3}$$

Numerous different dynamic system fuzzy model structures are described in literature [3, 4, 5, 6, 7, 8, 9]. Mentioned methods require at least partial information about the control system structure and parameters. Proposed methods in this article uses only experimental measured data from controlled system. Creating of the fuzzy model for dynamic system (1) was based on its discrete form description according to equation (2). It generally applies that a change in the state of a system (1) at a given moment depends on its input and on its state in the preceding moment. Its discrete description (relation between state **x** and input **u**) can then be described by means of fuzzy rules in the following form:

IF $\mathbf{u}_k$ is ... and $\mathbf{x}_k$ is ... THEN $\Delta\mathbf{y}_k$ is .. (4)

where index k denotes the value of the variable in the kth sampling step and the terms $\mathbf{u}_k$ is ... , $\mathbf{x}_k$ is ... , $\Delta\mathbf{y}_k$ is ... represent the relevant fuzzy definition areas of the given variable. Based on the above we can set up a block diagram of the dynamic system fuzzy model (Fig. 1), where block z^{-1} represents the delay by one sampling step.

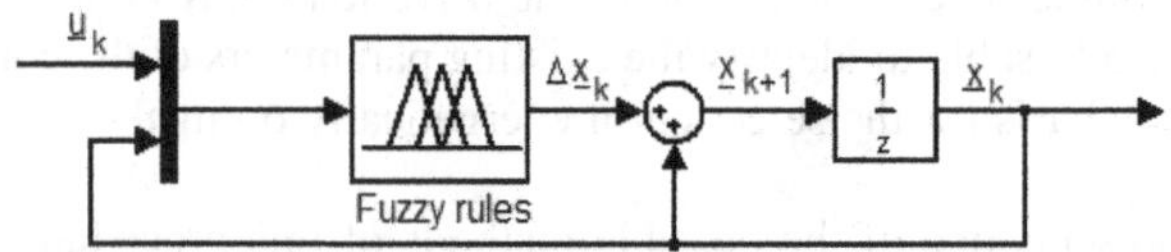

Fig. 1. Block diagram of dynamic system fuzzy model

In order to set up a fuzzy model we need to accumulate a database of appropriate input/output signals from the system, e.g. in a configuration shown in Fig.2. The fuzzy model will thus be designed on basis of the measured relations between [$\mathbf{u}_k$, $\mathbf{x}_k$] → $\Delta\mathbf{x}_k$, and the particular input signal values of the system input will be such that will cover the whole range of possible inputs and by that the whole working space of possible system states.

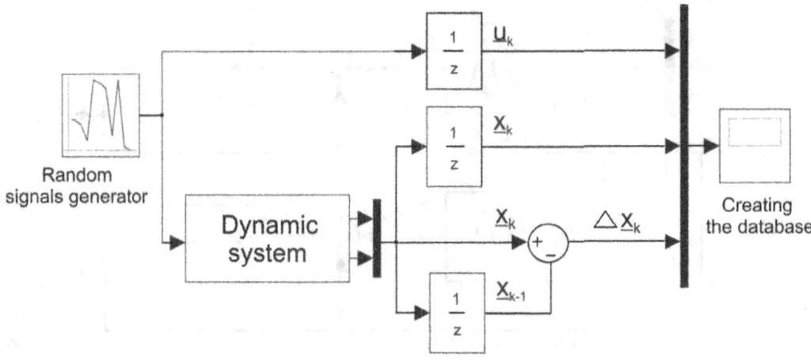

Fig. 2. Creating the database for fuzzy model setup

With the measured database it is possible to identify the system´s fuzzy model FIS structure that will best describe the measured relations. In the MATLAB programme package e.g. the Fuzzy Toolbox and its tool *Anfisedit* can be exploited for this purpose.

Example 1: We will consider a separately excited DC motor with parameters $K_A = 0.625\ \Omega^{-1}$, $T_A = 0.01$ s, $c\varphi = 0.7$ Vs, $J = 0.03$ kgm^2 as a 2nd order dynamic system. If we select state variables of the motor $x_1 = \omega$ (motor angular speed) and $x_2 = I_A$ (motor armature current), then the matrices of its state description will be the following:

$$\mathbf{A} = \begin{bmatrix} 0 & 22.22 \\ -43.75 & -100 \end{bmatrix} \quad \mathbf{B} = \begin{bmatrix} 0 \\ 62.5 \end{bmatrix} \tag{5}$$

Considering a fan-type load torque $M_z = k \,.\, x_1^2$ (k=0.0001), the response of this drive to connection of input voltage $U_A = 30$V is shown in Fig. 3

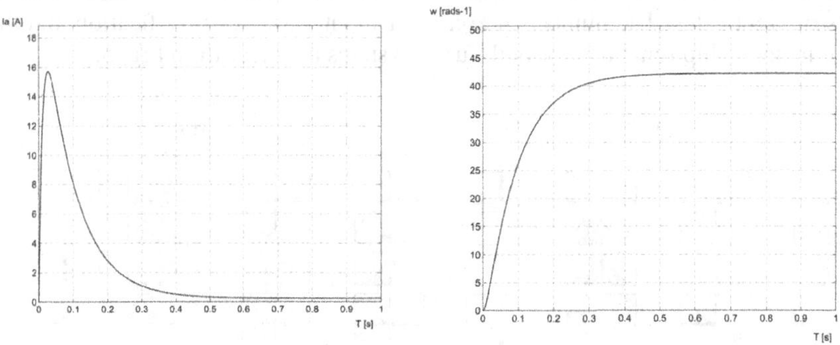

Fig. 3. DC motor response to voltage jump on armature

The structure of the fuzzy model of a DC drive representing a 2nd order dynamic system is shown in Fig. 4. The fuzzy model consists of two fuzzy subsystems for the individual state variables of the drive and each fuzzy subsystem has three inputs and one output.

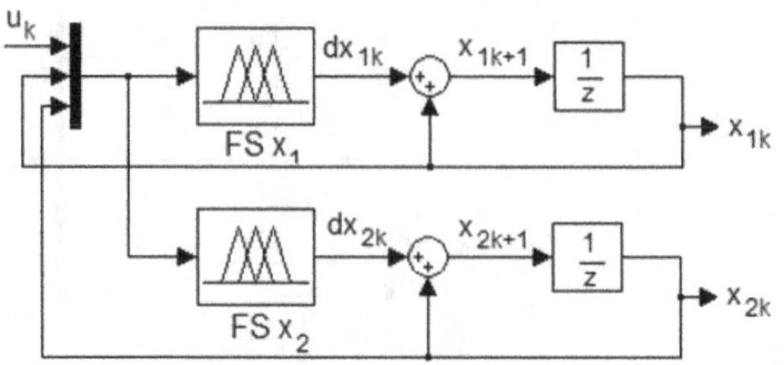

Fig. 4. Fuzzy model of DC drive

It is clear that in order to create a fuzzy model for the entire operational area of the drive it is necessary to measure the values for Table 1 in all of its potential states. In our case we divided the range of input voltage from 0 to 220 V into 10 levels and for the database for fuzzy model generation we measured the required values of state variables for all transitions between these values.

Table 1. Structure of table of values required for creating the fuzzy model

Data for FS x_1				Data for FS x_2			
$\mathbf{u_k}$	$\mathbf{x_{1k}}$	$\mathbf{x_{2k}}$	$\mathbf{dx_{1k}}$	$\mathbf{u_k}$	$\mathbf{x_{1k}}$	$\mathbf{x_{2k}}$	$\mathbf{dx_{2k}}$
90.00	123.58	2.18	0.00	90.00	123.58	2.18	0.00
130.00	123.58	2.18	6.10	130.00	123.58	2.18	20.33

Using the Anfisedit tool in the Matlab programme, two Sugeno type static fuzzy systems were created using the measured data database (see Fig. 4). Each of them consists of eight rules. The number of zones for input variables fuzzification and the shape of membership functions for both fuzzy systems are clear from Fig. 5.

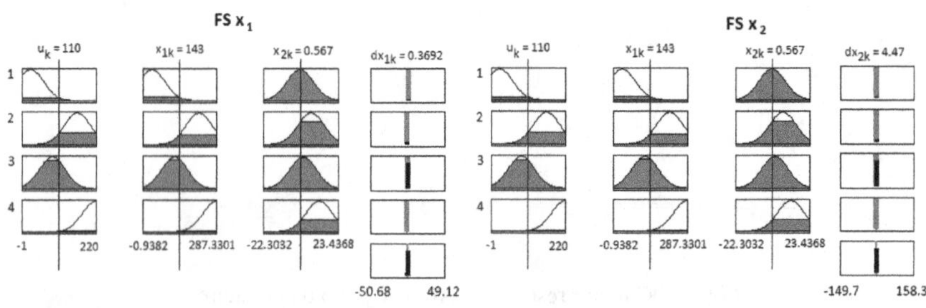

Fig. 5. Structure of fuzzy subsystems FS1 and FS2 (fuzzification and rules)

The comparison of the DC drive and its fuzzy model responses to the chosen voltage jump $U_A = 50$ V is presented in Fig. 6. It shows a practical correspondence of the model and the drive in dynamic states.

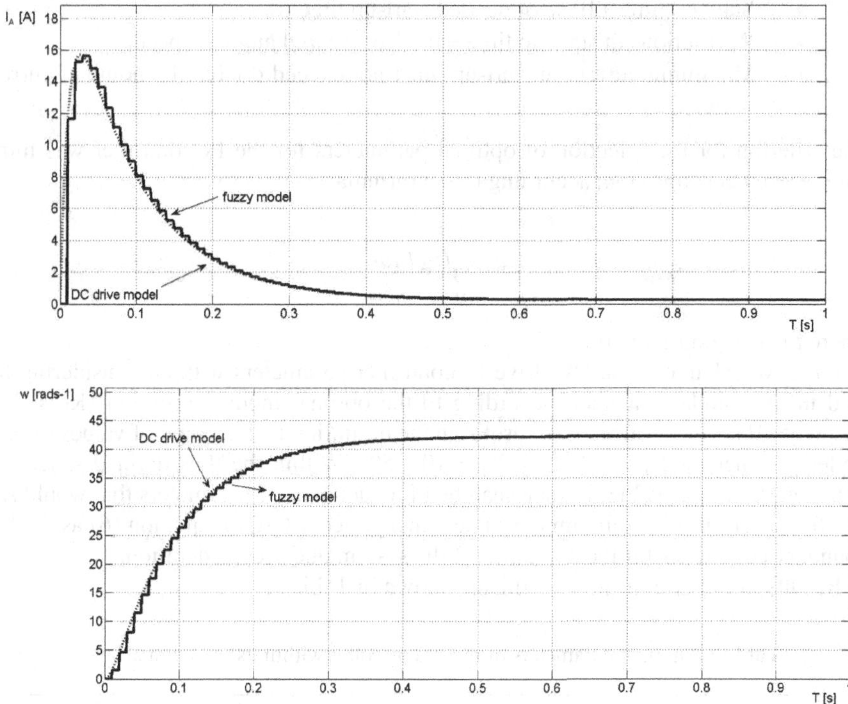

Fig. 6. Comparison of the dynamics of the DC drive and its fuzzy model

3 Optimisation of Drive Controller Parameters

Following the design and the verification of the dynamic system fuzzy model for the system´s operational space, it can be used for establishing the optimal parameters of the selected controller, most often of the PI type.

First we must define the space for the range of controller parameters. One of the ways, in case we have analytical knowledge of the system, is to calculate them in the standard manner and thus obtain information on their initial values, around which we then define their range. For thus calculated controller parameters we then define the energetic consumption of the system motion, with special attention on whether there are also other controller parameters that would be energetically optimal. For this purpose we have to scan the defined set of parameters in cycles and for each given concrete pair of parameters and concrete required input value calculate the energetic consumption of the motion of the system. The data that do not meet the determined boundary conditions are then excluded, and, at the same time, the parameters at which the motion is least energetically demanding are selected. The said algorithm can be implemented for example by means of an m-file created in MATLAB.

Example 2: For the DC drive in Example 1 we defined the following boundary conditions for the dynamics of its motion:

- Start-up time ±10 % of desired start-up time (0.5 s),
- Revolutions in start-up time ±10 % of desired angular speed,
- Maximum current at start-up must not exceed double the nominal current 2*In.

The criterion for the selection of optimal parameters for the PI controller was minimum energy consumption, according to the formula:

$$E = \int_0^t U_A I_A dt \tag{6}$$

where *t* is the simulation time.

First, we calculated the DC drive PI controller parameters without considering the load in the standard manner, according to the optimal module criterion ($K_P = 3.15$ a $K_I = 35.01$). These values were useful in the definition of the range of values of controller parameters (K_P = 1:0.05:4, K_I = 10:1:50). Within the thus defined space, for each concrete input value we then searched for such pair of parameters that would satisfy the criterion for minimum consumed energy according to equation (6) as well as boundary conditions for the dynamics of the system under consideration.

Results of the optimisation m-file are shown in Table 2.

Table 2. Optimal parameters of the PI controller within its operational space

ω_z[rads^{-1}]	K_P	K_I	E_{opt} [J]	E[J]	Saving
20	1.5	12	29.98	39.09	30.38 %
40	1.5	12	61.83	43.30	42.79 %
60	1.55	12	94.56	72.67	30.12 %
80	1.6	12	147.00	118.34	24.21 %
100	1.6	12	156.03	197.07	26.83 %
120	1.6	13	233.40	288.40	23.56 %
140	1.5	14	351.45	410.50	16.80 %
160	1.5	14	469.08	569.00	21.30 %
180	1.55	13	656.04	770.20	17.40 %
200	1.5	14	884.7	1020.10	15.30 %
220	1.5	14	1134.03	1356.30	19.60 %

From the table above it follows that PI controller parameters K_P, K_I (with regard to the fact that this is a system with non-linear load) have to be adjusted according to the magnitude of the desired drive angular speed value. On basis of the measured relations between $\omega_ž \rightarrow K_P$ and $\omega_ž \rightarrow K_I$ in Table 2, using the Anfisedit tool in Matlab, we can create two simple static Sugeno type fuzzy systems (FS K_P and FS K_I), the structure of which (number of rules, fuzzification method and membership functions shape) are shown in Fig. 7.

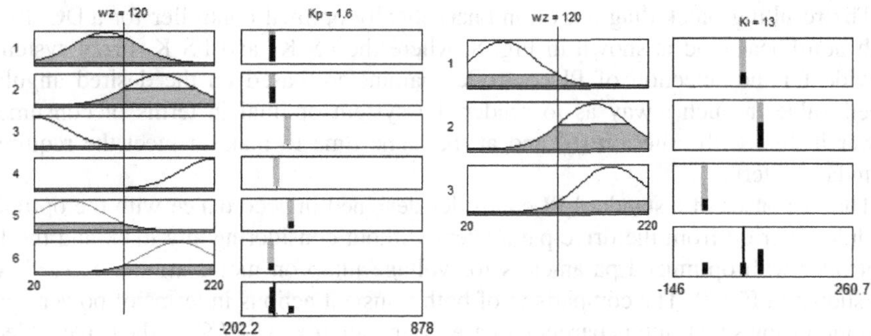

Fig. 7. Structure of fuzzy systems K_P and K_I

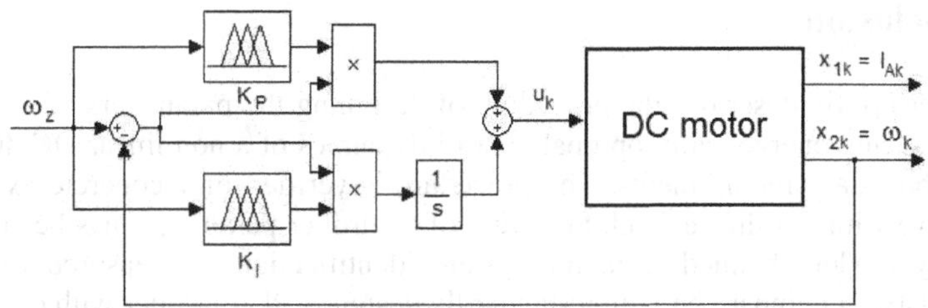

Fig. 8. Optimal fuzzy controller of DC drive with non-linear load

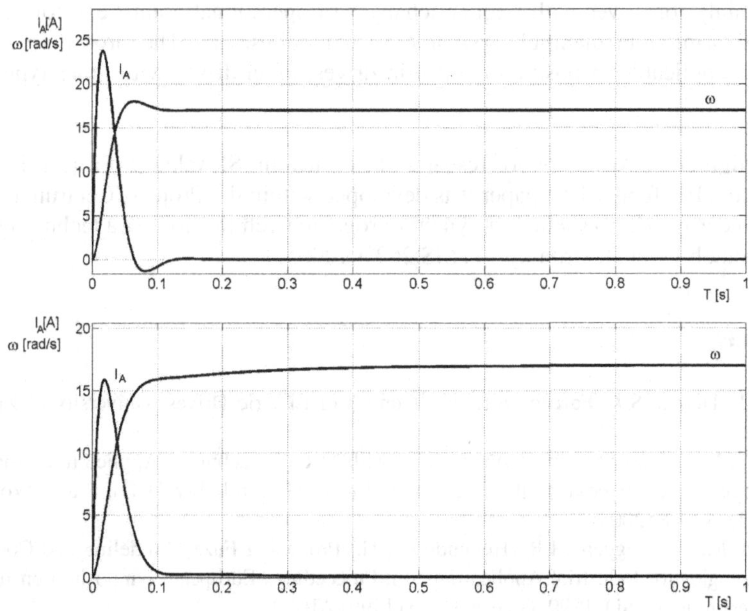

Fig. 9. Responses of a drive with standard and optimized PI controller for voltage jump on motor armature U_A=30V

The resulting block diagram of an energetically optimal controller for a DC drive with non-linear load is shown in Fig. 8, where the FS K_P and FS K_I fuzzy systems provide for the selection of PI controller parameters based on the desired angular speed value in such a way as to render the system optimal in terms of consumed power in line with criterion (6) and at the same time to make it meet the required boundary criteria.

The responses of a standard PI controller designed in accordance with the optimal module criterion from the drive parameters without considering any load, and the PI controller with optimized parameters for voltage jump on motor armature U_A=30V, are shown in Fig. 9. The comparison of both transient actions in terms of power consumption shows that in this particular case saving approx. 31.27 % of the drive´s electric power input is possible.

4 Conclusion

The paper briefly describes the procedure of designing the parameters of a controller that will secure energetically optimal control dynamics of a non-linear DC drive with defined boundary requirements. The procedure is verified on a concrete example of a DC drive with non-linear load. Selection of controller parameters has been done by the fuzzy model obtained from the system identification of measured inputs and outputs data. In comparison with a standardly designed PI controller with constant parameters for the whole operational space of the drive, in the concrete example presented it is possible to save as much as 15.30 % of electric power at each dynamic motion of the drive. This dynamic system controller design method will therefore be suitable mainly for drives with frequent changes of operational points, e.g. for drives of manufacturing line manipulators, of electro mobiles, etc. The procedure is of course also applicable in position control in drives, or in drives with other types of motors.

Acknowledgments. We support research activities in Slovakia / Project is co-financed from EU funds. This paper was developed within the Project "Centrum excelentnosti integrovaného výskumu a využitia progresívnych materiálov a technológií v oblasti automobilovej elektroniky", ITMS 26220120055.

References

1. Vittek, J., Dodds, S.J.: Forced Dynamics Control of Electric Drives. University of Zilina (2003)
2. Krishan, M.M.: Fuzzy Sliding Mode Control with MRAC Technique Applied to an Induction Motor Drives. International Review of Automatic Control (IREACO), Praise Worthy Prize 1(1), 42–48 (2008)
3. Babuška, R., Verbruggen, H.B., Hellendoorn, H.: Promising Fuzzy Modeling and Control Methodologies for Industrial Applications. In: Proceedings European Symposium on Intelligent Techniques ESIT 1999, Greece, Crete (1999) AB-02
4. Takagi, T., Sugeno, M.: Fuzzy identification of systems and its application to modeling and control. IEEE Trans. Systems, Man and Cybernetics 15, 116–132 (1985)

5. Brandstetter, P., Stepanec, L.: Fuzzy Logic in Vector Controlled Induction Motor Drive. In: Proc. of EPE 2003 Conference, Toulouse, France (2003)
6. Bourebia, O., Belarbi, K.: Fuzzy Generalized Predictive Control for Nonlinear Systems with Coordination Technique. International Review of Automatic Control (IREACO), Praise Worthy Prize 2(1), 169–176 (2008)
7. Babuska, R.: Fuzzy Modeling for Control. Kluwer Academic Publishers, Boston (1998)
8. Xu, Y.L.: Fuzzy model identification and self-learning for dynamic systems. IEEE Trans. Systems Man Cybernet., SMC-17 (4) (1997)
9. Žilková, J., Timko, J., Kover, S.: DTC on Induction motor Drive. Acta Technica CSAV 4(56), 419–431 (2011)

Optimal Input Vector Based Fuzzy Controller Rules Design

Pavol Fedor, Daniela Perdukova, and Zelmira Ferkova

Technical University of Kosice, Department of Electrical engineering and mechatronics, Letná 9, 042 00 Kosice, Slovak Republic
daniela.perdukova@tuke.sk

Abstract. The paper deals with the method of design of a fuzzy controller the rules of which are based on generating the optimal input vector using a genetic algorithm. The method is first demonstrated on a simple linear system and is then applied to the start-up of a drive with a three-phase asynchronous motor with constant torque, representing a strongly nonlinear fifth order dynamic system. The proposed controller is verified through simulation using the MATLAB software package. Achieved results present a simple applicability of this proposed procedure for a wide class of nonlinear dynamic black-box systems.

1 Introduction

Fuzzy controllers of mechatronic systems have been finding wider application in technological practice recently thanks to their simple and robust character [1, 2, 3, 4, 5]. They usually have a PID-type structure and the core of their design consists in specifying the rules and membership functions for the particular system controlled [6]. The application of general „metarules“ is not always suitable, especially where higher order nonlinear systems are concerned [7]. If, however, there exists for the system controlled a suitable input signal time sequence that will achieve the control objective in an optimal manner, it is possible to simply design a fuzzy controller for this system on basis of this sequence (of the optimal input control vector of inputs). This vector can be found, for example, with the help of genetic algorithms.

The paper tests the above mentioned method of fuzzification and rules proposal for a PI-type fuzzy controller and its application in a simple linear system, and for a P-type controller in a drive with an asynchronous motor. The quality of the fuzzy controller designed using the above mentioned method is verified in the paper by simulations in MATLAB, followed by the presentation of the results achieved.

2 PI-TYPE Fuzzy Controller Design Method

A PI-type discrete controller is generally described by the equation:

$$u_k = u_{k-1} + q_0 e_k + q_1 e_{k-1} \tag{1}$$

where u_k, u_{k-1} are values of controller output in the relevant sampling steps, e_k, e_{k-1} are values of the control error, and q_0 ,q_1 are parameters of controller . From these follows its possible structure, shown in Fig. 1 [6].

Á. Herrero et al. (Eds.): Int. JointConf. CISIS'12-ICEUTE'12-SOCO'12, AISC 189, pp. 371–380.
springerlink.com

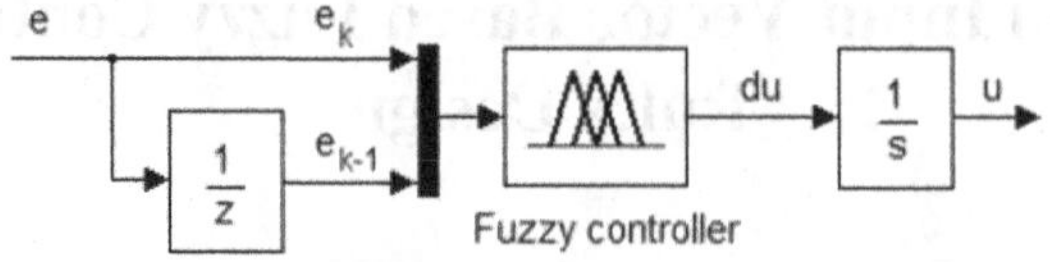

Fig. 1. Block diagram of dynamic system fuzzy model

Its discrete characteristic can then be described by means of the following fuzzy rules:

$$\textbf{IF } e_k \textbf{ is ... AND } e_{k-1} \textbf{ is ... THEN } du_k \textbf{ is ..} \quad (2)$$

where the formulations e_k is ... , e_{k-1} is ... and du_k is ... represent the relevant fuzzy areas of the given variables.

The design procedure of the fuzzy controller for the concrete controlled system can be summarized into the following three steps:

1. *Finding the optimal sequence of input values*

The areas and rules for a fuzzy controller can be found in the appropriate database among the appropriate triplets $[e_k\ e_{k-1}\ du_k]$ which can be collected from the optimal transition of the controlled system into the desired state. This transition corresponds with a particular sequence of input signal values (hereafter input vector $\mathbf{du}_{opt}$) that best complies with the selected optimality criterion. We chose the optimality criterion in the next form of the integral of the quadratic deviation of the system output y from the desired value w:

$$J(e) = \int (w(t) - y(t))^2 \mathrm{d}t \quad (3)$$

Searching for the $\mathbf{du}_{opt}$ vector requires repeated application of various input vectors to the controlled system (or to its model) and the evaluation of the criterion (3) for each one. The large number of possible input sequences is apparently given by the selected input digitalisation in both the time and value axes and it rapidly rises with the density of this digitalisation. For this reason it is advisable to use the genetic algorithm method in the search for vector $\mathbf{du}_{opt}$, as this method substantially speeds up the process.

2. *Finding the database of optimal data*

If we apply the optimal time sequence of inputs $\mathbf{du}_{opt}$ to the controlled system, we obtain the database of optimal data consisting of triplets $[e_k\ e_{k-1}\ du_k]$.

3. *Designing the fuzzy controller from the optimal data database*

From the obtained triplets $[e_k\ e_{k-1}\ du_k]$ of optimal data it is possible to design a concrete fuzzy controller of various types using standard procedures of clustering the data into significant clusters and describing them by means of rules.

Example 1: We need to design a fuzzy controller for a first order dynamic system with transition:

$$\mathrm{F(s)} = 1/(\mathrm{s}+1) \quad (4)$$

1. For finding the optimal input sequence $\mathbf{du}_{opt}$ we shall use the diagram shown in Fig. 2

For the selected input sequence vector **du** as shown in Fig. 3a, the time courses of the individual variables are shown in Fig. 3b (system output variable) and Fig. 3c (control deviation).

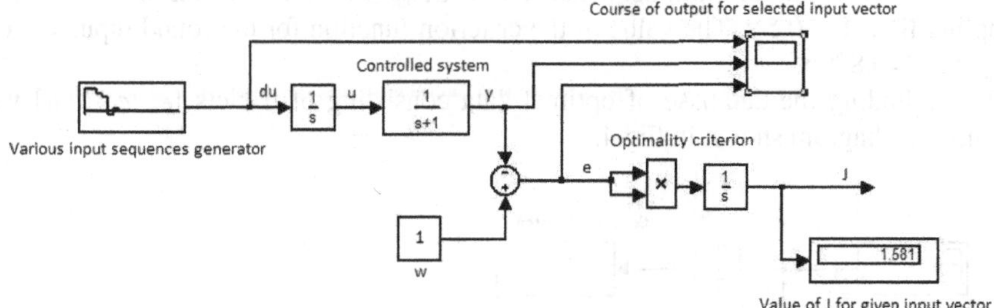

Fig. 2. Block diagram for finding vector $\mathbf{du}_{opt}$

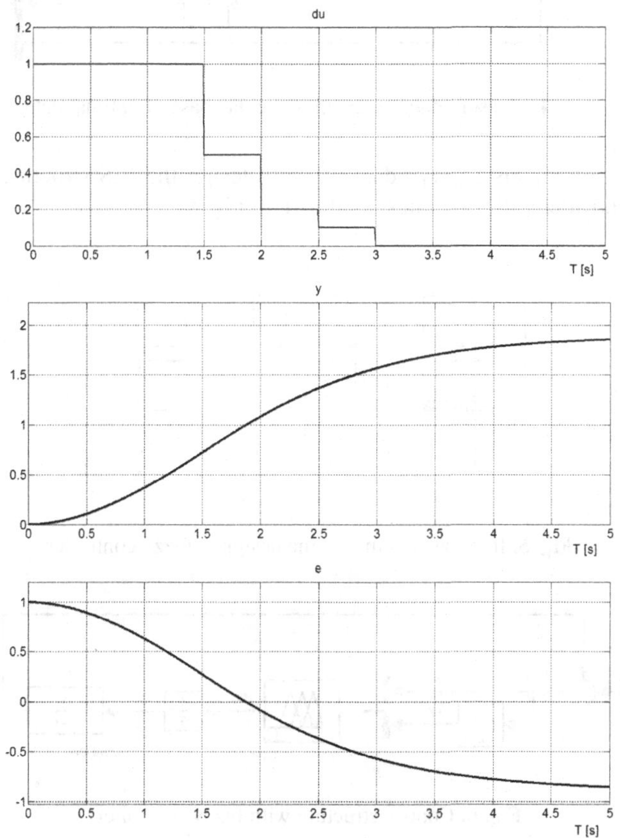

Fig. 3. Time courses of du, y, e

The above courses indicate that this particular adjustment of input sequence **du** is not suitable and the value of its criterion function J=2.187 is too high.

The optimal input vector $\mathbf{du}_{opt}$ can be found using the genetic algorithm method. The genetic algorithm was complemented by the direct search method (hybrid function). We used the criterion function J as the suitability measure (eq.3).

Using the genetic algorithm method we found for dynamic system (1) and desired value w = 1 the optimal input sequence vector $\mathbf{du}_{opt}$ = [1 1 1 -0.8 -0.2 0 0 0 ….] for sampling time T = 0.5 s. The value of the criterion function for the found input vector $\mathbf{du}_{opt}$ was J=0.8896.

2. For finding the database of optimal data consisting of triplets [e_k e_{k-1} du_k] we will use the diagram shown in Fig.4.

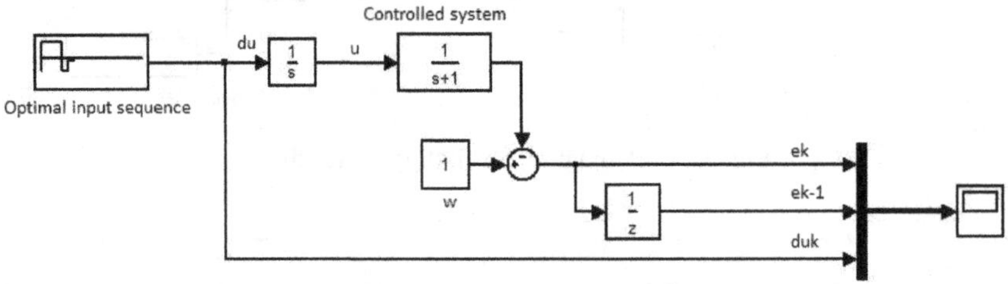

Fig. 4. Measure data for creating the database of optimal data

3. On basis of triplets [e_k e_{k-1} du_k] we will design the FIS structure of a Sugeno type fuzzy controller with two rules, as shown in Fig. 5

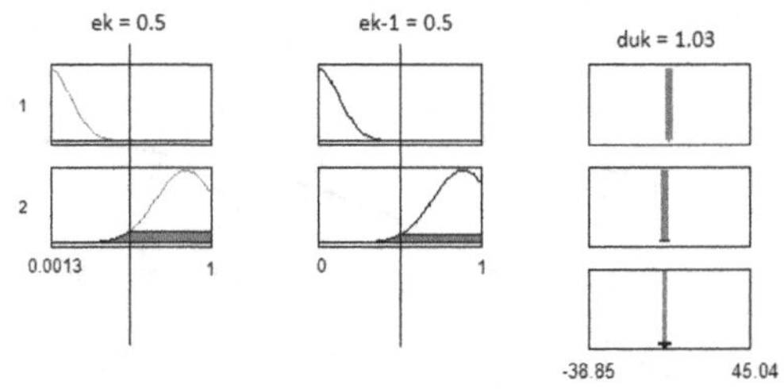

Fig. 5. Internal structure of the designed fuzzy controller

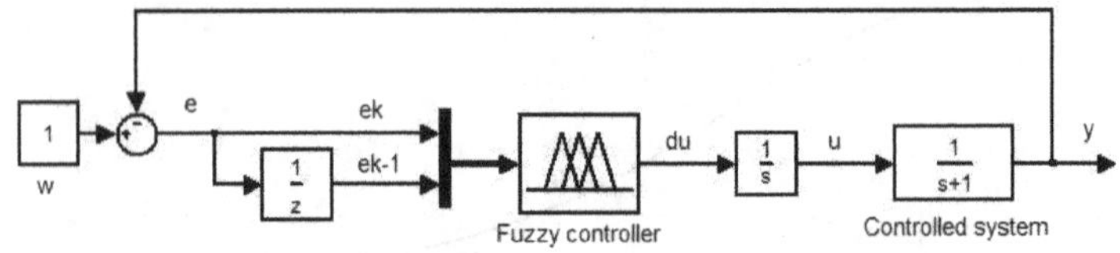

Fig. 6. Control structure with fuzzy controller

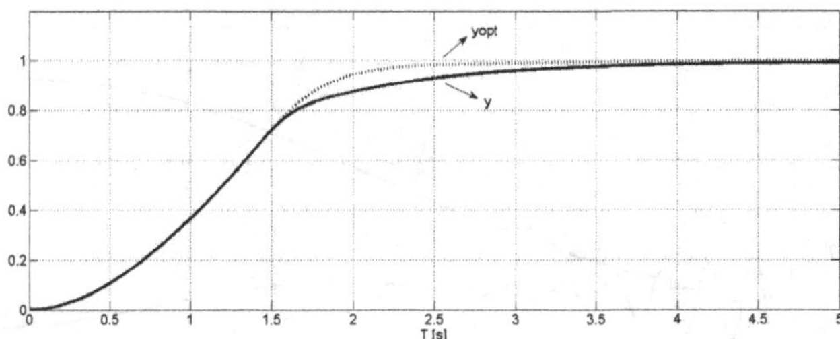

Fig. 7. The dynamics of the optimal and the designed control

The control circuit using the above designed fuzzy controller is shown in Fig. 6. The quality of control of system (1) output variable y using the fuzzy controller compared to its optimal course y_{opt} for input $\mathbf{du}_{opt}$ is demonstrated in Fig. 7.

3 Design of Fuzzy Torque Controller for an Asynchronous Motor Drive

The method described above was applied in an asynchronous motor drive. A three-phase asynchronous motor can be described according to general theory of electric machines in a two-phase right-angled system by a set of five differential equations. If the related system rotates against the stator with speed ω_1 (usually marked by coordinates x, y), the following equations apply for a squirrel-cage motor (see [8, 9]):

$$u_{sx} = R_s i_{sx} + \frac{d\psi_{sx}}{dt} - \omega_s \psi_{sy}$$

$$u_{sy} = R_s i_{sy} + \frac{d\psi_{sy}}{dt} - \omega_s \psi_{sx}$$

$$0 = R_r i_{rx} + \frac{d\psi_{rx}}{dt} - \omega_2 \psi_{ry}$$

$$0 = R_r i_{ry} + \frac{d\psi_{ry}}{dt} + \omega_2 \psi_{rx}$$

$$M_e = \frac{3}{2} p\,(\psi_{ry} i_{rx} - \psi_{rx} i_{ry}) \tag{5}$$

Mathematical symbols:

i_{sx}, i_{sy}	components of stator current space vector i_s
i_{rx}, i_{ry}	components of rotor current space vector i_r
u_{sx}, u_{sy}	components of stator voltage space vector u_s
ω_m	the motor mechanical angular speed
ω_1	angular frequency of the stator voltage
p = 2	number of pole pairs
ω_2	slip angular speed $\omega_2 = \omega_1 - \omega_m$
R_s, R_r	stator and rotor phase resistance
Ψ_s, Ψ_r	stator and rotor magnetic flux
M_e	electrical motor moment

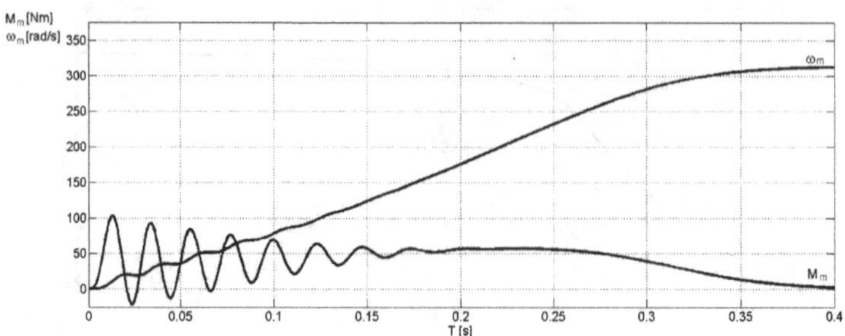

Fig. 8. Direct connection of drive with AM

Let us assume that this motor is powered from a standard (not vector) static frequency converter, the input to which is the desired frequency of the stator voltage vector, and the outputs are the segments of voltages on the individual stator phases. The converter internally secures maintaining of a constant ratio of stator voltage and frequency U_1/ω_1. When connected directly to the power supply network the motor shows a large increase of torque and also of current (Fig 8).

Let the aim of the torque controller design be to adjust the slip ω_2 (i.e. the difference of $\omega_1 - \omega_m$) according to the desired torque value. The structure of the controlled system will be as shown in Fig. 9.

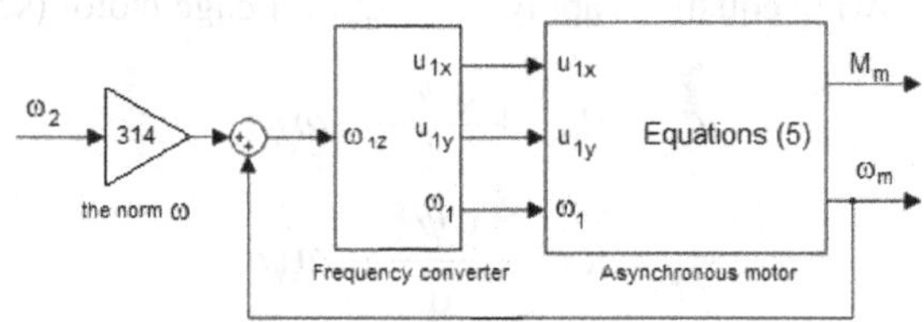

Fig. 9. Structure of the controlled drive

This aim can be achieved by finding the optimal input signal sequence that generates the course of the input stator voltage and frequency by using the genetic algorithm method (see Example 1) for the chosen criterion function according to equation (3). The course of the torque and angular velocity for one such optimal sequence, which every 50 ms adjusts the values of the input vector gradually to [0.165 0.1450.1550.120.1050.095 0 0 0 0 0 0] for the desired torque M_z=30 Nm is demonstrated in Fig. 10.

The said optimal input signal sequence was used for generating the database of triplets $[\omega_{mk}\ \omega_{mk-1}(\omega_1 - \omega_m)_k]$ for the design of the P-type fuzzy controller.

Using the Anfisedit tool in the Matlab programme the database of measured data $([\omega_{mk}\ \omega_{mk-1}(\omega_1 - \omega_m)_k])$ was used in the design of a static Sugeno type fuzzy system, which comprised three rules. The number of zones for fuzzification of the input values and the shape of membership functions for the fuzzy system mentioned is clear from Fig. 11.

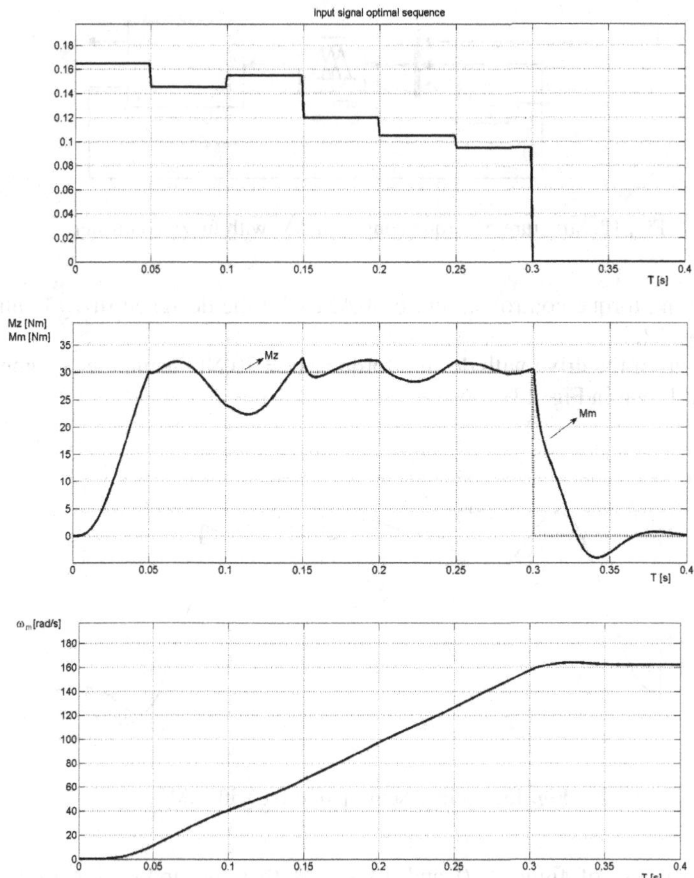

Fig. 10. Optimal start-up of drive with AM

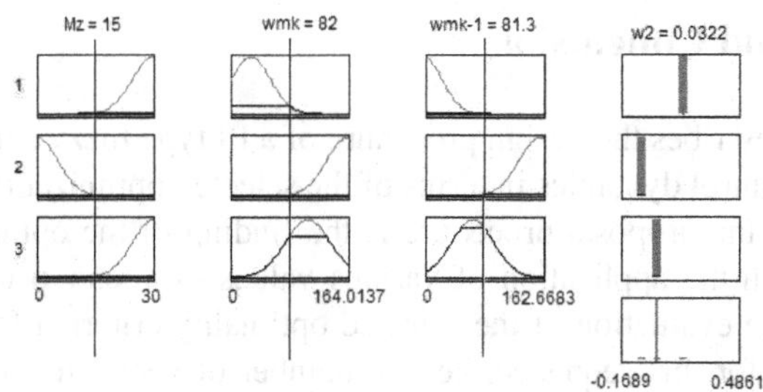

Fig. 11. Structure of AM fuzzy controller

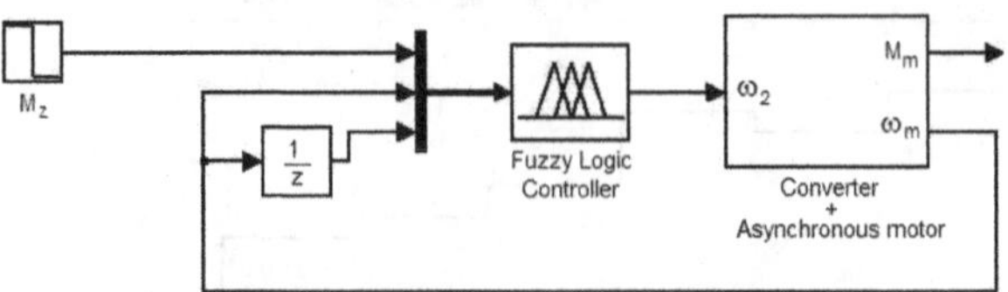

Fig. 12. Structure of torque control in AM with fuzzy controller

The resulting torque control structure of AM with the designed fuzzy controller is shown in Fig. 12.

The start-up of the drive with desired torque Mz = 30 Nm using the designed fuzzy controller is shown in Fig. 13.

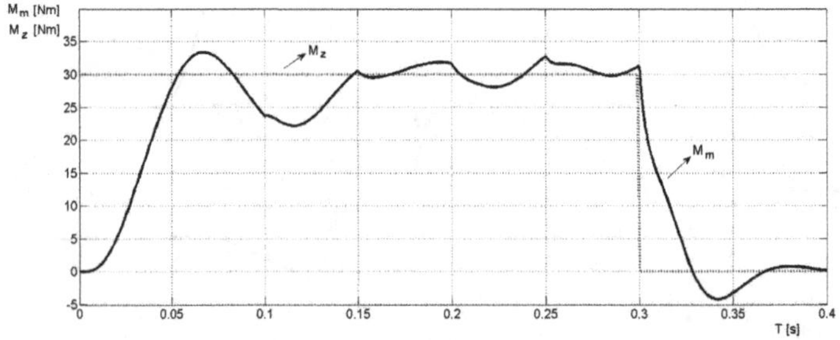

Fig. 13. Optimal start-up of drive with AM

The comparison of figures 10 and 13 shows that the application of the fuzzy controller resulted in the achievement of the desired torque responses of the AM at start-up.

4 Discussion and Conclusion

The paper briefly describes the design procedure of a PI type fuzzy controller that will establish optimal control dynamics in terms of the selected optimization criterion.

The first step of the proposal procedure is the finding of the optimal sequence of input values through the application of various values of vector **u** to the controlled system input and the evaluation of the selected optimality criterion for each of them. The sampling time for this sequence, i.e. the number of vector **u** samples should be determined in terms of the Shanonn-Kotelnik Theorem so that there are at least 10 of them for the whole transition activity of the closed circuit (e.g. in the example with the asynchronous motor which settles in approx. 0.5 s the sampling time of the input sequence was chosen as 50 ms). With the number of samples of the desired vector $\mathbf{u}_{opt}$ also the complexity of the search for it significantly increases, because the number of input vectors that need to be applied to the controlled system input increases. In case a computer model (at least approximate) of the controlled system is available, the whole process of searching for $\mathbf{u}_{opt}$ can be fully automated using the genetic algorithm method, as the case was in the examples in this paper. In case we have to deal with a real controlled system, we first either have to generate some type of its computer model from data measured through experiment, or draw on the experience of experts for finding and verifying some suitable sequence of **u** inputs that would become the basis for constructing a fuzzy controller with similar behavior.

In the second step of the proposal procedure, the input sequence $\mathbf{u}_{opt}$ is applied to the controlled system and through measurements of suitable variables a database is formed for the fuzzy controller design. The selection of variables for the database and the time of their sampling depend on the specific aims of the control, on the measurability of the controlled system variables, etc. It is possible to measure various different types of databases (ref. the various table forms in the paper) and design various fuzzy controllers for the given system. In all the cases, however, this task can again be fully automated.

The third step of the proposal procedure involves the design of fuzzy controller parameters from the obtained database, keeping in mind that the fuzzy controller is to behave in the same way as the optimal input vector $\mathbf{u}_{opt}$ and will thus ensure an optimal course of the selected variables of the controlled system. This step involves the use of standard clustering methods and computer tools and can also be fully automated.

The proposed procedure is verified on a concrete example of an asynchronous motor drive. Results of simulation experiments show that the controller, in spite of its simplicity and the uncomplicated computer oriented design procedure used, enables considerable improvement of the control circuit dynamic properties also in case of strongly nonlinear higher order controlled systems.

Acknowledgments. The financial support of the Slovak Research and Development Agency under the contract No. APVV 0138-10, is acknowledged.

References

1. Takagi, T., Sugeno, M.: Fuzzy identificationof systems and its application to modeling and control. IEEE Trans. Systems, Man and Cybernetics 15, 116–132 (1985)
2. Guillemin, P.: Fuzzy logic applied to motor control. IEEE Trans. Ind. Appl. 32(1), 51–56 (1996)
3. Li, W., Chang, X.G., Farrell, J., Wahl, F.M.: Design of an enhanced hybrid fuzzy P+ID controller for a mechanical manipulator. IEEE Trans. Syst., Man, Cybern., B, Cybern. 31(6), 938–945 (2001)
4. Barrero, F., Gonzalez, A., Torralba, A., Galvan, E., Franquelo, L.G.: Speed control of induction motors using a novel fuzzy sliding-mode structure. IEEE Trans. Fuzzy Syst. 10(3), 375–383 (2002)
5. Tsourdos, A., Economou, J.T., White, A.B., Luk, P.C.K.: Control design for a mobile robot: A fuzzy LPV approach. In: Proc. IEEE Conf. Control Applications, Istanbul, Turkey, pp. 552–557 (2003)

6. Žalman, M., Jovankovič, J.: Intelligent servosystems. STU, Bratislava (2007)
7. Babuska, R.: Fuzzy Modeling for Control. Kluwer Academic Publishers, Boston (1998)
8. Brandstetter, P.: A.C. Controlled Drives - Modern Control Methods. VSB-Technical University of Ostrava (1999)
9. Vittek, J., Dodds, S.J.: Forced Dynamics Control of Electric Drives. University of Zilina, Zilina (2003)

Correlation Methods of OCR Algorithm for Traffic Sign Detection Implementable in Microcontrollers

Radim Hercik, Roman Slaby, Zdenek Machacek, and Jiri Koziorek

VSB - Technical University of Ostrava

Abstract. This paper focuses on the correlation methods applicable for the recognition system of speed limit traffic signs and correlation methods comparison. The correlation method is one possible manner of the OCR algorithm (Optical Character Recognition) used to determine the degree of similarity between the input matrix and the defined pattern matrix. The presented correlation methods are verified using the proposed comparison algorithm, where the output data are evaluated by statistical methods of the exploratory statistic data analysis. The part of the recognition system for OCR algorithm proceeds is very time consuming and the limitation of microcontroller type depends on frequency instruction processing. High accuracy of the recognition system can be achieved by increasing the resolution of camera system, by segmentation methods of input image signal, correlation method type.

Keywords: image processing, correlation method, Statistical method, OCR algorithm.

1 Introduction

The automotive industry is one of the greatest area for implementing research and developing applications. The automotive companies motivation is to attract the customers and offer the variation of its products. It is reason for support the development to facilitate orientation in unfamiliar surroundings or to alert drivers on the possible risk occurrences. The presented paper is focused on the possible risk of driver oversight of traffic signs and specified speed limit. This recognition system can be realized for extension of GPS navigation system. The developed recognition system methods are implemented for the traffic signs identification and the speed limit detection used OCR algorithm [4]. The result of the presented algorithm methods are degree of similarity of recognized value of the speed limit from the image signal. The used patterns of vertical traffic signs and inside limits digits (dimension, symbols, font characters) are defined by European Standard EN12899-1: 2001. These patterns are compared with image signals from camera on the road by correlation methods implemented for OCR algorithm.

2 Recognition System Description for Traffic Signs Detection by Correlation Image Processing Methods

The recognition scene is scanned by a camera located in front of the car. In the first step, the scanned image is proceed by image segmentation, which is represented by

Á. Herrero et al. (Eds.): Int. JointConf. CISIS'12-ICEUTE'12-SOCO'12, AISC 189, pp. 381–389.
springerlink.com

the set of methods for image diversification on the parts with similar characters. After detection of area with speed limit digits, the area is proceeded by correlation methods to determine the degree of similarity of data patterns stored in the database. The block diagram illustrates the sequence with the image processing and it is shown in Fig. 1.

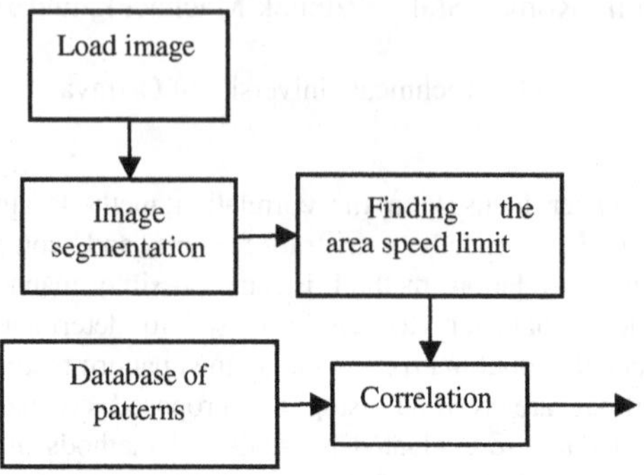

Fig. 1. The block diagram with the image processing sequence

The implemented patterns database includes alphanumeric digits, where their font is specified by the European Standard presented above. The result of the correlation methods is a correlation parameter *p*, which indicates the degree of similarity. This parameter *p* is defined in range from 0 to 1, where value 0 means the minimal similarity of the input matrix and specified pattern and the value 1 means the maximal similarity of input image and pattern comparison [3]. The highest consensus of recognized value determines the recognized speed limit of the traffic sign in input image. The correlation method type can significantly affect results of the recognition system. The presented correlation methods are suitable for microcontroller implementation, because of the elementary algorithms.

3 Direct Correlation Function Method

The direct correlation function method (method 1) calculates the similarity ratio p between matrix A and matrix B, where the matrixes size are the same dimension units. The correlation method can be applied only to binary data matrix. Coefficient p of similarity is calculated by equation (1).

$$p = \frac{\sum_{x}\sum_{y}\left(A_{(x,y)} \times B_{(x,y)}\right)}{x \cdot y} \tag{1}$$

where $A_{(x,y)} \times B_{(x,y)}$ is an exclusive logical sum (XNOR) [4].

4 Correlation Coefficient Function Method

The correlation coefficient function method (method 2) calculates the correlation coefficient p between matrix A and matrix B, where the matrixes size are the same dimension units.. The correlation coefficient p is calculated by equation (2) [2].

$$p = \frac{\sum_x \sum_y \left(A_{(x,y)} - \overline{A}\right) \cdot \left(B_{(x,y)} - \overline{B}\right)}{\sqrt{\left(\sum_x \sum_y \left(A_{(x,y)} - \overline{A}\right)^2\right) \cdot \left(\sum_x \sum_y \left(B_{(x,y)} - \overline{B}\right)^2\right)}} \tag{2}$$

where $\overline{A}$ is the arithmetic average of all elements of the matrix A and $\overline{B}$ is the arithmetic average of all elements of the matrix B.

5 Descript of the Comparison Methodology of Implemented Correlation Methods

The comparison of the correlation methods suitability is based on the statistical methods of exploratory statistic data analysis. Evaluation is obtained using the matrix algorithm implemented in mathematical programming environment MATLAB, which is used to sophisticate calculation, especially for research purpose, for modeling, simulation and signal processing. There is utilized Statistic Toolbox, which supports a wide range of tasks statistical calculations and algorithms [1].

There are presented obtained results from comparison algorithms, with calculation structure, which is shown by flowchart in Figure 2. In the first step, the comparison algorithm initializes specific variables and it reads each input matrix. Each input real image matrix is compared with the patterns by correlation methods, where the correlation coefficient indicates the degree of their similarity. The chosen scanned digit is given by the result of the chosen maximal computed correlation coefficient.

The comparison of the correlation methods is realized on the input images, which are represented by database of the scanned camera images (part with digit). The matrixes of database of alphanumeric digit symbols are shown in Fig. 3 with order number in Table 1.

The various correlation methods need matrix digit patterns which are shown in Fig. 4. The digit patterns represent the alphanumeric characters from 0 to 9, which are located in the database. The font type is realized by the European Standard.

alization of
ınd array data

for x = 1:NumberInput

input
image

for y = 1:Numberpattern

load pattern

calculating similarity
p1 for Method 1

calculating similarity
p2 for Method2

if
p1>data(x,1)

lata(x,1)=p1

if
p2>data(x,2)

data(x,2)=p2

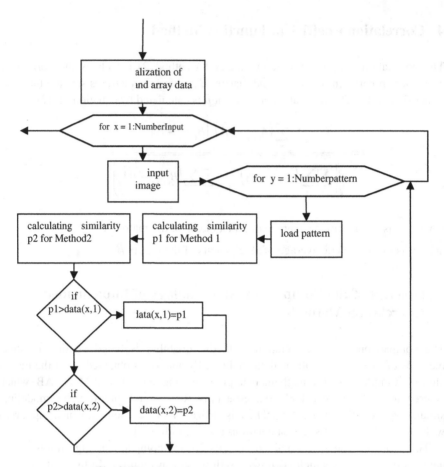

Fig. 2. Flowchart of comparison algorithms structure

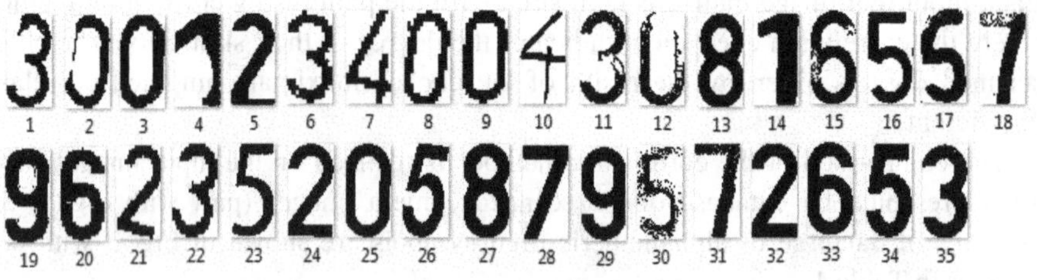

Fig. 3. Database of input digits from the scaned camera images

1234567890

Fig. 4. Matrix digit patterns in database

6 Statistical Methods for Comparison of the Correlation Function Methods

There are presented statistical calculations for correlation function methods comparison, where the results can lead to choose of the better method type. There are presented measured data of 35 images of the limit speed digits. Next column presents result of successful recognition of digit and there is shown the correlation coefficient for each correlation method in Table 1. There are label the values of bad identified digit with presented recognition algorithm.

The arithmetic mean $\overline{p}$ is a known diameter used for the quantitative variable. There is presented the algorithm and results from given example.

$$\overline{p_1} = \frac{\sum_{i=1}^{n} p_{1i}}{n} = 0{,}8238 \tag{3}$$

$$\overline{p_2} = \frac{\sum_{i=1}^{n} p_{2i}}{n} = 0{,}6594 \tag{4}$$

where p_{1i}, p_{2i} are individual values of correlation function of detected and pattern image. The value n is number of all digits from the scanned camera images, which are placed in table 1.

There are presented the results of arithmetic mean computation $\overline{p}$ of compared correlation methods, where the direct correlation function method is more visible compare to the correlation coefficient function method.

The variance s^2 of correlation coefficients is calculated from dispersion derived from the average of the the squared deviations from the arithmetic mean. It is determined by the relations (5) (6).

$$s_1^2 = \frac{\sum_{i=1}^{n}\left(p_{1i} - \overline{p_1}\right)^2}{n-1} = 0{,}0097 \tag{5}$$

$$s_2^2 = \frac{\sum_{i=1}^{n}\left(p_{2i} - \overline{p_2}\right)^2}{n-1} = 0{,}0372 \tag{6}$$

The standard deviation s is defined as the positive square root of sample variance, and it id defined by equations (7), (8).

$$s_1 = \sqrt{s_1^2} = \sqrt{\frac{\sum_{i=1}^{n}(p_{1i} - p_1)^2}{n-1}} = 0{,}0985 \tag{7}$$

$$s_2 = \sqrt{s_2^2} = \sqrt{\frac{\sum_{i=1}^{n}(p_{2i} - p_2)^2}{n-1}} = 0{,}1928 \tag{8}$$

Competitive disadvantage of the variance and standard deviation is the fact that they do not compare the variability of the variables expressed in different units, but this is not our case. This deficiency, however, but solves the coefficient of variation.

The coefficient of variation V_x is a characteristic variation of the probability distribution of random variables. As shown in equation (9), (10) it can be determined only for variables that takes only positive values. Coefficient of variation is dimensionless.

$$V_{1x} = \frac{s}{\overline{p_1}} = 0{,}1196 \tag{9}$$

$$V_{2x} = \frac{s}{\overline{p_2}} = 0{,}2923 \tag{10}$$

The quantile correlation coefficients are values in the statistics, which divides the sample into several equal parts, or characterize the position of each value in the variable. Quantiles are resistant to outlying observations. For determining the quantile is necessary to hold the sample size and from the smallest to largest. The lower quartile is 20% quantile dividing data file so that 25% of the values are smaller than this quartile. Statistical population divides into two equally large sets of numerical median. For an even number of values of the sample median is uniquely defined as the median with and consider any number between two intermediate values, including those values. Mostly, there is specified average between the two middle values. The upper quantile divides the data set so that 75% of the values are smaller than this quantile [5].

Quantile computation for correlation method 1:

$$kv_{25(p1)} = 0{,}7701 \doteq 1$$

$$kv_{50(p1)} = 0{,}8433 \doteq 1$$

$$kv_{75(p1)} = 0{,}9165 \doteq 1$$

Quantile computation for correlation method 2:

$$kv_{25(p2)} = 0{,}5641 \doteq 1$$

$$kv_{50(p2)} = 0{,}7022 \doteq 1$$

$$kv_{75(p2)} = 0{,}8403 \doteq 1$$

IQR - Interqantile range of correlation coefficients is defined as the difference between the upper and lower quartile (11) (12).

$$IQR_{p1} = x_{(p1)0,75} - x_{(p1)0,25} = 0,1464 \doteq 0 \tag{11}$$

$$IQR_{p2} = x_{(p2)0,75} - x_{(p2)0,25} = 0,2762 \doteq 0 \tag{12}$$

MAD of the correlation coefficients is determined on the base of the arrangement of sample size and by determination of the median. For a univariate data set *X1, X2, ..., Xn*, the MAD is defined as the median of the absolute deviations from the data's median (13)

$$MAD = median_i\left(\left|X_i - median_j(X_j)\right|\right) \tag{13}$$

that is, starting with the residuals (deviations) from the data's median, the MAD is the median of their absolute values. The next step is to calculate the absolute values of deviations from the median value for each sorted sample. Furthermore, there is determine the median from sorted absolute deviations.

$$MAD_{p1} = 0,0813$$

$$MAD_{p2} = 0,1532$$

Box plots have been used since 1977 for a graphical representation of the shape of the distribution, the mean and variability. Graphic version of this graph is slightly different in various applications. The horizontal line inside the box indicates the median and box boundaries are defined by upper and lower quantile [2],[5]. Outliers are shown as isolated points. Box plots of both correlation methods are shown in Fig. 5.

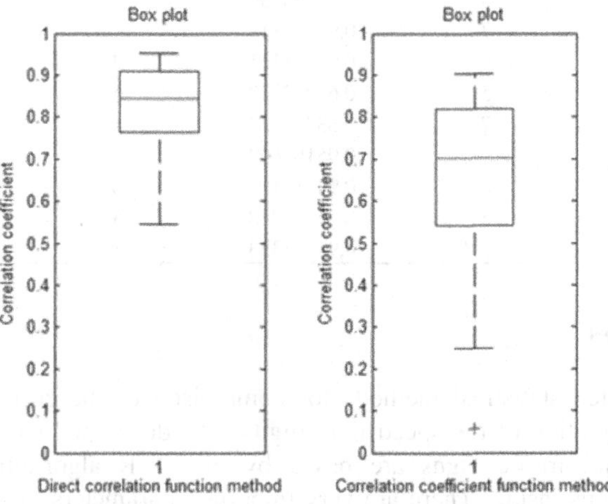

Fig. 5. Box plot of the direct correlation function method and box plot of the correlation coefficient function method

Table 1. Examples of calculated data which represents recognition results of speed limit digits

NO.	The real digit represented speed limit	Direct correlation function method		Correlation coefficient function method	
		Recognized digit	Correlation coefficient p_1	Recognized digit	Correlation coefficient p_2
1	3	3	0,80222222	3	0,62527363
2	0	3	0,77388889	0	0,56973409
3	0	0	0,88277778	0	0,77196623
4	1	1	0,88944444	1	0,80018254
5	2	2	0,92888889	2	0,86661210
6	3	3	0,84333333	3	0,70215603
7	4	4	0,72611111	4	0,44248296
8	0	0	0,78055556	0	0,60654130
9	0	0	0,85111111	0	0,72780759
10	4	4	0,54555556	5	0,05858087
11	3	3	0,83722222	3	0,69108661
12	0	0	0,70500000	0	0,51800402
13	8	8	0,89777778	8	0,78251524
14	1	1	0,83888889	1	0,67762109
15	6	6	0,56111111	6	0,73502759
16	5	5	0,79722222	5	0,62503498
17	5	5	0,76166667	6	0,53333959
18	7	7	0,71944444	7	0,41079431
19	9	9	0,87277778	9	0,73243781
20	6	6	0,91388889	6	0,82434668
21	2	2	0,76888889	2	0,56832326
22	3	3	0,79444444	3	0,59696663
23	5	5	0,72000000	5	0,49289727
24	2	2	0,91388889	2	0,83652378
25	0	0	0,89111111	0	0,78898813
26	5	5	0,72055556	5	0,45383958
27	8	8	0,93055556	8	0,84925183
28	7	7	0,95111111	7	0,89909460
29	9	9	0,91611111	9	0,82512159
30	5	5	0,68222222	6	0,46145130
31	7	7	0,63722222	1	0,24782331
32	2	2	0,95166667	2	0,90329306
33	6	6	0,94611111	6	0,88745634
34	5	5	0,86777778	5	0,73161168
35	3	3	0,91833333	3	0,83648054

7 Conclusions

There are presented statistical methods for comparison of the correlation function methods for recognition of the speed limit digits. The developed recognition system of the speed limit traffic signs are based by the OCR algorithms respectively correlation functions inside. There are very important parameters of processing time and accuracy and correctness and recognition success. The accuracy of recognition depends on the choice of methods for OCR algorithm, in our case the correct choice of the correlation method. The suitability of various methods for recognition systems are tested on the basis of statistical methods of exploratory data analysis statistics.

From the presented calculated values there is evident that the sample variance for the calculation of the correlation coefficient is much higher than for the direct correlation function method. This means that this function does not provide stable results and leads to much greater differences in accuracy. Using the method of calculation of correlation coefficients is not guaranteed to improve the accuracy of the recognition system.

In all evaluated parameters the direct correlation function method (method 1) reached better and more stable results. It is therefore preferable to implement into OCR algorithms.

Acknowledgments. The work was supported by project SP2012/111 - Data collection and data processing of large distributed systems II. Grand – aided student R. Hercik, Municipality of Ostrava, Czech Republic.

References

1. Ozana, S., Machacek, Z.: Implementation of the Mathematical Model of a Generating Block in Matlab&Simulink Using S-functions. In: The Second International Conference on Computer and Electrical Engineering ICCEE. Session 8, pp. 431–435 (2009)
2. Hlavac, V., Sedlacek, M.: Zpracování signálu a obrazu, 255 p. BEN, Praha (2007) ISBN 978-80-01-03110-0
3. Gibson, J.D.: Handbook of Image & Video Processing, 891 p. Academic Press, London (2000)
4. Machacek, Z., Hercik, R., Slaby, R.: Smart User Adaptive System for Intelligent Object Recognizing. In: Nguyen, N.T., Trawiński, B., Jung, J.J. (eds.) New Challenges for Intelligent Information and Database Systems. SCI, vol. 351, pp. 197–206. Springer, Heidelberg (2011)
5. Bris, R.: Exploratorní analýza proměnných. Ostrava. Scriptum. VŠB-Technical University of Ostrava
6. Krejcar, O., Jirka, J., Janckulik, D.: Use of Mobile Phone as Intelligent Sensor for Sound Input Analysis and Sleep State Detection. Sensors 11, 6037–6055 (2011)

Takagi-Sugeno Fuzzy Model in Task of Controllers Design

Jana Nowaková, Miroslav Pokorný, and Martin Pieš

VŠB-Technical University of Ostrava,
Faculty of Electrical Engineering and Computer Science,
Department of Cybernetics and Biomedical Engineering,
17. listopadu 15/2172, 708 33 Ostrava Poruba, Czech Republic
{jana.nowakova,miroslav.pokorny,martin.pies}@vsb.cz

Abstract. The designing of PID controllers is in many cases a discussed problem. Many of the design methods have been developed, classic (analytical tuning methods, optimization methods etc.) or not so common fuzzy knowledge based methods. In this case, the amount of fuzzy knowledge based methods is extended. New way of designing PID controller parameters is created, which is based on the relations of Chien, Hrones and Reswick design method (the modification of Ziegler-Nichols step response method). The proof of efficiency of the proposed method and a numerical experiment is presented including a comparison with the conventional Chien, Hrones and Reswick method (simulated in the software environment Matlab-Simulink). It is defined a class of systems for which - using the new proposed method - the settling time is shorter or the settling time is nearly the same but without overshoot, which could be also very useful.

Keywords: Expert system, knowledge base, PID controller, Ziegler-Nichols design methods, Chien, Hrones and Reswick design method, fuzzy system, feedback control.

1 Introduction

As it is written in [1] two expert system approaches exist. The first one, the fuzzy rule base way for controlling processes for which suitable models do not exist or are inadequate. The rules substitute for conventional control algorithms. The second way originally suggested in [2] is to use an expert system to widen the amounts of classical control algorithms. So this paper extends the number of methods of the second way of design.

The system of designing a PID controller (its parameters) with a knowledge base is created which is built on know-how obtained from the Chien, Hrones and Reswick design method which is a of the Ziegler-Nichols step response design method. The system created in this way is determined to the design parameters of a classical PID controller, which is considered in closed feedback control.

Á. Herrero et al. (Eds.): Int. JointConf. CISIS'12-ICEUTE'12-SOCO'12, AISC 189, pp. 391–400.
springerlink.com

2 Ziegler-Nichols and Chien, Hrones and Reswick Design Methods

In 1942 Ziegler and Nichols presented two classical methods for the identification of the parameters of PID. Both are based on determination of process dynamics. The parameters of the controller are expressed as a function by simple formulas [3],[4].

The transfer function of the PID controller using these methods and its modification is expressed as

$$G_R(s) = K\left(1 + \frac{1}{T_i\,s} + T_d\,s\right). \tag{1}$$

Furthermore, only one of Ziegler-Nichols' method will be used and its modification - Chien, Hrones and Reswick method.

2.1 The Chien, Hrones and Reswick Method

Chien, Hrones and Reswick (CHR) modified the step response method to achieve a better damped closed-loop system. There are two variants - the setpoint response method and load disturbance response method. And both of these variants have two tuning criteria - the quickest response without overshoot and for the quickest response with a 20% overshoot [3].

For the load disturbance response method the parameters of the controllers are expressed as functions of two parameters Table 1 as the Zeigler-Nichols step response method. The variant for a 20% overshoot is quite similar to the Ziegler-Nichols step response method [3].

Table 1. PID controller parameters obtained from the CHR load disturbance response method [3]

	0% overshoot			20% overshoot		
	K	T_i	T_d	K	T_i	T_d
PID	$0.95/a$	$2.4L$	$0.42L$	$1.2/a$	$2L$	$0.42L$

For the setpoint response method more than two parameters are used - a and L, but it is needed to use a time constant delay time D, Table 2.

For understanding it is important to explain constants a, L and D and how to obtain these constants from the unit step response of the controlled system.

The dead time L has the same importance as in the Ziegler-Nichols step response design method as well the constant a and are obtained so that the tangent is drawn at the maximum of the slope of the unit step response Figure 1. The intersection of the tangent with axis y and the distance of this intersection and the beginning of the coordinate axis determine parameter a. The dead time

Table 2. PID controller parameters obtained from the CHR setpoint response method [3]

	0% overshoot			20% overshoot		
	K	T_i	T_d	K	T_i	T_d
PID	$0.6/a$	D	$0.5L$	$0.95/a$	$1.4D$	$0.47L$

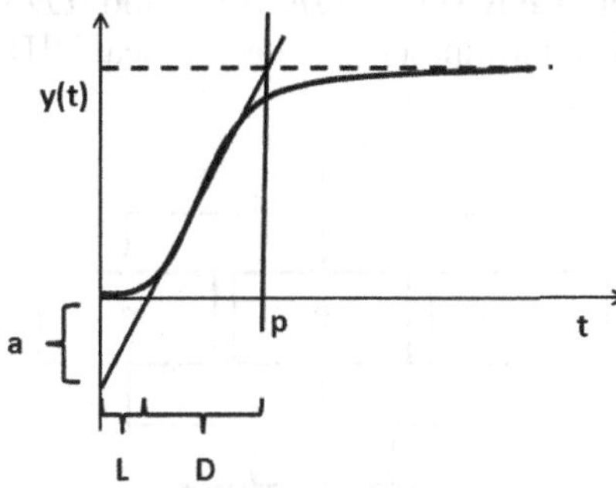

Fig. 1. Unit step response of the controlled system with the representation of parameter a, the dead time L and the delay time D [3],[5]

L is the distance of the intersection of the tangent with time axis (axis x) and also the beginning of the coordinate axis [3], [4]. The delay time D is the distance between the intersection of the tangent and the time axis and the point p on the time axis. The point p is created so that in the intersection of the tangent and the maximum of the unit step response of the controlled system is the perpendicular to the time axis raised and the intersection of the perpendicular and the time axis is the point p [3]. Point p could be also determined in another way; the difference is that the intersection of the tangent is done with 63% of the maximum of the unit step response of the controlled system. The remaining procedure is the same. In this case the first way of achieving point p is chosen. [3],[5]

3 The Description of the Controlled System

The first task is to describe the controlled system. It could be defined in many ways; in this case the controlled system is defined by a transfer function. The determining parameters are constants A, B and C from the denominator of the transfer function of the controlled system in the form

$$G_S(s) = \frac{1}{As^2 + Bs + C}. \tag{2}$$

The constants *A*, *B* and *C* are the inputs variables of the expert system which as it is explained furthermore, is used for identifying the parameters of a classical PID controller.

4 Expert System of Identification of Parameters of a PID Controller

The constants *A*, *B* and *C* as the inputs of the expert system are represented in Figure 2 so as the outputs *KKNOW*, *TIKNOW* and *TDKNOW* which are also constants and are used as parameters of the classical PID controller.

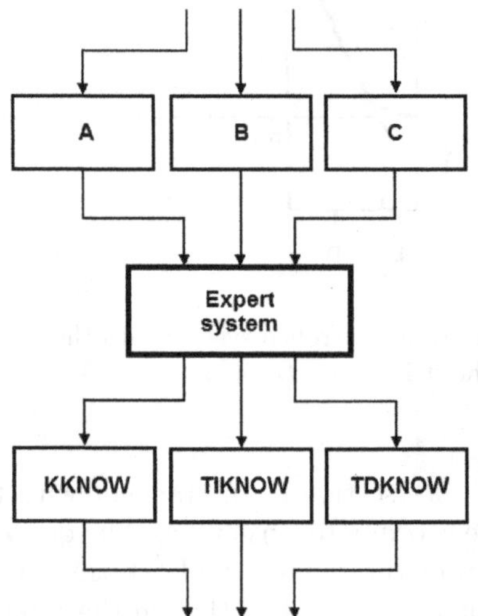

Fig. 2. The representation of inputs and outputs of the created expert system.

4.1 Description of Linguistic Variables

The inputs of expert system *A*, *B* and *C* are the linguistic variables expressed by fuzzy sets [6], for each linguistic variable by three linguistic values - small (S), medium (M) and large (L) Figure 3, 4, 5. The constant *A* can be from the range from 0 to 22, the *B* from 0 to 20 and the *C* from 0 to 28.
The membership functions of all linguistic values have a triangular shape. The shape could be described in three numbers. The first one is the point of the triangle where the membership degree is equal to zero, the same as the third number, and second number is the point, where the membership degree is equal to one.

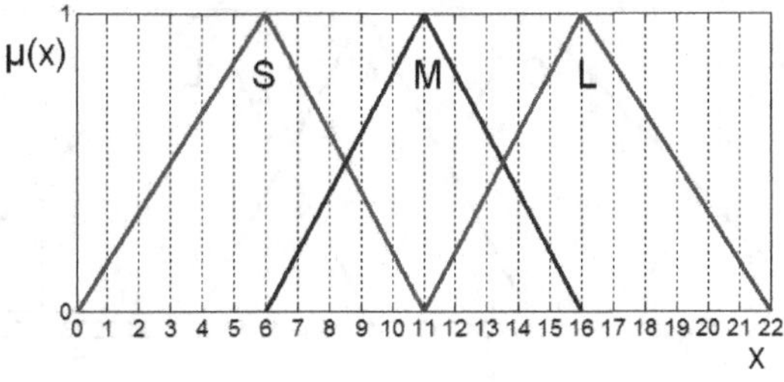

Fig. 3. The shape of the membership function of linguistic values for input linguistic variable A

Description of the shape of the linguistic values of the linguistic variable A:

- Small (S) - [0 6 11],
- Medium (M) - [6 11 16],
- Large (L) - [11 16 22].

So according to the description of the importance of three numerical descriptions, the membership function of linguistic value M of input linguistic variable A has the shape of an isosceles triangle, and shape of the membership functions for linguistic value S and L is a general triangle.

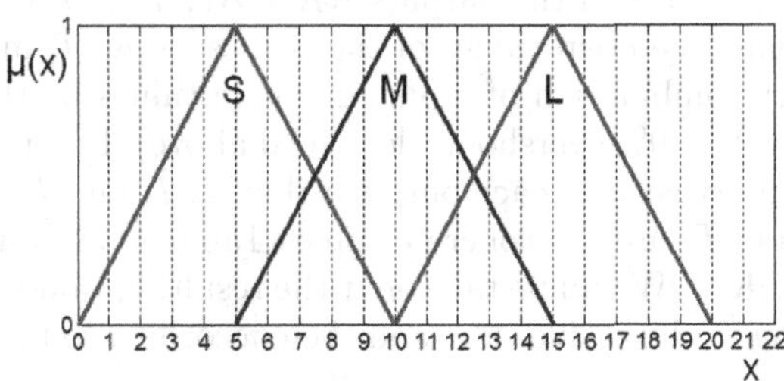

Fig. 4. The shape of the membership function of linguistic values for input linguistic variable B

Description of the shape of the linguistic values of the linguistic variable B:

- Small (S) - [0 5 10],
- Medium (M) - [5 10 15],
- Large (L) - [10 15 20].

Membership functions of all linguistic values of linguistic variable B have the shape of a isosceles triangle.

Description of the shape of the linguistic values of the linguistic variable C:

- Small (S) - [0 9 14],
- Medium (M) - [9 14 19],

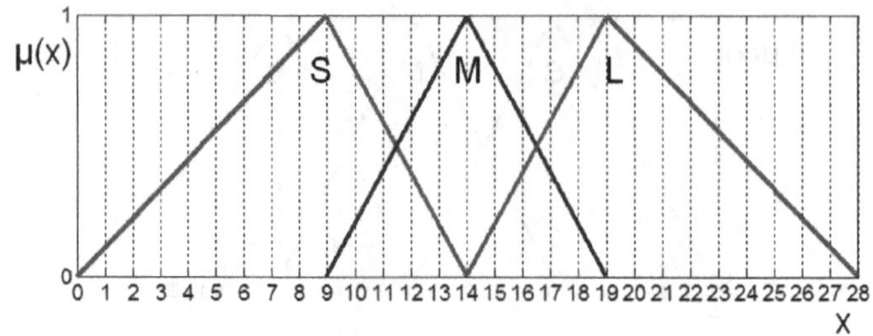

Fig. 5. The shape of the membership function of linguistic values for input linguistic variable C

– Large (L) - [14 19 28].

The membership function of linguistic value M of input linguistic variable C has the shape of a isosceles triangle, and the others are general triangles.

Outputs of the expert system are constants *KKNOW*, *TIKNOW* and *TDKNOW* which are parameters of the classical PID controller with a transfer function in a form

$$G_R(s) = KKNOW\left(1 + \frac{1}{TIKNOW \cdot s} + TDKNOW \cdot s\right). \tag{3}$$

For determining the values of the outputs *KKNOW*, *TIKNOW* and *TDKNOW* it is needed to define the crisp values of the constants K, T_i and T_d, which are identified for every combination of input linguistic values by the CHR setpoint response method with a 0% overshoot (described above). In total, 27 crisp values (a single element fuzzy set) for each output value K, T_i and T_d are created. The constants K, T_i and T_d are the concrete values Table 3. The outputs *KKNOW*, *TIKNOW* and *TDKNOW* can be taken but the resulting values of those outputs are given according to the active rules and then according to the weighted mean.

4.2 Knowledge Base of the Fuzzy System

The core of the knowledge base is the definition of the linguistic IF-THEN rules of the Takagi-Sugeno type [2]:

$$R_r\text{: If } (x_A \text{ is } A_r)\ \&\ (x_B \text{ is } B_r)\ \&\ (x_C \text{ is } C_r) \text{ then}$$

$$\text{then } (KKNOW_r = K_r)\ \&\ (TIKNOW_r = T_{i_r})\ \&\ (TDKNOW_r = T_{d_r}),$$

where $r = 1, 2, \cdots, R$ is a number of the rule.

The fuzzy conjunction in antecedent of the rule is interpreted as a minimum. The crisp output is determined using the weighted mean value [2], where $\mu_{A_r}(x)$ is the membership degree of the given input x.

Table 3. Crisp values of the constants K, T_i and T_d

r	K	T_i	T_d
1	19.792	1.271	0.173
2	27.085	0.948	0.147
3	34.039	0.775	0.130
4	34.381	1.752	0.138
5	43.675	1.250	0.120
6	52.304	0.998	0.109
7	53.384	2.250	0.114
8	64.626	1.565	0.102
9	74.895	1.228	0.093
10	16.688	1.556	0.252
11	23.451	1.174	0.210
12	29.962	0.972	0.185
13	26.210	2.029	0.209
14	34.464	1.475	0.180
15	42.220	1.192	0.161
16	38.102	2.516	0.178
17	47.820	1.784	0.157
18	56.806	1.418	0.142
19	15.293	1.782	0.315
20	21.799	1.356	0.261
21	28.096	1.128	0.229
22	22.694	2.249	0.268
23	30.440	1.653	0.228
24	37.773	1.346	0.203
25	31.707	2.731	0.233
26	40.679	1.959	0.202
27	49.036	1.569	0.182

Each output is defined as

$$KKNOW = \frac{\sum_{r=1}^{R}\left\{\min\left[\mu_{A_r}\left(x_A^*\right), \mu_{B_r}\left(x_B^*\right), \mu_{C_r}\left(x_C^*\right)\right]\right\} KKNOW_r}{\sum_{r=1}^{R}\min\left[\mu_{A_r}\left(x_A^*\right), \mu_{B_r}\left(x_B^*\right), \mu_{C_r}\left(x_C^*\right)\right]}, \quad (4)$$

$$TIKNOW = \frac{\sum_{r=1}^{R}\left\{\min\left[\mu_{A_r}\left(x_A^*\right), \mu_{B_r}\left(x_B^*\right), \mu_{C_r}\left(x_C^*\right)\right]\right\} TIKNOW_r}{\sum_{r=1}^{R}\min\left[\mu_{A_r}\left(x_A^*\right), \mu_{B_r}\left(x_B^*\right), \mu_{C_r}\left(x_C^*\right)\right]}, \quad (5)$$

$$TDKNOW = \frac{\sum_{r=1}^{R}\left\{\min\left[\mu_{A_r}\left(x_A^*\right), \mu_{B_r}\left(x_B^*\right), \mu_{C_r}\left(x_C^*\right)\right]\right\} TDKNOW_r}{\sum_{r=1}^{R}\min\left[\mu_{A_r}\left(x_A^*\right), \mu_{B_r}\left(x_B^*\right), \mu_{C_r}\left(x_C^*\right)\right]}, \quad (6)$$

where x_A^*, x_B^*, x_C^* are concrete values of inputs variables. So the output of the system is the crisp value and defuzzification is not required.

In total, $r = 27$ rules of the Takagi-Sugeno fuzzy model are used:

1. If$(A \text{ is } S)\&(B \text{ is } S)\&(C \text{ is } S)$then$(KKNOW_r = K_1)\&(TIKNOW_r = T_{i_1})\&(TDKNOW_r = T_{d_1})$

2. If$(A \text{ is } S)\&(B \text{ is } S)\&(C \text{ is } M)$then$(KKNOW_r = K_2)\&(TIKNOW_r = T_{i_2})\&(TDKNOW_r = T_{d_2})$
3. If$(A \text{ is } S)\&(B \text{ is } S)\&(C \text{ is } L)$then$(KKNOW_r = K_3)\&(TIKNOW_r = T_{i_3})\&(TDKNOW_r = T_{d_3})$
4. If$(A \text{ is } S)\&(B \text{ is } M)\&(C \text{ is } S)$then$(KKNOW_r = K_4)\&(TIKNOW_r = T_{i_4})\&(TDKNOW_r = T_{d_4})$
5. If$(A \text{ is } S)\&(B \text{ is } M)\&(C \text{ is } M)$then$(KKNOW_r = K_5)\&(TIKNOW_r = T_{i_5})\&(TDKNOW_r = T_{d_5})$
6. If$(A \text{ is } S)\&(B \text{ is } M)\&(C \text{ is } L)$then$(KKNOW_r = K_6)\&(TIKNOW_r = T_{i_6})\&(TDKNOW_r = T_{d_6})$
7. If$(A \text{ is } S)\&(B \text{ is } L)\&(C \text{ is } S)$then$(KKNOW_r = K_7)\&(TIKNOW_r = T_{i_7})\&(TDKNOW_r = T_{d_7})$
8. If$(A \text{ is } S)\&(B \text{ is } L)\&(C \text{ is } M)$then$(KKNOW_r = K_8)\&(TIKNOW_r = T_{i_8})\&(TDKNOW_r = T_{d_8})$
9. If$(A \text{ is } S)\&(B \text{ is } L)\&(C \text{ is } L)$then$(KKNOW_r = K_9)\&(TIKNOW_r = T_{i_9})\&(TDKNOW_r = T_{d_9})$
10. If$(A \text{ is } M)\&(B \text{ is } S)\&(C \text{ is } S)$then$(KKNOW_r = K_{10})\&(TIKNOW_r = T_{i_{10}})\&(TDKNOW_r = T_{d_{10}})$
11. If$(A \text{ is } M)\&(B \text{ is } S)\&(C \text{ is } M)$then$(KKNOW_r = K_{11})\&(TIKNOW_r = T_{i_{11}})\&(TDKNOW_r = T_{d_{11}})$
12. If$(A \text{ is } M)\&(B \text{ is } S)\&(C \text{ is } L)$then$(KKNOW_r = K_{12})\&(TIKNOW_r = T_{i_{12}})\&(TDKNOW_r = T_{d_{12}})$
13. If$(A \text{ is } M)\&(B \text{ is } M)\&(C \text{ is } S)$then$(KKNOW_r = K_{13})\&(TIKNOW_r = T_{i_{13}})\&(TDKNOW_r = T_{d_{13}})$
14. If$(A \text{ is } M)\&(B \text{ is } M)\&(C \text{ is } M)$then$(KKNOW_r = K_{14})\&(TIKNOW_r = T_{i_{14}})\&(TDKNOW_r = T_{d_{14}})$
15. If$(A \text{ is } M)\&(B \text{ is } M)\&(C \text{ is } L)$then$(KKNOW_r = K_{15})\&(TIKNOW_r = T_{i_{15}})\&(TDKNOW_r = T_{d_{15}})$
16. If$(A \text{ is } M)\&(B \text{ is } L)\&(C \text{ is } S)$then$(KKNOW_r = K_{16})\&(TIKNOW_r = T_{i_{16}})\&(TDKNOW_r = T_{d_{16}})$
17. If$(A \text{ is } M)\&(B \text{ is } L)\&(C \text{ is } M)$then$(KKNOW_r = K_{17})\&(TIKNOW_r = T_{i_{17}})\&(TDKNOW_r = T_{d_{17}})$
18. If$(A \text{ is } M)\&(B \text{ is } L)\&(C \text{ is } L)$then$(KKNOW_r = K_{18})\&(TIKNOW_r = T_{i_{18}})\&(TDKNOW_r = T_{d_{18}})$
19. If$(A \text{ is } L)\&(B \text{ is } S)\&(C \text{ is } S)$then$(KKNOW_r = K_{19})\&(TIKNOW_r = T_{i_{19}})\&(TDKNOW_r = T_{d_{19}})$
20. If$(A \text{ is } L)\&(B \text{ is } S)\&(C \text{ is } M)$then$(KKNOW_r = K_{20})\&(TIKNOW_r = T_{i_{20}})\&(TDKNOW_r = T_{d_{20}})$
21. If$(A \text{ is } L)\&(B \text{ is } S)\&(C \text{ is } L)$then$(KKNOW_r = K_{21})\&(TIKNOW_r = T_{i_{21}})\&(TDKNOW_r = T_{d_{21}})$
22. If$(A \text{ is } L)\&(B \text{ is } M)\&(C \text{ is } S)$then$(KKNOW_r = K_{22})\&(TIKNOW_r = T_{i_{22}})\&(TDKNOW_r = T_{d_{22}})$
23. If$(A \text{ is } L)\&(B \text{ is } M)\&(C \text{ is } M)$then$(KKNOW_r = K_{23})\&(TIKNOW_r = T_{i_{23}})\&(TDKNOW_r = T_{d_{23}})$
24. If$(A \text{ is } L)\&(B \text{ is } M)\&(C \text{ is } L)$then$(KKNOW_r = K_{24})\&(TIKNOW_r = T_{i_{24}})\&(TDKNOW_r = T_{d_{24}})$
25. If$(A \text{ is } L)\&(B \text{ is } L)\&(C \text{ is } S)$then$(KKNOW_r = K_{25})\&(TIKNOW_r = T_{i_{25}})\&(TDKNOW_r = T_{d_{25}})$
26. If$(A \text{ is } L)\&(B \text{ is } L)\&(C \text{ is } M)$then$(KKNOW_r = K_{26})\&(TIKNOW_r = T_{i_{26}})\&(TDKNOW_r = T_{d_{26}})$
27. If$(A \text{ is } L)\&(B \text{ is } L)\&(C \text{ is } L)$then$(KKNOW_r = K_{27})\&(TIKNOW_r = T_{i_{27}})\&(TDKNOW_r = T_{d_{27}})$

5 Verification of the Created System

For the verification the controlled system is selected with a transfer function in the form

$$G_S(s) = \frac{1}{1s^2 + 1s + 2}. \tag{7}$$

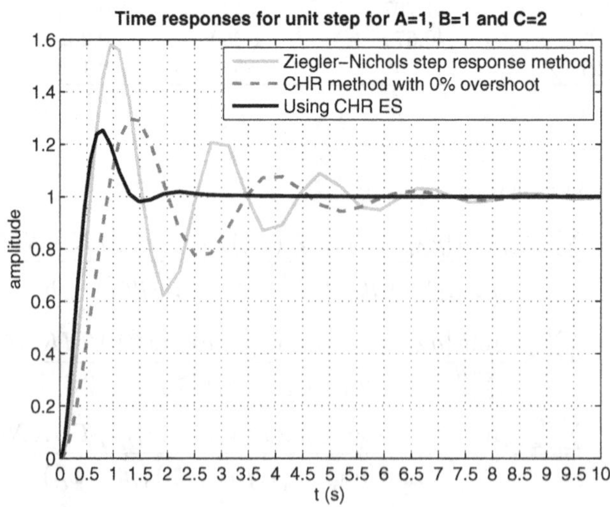

Fig. 6. Time responses of the closed feedback loop with PID controller with parameters determined using the created a expert system, the CHR setpoint response method with 0% overshoot and the Ziegler-Nichols step response method for a unit step

Table 4. Settling time

Settling time (3 % standard) (s)	
CHR step response method with 0% overshoot	Using expert system
5.6	1.2

so the variables A, B and C are small - "pure" small. For "pure" small variables there are determined parameters of PID controller, the same for all "pure" variables A, B and C, no matter their concrete values. It can be considered as a classic robust PID controller, but with parameters determined by a non-convetional method.

For this system three PID controllers are determined, one using only the CHR setpoint response method with 0% overshoot, one using the Ziegler-Nichols step response method (only for visual comparison) and the third one using the created expert system. All these PID controllers are inserted into a closed feedback loop with an appropriate system and the time response is assessed for the unit step of both closed feedback loops. The timing is displayed in time response of closed feedback loop with the PID controllers with parameters determined using the mentioned methods for a unit step Figure 6.

For evaluation of a time response the 3 % standard deviation from the steady-state value is chosen.

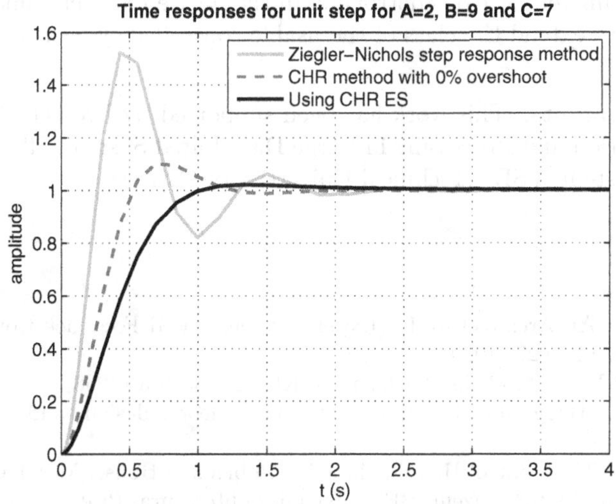

Fig. 7. Time responses of the closed feedback loop with PID controller with parameters determined using the created expert system, the CHR setpoint response method with 0% overshoot and the Ziegler-Nichols step response method for a unit step.

For our selected controlled system the difference in the settling time as a time response for the unit step is 4.4 second Table 4. So for this concrete example the difference is approximately 75 %.

For second experiment the controlled system is selected with a transfer function in the form

$$G_S(s) = \frac{1}{2s^2 + 9s + 7}. \tag{8}$$

In this experiment a very noticeable difference is in the overshoot. Using the PID controller tuned by using the expert system the timing is without overshoot Figure 7 (the overshoot under 3 % is considered as steady-state value [3]). The settling time is according to the 3 % standard deviation from the steady-state value nearly the same.

6 Conclusion

The created expert systems for determining the parameters of a classic PID controller have to been verified using simulation for a selected controlled systems in MATLAB Simulink. It was verified and it was found that it is very usable for systems with transfer function in shape (2), where the parameter A, B and C are rather small, then the settling time is much shorter compared with the CHR setpoint response method with 0% overshoot. For systems with the same transfer function where the parameter A is rather small, parameter B rather medium or large and parameter C small, medium or large, the settling times are nearly the same, but without overshoot, which could be also very useful. The fuzzy expert system which was created is the new non-conventional tool for design the parameters of PID controllers. In the future research some new rules will be added to extend the class of the usable controlled systems types.

Acknowledgements. This work has been supported by Project SP2012/111, "Data Acquisition and Processing in Large Distributed Systems II", of the Student Grant System, VSB - Technical University of Ostrava.

References

1. Arzen, K.-E.: An Architecture for Expert System Based Feedback Control. Automatica 25(6), 813–827 (1989)
2. Babuška, R.: Fuzzy Modeling for Control. Kluwer, Boston (1998)
3. Astrom, K.J., Hagglund, T.: PID Controllers: theory, design, and tuning, USA (1995)
4. Levine, W.S.: The Control Handbook. Jaico Publishing House, Mumbai (1999)
5. Balátě, J.: Automatické řízení. BEN - technicka literatura, Praha (2003)
6. Nowaková, J., Pokorný, M.: On PID Controller Desing Using Knowledge Based Fuzzy System. Advances in Electrical and Electronic Engineering 10(1), 18–27 (2012)

Real-Time Estimation of Induction Motor Parameters Based on the Genetic Algorithm

Petr Palacky, Petr Hudecek, and Ales Havel

VSB-Technical University of Ostrava, Department of Electronics,
17. listopadu 15/2172, 70833 Ostrava-Poruba, Czech Republic
petr.palacky@vsb.cz

Abstract. This article shows one of many ways how to identify the parameters of the IM in real time. There is used the theory of genetic algorithms for IM parameters identification. The introduction describes why the problem is discussed. Next chapters show induction motor's dynamic model and the principle and way how to implement the IM's parameters identification. Theory of used genetic algorithm and experimental results are demonstrated in the end of this article. The conclusion describes the potential use of this method and discusses further development in the field of real time estimation of induction motor's parameters.

1 Introduction

The development of a high-performance drive system is very important in industrial applications. The Induction Motors (IM), thanks to their well known advantages of simple construction, reliability, ruggedness and low cost, are now probably the most widely used machines in the high-performance drive systems. The IMs are often supplied by frequency converters for achieving better performance.

Frequency converters commonly use complex control strategies like FOC (Field Oriented Control) or DTC (Direct Torque Control). Both of these control techniques are highly dependent on correct estimation of motor flux linkage. It's generally well known, that FOC is very sensitive to variation of rotor time constant T_R, while DTC is similarly sensitive to variation of stator resistance R_S. However, any inaccuracy in evaluation in both of these control strategies can cause wrong value of magnetic flux (both amplitude and angle) and electromagnetic torque and therefore it isn't possible to achieve a correct field orientation [1], because the torque capability of drive degenerates and instability phenomena can even occur. Thanks to the use of new fast electronic devices and new DSPs with floating point is in these days possible to apply more complex strategies and methods for IM's parameters estimation, as for example genetic algorithms.

The genetic algorithm (GA) is a search technique used in many fields like computer science, to find accurate solutions to complex optimization or searching problems. The basic concept of GA is to emulate the evolution processes in natural system following the principles, which were first time formulated by Charles Darwin in the Survival of the fittest. The advantage of GA is that it is a very flexible and intuitive approach to optimization. The GA solutions present a higher probability of converging to local optima compared with traditional gradient based methods. More

Á. Herrero et al. (Eds.): Int. JointConf. CISIS'12-ICEUTE'12-SOCO'12, AISC 189, pp. 401–409.
springerlink.com

recently, research works have appeared in the scientific literature about the use of GA for control design in power electronics, drives and in general identification structures.

This paper deals with the real-time estimation procedure of IM's electrical parameters based on the GA, which uses IM's stator currents and voltages, DC link voltage, and velocity of rotor as input data. The procedure output gives the estimated parameters of stator and rotor resistances and stator inductance.

2 Mathematical Model of Induction Motor

The mathematical model of an induction motor is described in the stator coordinate system under usual assumptions. [2]

$$\frac{d\omega_m}{dt} = \frac{3}{2} \cdot \frac{p \cdot L_h}{J_C \cdot L_R} \cdot \left(i_{S\beta} \cdot \psi_{R\alpha} - i_{S\alpha} \cdot \psi_{R\beta}\right) - \frac{\tau_L}{J_C} \tag{1}$$

$$\psi_{R\alpha} = \int \left(\frac{M}{T_R} \cdot i_{S\alpha} - \frac{\psi_{R\alpha}}{T_R} - p \cdot \omega_m \cdot \psi_{Rb} \right) dt \tag{2}$$

$$\psi_{R\beta} = \int \left(\frac{M}{T_R} \cdot i_{S\beta} - \frac{\psi_{R\beta}}{T_R} + p \cdot \omega \cdot \psi_{R\alpha} \right) dt \tag{3}$$

$$\frac{di_{S\alpha}}{dt} = \frac{1}{\sigma \cdot L_S} \cdot u_{s\alpha} - \gamma \cdot i_{S\alpha} + \frac{\beta}{T_R} \cdot \psi_{R\alpha} + p \cdot \beta \cdot \omega_m \cdot \psi_{R\beta} \tag{4}$$

$$\frac{di_{S\beta}}{dt} = \frac{1}{\sigma \cdot L_S} \cdot u_{s\beta} - \gamma \cdot i_{S\beta} + \frac{\beta}{T_R} \cdot \psi_{R\beta} - p \cdot \beta \cdot \omega_m \cdot \psi_{R\alpha} \tag{5}$$

$$\beta = \frac{M}{\sigma \cdot L_S \cdot L_R} \tag{6}$$

$$\gamma = \frac{R_S}{\sigma \cdot L_S} + \frac{M^2 \cdot R_R}{\sigma \cdot L_S \cdot L_R^2} \tag{7}$$

Description of used variables and symbols:

$\psi_{R\alpha}, u_{S\alpha}, i_{S\alpha}$ - alpha components of rotor flux, stator voltage and current

$\psi_{R\beta}, u_{S\beta}, i_{S\beta}$ - beta components of rotor flux, stator voltage and current

L_S, L_R, L_h - stator, rotor and main inductances

ω_m, σ, p - mech. angular speed, leakage factor and number of pole pairs

J_C, T_R, M - total moment of inertia, mech. time constant, mutual inductance

R_S, R_R - stator and rotor resistances (rotor resistance is oriented to stator)

τ_L - the load torque

3 The Identification Methodology Based on Genetic Algorithm

By the modification of previous equations (1) to (7) we get the new relation (8) with the variables K_1 to K_5, which consequently leads to electrical parameters of IM.

$$K_1 i_{S\alpha} + K_2 u_{S\alpha} + K_3 p\omega_m i_{S\beta} + K_4\left(\frac{du_{S\alpha}}{dt} + p\omega_m u_{S\beta}\right) + K_5 \frac{di_{S\alpha}}{dt} = \frac{d^2 i_{S\alpha}}{dt^2} + p\omega_m \frac{di_{S\beta}}{dt} \tag{8}$$

$$K_1 = -\frac{R_S}{\sigma L_S T_R},\ K_2 = \frac{1}{\sigma L_S T_R},\ K_3 = -\frac{R_S}{\sigma L_S},\ K_4 = \frac{1}{\sigma L_S},\ K_5 = -\frac{R_S L_R + R_R L_S}{\sigma \cdot L_S \cdot L_R} \tag{9}$$

Electrical parameters of IM are given by:

$$R_S = -\frac{K_3}{K_4},\ R_R = \frac{K_3 - K_5}{K_4},\ L_S = \frac{K_3 - K_5}{K_2} \tag{10}$$

To describe, how the GA actually works, is used the relation (8). Let's consider the following simplistic rule. We know that IM parameters vary only in a certain range of values such as:

$$1\Omega \le R_S \le 10\Omega,\ 1\Omega \le R_R \le 10\Omega,\ 0.1H \le L_S \le 1H,\ 0.1H \le L_h \le 1H \ldots \tag{11}$$

Ranges of course can also expand or use other parameters, if needed. Substituting the upper and lower limits in relations (9), we will find the boundaries between which the variable K could fluctuate. The next procedure can be summarized in four steps:

1) **Initialize the population -** Is needed to create a zero population, which is composed from randomly generated individuals. Specifically, an individual is composed of five variables K_1 to K_5 randomly generated within the range specified above.

2) **Calculate the Fitness of New Individuals** - The fitness of new individuals has to be calculated. Then are selected a few individuals with high fitness from the population. In this case, the best marked individual is the individual whose K-variables have the difference between the right and left side of the equation (8) as much as possible close to zero.

3) **Termination Criteria** – If the stop condition isn't fulfilled, it is possible to continue directly to point **4**). Stopping condition may be a number of passes through the loop. This condition affects the speed and accuracy of determination of the K-variables. Individual with highest fitness is the main algorithm output and represents the best found solution.

4) **Creating the Next Generation** - To creation of the next generation are used two basic genetic operators. The mutation and the crossing. In some cases it may be useful to keep copies of the parents for the next generation. Specifically, the new population is generated as follows:

   - *crossing* - Swapping parts (K-variables) of a few individuals among them. It is necessary to ensure that the variable, which goes through the swapping procedure, must have the same index (not for example K_1 with K_5).
   - *mutation* – It is a random change of parts of a few individuals. Again, it is necessary to ensure that mutations for example in the K_1 individual were in appropriate range.
   - *reproduction* - The remaining individuals are copied unchanged.

Required values of $u_{S\alpha}$, $u_{S\beta}$, $i_{S\alpha}$, $i_{S\beta}$ are obtained in the SW part from a vector control structure, which is shown on Fig.1. The **Voltage inverter** supplies an induction motor (**M**). Phase currents are measured by current sensors. Position of the rotor is measured at the **Incremental encoder** (IE). True value of mechanical speed ω_m and the rotor angle ε are then evaluated in the **Position and speed estimator** block. The values in the three-phase stator coordinate system $[a, b, c]$ are transformed into two-axis stator coordinate system $[\alpha, \beta]$ in the **T3/2** block. In the block **α,β to x,y**, vector components are transformed to the oriented two-axis rotating coordinate system $[x, y]$. For vector rotation of components of stator current $i_{s\alpha}$, $i_{s\beta}$ to the oriented coordinate system $[x, y]$ is used the variable γ, which is calculated in the **Magnetizing current estimator** block. The components i_{sx}, i_{sy} serve as feedback variables for the current PI controllers $\mathbf{R_{isx}}$, $\mathbf{R_{isy}}$. The control of magnetizing current (or magnetic flux) is realized by a PI controller $\mathbf{R_{im}}$. The block processes a deviation between the desired magnetizing current i_m* and value i_m calculated in the magnetizing current estimator block. Angular speed control provides PI regulator $\mathbf{R_\omega}$, which handles the difference between a speed command ω_m* and speed ω_m, which is evaluated as derivation of rotor angle epsilon. The control voltages u_{sa}, u_{sb}, u_{sc} for **PWM generator** block are obtained from the components u_{sx}*, u_{sy}* by means of the **x,y to α, β** block and subsequent transformation block **T2/3**. The PWM modulation generates pulses for IGBT power transistors in the voltage inverter. In **Voltage Calculation** block are calculated real values of phase voltages in two-axis stator coordinate system $[\alpha, \beta]$. These voltages are corresponding to the control voltages u_{sa}, u_{sb}, u_{sc} and DC link inverter voltage. **IM's parameter estimator** contains genetic algorithm which is able to estimate IM's parameters in real time.

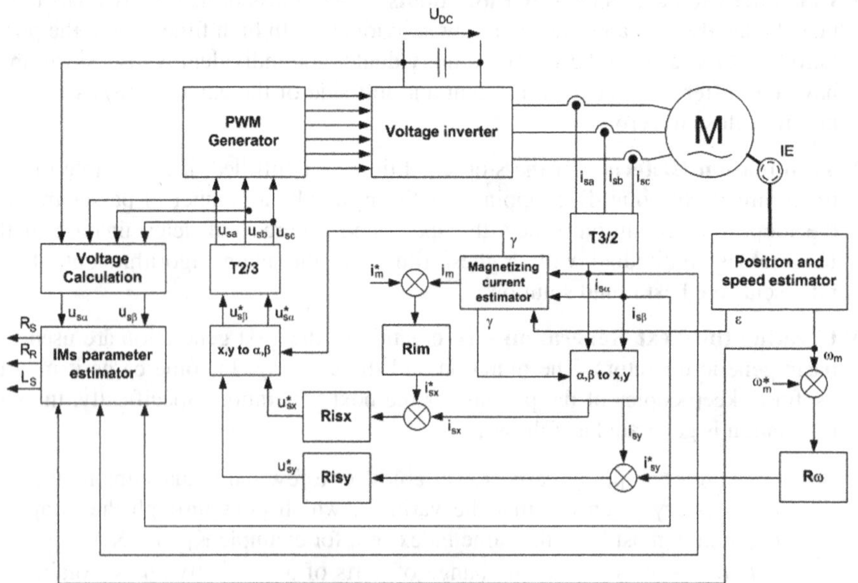

Fig. 1. Block structure of the used vector control with IM's parameter estimator

Detail block structure of IM's parameter estimator in Matlab-Simulink is shown on Fig. 2.

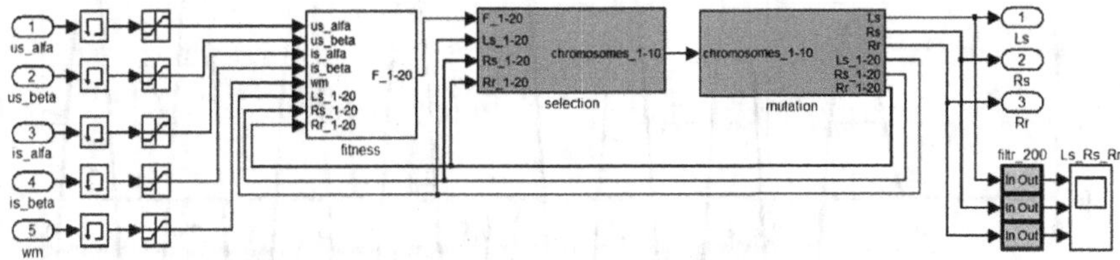

Fig. 2. Block structure of the used genetic algorithm in Matlab Simulink

4 Simulation Results

In all simulations was used the model of real three-phase IM type CANTONI Sg100L-4A, which has the value of stator resistance $R_S = 2.78\Omega$ and a value of rotor resistance $R_R = 2.84\Omega$. Value of stator inductance is $L_S = 0.319H$. The IM was supplied by a model of indirect frequency converter in wiring which consists of a DC voltage source at the input and the voltage inverter with IGBT transistors on the output. DC-link voltage was set to $U_d = 300V$. The control of output voltage was made by SPWM. The frequency and amplitude of the triangle signal was $f_p = 5kHz$ and $U_{pmax} = \pm 1V$. The incremental encoder model is based on the real encoder type ERN 420/TTL with 2048 pulses/rev. The total moment of inertia was set at $J_C = 0.043kgm^2$.

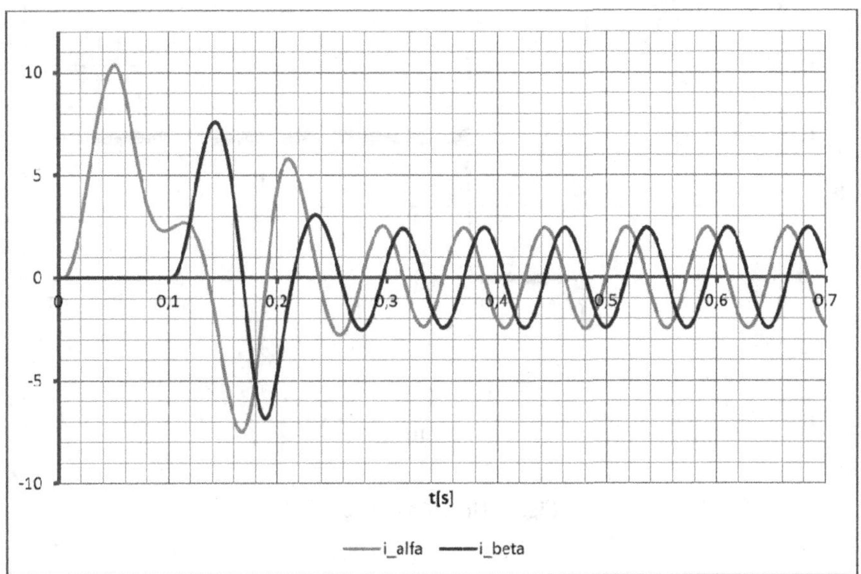

Fig. 3. Stator currents in two-axis stator coordinate system, which are required for IM's parameter estimation. Values are in amperes.

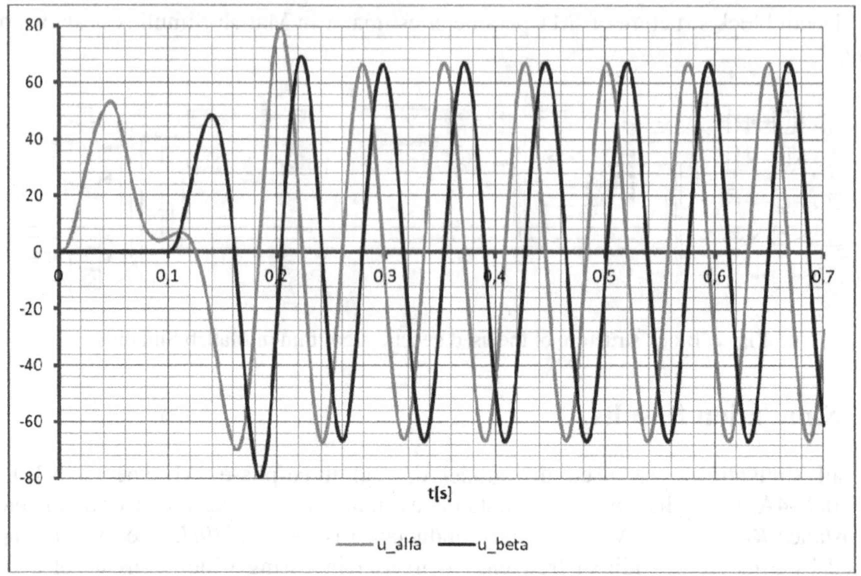

Fig. 4. Stator voltages in two-axis stator coordinate system, which are required for IM's parameter estimation. Values are in volts.

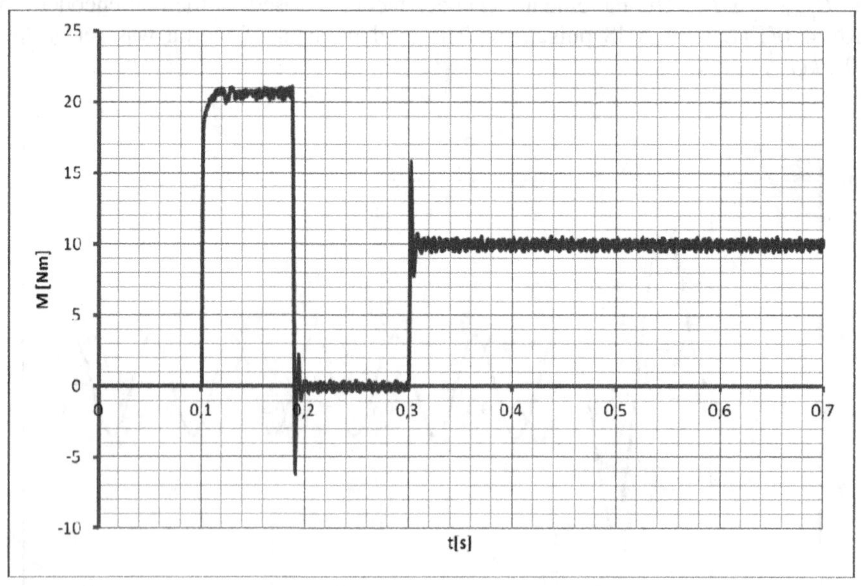

Fig. 5. The IM's torque

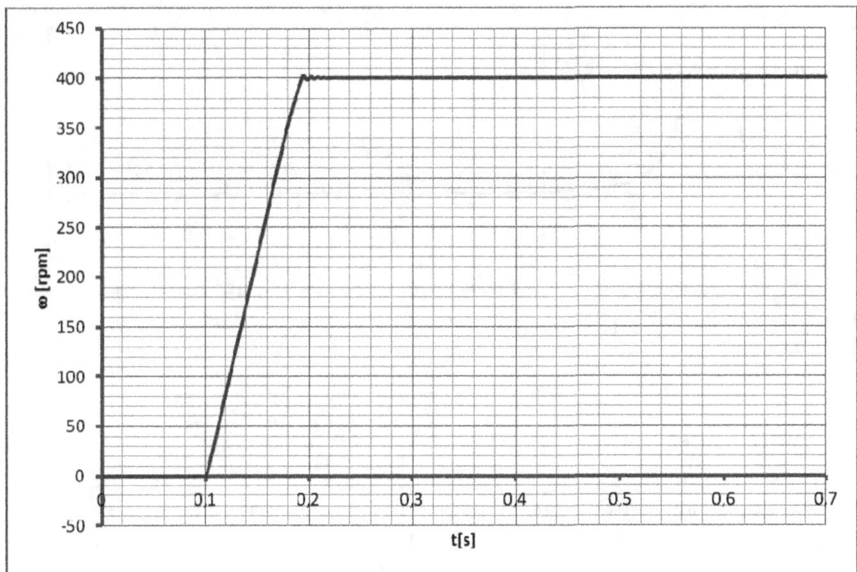

Fig. 6. The IM's mechanical speed

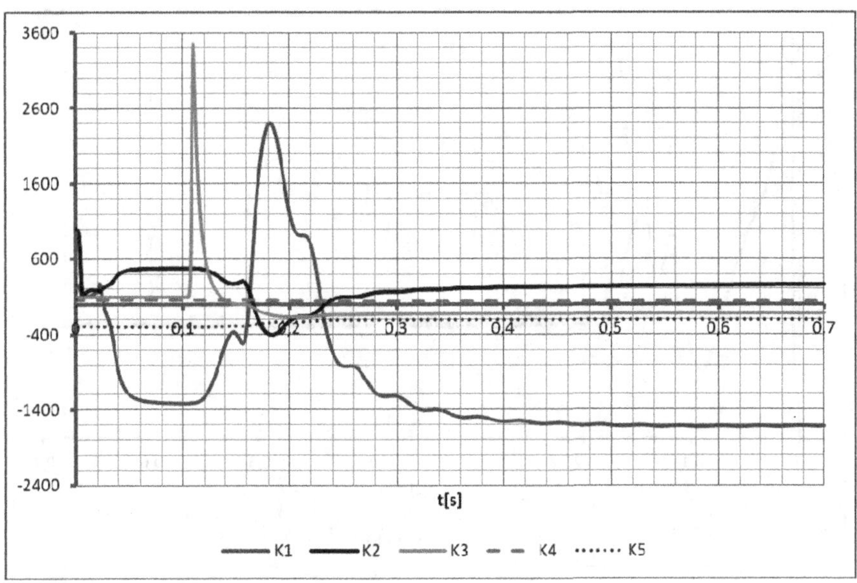

Fig. 7. The K-variables time behaviour

Figures 3, 4 and 6 show the input signals to IM's parameter estimator with GA (see Fig.1). Figures 5 and 6 show the time behaviour of IM and Fig. 7 shows the time behaviour of K-variables from equation (9).

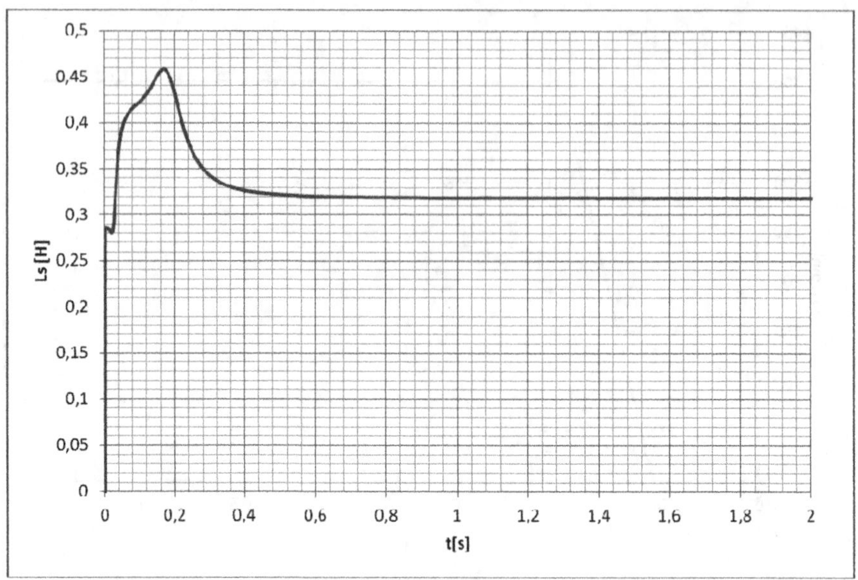

Fig. 8. The transient behaviour of L_S estimated with using of genetic algorithm

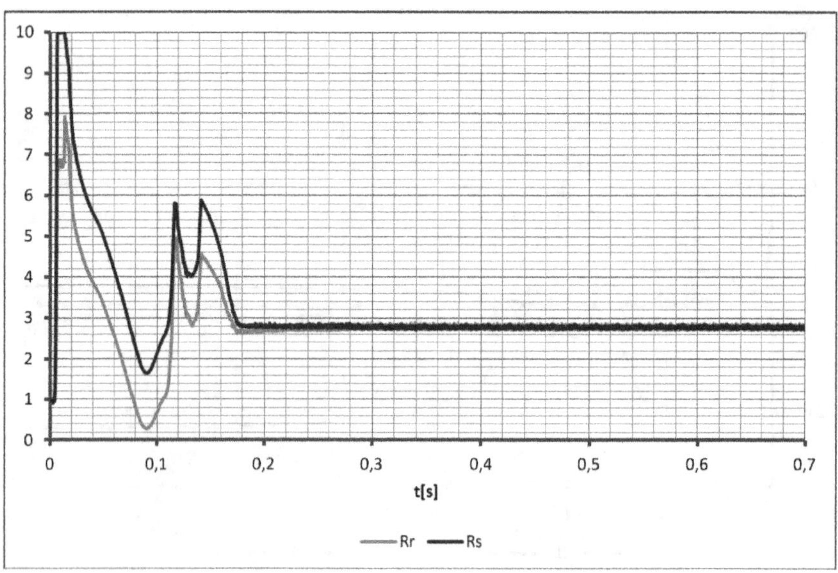

Fig. 9. The transient behaviour of R_S and R_R estimated with using of genetic algorithm. Values are in ohms

From the Fig. 8 and Fig. 9 is evident that the estimation of R_S, R_R and L_S returns true values. Identification isn't accurate in the area of excitation of IM (time 0 to 0.1s) and during the starting of IM (time 0.1 to 0.2s), because parameter estimation is very dependent on the mechanical speed ω_m. The parameter estimation has the best precision in constant mechanical speed. On the same figures is evident, that the GA estimator has a large computing inertia, because the steady values of IM's electrical parameters are available from the time 0.5s (as is evident on the Fig.7). This is one of the biggest disadvantages of genetic algorithms in this application.

5 Conclusion

Presented method can be used in modern high performance electric drives. This is evident from Fig. 8 or Fig. 9 and the whole chapter 4. The main reason why it is necessary to estimate the IM's parameters in real time, is to get the optimal time behaviour of the IM during the dynamic processes, because the adaptive values of the electrical parameters are needed for the PI controllers in modern control strategies. The other reasons are described in the introduction. Besides, for identifying the variables with GA, it is possible to detect electrical variables with using of the other methods, such are least squares methods (Total Least Squares, Ordinary Least Squares, Recursive Least Squares) or the theory of neural networks (see literature [2, 3]).

Acknowledgments. The research described in this paper was supported by SGS project SP2012/126 - Reversible Converters for Electric Energy Accumulation.

References

1. Leonhard, W.: Control of Electrical Drives. Springer, Berlin (2001) ISBN 3-540-41820-2
2. Cincirone, M., Pucci, M., Cincirone, G., Capolino, G.A.: A New Experimental Application of Least-Squares Techniques for the Estimation of the Parameter of the Induction Motor. IEEE Trans. on Ind. Applications 39(5), 1247–1255 (2003)
3. Chiasson, J.: Modeling and High-Performance Control of Electric Machines. IEEE Press, USA (2005) ISBN 0-471-68449-X
4. Ferková, Ž., Zboray, L.: Contribution to parameter identification of an induction motor by genetic algorithms. Acta Electrotechnica et Informatica 5(2), 1–5 (2005) ISSN 1335-8243
5. Brandstetter, P., Chlebis, P., Palacky, P.: Direct Torque Control of Induction Motor with Direct Calculation of Voltage Vector. Advances in Electrical and Computer Engineering 10(4), 17–22 (2010) ISSN 1582-7445
6. Brandstetter, P.: Střídavé regulační pohony – moderní způsoby řízení. In: VŠB-TUO, Ostrava, Czech Republic (1999) ISBN 80-7078-668-X
7. Bose, B.K.: Modern Power Electronics and AC drives. Prentice Hall PTR, Upper Saddle River (2002) ISBN 0-13-016743-6
8. Vas, P.: Sensorless Vector and Direct Torque Control. Oxford University Press, USA (1998) ISBN 0-19-856465-1

Comparison of Fuzzy Logic Based and Sliding Mode Based Controller for IM Drive Fed by Matrix Converter

Jan Bauer and Stanislav Fligl

Czech Technical University in Prague, Faculty of Electrical Engineering,
Department of Electric Drives and Traction, Technicka 2,
Prague 6, 166 27, Czech Republic
{bauerja2,xfligl}@fel.cvut.cz

Abstract. Modern electric drive can be comprehended as electro mechanic energy changer. The own change of electric energy to the mechanic energy is done by the electric motor. It needs for the demanded conversion proper voltage, currents or frequency on its terminals. In case of higher demands, like high dynamics of the motor is required, the converter is usually connected in front of the motor. Due to their robustness and durability recently induction machines are used in modern regulated drives. To gain maximal performance from the induction machine is the research also focused on the so called all silicon converter solutions like matrix converter. From the point of control algorithm the the Direct Torque Control (DTC) is one of the most common high performance induction motor control methods nowadays. This paper compares two modern controller approaches for realization of DTC IM drive fed by matrix converter.

Keywords: Matrix Converter, Induction Machine, Direct Torque Control.

1 Introduction

The area of the regulated AC drives is very wide. But most of them are based on the induction machines and frequency converters. Till now indirect frequency converters are the spread ones, but they have several disadvantages. In order to push the limits of the IM drives and reach maximal efficiency the research in this field focuses also to other converter topologies like multilevel inverters or so called all silicon solutions. One of them is matrix converter [1], [2]. The matrix converter got the name because of its switches can be formed into two dimensional matrix.

The matrix converter does not include passive accumulation element in DC-link, thus uses 9 bidirectional switches to transfer voltage from its input to the output. Compared to the indirect frequency converters it has some attractive features:

- It has relative simple power circuit, indirect frequency converters comprises of rectifier part, inverter part and DC-link with passive accumulation element.
- The features of the matrix converter are same as VSI with active front end on its input. But the drawback is, that because of the DC-link absence, the output voltage amplitude is limited to 86,6% of the input voltage amplitude when we wish to maintain sinusoidal input currents.

Á. Herrero et al. (Eds.): Int. JointConf. CISIS'12-ICEUTE'12-SOCO'12, AISC 189, pp. 411–420.
springerlink.com

The matrix converter is not the topic of the paper, however some of its capabilities are used in the controller design. The modulation strategy employed and control algorithm are very important parts of the converter's controller. In a general case, common modulation strategies from indirect frequency converters cannot be used because of the absence of a DC-link [2]. Therefore, modulation strategies based on a virtual DC-link space vector modulation were developed. Such an approach helps to reduce the modulation complexity and enables the reuse of traditional modulation strategies (after slight modification). At the same time, it works as a dependency injection. In this paper phase control of the input current in combination with the sliding mode or fuzzy control of the output voltage is presented. In other words the input of the converter will be controlled to consume sinusoidal current that will produce demanded voltage in virtual DC-link. The virtual DC-link voltage will be than switched, as one of the 6 vector from Tab. 1, to the output without any modulation. This enables us to overcome the traditional limitations of the output voltage without resigning from the input current control.

Table 1. Virtual inverter output voltage vectors

u_x	u_0	u_1	u_2	u_3	u_4	u_5
u_α	1	0,5	-0,5	-1	-0,5	0,5
u_β	0	√3/2	√3/2	0	-√3/2	-√3/2

The input current modulation index can be calculated from the power balance between the input and output of the converter.

$$\underline{u}_{out} = \mathrm{m}_U \cdot \mathrm{m}_I \cdot \underline{u}_{in} \tag{1}$$

The modulation index of the virtual rectifier can be expressed as:

$$m_I = \frac{\frac{|i_{in}|}{k}}{\frac{\sqrt{3}}{2}\sqrt{3} i_{DC}} = \frac{|i_{in}|}{i_{DC}} = \sqrt{3}\frac{|u_{out}|}{|u_{in}|} \tag{2}$$

which describes current modulation index that must be set for the virtual rectifier to obtain proper output voltage.

Currently the Field Oriented Control (FOC) [3] and the Direct Torque Control (DTC) [4] are the most common high performance induction motor (IM) control methods from the point of view of control algorithm. In the FOC, control of the torque and flux is carried out through the decoupled control of the stator current - torque and flux-producing components are controlled separately [3]. Because it operates in field coordinates, it requires coordinate transformation between the stationary and rotating flux coordinates. In the DTC [4], [5], the torque and flux are controlled directly by switching suitable voltage vectors, which is relatively close to a sliding mode or fuzzy logic control.

2 Induction Machine Reaction Analysis

The Γ-equivalent circuit (Fig. 1) will be used in this paper to describe the machine's behavior. The advantage of this circuit is a simplification consisting in fusion of the rotor and stator inductances into one inductance on the rotor side without loss of information.

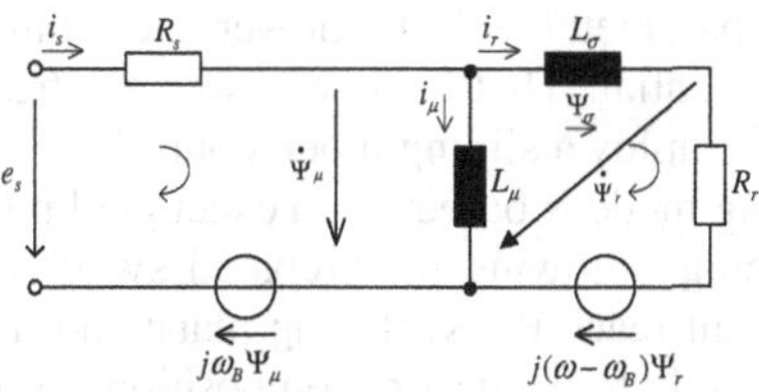

Fig. 1. Γ-circuit of an IM

The circuit can be mathematically described by two equations for the stator and rotor flux. The information about rotor flux can not be easily obtained. However [6] shows that any combination of fluxes or currents can be used to describe the circuit. This intuitively leads to select easily measurable stator variables. This choice forms an equation system (3)

$$\begin{pmatrix} \dot{\underline{\Psi}}_{\mu}(t) \\ \dot{\underline{i}}_{s} \end{pmatrix} = \mathbf{A}_{s} \begin{pmatrix} \Psi_{\mu}(t) \\ i_{s}(t) \end{pmatrix} + \mathbf{b}_{s}(t) \tag{3}$$

Here can be proposed first simplifying assumption: that the machine state variables will be expressed in coordinate system that is rotating with the speed of flux. Therefore $\omega_B = \omega_1$, $\Psi_d = \Psi$ and $\Psi_q = 0$. Using this dynamically rotating coordinate system the original 4th order system will be reduced. The Ψ_q component of the flux will always be zero. The $\mathbf{A_s}$ and $\boldsymbol{b}$ from (3) will be reduced to the 3rd order system (4).

$$\begin{pmatrix} \frac{d\psi_{\mu d}}{dt} \\ \frac{di_{sd}}{dt} \\ \frac{di_{sq}}{dt} \end{pmatrix} = \begin{pmatrix} 0 & -R_s & 0 \\ \sigma_2 & \sigma_1 & \omega_B - \omega \\ -\frac{\omega\sigma_3}{L_\mu L_\sigma} & \omega - \omega_B & -\sigma_1 \end{pmatrix} \cdot \begin{pmatrix} \psi_{\mu d} \\ i_{sd} \\ i_{sq} \end{pmatrix} + \begin{pmatrix} 1 & 0 \\ \frac{L_\mu + L_\sigma}{L_\mu L_\sigma} & 0 \\ 0 & \frac{L_\mu + L_\sigma}{L_\mu L_\sigma} \end{pmatrix} \cdot \begin{pmatrix} e_{sd}(t) \\ e_{sq}(t) \end{pmatrix} \tag{4}$$

where $\sigma_1 = -\frac{R_r L_\mu + R_s L_\mu + R_s L_\sigma}{L_\mu L_\sigma}$, $\sigma_2 = \frac{R_r}{L_\mu L_\sigma}$, $\sigma_3 = L_\mu + L_\sigma$.

The torque produced by the IM can be then calculated according to

$$M_E = \frac{3}{2} p(\Psi \times i) = \frac{3}{2} p(\Psi_d i_q - \Psi_q i_d) = \frac{3}{2} p(\Psi_d i_q) \tag{5}$$

Because we aim to design a torque controller, we will set Ψ_d and i_q as a plane and all investigations will be related to it. The last state variable current i_d remains uncontrolled but from the [7] it follows that it can be controlled vicariously by proper selection of the input voltage vector or by limiting the area of operation.

3 Sliding Mode Controller Design

The sliding mode theory comes from the systems that naturally included a kind of relay [8]. In such systems the use of traditional theories is often far from optimal. On the other hand there are many areas and systems that do not include any kind of relay and where the sliding mode control can be successfully implemented and used bringing its robustness and well defined response to effect major changes to the target value of the controlled parameter [8]. In power electronics is existence of non-linearity and switching essential. That is why power electronics and drives are very suitable areas in which to employ a sliding mode control.

In most cases, the sliding mode is based on an exact guiding of the system state space vector via suitable switching following the prepared switching surface. However, the state vector has to first of all reach the switching plane and then the state space vector will move more or less directly to the destination position. In order to define criteria for the best control law, we could say: The aim is to move from any current system state (defined by system state variables) to any target system state in the shortest time possible. This can be simplified as the task to find the shortest route to the target point.

The induction machine is a 4th order system and which has two action inputs that that can be excited by six vectors with predefined direction and value [9]. The possible directions of the vector are given by the inverter construction and the amplitude is given by the DC-link voltage; it cannot be changed quickly, or not at all (for a matrix converter it could be done in a virtual rectifier by adding a switching step). Moreover, depending on the coordinate system selection the destination point (or the curve) might be steadily moving. In other words, the aim of the control is not to achieve a certain point in the state space but to produce a never-ending movement that delivers the required torque as a side effect.

Existence and stability of the system can be evaluated by analyzing the state space description of the induction motor in order to find limitation boundaries where the available action vectors can no longer produce the required movement in all directions. In other words, the boundary can be defined as system state where the time derivative of one particular state variable is equal to zero and at the same time the other available vectors deliver the same sign (the state value reaches its limit and can no longer be increased).

Typically for IM drive, the flux is kept constant and the torque control is actualized via changes in the orthogonal current (i_q), in this case, the sliding line orthogonal to flux axis. There are typically two movements: upwards if the torque is increased, and downwards if the torque is decreased. Both cases are symbolically depicted in the following figure:

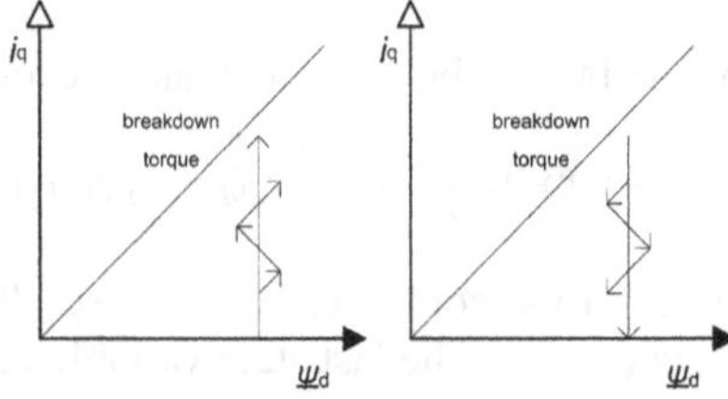

Fig. 2. Sliding mode torque control (increasing and decreasing torque)

The sliding mode increasing the torque exists if there is always a pair of voltage vectors, both increasing the i_q, where the first one causes a positive and the latter one a negative torque derivation and vice versa.

3.1 Design of the Controller

Based on the analysis presented a sliding mode controller for a drive fed by a matrix converter was designed. Input for the regulation is desired flux Ψ_d^* and desired torque M^*. From these two values a current component i_q^* is calculated that is required for the torque generation. Values of Ψ_d^* and i_q^* are then compared with the actual values of Ψ_d and i_q and according to the result of this comparison the voltage vector that should be generated is selected to produce the desired movement of the operation point of the machine. This is illustrated in Fig. 3.

The directions of voltage vectors that are demanded were selected based on the previous analysis of the reaction to the voltage vector [7]. From the analysis it followed that the decrease of the torque is faster than its increase, that is why we choose to boost the increase of the i_q component when an increase in the torque is required by demanding the voltage vector with an angle of 60 degrees. And vice versa, we reduce the decrease of the i_q component when the decrease of the torque is required by demanding the voltage vector with an angle of -45 degrees.

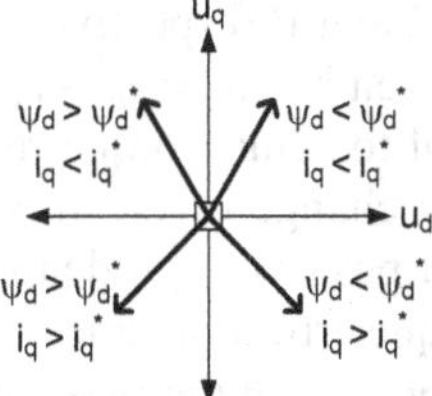

Fig. 3. Voltage vector direction selection

The most suitable voltage vector available at the converter output is then selected by searching for the maximum of scalar multiplication of the demanded vector and the available converter's output vectors (Tab. 1.).

4 Fuzzy Controller Design

The idea of the fuzzy controller is to design practical controller based on the qualitative knowledge of the system [9]. In the transferred meaning the fuzzy control then incorporates the experience and intuition of the designer. It does not need accurate mathematical model of the controlled system. That is why the fuzzy control perfectly fits for the systems with high degree of uncertainty or complex dynamics. Controller based on the linguistic information has advantages like robustness and model free rules based algorithm [9]. Internal structure of the designed controller is in Fig. 4. Input numerical variables are transformed into linguistic ones (fuzzification). Then

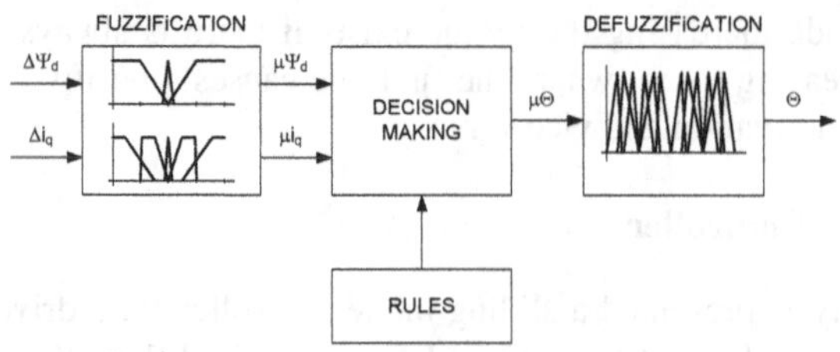

Fig. 4. Fuzzy controller block diagram

based of the predefined decision table and logical linguistic rules the decision making block decides the value of the output. The linguistic output value of the decision block is then transformed back to the numeric output variable (defuzzification).

Using of reduced model of the IM (4) and (5) for the calculation of the machine's torque leads to the fuzzy controller with two inputs and one output [10]. First input variable is flux error, second is torque error, but as shown in (5) the torque depends on the stator current component i_q the current component error is used instead. The output of the controller is then required voltage vector angle that will push the operation point to the desired one. The flux linkage error is given by $\Delta\Psi = \Psi_d^* - \Psi_d$. Where Ψ_d^* is the demanded value of stator flux and Ψ_d is actual stator flux. Because the changing of the flux in the IM is relatively slow only three linguistic values, negative, zero and positive denoted as N, Z and P respectively are used to fuzzify flux linkage error domain. The torque error given by $\Delta i_q = i_q^* - i_q$. Where i_q^* is proportional to desired torque and i_q is proportional to actual torque. Because the change of the torque of the machine is faster then the change of the flux five linguistic values, negative high, negative, zero, positive and positive high denoted as NH, N, ZE, P and PH respectively are used to fuzzify torque error domain.

Each fuzzy rule can be described using the two input linguistic variables $\Delta\Psi_d$, Δi_q and one output linguistic variable that represents desired angle of the converter's output voltage [9]. The ith IF-THEN fuzzy rule R_i of can be expressed as:

$$\text{IF } \Delta\Psi_d \text{ is } A_i, \Delta i_q \text{ is } B_i \text{ THEN } \theta \text{ is } C_i$$

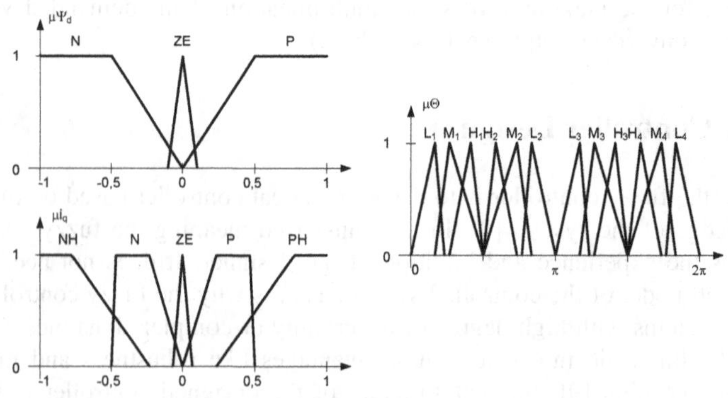

Fig. 5. Fuzzy controller membership functions

Where A_i, B_i and C_i are fuzzy membership function values obtained from μ_A, μ_B and μ_C membership functions of inputs and output linguistic variables of the ith rule. In this paper, Mamdani's fuzzy rule base model is used to perform the function of proposed control algorithm.

The output membership function is designed based on the knowledge of machine reaction analysis to voltage vector [OPTIM].

Table 2. Fuzzy Rules for Output Voltage Angle Selection

$\Delta\Psi_d \setminus \Delta i_q$	NH	N	ZE	P	PH
N	L_3	L_3	L_2	M_2	H_2
ZE	L_4	L_4	L_2/L_1	M_1	H_1
P	M_4	L_4	L_1	H_1	H_1

Based on this we can design the output membership function as shown Fig. 5. The defuzzified output then represents the angle of the voltage vector that should be generated in order to produce fastest movement of the machine's current operation point to the desired one. The output of the fuzzy controller is then used as input for selector of the nearest voltage vector available in the virtual inverter (Tab. 1).

5 Simulation Results

The models of the IM drive fed by matrix converter were created in Matlab/Simulink. Simplified block diagram of the controller is in Fig. 6. From the measured values of voltages and currents are recalculated quantities, that cannot be measured directly - the motor flux Ψ and the angle of the flux phasor φ. Based on the knowledge of these two variables the current values of the Ψ_d and i_q can be expressed. The most suitable voltage vector available is then selected by searching for the maximum of scalar multiplication of the demanded vector and the available converter's output vectors (Tab. 1.).

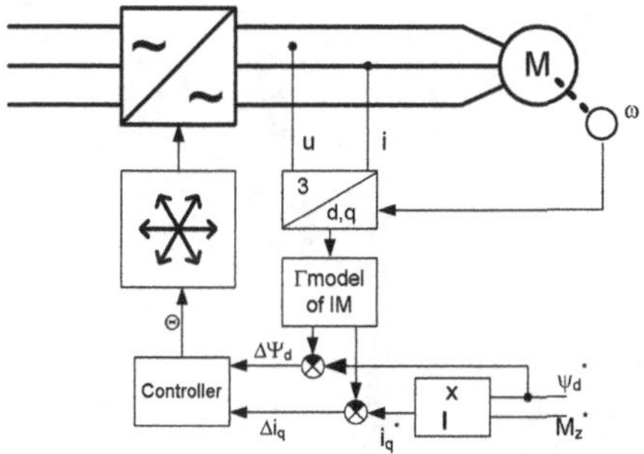

Fig. 6. Block diagram of the drive controller

The simulation results for both controllers are depicted in Fig. 7. It shows the results of the simulation with the proposed control strategy for the desired values of Ψ_d = 0.84 Wb and a different load torque. The red line shows the desired torque and the green line represents the real torque. It can be seen that the controller tracks the desired value very quickly and without any oscillation in transitions.

In Fig.8 the trajectory of the controlled variables in a polar coordinate system is shown. It can be seen that the machine's flux firstly reaches its desired value and then starts chattering around the desired flux line, and the orthogonal current component is adjusted according to the mechanical torque needed. Figure 9 shows the voltage vectors that are generated by the MC. It can be seen that the voltage vectors are scattered around the desired vectors +/- 60° and +/- 45°, however the fuzzy controller produces voltage vectors more unfolded. It is advantageous mainly when the torque reaches the desired point and strong chattering in the direction of i_q axis of the sliding mode controller appears.

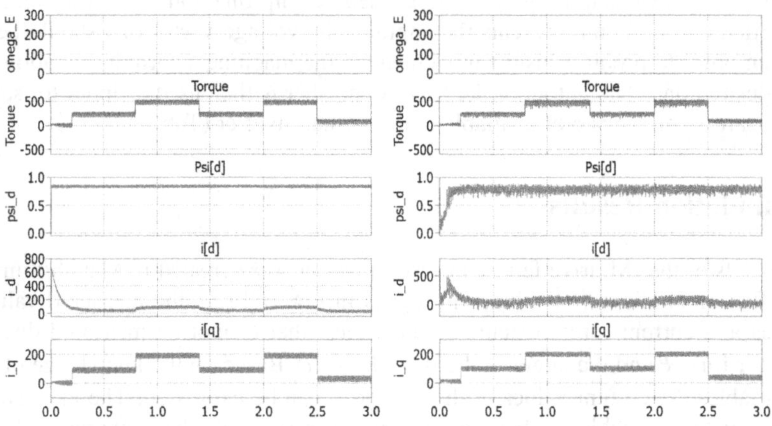

Fig. 7. Simulation result - torque steps sliding mode controller-left; fuzzy controller-right

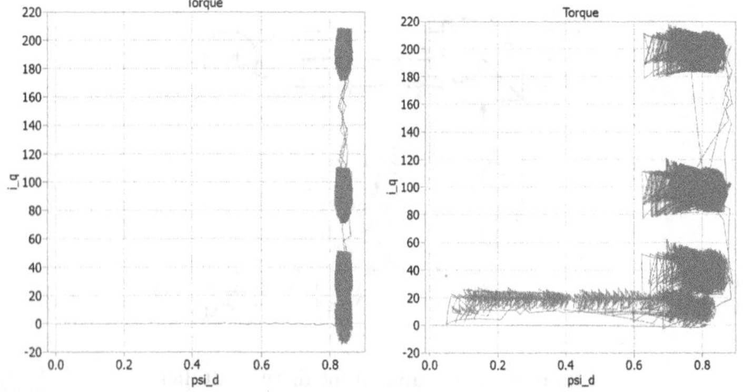

Fig. 8. Simulation result - controlled variables sliding mode controller-left; fuzzy controller-right

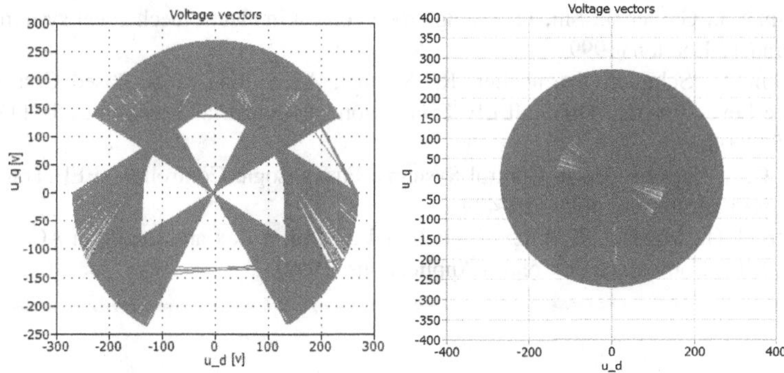

Fig. 9. Simulation result - selected voltage vectors sliding mode controller-left; fuzzy controller-right

6 Conclusion

A control strategy of the matrix converter based on a virtual DC-link concept and a generalized sliding mode or fuzzy control of the virtual inverter part has been presented. The simulation results look very promising and show that both types of control can operate simultaneously with the independent input sinusoidal modulation. The responses of the controllers to a change of the desired state are quick and without any overshoot. However, the real impact of the modulation can be evaluated first after the realization of the algorithm on the real converter prototype. Nevertheless, it is a solid basis for further investigation and improvement of the suggested control concept for a matrix converter induction motor drive system.

References

[1] Kolar, J.W., Friedli, T., Rodriguez, J., Wheeler, P.W.: Review of three-phase PWM AC–AC converter topologies. IEEE Transactions on Industrial Electronics 58(11), 4988–5006 (2011)

[2] Rodriguez, J., Rivera, M., Kolar, J.W., Wheeler, P.W.: A Review of Control and Modulation Methods for Matrix Converters. IEEE Transactions on Industrial Electronics 59(1), 58–70 (2012)

[3] Vas, P.: Sensorless vector and direct torque control. Oxford University Press, Oxford (1998)

[4] Depenbrock, M.: Direct Self Control of Inverter-Fed Induction Machines. Proceedings of the Part I/II, IEEE Trans. Systems Man. Cybernetics 20, 404–435 (1990)

[5] Brandstetter, P., Hrdina, L., Simonik, P.: Properties of selected direct torque control methods of induction motor. In: 2010 IEEE International Symposium on Industrial Electronics (ISIE), pp. 1456–1461 (2010)

[6] Fligl, S., Bauer, J., Vlcek, M., Lettl, J.: Analysis of Induction Machine T and Γ Circuit Coequality for Use in Electric Drive Controllers. In: 13th Int. Conference on Optimization of Electrical and Electronic Equipment, OPTIM 2012 (2012)

[7] Fligl, S., Bauer, J., Ryvkin, S.: Matrix Converter Sliding Mode Based Control for Induction Motor Drive. In: 13th Int. Conference on Optimization of Electrical and Electronic Equipment, OPTIM 2012 (2012)

[8] Utkin, V.I., Gldner, J., Shi, J.: Sliding mode control in electromechanical system. Taylor & Francis, London (1999)
[9] Ryvkin, S., Schmidt-Obermoeller, R., Steimel, A.: Sliding-Mode-Based Control for a Three-Level Inverter Drive. IEEE Transaction on Industrial Electronics 55(11), 3828–3835 (2008)
[10] Lee, C.C.: Fuzzy Logic in Control Systems: Fuzzy Logic Controller. IEEE Transactions on Power Electronics 3(4), 420–429
[11] Wilfried, H., Michael, K.: Fuzzy logic-based control of flux and torque in AC-drives. In: First IEEE Conference on Control Applications (1992)

Discrete-Time Markov Chains in Reliability Analysis-Case Study

Pavel Skalny and Bohumil Krajc

VSB-Technical University of Ostrava, 17. listopadu 15/2172,
Department of applied mathematics
{pavel.skalny,bohumil.krajc}@vsb.cz

Abstract. This paper presents reliability analysis drawn up for an industrial firm. The main goal of this paper is to estimate the probability of firms failure to satisfy an order to its industrial partners. The second aim is to quantify expected value of amount of manufactured products for specific time period. Discrete Markov chains- well-known method of stochastic modelling describes the issue. The method is suitable for many systems occurring in practice where we can easily distinguish various amount of states. The disadvantage of Markov chains is that the amount of computations usually increases rapidly with the amount of states. The Monte Carlo method was implemented to deal with the problem. Chebyshev's inequality was applied to estimate sufficient number of simulations.

1 Introduction

The reliability of production plays a fundamental role in an industrial sphere. Nowadays the reliability of industry process is on a high level. It increases by improving the quality of each component or by redundancy of the production process. Even though it is the top reliability process, there is still a chance of system failing or system proceeding limitedly. In our case we analyse the process which has no redundancy. Thus the information about the probability of the systems failures for certain time period t is very valuable.

The research presented here, was motivated by the practical problem. Analysed company was asked what the probability of production failure was. Knowledge of risk, that the order won't be delivered in time, is important for the partner's firm to establish sufficient goods supplies.

Our aim is to solve the problem by applying discrete time Markov chains (from now on we will understand Markov chain as a discrete time model) a well-known method of stochastic modelling. Nowadays Markov chains has various fields of application in physics, statistics, computer science etc. Markov chains were also effectively applied in stochastic and reliability modelling of industry process [1,2]. The advantage of Markov chains is that calculations of its results are very easy in principle. On the other hand the amount of computations usually increases rapidly with amount of states and time.

Many numerical algorithms were developed to deal with the problem of high amount of computations [3,4]. Another way how to deal with the problem is to utilize simulations method. In this paper we have chosen the second approach, so well known Monte Carlo method was applied [5,6,7]. The simulation approach was used because even in our case study the amount of possibilities was too large to be solved analytically.

Á. Herrero et al. (Eds.): Int. JointConf. CISIS'12-ICEUTE'12-SOCO'12, AISC 189, pp. 421–427.
springerlink.com

2 Markov Chains

Markov chain is a random process with a discrete time set $T \subset \mathbb{N} \cup \{0\}$, which satisfies so called Markov property. The Markov property means that the future evolution of the system depends only on the current state of the system and not on its past history.

$$P\{X_{n+1} = x_{n+1} | X_0 = x_0, \cdots, X_n = x_n\} = P\{X_{n+1} = X_{n+1} | X_n = x_n\}. \quad (1)$$

where:

$X_1, \cdots, X_n$ is a sequence of random variables. The index denotes certain time $t \in T$ $x_1, \cdots x_n$ is a sequence of states in time $t \in T$. As a transition probability p_{ij} we regard probability, that the system changes from the state i to the state j.

$$p_{ij} = P\{X_{n+1} = x_j | X_n = X_i\}. \quad (2)$$

Matrix P, where p_{ij} is placed in row i and column j, is for all admissible i and j called transition probability matrix.

$$P = \begin{pmatrix} p_{11} & p_{12} & \cdots & p_{1n} \\ p_{21} & p_{22} & \cdots & p_{2n} \\ \vdots & \vdots & \vdots & \vdots \\ p_{m1} & p_{m2} & \cdots & p_{mn} \end{pmatrix}. \quad (3)$$

Clearly all elements of the matrix P satisfy the following property:

$$\forall i \in \{1, 2, \cdots, m\} : \sum_{j=1}^{n} p_{ij} = 1. \quad (4)$$

As $v(t)$ we denote vector of probabilities of all the states in the time $t \in T$. As $v(0)$ we denote an initial vector. Usually all its values are equal to zero except the ith, which is equal to 1. It can be proved:

$$v(t) = v(0) \cdot P^t. \quad (5)$$

3 Monte Carlo Method

As F we denote a random variable, which presents the production of the factory within 8 hours. Let R denotes the desired production. In this chapter we will describe the estimation of a probability $p = P(F \geq R)$. The probability will be estimated by the Monte Carlo method.

Let us consider n independent simulations $F_1, F_2, \cdots, F_n$ of F. Let X be a binomial random variable, which represents an amount of successful simulations $X = |\{(F_k : F_k \geq R\}|$. This random variable can be considered as a sum of n independent alternative random variables with parameter p. Let us recall formulas for an expected value and a variance:

$$E(X) = np, \quad D(X) = np(1-p). \quad (6)$$

It follows immediately an expected value and a variance of a random variable $\frac{X}{n}$ from the formula (6):

$$E\left(\frac{X}{n}\right) = p, \quad D\left(\frac{X}{n}\right) = \frac{p(1-p)}{n}. \tag{7}$$

The random variable $\frac{X}{n}$ represents the mean success rate. Let us remind, that for every finite $\varepsilon > 0$ and every random variable Y with a finite expected value $E(Y)$ and a finite variance $D(Y)$ the well-known Chebyshev's inequality holds:

$$P\left(|Y - E(Y)| < \varepsilon\right) \geq 1 - \frac{D(Y)}{\varepsilon^2}. \tag{8}$$

Let us write $Y = \frac{X}{n}$. This implies:

$$P\left(|\frac{X}{n} - p| < \varepsilon\right) \geq 1 - \frac{p(1-p)}{n\varepsilon^2}. \tag{9}$$

Since $p(1-p) \leq \frac{1}{4}$ and

$$P\left(p > \frac{X}{n} - \varepsilon\right) \geq P\left(|\frac{X}{n} - p| < \varepsilon\right), \tag{10}$$

we can write the lower estimate of the probability p:

$$P\left(p > \frac{X}{n} - \varepsilon\right) \geq 1 - \frac{1}{4n\varepsilon^2}. \tag{11}$$

Let us show an example. We choose $\varepsilon = 10^{-2}$, $n = 10^6$. Therefore:

$$p > 10^{-6}\left(X - 10^4\right). \tag{12}$$

The previous result is valuable in a significance level $1 - \alpha = 0.9975$.
Let us note, that correctness of the calculation mentioned above depends on an independence of simulations. Thus the quality of simulation depends on the quality of a pseudo random number generator. In our research we have used the generator from the MATLAB program.

4 Estimation of Input Probabilities

The firm, which our reliability analysis is made for, is a medium size company specialized in thin layer coating of hard materials. The company has requested not to publish its name and not to publish important data of its production. Thus some information (for example production of certain machines) will be demonstrated as a relative variable.

Although the production process of the firm is much more sophisticated, only crucial part of the process will be analysed. We will calculate probabilities that the system fails or works under the condition that the rest of the process works properly.

The analysed part of a process consists of 4 machines, where each of them could work properly- state 1 or could be in failure -state 0. Since we have 4 machines we can distinguish $2^4 = 16$ different states. The huge data set (over three years) of fails and the length of repairs was granted for each machine.

We choose one hour interval as an appropriate period of each time step. The period of one hour was chosen, because it is approximately equal to the length of one coating process.

In the first data source, there was an information about the frequency of fails for each machine. The data were obtained from reports about the failure. We expect that the data of fails come from exponential distribution. Thus to estimate probability p_f that the system fails during one hour we calculated the expected value as an average length of period fail- Q:

$$p_f = \frac{1}{Q}. \tag{13}$$

In the data about length of periods between two fails, there were few outlier measurements. These values increased the arithmetic mean of the variable Q significantly. After the discussion with the production manager we removed the outlying measurements from the data. In the second data source file, there was information about the length of repair time. Using the maximum likelihood method we estimated the probability p_r which says that a machine will be repaired within one hour:

$$p_r = \frac{\Delta V}{V}. \tag{14}$$

where: V is an amount of all repairs that were realized. ΔV is an amount of all repairs that lasted less than one hour. To calculate the transition probability matrix P we calculated probabilities p_f, p_r for each machine. The calculated probabilities for given machines a, b, c, d are presented in the table 1.

Table 1. Calculated probabilities p_f and p_r

probability	a	b	c	d
p_f	0.104	0.0063	0.0048	0.0031
p_r	0.36	0.36	0.4	0.375

In the second step we have to define all of 2^4 states. As a state 1 we understand the situation that all machines work properly. As a state 16 we understand the situation that all machines fail. Some states are presented in the table 2 , where 1 denotes the certain machine works and 0 denotes the certain machine is in a failure.

When we calculated the elements of probability transition matrix, we took into an account the change of states of certain machine. For example certain machine remains in the state 1 with the probability $1 - p_f$, change of the state from 1 to 0 is related to probability p_f. So the probability $p_{3,15}$ that the system changes from state 3 to the state 15 is equal to:

$$p_{3,15} = 0.0104 \cdot (1 - 0.0063) \cdot 0.36 \cdot (1 - 0.0031) = 0.0037.$$

Table 2. Representation of states p_f and p_r

state	a	b	c	d
1	1	1	1	1
2	1	1	1	0
3	1	1	0	1
4	1	0	1	1
⋮	⋮	⋮	⋮	⋮
15	0	0	0	1
0	0	0	0	0

Then we have to define the production of certain machines. As it was written above the production had to be presented as a relative variable. Table 3 presents the real values of power for every machine.

Table 3. Production of machines

state	a	b	c	d
production	0.4	0.3	0.15	0.15

Now, using the previous table we can distinguish various levels of productions w_i for each state i. For example, according to the state 3:

$$w_3 = 10.4 + 10.3 + 00.15 + 10.15 = 0.85.$$

5 Monte Carlo Simulation

In this chapter we will describe how our simulation process works. We use the simulation to estimate the probability that the firm fails to fulfil the order, in a condition that the system begins in a state 1. We will demonstrate the computation for period of 8 hours with one hour time step. We choose the 8 hours period, because it corresponds to the length of a working day. Each one of million simulations follows steps according to the figure 1:

6 Calculating the Expected Value

The expected value of production F within 8 hours is a summary of all expected values for each time step $t \in \{1, 2, \cdots, 8\}$. In our case, the expected value $E(W)(t)$ of production for the certain time t is equal to:

$$E(W)(t) = \sum_{i=1}^{s} w_i v_i(t), \tag{15}$$

where w_1, w_2, $\cdots$, w_s are possible outcomes and $v_1(t)$,$v_2(t)$,$\cdots$,$v_s(t)$ are related elements of probability vector $v(t)$ from the formula (5) .

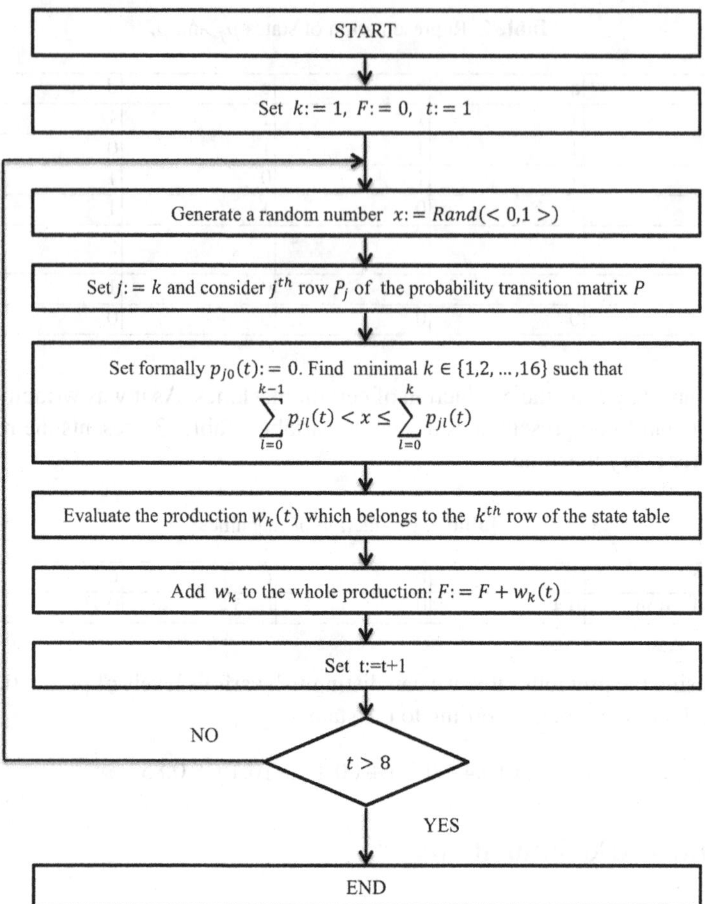

Fig. 1. Principle simulations

7 Results

In this chapter we will demonstrate the results of our application. At first we will present the outputs of the Monte Carlo method. In the table 4 , there are lower estimates of probabilities $P(F \geq R)$ that the production F will be less than the given parameter R. Since we have realized one million simulations, our results are valuable in the significance level 0.9975 (See formula (11)).

In the table 5 there are presented several variables of exploratory statistics of an arithmetic mean of production during the 8 hours period.

Table 4. Calculation of probability $P(F \geq R)$

R	8	7.6	7.2	6.4
$P(F \geq R)$	0.8	0.89	0.93	0.98

Table 5. Exploratory statistics of simulated production

mean	median	st.dev.	1. decile	2. decile
7.84	8	0.41	7.4	7.85

8 Conclusion

In this paper we have presented primeval reliability analysis of the industry firm. The analysis will be also processed as a report for the management of the firm. The probabilities of the failure will be also calculated for longer periods.

As we can see from the results in the previous chapter, probability the firm fails in delivering an order is quite little. We can expect that, if we would analyze the whole production process, the output probabilities of failure would be more pessimistic.

In the further research we want to analyze all the production process and develop a software application for collecting the input data and calculating the probabilities of failure and repair.

Acknowledgement. This work was supported by VSB-tichnical university Ostrava Grant No. TA 9302611

References

1. Moura, M.C.: Semi Markov decision process for determining multiobjective optimal condition-based replacement policies. In: ESREL (2000)
2. Koutras, M.V.: On a Markov chain approach for the study of reliability structures. Journal of Applied Probability (1999)
3. Marek, I.: Algebraic Schwarz methods for the numerical solution of Markov chaos. Linear Algebra and Its Applications (2004)
4. Stewart, W.: Introduction to the Numerical Solution of Markov Chains, 1st edn., 538 p. Princeton (1994)
5. Dubi, A.: Monte Carlo Applications in Systems Engineering, 1st edn., 266 p. Wiley (2000)
6. Bernd, A.: Markov Chain Monte Carlo Simulations and Their Statistical Analysis, World Scientific, Singapore (2004)
7. Beck, J.L.: Bayesian Updating of Structural Models and Reliability using Markov Chain Monte Carlo Simulation. Journal of Engineering Mechanics (2001)

Genetic Algorithm Based Optimization of MGP-FIR Current Reference Generator for Active Power Filters

Jakub Talla[1], Zdenek Peroutka[1], Seppo J. Ovaska[2], and Josef Stehlik[3]

[1] University of West Bohemia, Faculty of Electrical Engineering, Regional Innovation Center for Electrical Engineering (RICE), Pilsen, Czech Republic
talic@rice.zcu.cz, peroutka@ieee.org
[2] Aalto University, Department of Electrical Engineering, Finland
seppo.ovaska@tkk.fi
[3] University of West Bohemia, Faculty of Electrical Engineering, Department of Applied Electronics, Pilsen, Czech Republic
dzouzi@kae.zcu.cz

Abstract. This paper describes genetic algorithm (GA) based optimization of multiplicative general parameter finite impulse response filter (MGP-FIR) current reference signal generator for shunt-type active power filters. The MGP-FIR consists of two fixed subfilters with GA's preoptimized fixed-point coefficients sharing one common delay line and two general adaptive multiplicative parameters to avoid harmful phase shift during fundamental frequency fluctuations. Presented theoretical conclusions are verified by simulation results and experiments performed on a designed laboratory prototype of the single-phase active power filter.

1 Introduction

Our paper focuses on a control of shunt active power filter (APF) which is very popular research topic in the past decade due to power quality constraints in power grids with increasing content on non-linear loads [1]. The simplified scheme with the proposed filtering system is shown in Fig. 1. In principle, the current reference generator extracts the fundamental current directly from the load current. The estimated fundamental frequency component is then subtracted from the actual distorted load current. The resulted harmonics content should be then injected to the power line in the opposite direction in order to achieve harmonics cancellation. A key factor determining a quality of the active filter control is precise current reference generator. This paper investigates the current reference generator based on MGP-FIR with fixed-point coefficients. A serious problem of this filter is its optimal tuning. The aim of this paper is introduction of proposed genetic algorithm based optimization of MGP-FIR parameters. Our paper is organized into five sections. In section 2, the idea and the functionality of MGP-FIR filter is described. The adaptation mechanism is also discussed there. Section 3 explains GA optimization method suitable for the design of MGP-FIR filter, GA setting recommendation and the best designed filters. Section 4 documents the behaviour of the filter in a laboratory experiments with typical non-linear load (thyristor rectifier with RL load).

Á. Herrero et al. (Eds.): Int. JointConf. CISIS'12-ICEUTE'12-SOCO'12, AISC 189, pp. 429–438.
springerlink.com

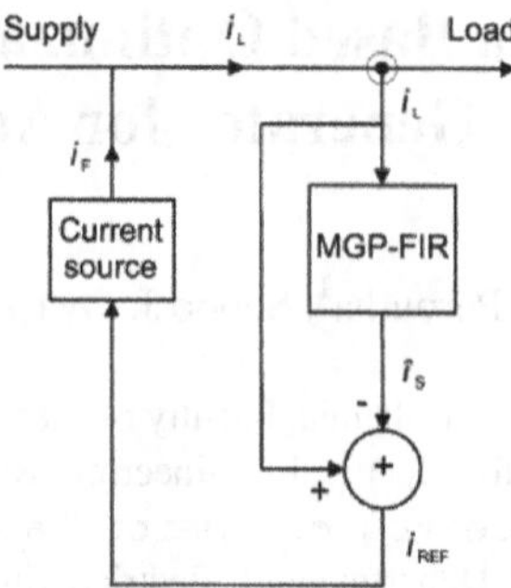

Fig. 1. Shunt active power filter using control based on MGP-FIR current reference generator

2 Predictive MGP-FIR Filter

The MGP-FIR filter was first introduced by Vainio and Ovaska in 2002 [2]. They proposed a simple and computationally efficient solution for adaptive bandpass filtering. The main advantages of the MGP-FIR are: Low computational burden, precise fundamental harmonic extraction with negligible phase delay and robust adaptation around the nominal frequency. The entire current reference estimator is shown in Fig. 2. Due to instability and sensitivity of adaptation (Widrow-Hoff least mean square) to the scaling of input signal it is necessary to normalize input signal (Komrska et al.[3]) which leads to noise transmition from input to output via normalization/denormalization. We improved the original version by normalization of adaptation constant to prevent this problem. The filter works with the sampling frequency 1.67 kHz. This frequency was chosen as a compromise between the filtering quality and the length of the filter needed. Therefore, a polyphase LaGrange interpolator is used here for increasing the sampling rate from 1.67 kHz to 10 kHz. The output of the interpolator is in the end subtracted from the incoming load current and the result is reference current for APF current regulator i_{ref} (Fig. 1.).

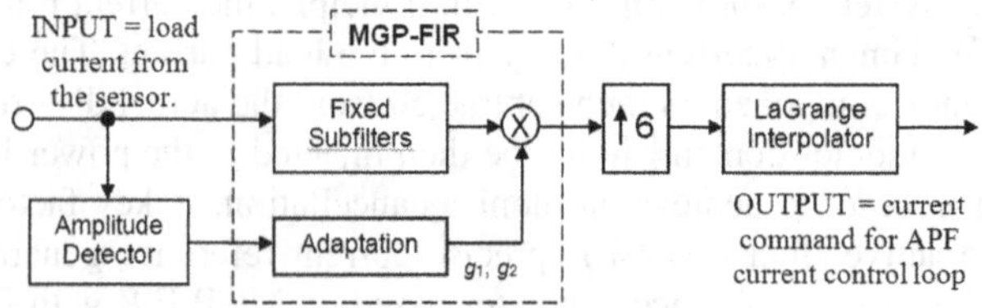

Fig. 2. Current reference generator based on MGP-FIR filter

2.1 MGP-FIR Filter Functionality

The MGP-FIR filter consists of two subfilters sharing one common delay line. The overall scheme is shown in Fig. 3. The delay line includes the fixed coefficients values $h_0, h_1, \ldots, h_{N-1}$. Each of the coefficients belongs either to the structure of the

Subfilter A or the Subfilter B. The subfilter outputs are scaled with the adaptive gain values of $g_1(n)$ and $g_2(n)$. The following equation forms the filter output:

$$y(n) = g_1(n) \overbrace{\sum_{k=0}^{N-1} h_A(k)x(n-k)}^{\text{Subfilter A}} + g_2(n) \overbrace{\sum_{k=0}^{N-1} h_B(k)x(n-k)}^{\text{Subfilter B}} \tag{1}$$

where $h_A(k)$ and $h_B(k)$, $k \in \{0,1,2,\ldots,N\}$, are the fixed subfilter coefficients, $g_1(n)$ and $g_2(n)$ are the multiplicative general parameters and $x(k)$ is the input sequence. Our MGP-FIR works as a p-step ahead predictor. According to this fact the general multiplicative parameters are updated as follows:

$$g_1(n+1) = g_1(n) + \mu \frac{x(n) - y(n-p)}{A^2} \overbrace{\sum_{k=0}^{N-1} h_A(k)x(n-k)}^{\text{Subfilter A}} \tag{2}$$

$$g_2(n+1) = g_2(n) + \mu \frac{x(n) - y(n-p)}{A^2} \overbrace{\sum_{k=0}^{N-1} h_B(k)x(n-k)}^{\text{Subfilter B}} \tag{3}$$

where $\mu < 1$ is the adaptation gain factor. The value of the adaptation gain affects the speed of convergence during the transients and the residual ripple of multiplicative parameters in the steady state. Residual ripple adds additional distortion to the output signal, so the compromise decision between the speed of adaptation and the desired output precision must be made.

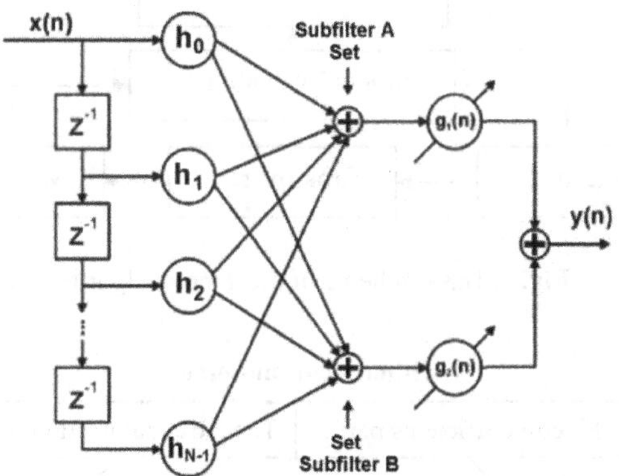

Fig. 3. MGP-FIR filter structure

3 Genetic Algorithm Based MGP-FIR Filter Optimization

The proposed filtering system is hard to design by conventional methods because of its time-domain adaptation behaviour. Therefore, Ovaska and Vainio [2] introduced simple and efficient evolutionary programming based algorithm for optimizing sub-filter coefficients. While this simple solution is highly suitable for the case of plus-minus one restricted coefficient values it is no longer practical for the real-valued or fixed-point coefficients. Consequently, we are introducing in this paper a simple optimization technique based on genetic algorithm [4] (Fig. 4.) suitable for this purpose. Generally, the genetic algorithm translates multi-dimensional problem into the one dimensional by forming one binary string. The binary string represents one member of population. Proper coding of the variables is necessary for the success. The coding for the fixed-point version of MGP-FIR is described in Fig. 5.

First, an initial random population of chromosomes is created. The chromosome consists of two sections. The first main part comprises of the fixed coefficient variables. The length of this part depends on the filter length and the desired coefficient precision. The second part is N-bits long, where the N is the filter length. The value of zero in the field means that the underlying coefficient in the fixed-point part pertains to the subfilter number 1 and vice versa.

In the next step the filtering ability of each chromosome (MGP-FIR filter) is verified with the fitness function and the results are saved in the memory. The selected members of the population are then randomly crossovered and mutated. Resulting new population of the size of $2 \times N$ consists of the newly created chromosomes (N offspring) and older chromosomes (N parents) whose fitness score is already known. Finally, the selection of N mates for reproduction is provided. This scheme is followed for certain number of iterations. After that the fittest member of the population represents the optimally designed filter.

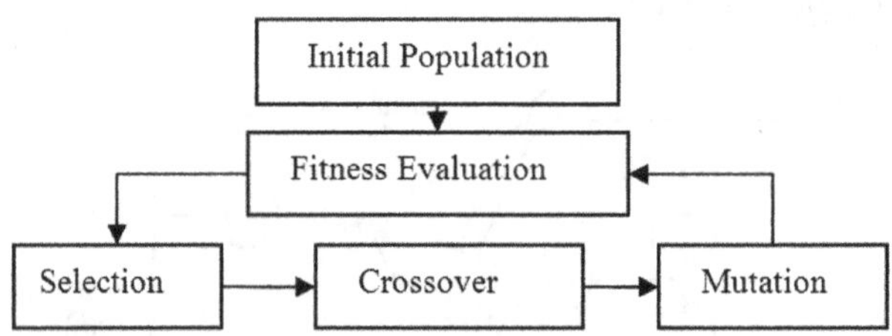

Fig. 4. Basic scheme of the genetic algorithm

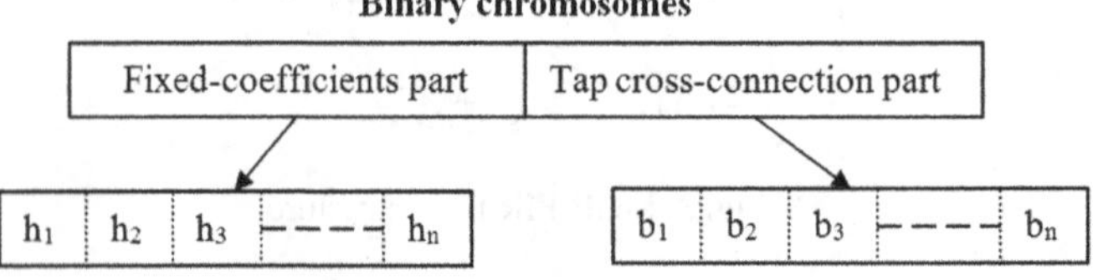

$h_1 \ldots h_n$ – fixed coefficients (integers)
$b_1 \ldots b_n$ – subfilter membership bit coding (0 – first, 1 – second subfilter)

Fig. 5. Employed binary chromosome representation

3.1 Fitness Function

The fitness function describes our requirements on the filter behaviour. The fitness function consists of two parts. The first part is designed to meet the following two criteria: (i) speed of convergence, and (ii) steady state output error. For this purpose the weighted Integral of Time Absolute Error (ITAE) was used. The error is multiplied by the sample number as we are putting more emphasis on the converged filter state:

$$ITAE = \sum_{n=0}^{N-1}\left(n \cdot \left| e_{(n)} \right| \right), \tag{4}$$

where $e_{(n)} = x_F - y_{(n-1)}$ is the error between the fundamental component x_F of the artificial signal and the one step delayed output of the filter $y_{(n-1)}$ and n is the sample number. The filter must be one step predictive due to the LaGrange interpolator on its output, which caused one step delay. The second part of the fitness function minimizes the white noise gain (NG) value. Because our filter changes its parameters during the time, the maximum value of NG is found during the tests and used in the fitness function.

$$NG(n) = \sum_{k=0}^{N-1} [g_1(n)h_A(k)]^2 + \sum_{k=0}^{N-1} [g_2(n)h_B(k)]^2 , \tag{5}$$

We are looking for the maximum value of the fitness function. Our fitness function is the same as originally proposed in [2].

$$ITAE = \sum_{n=0}^{N-1}\left(n \cdot \left| e_{(n)} \right| \right) \tag{6}$$

For optimizing frequency variations we are using three separate training signal sequences together 900 samples long. Each third consists of 300 samples and contains another fundamental frequency. In the power distribution network there should be fundamental frequency fluctuations at maximum around 2% (typically 0.5%) from the rated frequency. It matches the 49 and 51 Hz around the nominal 50 Hz. Thus in our case we are using the ITAE in the following form where all three signals are weighted separately:

$$ITAE = \sum_{n=0}^{299}\left(n \cdot \left| e_{49(n)} \right| \right) + \sum_{n=300}^{599}\left((n-300) \cdot \left| e_{50(n)} \right| \right) + \sum_{n=600}^{899}\left((n-600) \cdot \left| e_{51(n)} \right| \right) \tag{7}$$

We are using the training signal in the following form:

$$x(n) = \sin(\omega_F n) + \sum_{m \in \{3,5,7,9,11,13,15\}} 0.2 \cdot \sin(m\omega_F n) + whitenoise \tag{8}$$

where ω_F corresponds to the three fundamental frequencies 49, 50 and 51 Hz.

3.2 Selection

The population consists of N offspring and N parents. For a certain number of iterations parents are only compared directly with their own offspring. If the offspring is better than its parent it takes his place in the population otherwise is discarded. This is the preparation phase and all the members of the population are slowly improving their fitness score. Premature convergence to the local optimum is avoided in this way and the diversity in the population is preserved. The selection scheme is changed after the specified time to the natural selection. Only the best members from the whole population can survive at this second stage. The population then converges very fast to the final solution.

Fig. 6 compares the fitness evolution for the cases of pure natural selection with random reorder and our proposed hybrid selection approach. The natural selection case converges very fast to the point where all the members of the population are highly similar. Therefore, the population should be very large; otherwise it often leads to poor results. Our approach also uses this feature for the fast convergence, but it gives the algorithm better chance to find a more fit solution, because the initial conditions are always different. In this way, the size of the population can be reduced. Computational effort is also reduced, because the sorting of population is needed only at the end of the simulation and a smaller population will be sorted. Besides, the number of fitness evaluations is decreased.

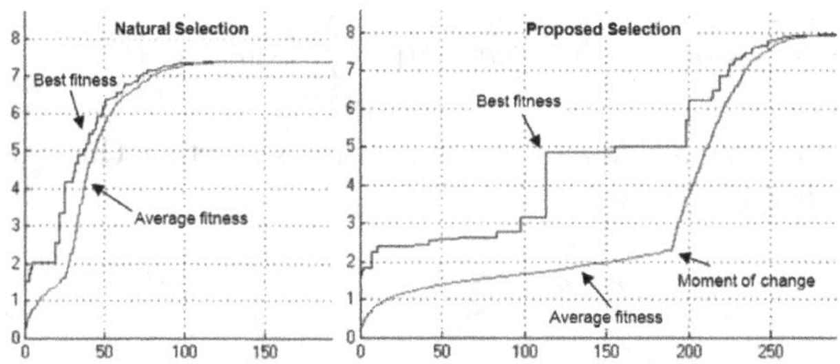

Fig. 6. GA fitness evolution: Comparison of natural selection and proposed hybrid selection

3.3 Crossover and Mutation

The parents are randomly reordered before the crossover is provided. We are using the multiple-point crossover method (Fig 7) applied on the whole chromosome. It is worth mentioning, that the single-point approach did not bring any successful results. For the filter lengths from 15 to 40 and the 16-bit coefficients resolution from 4- to 6-point crossover gave the best results.

The gene is selected for the mutation with the probability $P_{mut} \in (0,1)$. One bit is then randomly selected and negated.

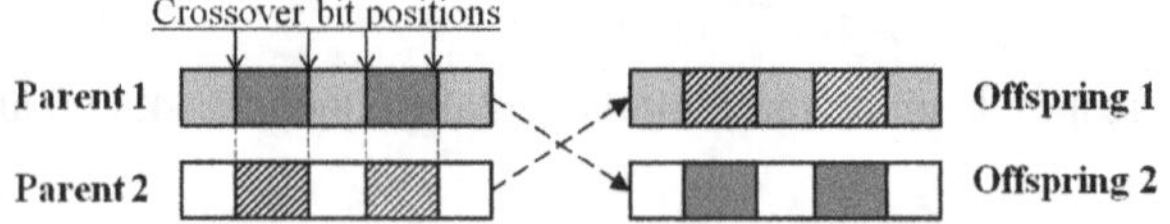

Fig. 7. Four-point crossover example

3.4 GA General Recommendation

We recommend the following parameters to be used with the GA optimization method in this particular application:

Mutation probability = 0.7
Crossover probability = 0.7
Crossover rate = 4

The moment of change for the Parent's fight algorithm depends on the population size. We were using 100 iterations before the end for the population sizes to 300. Above the 300 population higher number of iterations is recommended. We were using maximum 160 iterations from before the end of simulation for the population size of 1500. For the fixed-point version of MGP-FIR, the population size at least twice the binary chromosome length is recommended.

3.5 GA Optimization Results

Extensive optimization runs were performed with the genetic algorithm. The average results for the different filter lengths are included in the Table 1. Each test was run 20 times. One test run consisted of 500 iterations and the number of population was two times binary gene length. The total harmonic distortion of the training sequence according to (8) was 52%.

Table 1. GA validation results

Filter Length	16 μ =0.002		20 μ =0.0017		25 μ =0.0015		30 μ =0.001		35 μ =0.0007		40 μ =0.0005	
	Avg	*Dev*	*Avg*	*Dev*	*Avg*	*Dev*	*Avg*	*Dev*	*Avg*	*Dev*	*Avg*	*Dev*
Fitness	1,42	0,04	2,81	0,11	3,42	0,15	4,03	0,25	4,96	0,22	5,15	0,21
ITAE	4849	216	3225	220	3040	117	3079	94	2727	101	2679	133
NG	0,164	0,003	0,125	0,004	0,111	0,002	0,102	0,002	0,087	0,001	0,082	0,002
THD	4,14	0,22	1,91	0,04	1,91	0,04	1,80	0,12	1,69	0,04	1,64	0,04

Dev = Standard deviation

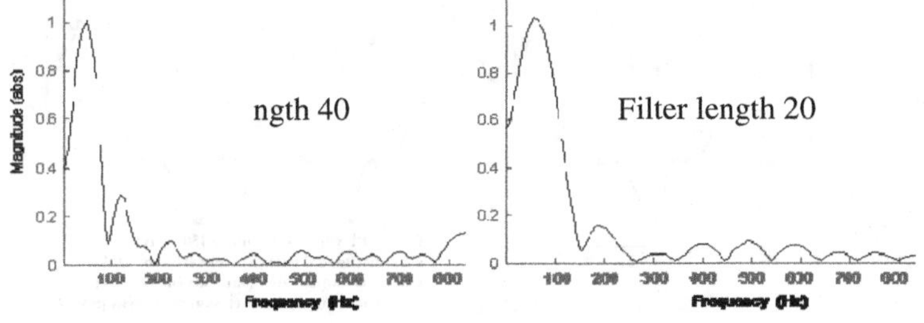

Fig. 8. Instantaneous magnitude responses for filter lengths 20 and 40

Fig 8 shows the instantaneous magnitude responses for the best filters of lengths 20 and 40. The filter length 20 is exactly the length needed for attenuation of the first problematic odd frequency. Consequently the zeros are equally distributed at the odd higher level harmonics.

4 Experimental Results

Proposed MGP-FIR was implemented using a 32-bit fixed-point digital signal processor TMS320F2812. APF current was controlled by hysteresis controller with 1.6 A hysteresis band. The configuration of the APF laboratory prototype with thyristor bridge rectifier representing non-linear load (test bed) is shown in Fig.9. Fig. 10 shows APF behavior under steady state conditions. The spectral analysis of the load and line current computed for the scenario in Fig. 10 is depicted in Fig. 11 and Fig. 12, respectively.

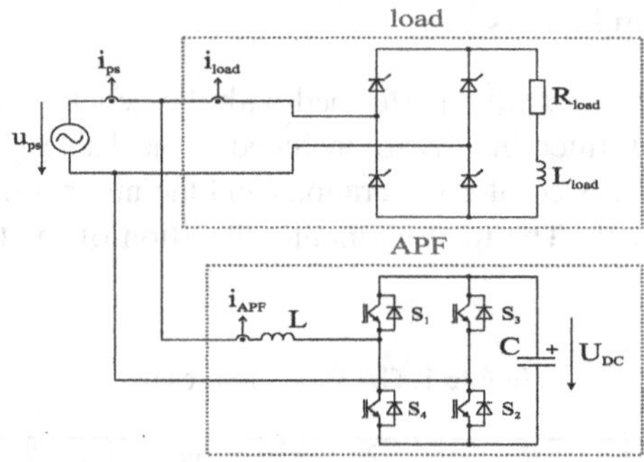

Fig. 9. Configuration of the active power filter (APF) laboratory prototype

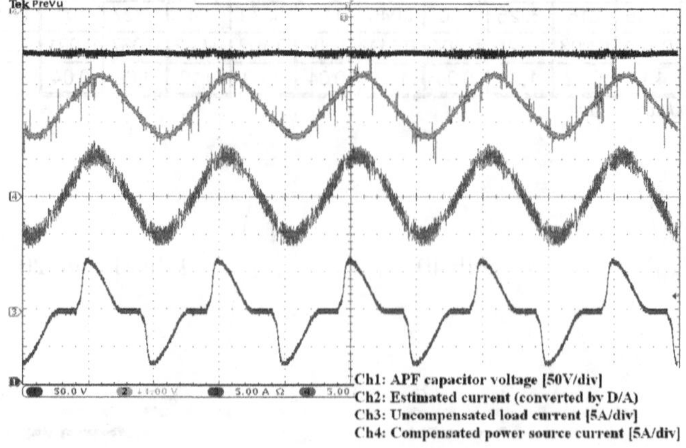

Fig. 10. Experimental results – steady-state conditions: grid voltage u_{ps} = 230V/50Hz, U_{dc}=450V, nonlinear load - thyristor rectifier with RL load (α=30°, R=45Ω, L=16mH).

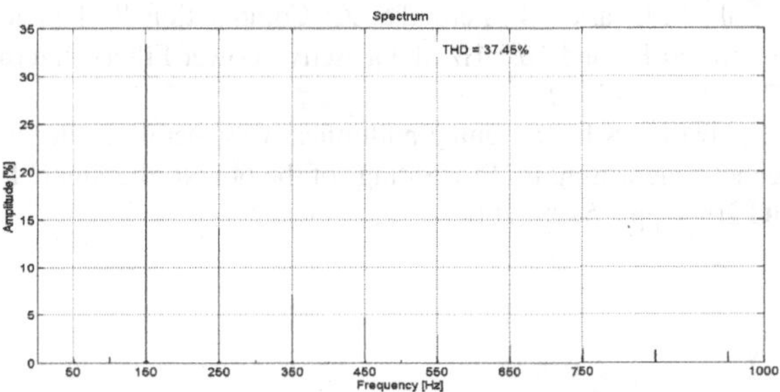

Fig. 11. Spectral analysis of load current i_{load} (Test scenario displayed in Fig. 10)

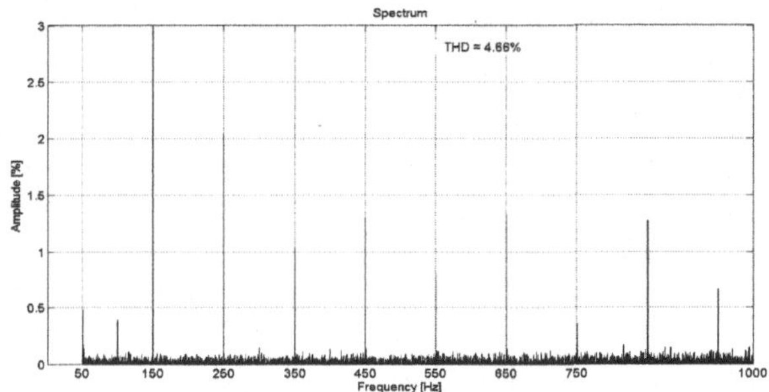

Fig. 12. Spectral analysis of line current i_{ps} (Test scenario displayed in Fig. 10)

5 Conclusion

This paper presented the new method based on genetic algorithm (GA) proposed for optimization of the real-valued multiplicative general parameter finite impulse response filter (MGP-FIR) current reference signal generator for shunt-type active power filters. The paper also explained general recommendation of GA optimization parameters. The MGP-FIR filter performance was verified by simulations and experiments made on developed single-phase shunt active power filter prototype controlled by hysteresis current control.

Acknowledgments. This research has been supported by the ERDF and Ministry of Education, Youth and Sports of the Czech Republic under project No. CZ.1.05/2.1.00/03.0094: Regional Innovation Centre for Electrical Engineering (RICE).

References

1. El-Habrouk, M., Darwish, M.K., Mehta, P.: Active power filters: A review. IEE Proc.-Electric Power Applications 147, 493–413 (2000)
2. Ovaska, S.J., Vainio, O.: Evolutionary-programming-based optimization of reduced-rank adaptive filters for reference generation in active power filters. IEEE Transactions on Industrial Electronics 51(4), 910–916 (2002)

3. Komrska, T., Žák, J., Ovaska, S., Peroutka, Z.: Computationally Efficient Current Reference Generator for 50-Hz and 16.7-Hz Shunt Active Power Filters. International Journal of Electronics 97(1), 63–81 (2010)
4. Martikainen, J., Ovaska, S.J.: Designing multiplicative general parameter filters using multipopulation genetic algorithm. In: Proceedings of the 6th Nordic Signal Processing Symposium, NORSIG 2004, pp. 25–28 (2004)

Correlation-Based Neural Gas for Visualizing Correlations between EEG Features

Karla Štěpánová, Martin Macaš, and Lenka Lhotská

Dept. of Cybernetics, FEL CTU in Prague
{stepakar,lhotska}@fel.cvut.cz, mmacas@seznam.cz

Abstract. Feature selection is an important issue in an automated data analysis. Unfortunately the majority of feature selection methods does not consider inner relationships between features. Furthermore existing methods are based on a prior knowledge of a data classification. Among many methods for displaying data structure there is an interest in self organizing maps and its modifications. Neural gas network has shown surprisingly good results when capturing the inner structure of data. Therefore we propose its modification (correlation - based neural gas) and we use this network to visualize correlations between features. We discuss the possibility to use this additional information for fully automated unsupervised feature selection where no classification is available. The algorithm is tested on the EEG data acquired during the mental rotation task.

1 Introduction

With an increasing amount of data and with our ability to detect more and more relevant features in the data, the number of extracted features is enormously high. Redundant features increase a classification error [1] and therefore feature selection is indispensable for the automatic data analysis.

Due to a combinatorial complexity it is impossible to test all combinations of features to select the best set of them. The overview of different models, search strategies, feature quality measures and evaluation can be found in [2]. Search strategies can be divided into the greedy methods such as sequential forward search (SFS [3]) and sequential backward search (SBS [4]), its modification [5,6] and randomized methods [7] (these include: genetic algorithms [8,?,10], Las Vegas approach [11] or random mutation hill climbing [12]). All of these method are heuristic and cannot guarantee optimal solution.

The upper mentioned methods are dependent on a prior classification of the data. But isn't there any other information in the features which can be used for unsupervised feature selection? Among features there is a mutual relationship which should be incorporated into the consideration. There are many statistical parameters which are able to describe relationship between features (eg. correlation or mutual information). Further we can focus on the time dependency where hidden Markov models or recurrent neural networks can be used. Among these parameters is the simplest one correlation which can give us the first insight into the structure of features. This insight can be further used for example for an automatic grouping of features when appropriate unsupervised clustering algorithm would be used. The automatic features grouping based on a k-means algorithm was proposed in [13]. But the k-means is very computationally demanding.

Á. Herrero et al. (Eds.): Int. JointConf. CISIS'12-ICEUTE'12-SOCO'12, AISC 189, pp. 439–446.
springerlink.com

Furthermore we don't know the number of feature groups apriori, so we should use some unsupervised clustering algorithm which can adaptively find the optimal number of clusters. For example, greedy gaussian mixture models with merging (gmGMM) algorithm proposed in [14] could be used.

How can we visualize the relationships between data? We can take inspiration from undirected graphs and especially from the social theory where social-network diagrams are used to represent relationships between social groups [15] (those include: force-directed diagram[16,?] or exponential random graph model [18]). These methods can be applied to any other data. As well the self organizing maps (SOM) or its modification can be used to find structures in data.

In our contribution we propose a correlation-based neural gas network (cNG) which is derived from the neural gas which was introduced by Martinetz in 1991 [19]. The modification is that neurons represent features and their connections represent correlations between these features. This network is in the final stage able to visualize relationships between features. We tested this network on the features extracted from the EEG signal. Further we suggest to use unsupervised clustering technique to detect ideal number of features which would be able to sufficiently describe data.

1.1 Neural gas

The neural gas algorithm was proposed by Martinetz and Schulten in 1991 [19]. This algorithm is a modification of WTA (winner-takes-all) algorithm [20] for vector quantization and was proposed to adaptively quantized a given set of input data. Other similar methods is for example SOM or its dynamic modification "growing neural-gas". Neurons in neural gas are not free, but are restricted by competition.

Neurons are initialized with random weights. In each iteration step one input data vector is presented to the network and the weights of the closest neurons are adapted so that their values would be closer to the input vectors' values. After predefined number of iterations k_{max} the algorithm stops. Its ability to effectively capture data structure depends on the number of neurons, variability in the input data and parameters describing the learning process (neigbourhood function G, learning ratio η and neigbourhood radius λ).

The algorithm works as follows:

Algorithm 1: Neural gas algorithm

---

input data D, λ_0, $\lambda_m in$, η_0, η_{min}, k_{max}
initialize set of neurons A with the uniformly distributed random weights w
for $k = 1 : k_{max}$ **do**
  Choose an input vector c from dataset D
  Sort neurons in A ascending by distance from input vector c
  update weights of each neuron i: $w(i,k+1) \leftarrow w(i,k) + \eta(k) * G(i,k) * (c - w(i,k))$
  where $G(i,k) = \exp(\frac{-k_i}{\lambda(k)})$ is a neigbourhood function
  k_i is the order of the i-th neuron in the sorted list
  $\lambda(k) = \lambda_0(\frac{\lambda_{min}}{\lambda_0})^{k/k_{max}}$ is a neigbourhood radius and
  $\eta(k) = \eta_0(\frac{\eta_{min}}{\eta_0})$ is a learning ratio
end for

The values of λ resp. η are decreasing in each step from the initial values λ_0 resp. η_0 to the final values λ_{min} resp. η_{min}. G is a decreasing function of i. While the update in weights is negligible for neurons with $k_i >> \lambda$ it is suggested to cut off these [19] to reduce computational time. In the neighbourhood function $G(i,k)$ the value of λ determines how many neurons will significantly change their weights during the adaptation step.

2 Correlation - Based Neural Gas (cNG)

We propose a novel correlation-based neural gas (cNG) algorithm. This approach is able to visualize relationships between data. The biggest difference between cNG and neural gas (NG) is that movement of neurons in cNG is accomplished based on their correlation with an input vector compared to the NG where are nodes moved according to their distance from this input signal.

Furthermore there is another big difference which lays in the fact that contrary to the standard neural gas we do not try to create some "codebook" which will be able to generalize the data and provide vector quantization, but we want to visualize relationships among given set of data. Therefore the number of neurons in a network is the same as the number of input data vectors and the neurons are initialized with the data values. In our case, the input vectors are vectors of features which were extracted from data. The coordinates of the neurons are chosen randomly. In our case we have randomly distributed them into the square grid which just ease the work with them.

The positions of nodes are updated in each adaptive step according to the same rules which are used in a neural gas. The only difference is that we choose to-be-moved neurons according to their correlations with an input vector.

In each cycle we go through all neurons in random order and choose each of them as an input vector.

In the final stage neurons represent features and connections between nodes represent correlations between these features.

The cNG algorithm is described bellow:

Algorithm 2: Correlation - based neural gas algorithm

input data D, λ_0, λ_{min}, η_0, η_{min}, k_{max}
create vector $rand_{vec}$ with randomized numbers (the highest number is the number of data vectors $nmbFeatures$)
initialize set of neurons NG so that for each neuron:
$ID = 1$
for $i = 1 : \lceil sqrt(nmbFeatures) \rceil$ **do**
  for $j = 1 : \lceil sqrt(nmbFeatures) \rceil$ **do**
    $NG(randvec(ID)).cord \leftarrow [ij]$
    $NG(randvec(ID)).values \leftarrow$ random data vector $D(randvec(i))$
    $NG(randvec(ID)).correl \leftarrow$ vector of correlations of $D(randvec(i))$ with other data vectors
    $ID = ID + 1$
    if $ID > nmbFeatures$ stop adding neurons
  end for
end for
$p = 1$
for $cyc = 1 : nmbOfCycles$ **do**
  for $k = 1 : nmbOfNeurons$ **do**
    Choose an input neuron $NG(sel)$ from set NG
    Sort neurons descending by correlations with the neuron $NG(sel)$
    update coordinates of each neuron i: $NG(i).cord \leftarrow NG(i).cord + \eta(p) * G(i,p) * (NG(sel).cord - NG(i).cord)$
    where $G(i,p) = \exp(\frac{-k_i}{\lambda(p)})$ is a neigbourhood function
    k_i is the order of the i-th neuron in the sorted list

$\lambda(p) = \lambda_0(\frac{\lambda_{min}}{\lambda_0})^{p/p_{max}}$ is a neigbourhood radius and
$\eta(p) = \eta_0(\frac{\eta_{min}}{\eta_0})$ is a learning ratio
$p \leftarrow p+1$
end for
end for

The neurons which have the highest correlation with the input vector (in our case feature vector) will be pulled to the input vector the most. On contrary the neurons with lower correlations will not move at all. The parameter λ determines how many of the most correlated neurons will be significantly moved. The best results will be achieved when the number of features will be N^2. In the case neurons will be initially distributed into the regular square grid.

3 Experimental

3.1 Data

The correlation-based neural gas algorithm (cNG) was tested on the features which were extracted from the EEG signal. The EEG signal was recorded from 19 electrodes, positioned under 10-20 system, during the mental rotation task. Subjects were solving 228 2D and 144 3D mental rotation tasks during the measurement. In the beginning there was one minute baseline with the open eyes and one minute with the eyes shut. A segmentation of the signal was performed so that each trial and time between trials was marked as a one segment. The segmentation divided signal into 837 time-serie values. For each segment and each electrode 44 features were extracted. These include statistical parameters, mean and maximum values of the first and second derivation of the samples and absolute/relative power for five EEG frequency bands (delta, theta, alpha, beta 1, beta 2, gamma). This give a total number of 736 features for all 19 electrodes, each feature vector contains 837 values (one value for each time segment).

3.2 Results

In the first stage we visualized correlations between 36 features from 1 electrode using 6x6 cNG network. The network was initialized as a square grid with the nodes corresponding to the features. A distance between neigbourhood nodes was 1. Adaptation steps of cNG network were performed. In each cycle were selected one by one all nodes as an input vector and presented to the network. Nodes were pushed towards the input vector according to the neigbourhood function, neigbourhood radius and learning ratio as described above in the Algorithm 2. The neigbourhood function in cNG depends on the correlation between these features (the higher the correlation, the bigger this function is). Evolution of the network during 20 cycles is visualized in Fig. 1. The used initial settings are stated.

The final network after 20 cycles is a diagram of features distributed in the space where the distances between them correspond to their correlation.

In Fig. 2 there are displayed connections of one selected feature with all other features. The thickness of the line is in accordance with the correlation between connected features.

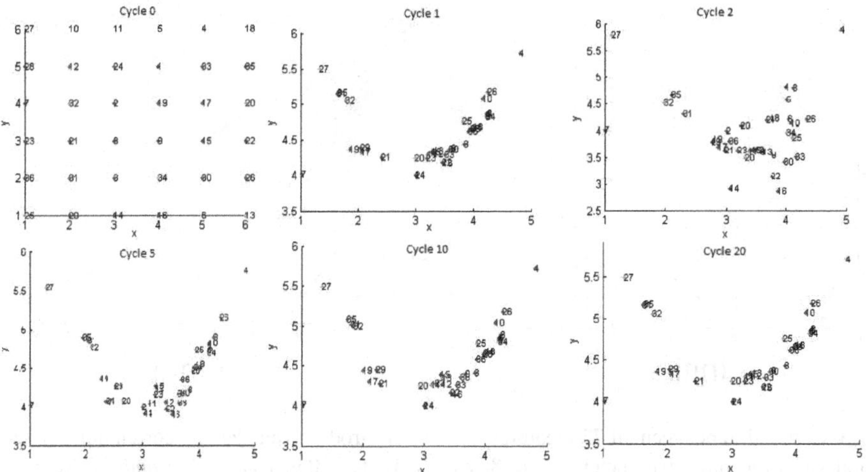

Fig. 1. There is displayed evolution of the cNG applied on the features from one electrode (Fp1), during 20 cycles. In each cycle all features (neurons) are presented to the network in the random order. Above is a visualized network at initial position (Cycle 0), at Cycle 1, 2, 5, 10 and 20 (final number of cycles). As can be seen the evolution proceed from regular square grid (with random features in the nodes) to the clusters of features. The initialization parameters for the network were: $\lambda_0 = 4$, $\lambda_{min} = 0.5$, $\eta_0 = 0.9$, $\eta_{min} = 0.3$, $nmbOfCycles = 20$.

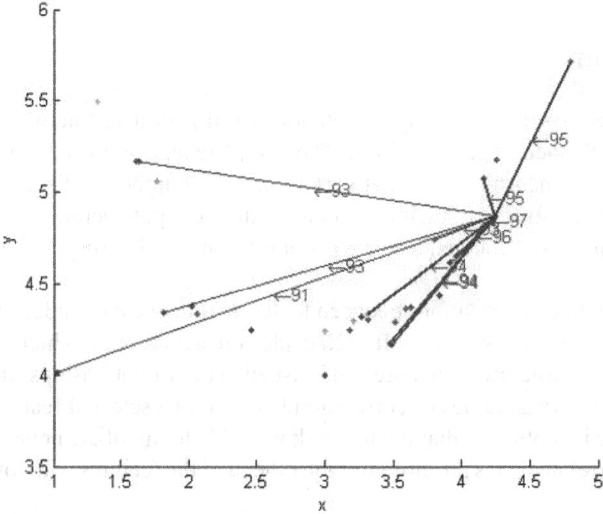

Fig. 2. There are visualized correlations between the feature number one and other features from the electrode Fp1. The more the blue line thick, the bigger correlation between 2 features is present. For some connections there is displayed the value of the correlation. As can be seen the weaker correlations are displayed as a longer and weaker line.

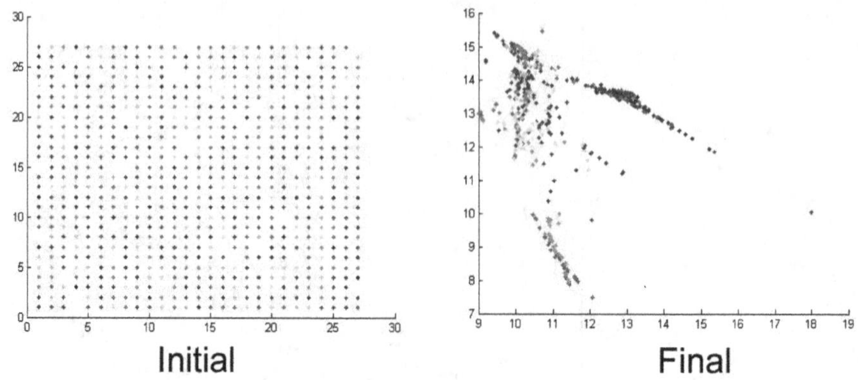

Fig. 3. Correlation between all 736 features from 19 electrodes visualized by correlation - based neural gas with initial parameters: $\lambda_0 = 8$, $\lambda_{min} = 1$, $\eta_0 = 0.9$, $\eta_{min} = 0.3$, $nmbOfCycles = 20$. There is displayed initial state when neurons in cNG are distributed into the regular square grid, and final state after 20 cycles. Different colors correspond to the features from different electrodes.

In Fig. 3 there is a visualization of 736 feature vectors from all 19 electrodes. Features from different electrodes have different colors. As can be seen by eye features from one electrode tend to form clusters even though these have been initialized randomly.

4 Discussion

We've developed a self-organizing correlation-based neural gas network, which is a modification of the neural gas algorithm. The complete algorithm is described in Algorithm 2. Number of neurons in our network is same as number of display data and the nodes are moved according to their correlation with the input vector.

The algorithm was applied on the data acquired from EEG during the mental rotation task.

We have visualized correlations between features from one electrode. As can be seen from the Fig. 1, network was able after 20 cycles create structure which is a representation of the inner structure of features and visualize their relationships. In Fig. 2 there are displayed numerical values of correlation between one selected feature and others. From this figure it is obvious that the network was able to distribute nodes (features) so that the high correlated ones are close to each other and the features with low correlation are farther.

We have also visualized data from all 19 electrodes using the same algorithm. We displayed features from different electrodes by different colors. The results of the network evolution during 20 cycles can be seen in Fig. 3. The network revealed that features from one electrode are in most cases highly correlated (much more than features from different electrodes). Therefore features from each electrode formed more or less separated clusters.

The unsupervised clustering algorithm could be applied to create feature groups automatically and suggest a set of features for unsupervised classification. The algorithm must be able to detect unknown number of clusters and have low computational demands. When the clustering analysis would be performed on all features we would detected as a selected subset of features the ones which are originated from different electrodes. Therefore the selection using unsupervised clustering is not applicable on all EEG features. It could be more useful to visualize structure of features from separate electrodes. When the correlation among the same features will be detected across all electrodes then these features are redundant and can be removed. Furthermore time information could be included.

The proposed algorithm can be used for visual inspection into the feature structure when deciding which features are correlated. This network is usable not only to visualize features' correlations but we can visualize as well other feature relationship parameters. Because this network is not constrained to the features as an input data, the application field is very wide.

Acknowledgement. This research has been supported by SGS grant No. 10/279/OHK3/3T/13, sponsored by the CTU in Prague, and by the research program MSM 6840770012 Transdisciplinary Research in the Area of Biomedical Engineering II of the CTU in Prague, sponsored by the Ministry of Education, Youth and Sports of Czech Republic.

References

1. Yu, L., Liu, H.: Efficient feature selection via analysis of relevance and redundancy. JMLR 5, 1205–1224 (2004)
2. Liu, H., Motoda, H.: Feature selection for Knowledge Discovery and Data Mining. Kluwer Academic Publishers, Boston (1998)
3. Whiteney, A.W.: A direct method of nonparametric measurement selection. IEEE Trans. Comp. 20, 1100–1103 (1971)
4. Marill, T., Green, D.M.: On the effectiveness of receptors in recognition systems. IEEE Trans. IT 9, 11–17 (1963)
5. Pudil, P., Novovièová, K.J.: Floating search methods in feature selection. Patt. Rec. Letters 15, 1119–1125 (1994)
6. Stearns, S.D.: On selecting features for pattern classifiers. In: 3rd Int. Conf. on Patt. Rec., pp. 71–75 (1976)
7. Motwani, R., Raghavan, P.: Randomized algorithms. Cambridge University Press, Cambridge (1995)
8. Goldberg, D.: Genetic Algorithms in Search, Optimization, and Machine Learning. Addison-Wesley, Reading (1989)
9. Mitchell, M.: An Introduction to Genetic Algorithms. MIT Press, Cambridge (1996)
10. Vafaie, H., DeJong, K.: Genetic algorithms as a tool for restructuring feature space representations. In: Proceedings 7th Int Conf. of Tools with AI. IEEE Computer Society Press, New York (1995)
11. Liu, H., Setino, R.: A probabilistic approach to feature selection. In: ML Proceedings 13th ICML, pp. 319–327 (1996)
12. Skalak, D.B.: Prototype and feature selection by sampling and random mutation hill climbing. In: ML Proceedings 11th ICML, pp. 293–301 (1994)

13. Yao, J.: Feature selection for fluoresence image classification. KDD Lab Proposal. Carregie Mellon University (2001)
14. Štìpánová, K., Vavreèka, M.: Improved initialization of clusters in greedy GMM for Neural modeling fields (submitted, 2012)
15. Wasserman, S., Faust, K.: Social Network Analysis: Methods and Applications. Cambridge University Press, UK (1994)
16. Eades, P.: A Heuristic for graph drawing. congressus Numerantium 42, 149–160 (1984)
17. Mutton, P.: Force directed layout of diagrams. PhD thesis. University of Kent, Canterbury (2005)
18. Robins, G., Pattison, P., Kalish, Y., Lusher, D.: An introduction to exponential random graph models for social networks. Social Networks 29, 173–191 (2007), doi:10.1016/j.socnet.2006.08.002
19. Martinetz, T., Schulten, K.: The "Neural gas" network learns topologies. Artificial Neural Networks 1, 397–402 (1991)
20. Heidemann, G.: Efficient Vector Quantization Using the WTA-Rule with Activity Equalization. Neural Processing Letters 13(1), 17–30 (2001)

Built-in Smartphone Accelerometer Motion Pattern Recognition Using Wavelet Transform

Jakub Jirka and Michal Prauzek

Department of Cybernetics and Biomedical Engineering,
VSB-Technical University of Ostrava,
17. listopadu 15, Ostrava-Poruba,
Czech Republic, Europe
{jakub.jirka1,michal.prauzek}@vsb.cz

Abstract. This paper is concerned with motion pattern recognition using built-in accelerometers inside of modern mobile devices - smartphones. More and more people are using these devices nowadays without using its full potential for user motion recognition and evaluation. As accelerometer magnitude level comparison is not sufficient for motion pattern recognition morlet wavelet based recognition algorithm is introduced and tested as well as its device implementation is described and tested. Set of basic motion patterns walking, running and shaking with the device is tested and evaluated.

1 Introduction

Smartphone devices are spreading more and more in this society [1] and these devices themselves contain one of the most sophisticated sensors able to detect user motion in the three dimensional space we are living in. This motion is always connected to the real world context and represented by numeric values in the smartphone. This context can represent normal physiological everyday motion pattern as walking, running, driving or it can also represent pathological motion pattern as for example epileptical seizures or fall. All of this information is vital either for person´s circadian rhythms [2] in case of physiological motion patterns or dangerous situation evaluation in case of pathological motion patterns.

This paper is concerned with context data extraction from device´s accelerometer sensor and motion pattern recognition. Device used in this paper is Google Nexus S with built-in accelerometer ST Microelectronics - LIS331DLH capable of measuring acceleration ±8g on 16bit ADC. Thus accelerometer gives sensitivity of 3,9mg/digit. Sampling frequency of this accelerometer varies from 23-26Hz in the normal setting but can go up to 50Hz in case of necessity.

Simple value threshold is not sufficient for motion pattern recognition as some of the movements like running shaking and fast walking have all similar amplitude. Thus time and frequency domain analysis is required. Based on the nature of accelerometer signal which is rather unpredictable and non-periodical, fast fourier transform had been skipped and wavelet transform was used to classify different events.

Á. Herrero et al. (Eds.): Int. JointConf. CISIS'12-ICEUTE'12-SOCO'12, AISC 189, pp. 447–455.
springerlink.com

2 Accelerometer Data Acquisition and Data Dimension Reduction

Dataset acquired from device was expressed by time series of three independent streams of data points from accelerometer axis X, Y, Z. This data express user acceleration in all direction. In most cases and use scenarios accelerometer data amplitude independent on direction is required. Thus in order of data dimension reduction and process simplification modulus of these three dimension values is calculated according to (2.1) as an absolute value of three axial data from the accelerometer.

$$Z_{ACC} = \sqrt{X^2 + Y^2 + Z^2} \tag{2.1}$$

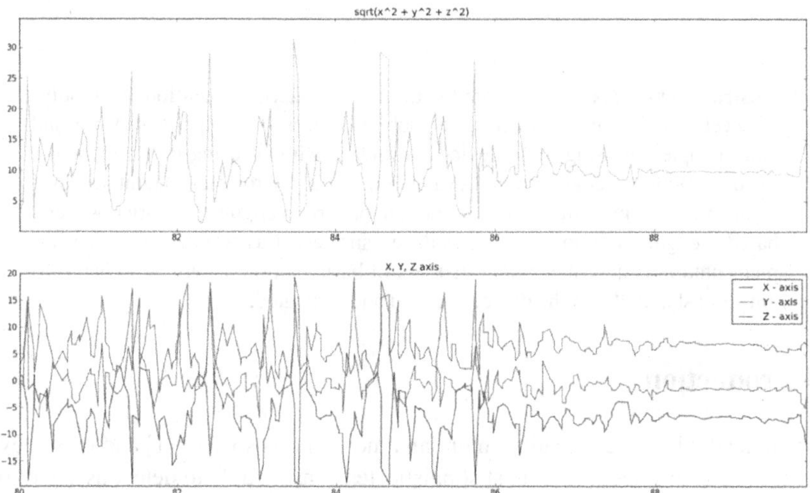

Fig. 2.1. Data modulus and original tri-axial data from walking signal

Data retrieved from the device´s accelerometer have frequency that varies between 23-26Hz which is sufficient for the most of the applications. Device is capable of frequency 50Hz in fastest setting. Three axial accelerometer is a very sensitive device capable of measuring acceleration ±8g on 16bit ADC, accelerometer giving final sensitivity of 3,9mg/digit.

3 Motion Patterns

Data measured by the device can be classified into motion patterns. These motion patterns described in this paper represents walking, running and shaking. The major attention is paid to the difference between running and shaking as it is same in the amplitude but different in time distribution as we can see on the figures (2.2) and (2.3). Walking and running is different in amplitude and thus not interesting for another processing in time or frequency domain.

This application is trying to put into context following activities (movement, shaking, pathological movements), thus this application is trying to create following equation motion pattern = contextual user action (movement). This request for real time processing gives a new request for minimal delay between data´s input data analysis and output

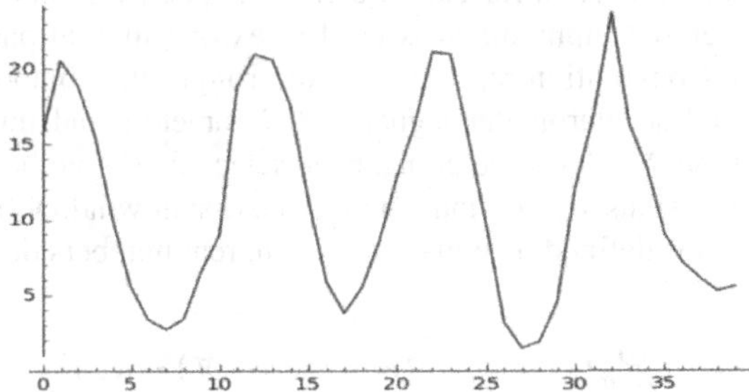

Fig. 2.2. Accelerometer modulus values from running

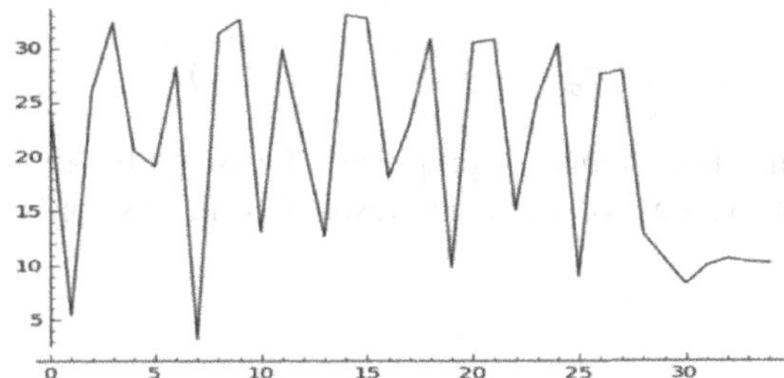

Fig. 2.3. Accelerometer modulus values from shaking

4 Current Research and Results

There is a lot of information described in medical magazines describing the problematic of motion pattern classification [3], [4]. Major differences between this application and applications described in medical articles are:

- Sensors used in medical area are fixed on patient's body and gather more exact information about patient movements. In our case the sensor is not fixed to a patient's body but they are built in the mobile device that is freely for instance in a pocket that produces a lot of disturbing signals
- In a medical research data are usually processed offline after recording. Solution purposed in this article is designed for processing in real time.

From the previous research of frequency analysis following conclusion are known:

- Different versions of furrier transform ale less effective from the reasons of bigger time-frequency wavelet flexibility depending on the specific signal
- There is not a single algorithm that would be able to classify any motion, there is a set of multiple algorithms that can be used

5 Wavelet Analysis

Thanks to furrier transform limitations for non-periodical signals wavelet analysis is used for accelerometer signal analysis and followed classification. Wavelet is a time limited signal (wavelet + window) that interfere with components of analyzed signals. Wavelet is specifically designed for classification of specific components and convolution of this wavelet and input signal is used to extract motion pattern information from the signal [5]. Convolution represents how strongly the components of wavelet presented in analyzed accelerometer signal are. Parameters and thus wavelet shape can be changed in time. Morlet wavelet had been selected as a mother wavelet.

Morlet wavelet is a sinus signal modulated by Gaussian window function of probability distribution and is defined by equation (5.1) in real numbers domain [6].

$$\Psi_\sigma(t) = c_\sigma \pi^{-\frac{1}{4}} e^{-\frac{1}{2}t^2} (\cos(\sigma t) - \kappa_\sigma) \quad (5.1)$$

$$\kappa_\sigma = e^{-\frac{1}{2}\sigma^2} \quad (5.2)$$

$$c_\sigma = (1 + e^{-\sigma^2} - 2e^{-\frac{3}{4}\sigma^2})^{-\frac{1}{2}} \quad (5.3)$$

σ defines change in a time-frequency properties of wavelet. Lesser the value better the time resolution for the cost of frequency resolution. In this application σ=5 is used shown on figure (5.1).

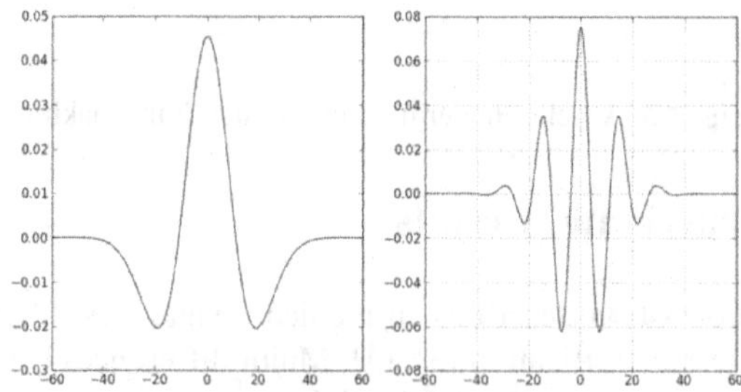

Fig. 5.1. Morlet wavelet with $\sigma = 1$ on left and σ=5 on right

Wavelet analysis performed on input signal is defined with equation (5.4) as convolution of input signal with given wavelet and implemented by fast wavelet transformation performing convolution of FIR filters bank. Filters in filter bank are created by the wavelet property change. Fast wavelet transform is using downsampling of mother wavelet samples changing its central frequency. Downsampling to the half number of samples gives twice the central frequency of original wavelet.

$$\omega[n] = \sum_{-\infty}^{+\infty} X[m] f[n-m] \quad (5.4)$$

6 Wavelet Algorithm Implementation for Accelerometer Data Analysis and Motion Pattern Extraction

Implemented application is using cascade of low pass H(scaling) and high pass G(wavelet) filters created as a quadrature mirrored filters [6] of scaled mother wavelet. Used scaled versions of mother wavelet are described in table (). Mother wavelet is defined from -2 to 2 with default step between samples 0,0125 displayed in figure (Fig 5.1).

Table 6.1. Scaled wavelets table that forms the filter bank

Wavelet ID	Time step between samples
w1	0,0125
w2	0,0250
w3	0,05
w4	0,1
w5	0,2
w6	0,4
w7	0,5

Scaled morlet wavelet is applied to the input signal as described in chapter (5). Output signal power is calculated according to the following equation (6.3).

$$p_n[k] = \frac{1}{N}\sqrt{\sum_{k-N+1}^{k} w_n[k]^2} \tag{6.3}$$

Where N is an average constant and in this application N = 20. Output signal used for classification is obtained with following steps:

a) Input signal is prepared and it must according to decimation rule have 2^n samples. If the signal does not comply with this rule, then signal is zero padded.
b) Coefficients of low pass and high pass filter are calculated from scaled morlet wavelet and high pass filter is obtained as quadrature mirror filter of the low pass filter
c) Convolution of input signal in time domain with coefficients of low pass and high pass filter is performed containing signal approximation coefficients and detail coefficients
d) Steps a-c are repeated for decimated (scaled) version of wavelet until list of all wavelets is exhausted

7 Wavelet Algorithm Testing Results

At first graph of signal that represents walking is displayed in figure (2.1). This signal's wavelet analysis is displayed in figure (7.1). This figure and all others that follow this result contains input data in a first subplot wavelet results in the second subplot and signal power spectrum in the third subplot.

7.1 Walking Test Results

Wavelet number w4 has the highest relevance to the change of the input signal with the step between the samples of 0,2. This gives in the mother wavelet given from -2 to 2 altogether 20 samples that equals to the time of 800ms with average accelerometer sampling frequency of 25Hz. That means that 1 step had been made in 800ms. This wavelet specify and represent both time and frequency information. Time information preservation while containing frequency information of w4 which is highlighted can be seen on figure (7.2).

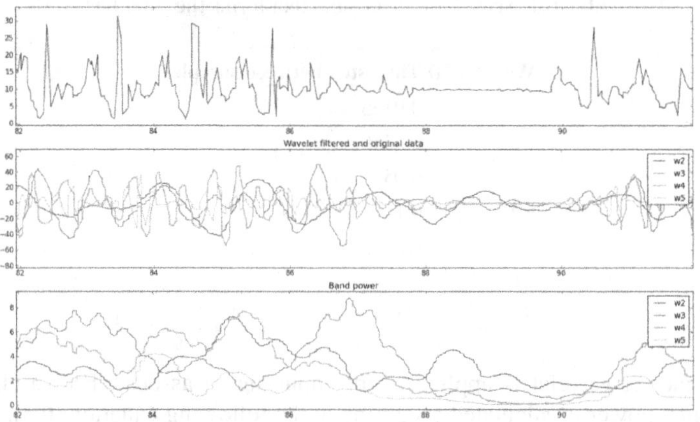

Fig. 7.2. Walking action data wavelet analysis with highlighted w4

7.2 Running Test Results

As seen from figures (7.2) and (7.4) there is not much difference in a frequency spectrum of running and walking both running and walking had happened in the same

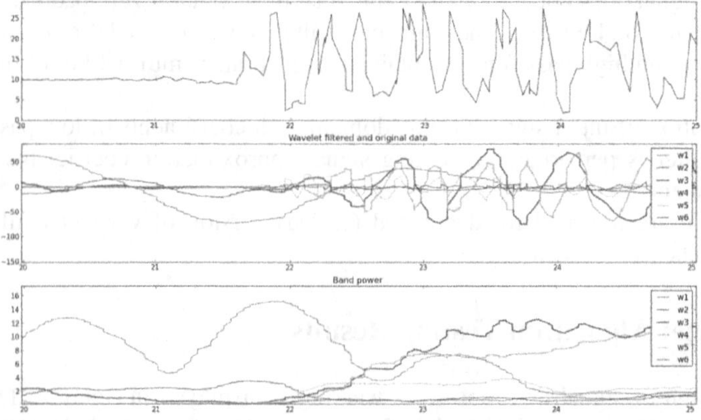

Fig. 7.4. Running action wavelet analysis with highlighted w3 and w4

pace around 1 step per second. But as stated previously there is a difference in a magnitude and thus power of these signals which is clearly visible from figures (7.2) and (7.4) where walking wavelet (Fig 7.2) band power has maximum energy of wavelets w3 and w4 within 8 and running wavelets w3 and w4 (Fig. 7.4) has maximum energy at 13. These values are clearly distinguishable from each other and it is possible to classify between this two movements.

7.3 Shaking Test Results

As expected energy of shaking had moved to higher frequency band and most time and frequency expressing wavelet is wavelet w6 (Fig. 7.6) that has step between samples 0,5 which gives us 8 samples in scaled wavelet from -2 to 2. This means at sampling frequency 25Hz that shaking was at frequency 3,125Hz. From this we can clearly extract the information about shaking from accelerometer signal.

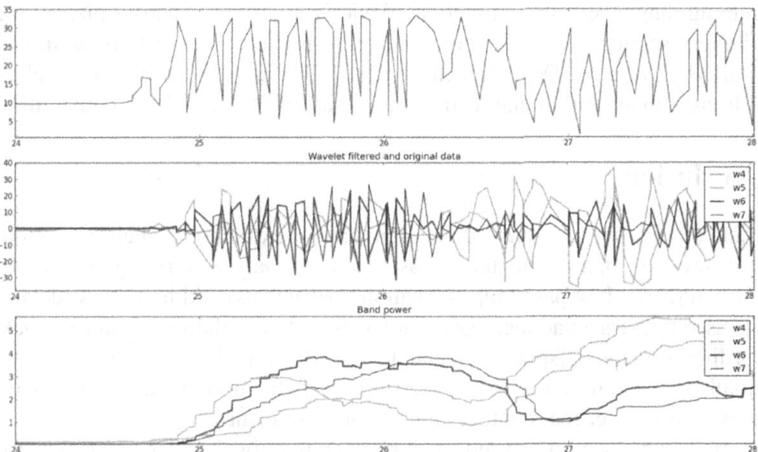

Fig. 7.6. Shaking action wavelet analysis with highlighted w4, w5 and w7

8 Test and Results Evaluation

Algorithm motion pattern extraction capability and correct classification had been tested in number of cases using three devices with built in accelerometers running android operating system. These devices are, Samsung Nexus S, Motorola Razr, LG Optimus One. Each motion pattern was tested 8 times always with different user. Tested parameters were motion pattern extraction success rate, with other variables of device type and device placement. These results are displayed in following table (8.1). Main tested parameter is average success rate of motion pattern detection expressing success rate over the number of 8 tests on 3 devices.

Table 8.1. Algorithm testing results

Motion pattern	Number of tested device	Device placement	Average success rate of motion pattern detection [%]
Walk	3	Pocket	83,3
Walk	2	Hand	62,5
Run	3	Pocket	91,6
Run	2	Hand	95,8
Shake	3	Pocket	58,3
Shake	2	Hand	91,6

9 Device Implementation Notes

Motion pattern extraction and detection algorithm was implemented in java and C programming languages into smartphone device running on Android platform. Performance hungry sections of application code were implemented in C programming language via NDK (Native Development Toolkit). These sections of code includes fast wavelet transform algorithm that renders into series of FIR filter bank convolutions.

10 Conclusion

This article purposes a solution to the motion pattern extraction and classification by the means of wavelet analysis method. Fast wavelet analysis forming a filter bank of quadrature mirrored low pass (approximation coefficients) and high pass (detail coefficients) is used to extract and identify motion pattern like walking, running or shaking.

According to results displayed in chapter 7 and 8 wavelets had proven to be appropriate mathematical tool for the accelerometer data motion pattern extraction and analysis with final user real contextual movement information.

Algorithm introduced and used in this paper is only the base algorithm for more sophisticated algorithm used for the classification of spastic epileptical seizures detection algorithm.

Acknowledgments. The work and the contribution were supported by the project: Ministry of Education of the Czech Republic under Project 1M0567 "Centre of applied electronics", and TACR TA01010632 "SCADA system for control and measurement of process in real time". This paper was also supported student grant agency SP 2011/22 "User's adaptive systems I." and student grant agency SP 2012/75 "User's adaptive systems II."

References

1. Brownlow, M.: Smartphone statistics and market share. Email marketing reports (June 2010)
2. Cagnacci, A., Elliott, J.A., Yen, S.S.: Melatonin: a major regulator of the circadian rhythm of core temperature in humans. The Journal of Clinical Endocrinology & Metabolism 75(2), 447–452 (1992)

3. Barralon, P., Vuillerme, N., Noury, N.: Walk Detection With a Kinematic Sensor: Frequency and Wavelet Comparison. In: Proceedings of the 28th IEEE EMBS Annual International Conference, New York City, USA, August 30-September 3 (2006)
4. Goupillaud, P., Grossman, A., Morlet, J.: Cycle-Octave and Related Transforms in Seismic Signal Analysis. Geoexploration 23, 85–102 (1984)
5. Mallat, S.: A Wavelet Tour of Signal Processing, 2nd edn., p. 5, 257. Academic Press (1999)
6. Schneider, K., Farge, M.: Wavelets: Mathematical Theory. Encyclopedia of Mathematical Physics, 426–438 (2006)
7. Vrána, J.: Quadrature mirror filter banks with sigma delta modulators. Doctoral thesis, VUT-Brno (2007)

Temporal Analysis of Remotely Sensed Data for the Assessment of COPD Patients' Health Status

Sara Colantonio and Ovidio Salvetti

Institute of Information Science and Technologies, ISTI-CNR
Via G. Moruzzi 1, 56124 Pisa – Italy

Abstract. In the last years, ICT-based Remote Patients' Monitoring (RPM) programmes are being developed to address the continuously increasing socio-economic impact of Chronic Obstructive Pulmonary Disease (COPD). ICT-based RPM assures the automatic, regular collection of multivariate time series of patient's data. These can be profitably used to assess patient's health status and detect the onset of disease's exacerbations. This paper presents an approach to suitably represent and analyze the temporal data acquired during COPD patients' tele-monitoring so as to extend usual methods based on e-diary cards. The approach relies on Temporal Abstractions (TA) to extract significant information about disease's trends and progression. In particular, the paper describes the application of TA to identify relevant patterns and episodes that are, then, used to obtain a global picture of patient's conditions. The global picture mainly consists of TA-based qualitative and quantitative features that express: (i) a characterization of disease's course in the most recent period; (ii) a summarization of the global disease evolution based on the most frequent pattern; and (ii) a profiling of the patient, based on anamnesis data combined with a summary of disease progression. The paper focuses on the description of the extracted features and discusses their significance and relevance to the problem at hand. Further work will focus on the development of intelligent applications able to recognize and classify the extracted information.

1 Introduction

In the last years, dedicated health programmes for Chronic Obstructive Pulmonary Disease (COPD) patients are being developed to provide a long-term care option mainly based on Remote Patients' Monitoring (RPM). The principal aim is to reduce disease impairment and exacerbation episodes, thus improving patients' quality of life. Several trials have demonstrated the real advantages of patients' tele-monitoring [1]. Such trials mainly consist in providing tele-assistance to patients via phone calls from clinical personnel, i.e., qualified nurses. During the periodical calls, nurses use to investigate patient's conditions by asking a number of questions. In the most structured cases, interviews are organized according to an *e-diary card*, which assigns a score to a number of health indicators. According to the scores computed at the current call and their comparison with previous ones, the caller assesses the patient's current status, understands if conditions are worsening and decides whether alert specialized clinicians or not [2-3].

Á. Herrero et al. (Eds.): Int. JointConf. CISIS'12-ICEUTE'12-SOCO'12, AISC 189, pp. 457–466.
springerlink.com

In the last years, advanced settings are emerging for RPM. These valuably rely on the deployment of an ICT-based sensorized infrastructure, which is used to continuously acquire and monitor key health indicators, within patients' long-stay settings. The advantages obtained are obvious: (*i*) the automatic acquisition and communication of all the sensed data; and (*ii*) the collection of a much larger and precise corpus of data (some examples above all: ECG signals acquired by wearable sensors, or data about air quality provided by some environmental sensor). Both these advantages pose the urgent demand for intelligent applications able to support clinical professionals in the investigation and correlation of the multi-parametric sensed data. Few attempts to tackle this problem have been recently reported in the literature for different chronic diseases [4-5]. A big challenge, in this frame, is the analysis of the time series of sensed data so as to automatically recognize patient's worsening signs and suitably alert care provisioners.

This paper addresses just this problem in the case of COPD patients monitored remotely. The main goal is to early identify the onset of disease exacerbation events, since this is, indeed, still an open problem. Actually, although different evidence based indications are known, the medical community has not yet reached a general consensus about the criteria on how to predict when patient's conditions are starting to deteriorate. The driving idea of the work is to suitably exploit, for the diagnosis, the time series of sensor data acquired during tele-monitoring, and extending e-diary based approaches that consider only the current values and some past reference values. Specifically, the work reported focuses on the problem of representing time series data, this means that we describe the extraction of significant features that can be easily managed, processed and understood by intelligent recognition or classification applications. Precisely, we use *Temporal Abstractions* (TA) [6] to define the criteria of temporal data analysis, extract relevant patterns or episodes and suitably summarize patient's disease course. The temporal information obtained is meant to be afterwards processed by a classifier, suitably trained to diagnose patient's health status. This will be done in the next phase of our research activity. The temporal analysis based on TA offers several advantages. Not by chance, TA have been extensively used in data mining within the clinical domain [7-8]. In this work, TA serve to transform time series into a high-level symbolic form that is easily understandable also by non technical users. This fosters valuable peculiarities of the proposed TA-based method with respect to other model-based approaches to time series analysis (e.g., Hidden Markov Models). Actually, the TA-based method can be (*i*) easily understood by clinicians, (*ii*) better compared and verified with known clinical knowledge about the disease; (*iii*) used within a further stage of knowledge discovery to have an insight into COPD pathological signs and symptoms that regulate the onset of an exacerbation. In the following, we firstly introduce the problem of time series representation and describe the application of TA-based analysis of sensed data acquired in COPD patients' monitoring. Then, we illustrate the features extracted from data time series so as to portray a global picture of patient's conditions. Finally, we conclude the paper by discussing the procedure followed to assess the feasibility of the method on synthetic data and present future works.

2 Temporal Analysis of Patients' Sensed Data

Formally, the problem can be stated as follows. We have a set of health indicators that are collected remotely about disease signs and symptoms, patients' behavior and activity, as well as environmental conditions. These correspond to a set of variables V_i, $i=1,\ldots,N$, measured in time series: $Meas_{it}=<v_{i0},\ldots, v_{it}>$ such that, for each V_i, we have a series of values v_{ij} acquired at time j, being $j=0$ the starting point and $t\in T$ the current time of acquisition. Globally, at each time t, we have a dataset of N time series, one for each variable:

$$D_{Mt} = \{Meas_{it}\}, i=1,\ldots,N, t=0,1,2,\ldots$$

The goal is to diagnose patient's status, i.e. predict the value of a variable S_t at each time $t\in T$. This means to define a function f: $D_{Mt} \rightarrow S_t$.

A data-driven approach can be envisaged to learn S directly from the measurement observations with a standard Machine Learning method, possibly enriched with some a priori knowledge elicited from clinicians. However, such a method should not be applied directly to raw time series, since this would raise several drawbacks: among others, the high dimensionality of the input space, which would be difficult to handle and the strong correlation among data, which will interfere with the learning method. It is, on the contrary, desirable to perform a data space transformation Θ: Θ: $D_{Mt} \rightarrow (D_{Mt})^*$ that maps each series $<v_{i0},\ldots, v_{it}>$ into a fixed-size feature vector ϕ_i that maintains the significant information hidden into the time series as much as possible. $(D_{Mt})^*$ is, then, represented by the collection of all the features vector computed at time t for each of the N variables. Temporal analysis is employed to define the function Θ and the vectors of features ϕ_i. In particular, the analysis consists in a combination of both TA qualitative features and quantitative statistical features. More specifically, we define a data space transformation Θ to express the following information that characterizes patient's current status: (*i*) a characterization of disease's course in the most recent period; (*ii*) a summarization of the global disease evolution based on the most frequent pattern; (*iii*) a profiling of the patient, based on both anamnesis data and a summary of disease progression.

In the next sub-section, we report how our approach adapts the main concepts of TA to COPD patients' monitoring, while in the next section, we describes how we obtain the Θ transformation, i.e. the vector of feature ϕ_i.

2.1 TA-Based Variable Representation

To describe our TA-based approach, we firstly introduce the variables sensed in time series, discuss how we treat different types of variable and, then, introduced the TA primitives used to represent them.

Variables Monitored. The variables V_i considered are those collected within CHRONIOUS, an EU IST Project aimed at developing a platform of services for RPM of chronic patients affected by COPD and chronic kidney diseases. More precisely, such variables express information about: (*i*) disease signs and symptoms; (*ii*) patient's disability; (*iii*) patient's mood and frailty; (*iv*) patient's real activity; and (*v*) environmental conditions (especially in terms of air quality). Some of these are automatically measured - on a daily basis, also twice a day in some cases - by body or environmental sensors, while others are obtained through periodic questionnaires answered by patients themselves or their relative caregivers. Thirty variables are, this way, obtained. Precisely, the variables acquired by questionnaire are: dyspnoea, ventilator interaction, cough, asthenia, sputum, sputum colour, wheeze, minute ventilation, sleep quality, and walk. The variables acquired by sensors are: body temperature, weight, heart rate, ECG, systolic and diastolic blood pressure, oxygen saturation (SpO2), respiratory rate, inspiratory time, expiratory time, respiratory asynchrony, tidal volume, coughs counter (#/min), patient activity, ambient temperature, ambient humidity, patient position (supine)/h, sleep quality (#position changes/h). Demographic variables are: age and depressive phenotype (HAD scale).

Currently, within CHRONIOUS, an e-diary card is defined to evaluate all these variables, according to a scoring pattern [9]. This means that a score is assigned to different ranges of the variable values or combination of variables values. See Table 1 for some examples of variable scoring. To assess patient's current health status, the scores of each card item are summed up and the result is compared to the sum computed at patient's *baseline* time (i.e., the value computed at patient's enrolment or after the recover from the last exacerbation). If the comparison exceeds a certain threshold, patient's conditions are considered worsened and an alert is sent. For a more detailed description of the card evaluation procedure, please refer to [9]. As it is evident, this procedure discards all the information hidden in the evolution of the different card items in between the baseline time and the current time of card evaluation. For this reason, we propose an approach to extend the e-diary card evaluation: such an approach retains the variables and the score pattern of the card, and evaluates patient's status according to all the information hidden in the temporal series of the different variables.

Table 1. The scoring pattern of the e-diary card for some of the monitored variables

Item \ score	0	1	2	3	4
Sputum	No need for sputum	Moderate	Copious	Very copious	Unbearable
Syst. Blood Pr.	120	120-140	140-160	140-160	>190
Patient Activity (#steps/day)	>2000	< 2000 > 1800	< 1800 > 1000	< 1000 > 500	<500

As already stated, variables are acquired with different frequency, i.e. some once or twice a day, while others on a regular basis of two or three days. For those variables acquired twice a day, we introduce a couple of other variables: for instance, blood pressure is measured both early in the morning and in the middle afternoon, so we consider a variable *BP* split into *BPm* for the morning measures and *BPa* for the afternoon measures. Besides the thirty variables of the card, we also introduce variables related to patient's anamnesis (e.g., presence of comorbities) and therapy (e.g., medication taken). All these variables belong to three main categories: (*i*) numerical, e.g., systolic/diastolic blood pressure or temperature; (*ii*) categorical, e.g., cough, sputum or depressive phenotype, which assume textual, qualitative values; (*iii*) Boolean, e.g., assumption of antipyretics, which can be true or false.

TA Representation. For all these variables, we use TA to represent the content of the corresponding time series. More specifically, each time series $Meas_{it}$=< $v_{i0},\dots, v_{it}$> is transformed into an *interval-based* representation as a sequence of a certain number *n* of:

$$<\pi_{i1}[s_{i1}: e_{i1}], \dots, \pi_{in}[s_{in}, e_{in}]>$$

where $\pi_{ik} \in \mathcal{A}$ is a *primitive* abstraction pattern for the variable V_i that lasts from the starting time s_{ik} (s_{i1}=0 $\forall i$=1,..,*N*) to the ending time e_{ik} (e_{in}=*t* $\forall i$=1,..,*N*), and $\mathcal{A}$ is a TA alphabet that contains the finite number of permitted abstraction *primitives.* Our approach customizes this alphabet introducing dedicated primitives. More precisely, for Boolean variables, we introduce simple primitive patterns of the alphabet $\mathcal{A}_B$ = {*ON*, *OFF*}, for example $ON_{antipyretics}[s{:}e]$ indicates that patient has being taken antipyretics during the time interval [*s*:*e*]. For numerical and categorical variable, we consider abstraction primitives that regard the value, i.e. the *state*, of the variable, and its evolution, i.e. the *trend.* In addition to these, we also introduce a *Peak* primitive. All these primitives are described in detail in the following. In addition, we introduce the concept of *Relations* among primitives and two custom patterns, i.e. *Combined* and *Complex* patterns, which are all described hereafter.

State Primitives. The state primitive patterns are defined extending the scoring pattern of the e-diary card: the patterns considered are *Severely Low* (*SL*), *Very Low* (*VL*), *Low* (*L*), *Mildly Low* (*ML*), *Normal* (*N*), *Mildly High* (*MH*) *High* (*H*), *Very High* (*VH*), and *Severely High* (*SH*), i.e.$\mathcal{A}_S$ = {*SL*, *VL*, *L*, *ML*, *N*, *MH*, *H*, *VH*, *SH*}. Table 2 shows how these states are computed starting from the e-diary card scores.

Table 2. The *State* patterns descending from the e-diary card scores for two variables

Item \ score	SL	VL	L	ML	N	MH	H	VH	SH
Systolic Blood Pressure	<80	80-90	90-100	110-120	120	120-140	140-160	160-180	>180
Patient Activity (#steps /day)	<500	< 1000 > 500	< 1800 > 1000	< 2000 > 1800	>2000 <2100	>2100 <2200	>2200 <2300	>2300 <2400	>2400

Trend Primitives. For the trend abstractions, we considered the usual patterns *Decreasing* (*D*), *Steady* (*S*) and *Increasing* (*I*), i.e., $\mathcal{A}_T$ = {*D*, *S*, *I*}. These are computed with an "online" version of the sliding windows algorithm for time series segmentation [10]. The sliding windows algorithm approximates the time series with a piece-wise linear function, whose segments are determined so as to minimize an interpolation error. We define our version "online" since it is designed to be applied while data are coming: the segmentation is performed each time a new value arrives at the current time *t*. The algorithm works by trying to conjunct the currently arrived point with the starting point of the last segment, if the interpolation error exceeds a predefined bound, a new segment is started, otherwise the point is added to the last segment. The abstractions are determined from the slopes of the fitted segments (see Fig. 1 for an example of segmentation).

Peak Primitive. We also consider the primitive pattern *Peak* (*P*), $\mathcal{A}_P = \{P\}$ which correspond to a local maximum of a variable. Peaks are important to indicate that some peculiar situation is happening.

Relations. A simple *relation* pattern is introduced besides primitives. More precisely, we define the relation "*change state to*", expressed with the symbol "-", i.e. $\mathcal{R}^- \subseteq \mathcal{A}_S \times \mathcal{A}_S$. This represents those periods during which state transitions are recorded, e.g. "*H changes to VH*" ≡ (*H-VH*).
Starting from the primitive patterns, *combined* and *complex* patterns are defined so as to represent two kinds of co-occurring patterns.

Combined Patterns. A *combined* pattern is used to postulate that a trend pattern co-occurs with a state or a relation pattern or that a peak pattern co-occurs with a state pattern. This way, each segment of the time series has a single, condensed representation. Combined patterns are obtained as $\mathcal{A}^C \equiv (\mathcal{A}_T \times (\mathcal{A}_S \cup \mathcal{R}^-)) \cup (\mathcal{A}_P \times \mathcal{A}_S)$. For instance, (*I, N-MH*)[s:e] is a combined pattern associated to the time period [s:e], i.e., a segment of the time series, during which the variable is increasing and its state changes from normal to mildly high.

Complex Patterns. *Complex* patterns are introduced to indicate when two TA patterns of two different variables co-occur. In particular, we consider only the co-occurrence between a combined pattern of a numerical or categorical variable with a primitive pattern of a Boolean variable: $\mathcal{A}^O \equiv \mathcal{A}^C \times \mathcal{A}_B$. Complex patterns for two generic variables V_1 and V_2 are expressed as $(\pi_{1k} \wedge \pi_{2h})[s{:}e]$. An example of complex pattern is $((I, N\text{-}MH)_{temperature} \wedge \text{ON}_{antipyretics})$[s:e]. Examples of patterns are reported in Fig. 1.

Trend Primitives. For the trend abstractions, we considered the usual patterns *Decreasing* (*D*), *Steady* (*S*) and *Increasing* (*I*), i.e., $\mathcal{A}_T = \{D, S, I\}$. These are computed with an "online" version of the sliding windows algorithm for time series segmentation [10]. The sliding windows algorithm approximates the time series with a piecewise linear function, whose segments are determined so as to minimize an interpolation error. We define our version "online" since it is designed to be applied while data are coming: the segmentation is performed each time a new value arrives at the current time *t*. The algorithm works by trying to conjunct the currently arrived point with the starting point of the last segment, if the interpolation error exceeds a predefined bound, a new segment is started, otherwise the point is added to the last segment. The abstractions are determined from the slopes of the fitted segments.

Peak Primitive. We also consider the primitive pattern *Peak* (*P*), $\mathcal{A}_P = \{P\}$ which correspond to a local maximum of a variable. Peaks are important to indicate that some peculiar situation is happening.

Relations. A simple *relation* pattern is introduced besides the primitive ones. More precisely, we define the "*change state to*" relation, expressed with the symbol "-", i.e. $\mathcal{R}^- \subseteq \mathcal{A}_S \times \mathcal{A}_S$. This represents those periods during which state transitions are recorded, e.g. "*H changes to VH*" ≡ (*H-VH*).

Starting from the primitive patterns, *combined* and *complex* patterns are defined so as to represent two kinds of co-occurring patterns.

Combined Patterns. A *combined* pattern is used to postulate that a trend pattern co-occurs with a state or a relation pattern or that a peak pattern co-occurs with a state pattern. This way, each segment of the time series has a single, condensed representation. Combined patterns are obtained as $\mathcal{A}^C \equiv (\mathcal{A}_T\times(\mathcal{A}_S \cup \mathcal{R}^-)) \cup (\mathcal{A}_P\times\mathcal{A}_S)$. For instance, *(I, N-MH)*[s:e] is a combined pattern associated to the time period [s:e], i.e., a segment of the time series, during which the variable is increasing and its state changes from normal to mildly high.

Complex Patterns. *Complex* patterns are introduced to indicate when two TA patterns of two different variables co-occur. In particular, we consider only the co-occurrence between a combined pattern of a numerical or categorical variable with a primitive pattern of a Boolean variable: $\mathcal{A}^O \equiv \mathcal{A}^C \times\mathcal{A}_B$. Complex patterns for two generic variables V_1 and V_2 are expressed as $(\pi_{1k} \wedge\pi_{2h})[s{:}e]$. An example of complex pattern is $((I, N\text{-}MH)_{temperature} \wedge \mathrm{ON}_{antipyretics})$[s:e].
Examples of patterns are reported in Fig. 1.

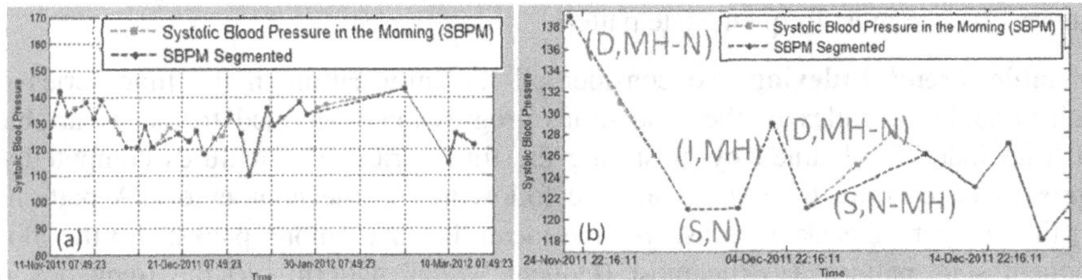

Fig 1. The left hand shows the time series of the variable Systolic Blood Pressure in the morning (SBPM) and its segmentation. The right hand reports a zoom on a selected period and the TA patterns annotated on the different segments.

2.2 Patient's Current Status Summarization

According to the temporal analysis introduced in the previous section, the time series of each variable are abstracted into TA sequences of the following types:

$$Meas_{it}=<v_{i0},\ldots, v_{it}> \rightarrow<\pi_{i1}[s_{i1}: e_{i1}], \ldots, \pi_{in}[b_n, e_n]>;\ s_1=0 \text{ and } e_n=t$$

where $\pi_k\in \mathcal{A}^C \cup \mathcal{A}^O$ for each k=1,…,n, i.e., each pattern can be either a combined or a complex pattern. Such a transformation is performed "online", that is each time new data arrive, the time series are extended of one point added at time t+1, and their segmentation is refined just around that new point. The last part of the abstracted sequence is updated consequently. This means that the update can consist either in the extension of the last pattern $\pi_{in}[b_{in}, e_{in}]$ with $e_{in}=t+1$ or in the introduction of a new pattern $\pi_{i(n+1)}[b_{i(n+1)}, e_{i(n+1)}]$ with $b_{i(n+1)}= e_{in}$ and $e_{i(n+1)}=t+1$.
Starting from this first transformation, time series are further processed to extract qualitative TA and quantitative features. For each variable V_i, a vector of features is extracted organized in three groups of data, which describe the following abstract content:

$\forall V_i$, i=1,..,N, ϕ_i = [*Recent Variable Course* features; *Variable Trend Index* features; *Patient's Typical Profile for the Variable* features]

Recent Variable Course. Since disease signs and symptoms evolve in definite periods of time, the current patient's health status likely depends on the progression of his/her conditions in the last monitoring period. For this reason, the diagnosis is based on an evaluation of how the disease course is evolving in the last, definite time window. This means that for each variable, the first group of features consists of the sequence of TA patterns that, starting from t, goes back to t-δ. To improve accuracy, quantitative features are associated to each of the PA patterns of the time window. Precisely, we include the last value of the segment for patterns of type *Increasing* and *Decreasing*, and the maximum value of the variable in that segment for *Stable* patterns. For all the patterns, the slope of the approximating segment is also inserted. Since the number of patterns can vary from window to window, also a maximum number of patterns η is established. This way, the length of the feature vector is maintained fixed. When the number of patterns within the δ-long window exceeds η, the last η-1 TA patterns are included in the feature vector. The ηth feature is computed as a summary of the window time not covered by the patterns selected. Such a summary is obtained by computing the following quantitative features: (*i*) the number of trend patterns; (*ii*) the most frequent state pattern.

Variable Trend Indexing. To consider all the information in the time series of each variable, an index of the time series progression is defined for each variable. Such an index is obtained by clustering the time series. The features computed to represent each entire time series are the following: (*i*) the number of TA patterns; (*ii*) the number of peak patterns; (*iii*) the mean distance among peaks; (*iv*) the most frequent state pattern; (*v*) the most frequent "*long*" pattern; (*vi*) the frequency of pattern changes. Peaks are considered important information to understand if that variable is subject to a frequent or sporadic suddenly increase. *Long* patterns correspond to a sequence of three patterns [(I, H-VH)$_{ik}$[e_{ik},s_{ik}]; (S,VH)$_{i(k1+1)}$[$e_{i(k+1)}$,$s_{i(k+1)}$]; D(VH-MH)$_{i(k+2)}$[$e_{i(k+2)}$,$s_{i(k+2)}$] with $s_{i(k+1)} = e_{ik}$ and $s_{i(k+2)} = e_{i(k+1)}$. The frequency of such patterns is easily computed in the online feature extraction process, by maintaining a frequency table for the possible combinations of patterns. This way, any time a new pattern is introduced at the end of the series segmentation, it is append with the last two patterns and the frequency table of the resulting long pattern is updated. All these features are passed to a *Self Organizing Map* (SOM) model which returns an index (the number of the winning neuron) used as the trend index of the variable.

Patient's Variable Profiling. Finally, each variable is profiled for the considered patient. Variable profiling is based on the following data: (*i*) the most frequent *long* pattern; (*ii*) the sequence of TA patterns before, during and after the last exacerbation event. Any time an exacerbation is detected, the evolution of each monitored variable is tracked and suitably stored in patient's profile so as to be used as reference for the next deterioration events. The time interval considered around the exacerbation is fixed to 5 day according to clinicians' suggestions. The variable profiles are meant to be added to the general profile of the patient. This relies on the definition of a patient's *phenotype*, which characterizes the disease manifestation for specific patients. The phenotypes are defined by combining the values of the following parameters: (*i*) the BODE index; (*ii*) smoking habits; (*iii*) physical activity carried out by the patient; and (*iv*) patient's attitude to develop exacerbations, as suggested by clinicians [9].

3 Discussion and Conclusions

Summarizing, the paper presents an approach to analyze time series of data variables collected during COPD patients' tele-monitoring. The analysis results in the transformation of the series into a set of features vectors, one for each variable, of finite and fixed length. Such vectors can be more easily processed and, most importantly, it expresses fundamental information about the course of each variable. The method presented is innovative since relies on a combination of both TA qualitative and quantitative features and these are meant to be used for online classification, i.e. diagnosis of patient's status at the current state point of the time series dataset (contrary to the standard use of TA for data mining).

Currently, we verified the practicability of the approach in simulated cases created with clinicians' support. More precisely, synthetic time series were created to simulate peculiar cases that can be detected: (*i*) patients with a rather stable course of variable (called *stable*); (*ii*) patients with a very peaky course of some variables such as heart rate and blood pressure, but with an overall clinical picture that clinicians consider stable (called *peaky*); (*iii*) patients with variable conditions and unstable time series that clinicians deem to be monitored more closely (called *tricky*). The analysis proposed allows us to distinguish these peculiar differences, thanks precisely to variable indexing and patient's profiling. For example, Fig. 2(a) shows the result of the SOM trained for clustering systolic blood pressure: it is a 10x10 map trained with the Kohohen's training algorithm (MATLAB implementation).

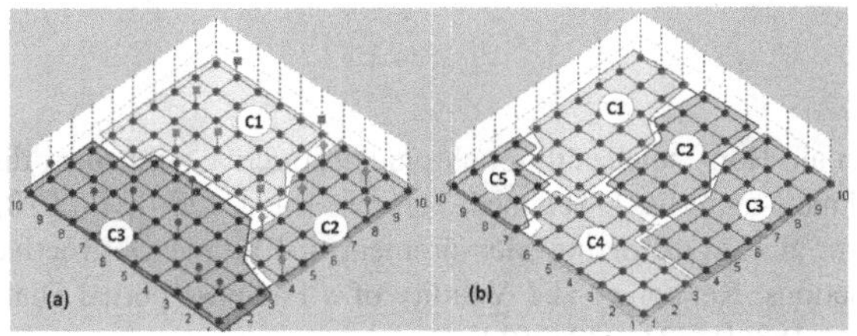

Fig 2. Clustering result for two 10x10-lattice SOMs trained to index systolic blood pressure. (*a*) Results of *in vitro* tests: cluster C1 corresponds to *stable* patients, C2 to *peaky* and C3 to *tricky*. (b) Preliminary results for *in vivo* tests, the clusters are being interpreted with clinicians' support.

The lattice arrangement after training showed clearly a three cluster structure as illustrated in the picture. In addition, for the different types of patients, clinicians suggested how to simulate some exacerbation events. These mainly consisted in the slow deterioration of some known variable, such oxygen saturation and respiratory rate. The TA-based approach to variable course representation clearly highlighted this behaviour of the variables, as shown in Fig. 3. Anyway, these were well-known situations created in vitro. The most interesting results are those that will be achieved in vivo, when the proposed method will allow novel patients' typologies and pathophysiological behaviour to be recognized and/or discovered. CHRONIOUS is currently in its validation phase and real patients' data are being collected. We are applying the method to these data. For each feature, a SOM has been trained on two

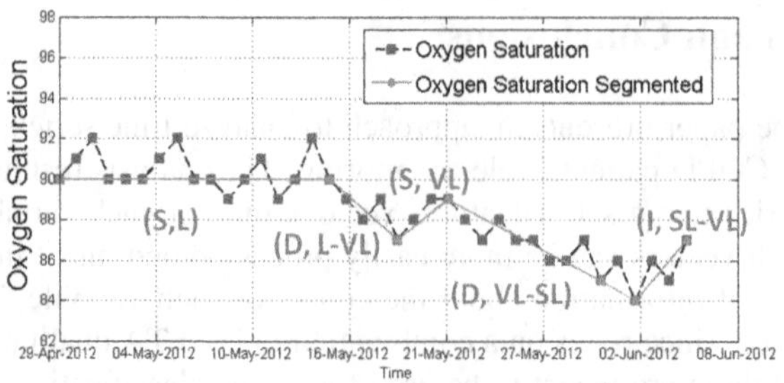

Fig 3. Time series for oxygen saturation that shows a clear deterioration: this behaviour is clearly identified by the two TA patterns (D, L-VL) and (D,VL-SL)

months data and the preliminary clustering is being used. Fig. 2(b) shows the newly obtained structure of the map lattice with two new clusters that we are interpreting with clinicians support. The further step of the research activity is to train a Machine Learning classifier that will be able to process the defined features and diagnose patient's status. Solutions based on classifiers' combinations and hybridization are being considered.

Acknowledgments. Authors would like to express their gratitude to the clinical team of the CHRONIOUS Project for their wise contribution and support.

References

[1] Polisena, J., et al.: Home telehealth for chronic obstructive pulmonary disease: a systematic review and meta-analysis. J. Telemed. Telecare 16(3), 120–127 (2010)

[2] Leidy, N.K., et al.: Standardizing Measurement of Chronic Obstructive Pulmonary Disease Exacerbations. Reliability and Validity of a Patient-reported Diary. Am. J. Respir. Crit. Care Med. 183(3), 323–329 (2011)

[3] Vijayasaratha, K., Stockley, R.: Reported and unreported exacerbations of COPD: analysis by diary cards. Chest 133, 34–41 (2008)

[4] Chiarugi, F., et al.: Decision support in heart failure through processing of electro- and echocardiograms. AIiM 50, 95–104 (2010)

[5] Basilakis, J., et al.: Design of a decision-support architecture for management of remotely monitored patients. IEEE Trans. Inf. Tech. Biomed. 14(5), 1216–1226 (2010)

[6] Shahar, Y.: A framework for knowledge-based TA. Artif. Intell. 90, 79–133 (1997)

[7] Batal, I., et al.: A Pattern Mining Approach for Classifying Multivariate Temporal Data. In: IEEE Int. C. on Bioinf and Biomed., Georgia (November 2011)

[8] Verduijn, M., et al.: Temporal abstraction for feature extraction: A comparative case study in prediction from intensive care monitoring data. AIiM 4(1), 1–12 (2007)

[9] Colantonio, S., et al.: A Decision Making Approach for the Remote, Personalized Evaluation of COPD Patients' Health Status. In: Proc. 7th Int. W. BSI 2012, Como, Italy, July 2-4 (2012)

[10] Keogh, E., Chu, S., Hart, D., Pazzani, M.: Segmenting Time Series: A Survey and Novel Approach. In: Data Mining in Time Series Databases. World Scientific (2003)

Encoding Time Series Data for Better Clustering Results

Tomáš Bartoň and Pavel Kordík

Faculty of Information Technology
Czech Technical University in Prague

Abstract. Clustering algorithms belong to a category of unsupervised learning methods which aim to discover underlying structure in a dataset without given labels. We carry out research of methods for an analysis of a biological time series signals, putting stress on global patterns found in samples. When clustering raw time series data, high dimensionality of input vectors, correlation of inputs, shift or scaling sensitivity often deteriorates the result. In this paper, we propose to represent time series signals by various parametric models. A significant parameters are determined by means of heuristic methods and selected parameters are used for clustering. We applied this method to the data of cell's impedance profiles. Clustering results are more stable, accurate and computationally less expensive than processing raw time series data.

1 Introduction

Clustering techniques partition objects into groups of **clusters** so that objects within a cluster are *similar* to one another and *dissimilar* to objects in other clusters. Similarity is commonly defined in terms of how "close" the objects are in space, based on a *distance function* [1].

In real-world application clustering can be misled by the assumption of uniqueness of the solution. In the field of unsupervised learning a single best solution might not exists. There could be a set of equally good solutions which show a dataset from different perspectives. Because of noise, intrinsic ambiguity in data and optimization models attempting to maximize a fitness function, clustering might produce misleading results. Most clustering algorithms search for one optimal solution based on a pre-specified clustering criterion. Usually the quality of the solution can be adjusted by setting up algorithm-specific parameters [2,3].

In the end, all the user cares about is the "usefulness" of the clustering for achieving his final goal [4]. This usefulness is vaguely defined, but can be easily evaluated by an expert who has background knowledge of the dataset and is able to verify the correctness of a result. Sometimes constructive criticism is too difficult, but negative examples (marking incorrect cluster assignments) are easy to find.

2 Problem Specification

A given dataset contains time series data that reflect the biological activity of cells placed on a plastic plate with 96 separated wells. On the bottom of each well an electrode was placed, which measured impedance while a small electric current was going

Á. Herrero et al. (Eds.): Int. JointConf. CISIS'12-ICEUTE'12-SOCO'12, AISC 189, pp. 467–475.
springerlink.com

through. These datasets were measured at IMG CAS[1]. The goal is to find a grouping of samples in a way which would put similar response patterns into the same groups. The similarity is defined rather by the visual similarity of the curves than the absolute values of the curves. The measured quantity is called *Cell Index* and it is a proportional variable displaying a change of impedance compared to the initial state. The concentration of samples has a huge influence on the value of Cell Index. However the shape of curve stays the same. The signals are not periodical and all of them should converge to zero in the end. A sample input is shown at Figure 1.

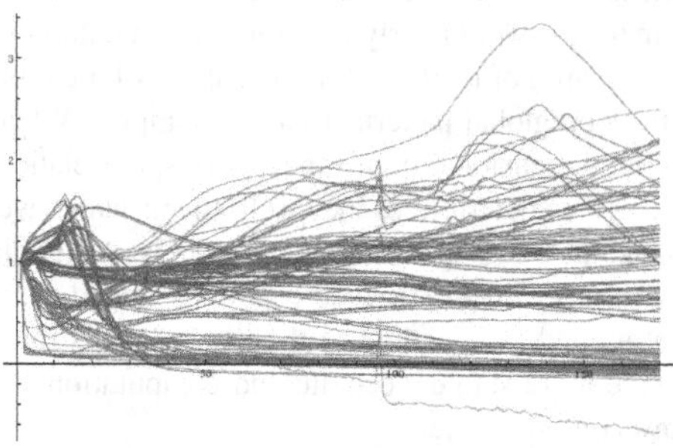

Fig. 1. This chart represents an example of an input dataset with 112 samples – on the y axis Cell Index is shown, which captures number of surviving cells in specific well. The x axis displays time in hours, at the beginning a reactive compound was added. This caused a significant decrease (cells died) in some wells, other samples survived.

The time series signal starts at the moment when a reactive compound is added. The most interesting samples are those, which are resistant to the added virus or capable of recovery after a couple of hours. The samples that are decreasing quickly could be used as a reference group but they are not really perspective for further research.

To sum up, what really matters are the global trends and possibility to discover new, unknown patterns. Which is also the reason why we used an unsupervised learning method.

3 Clustering Time Series

Time series include a huge variety of data that is processed in the areas of medicine [5], biology [6], speech recognition [7], financial market analysis and many other fields. We can distinguish several categories of time series data by their character: **horizontal** (data fluctuate around a constant value), **trend** (contains a visible global trend of increase or decrease), **seasonal** (the trend in data repeats periodically) and **cyclical** (rises and falls repeat without a fixed period).

[1] Institute of Molecular Genetics, Academy of Sciences Czech Republic `http://www.img.cas.cz/`

Some time series datasets could include a combination of these patterns. It is quite clear that some algorithms perform better on a specific type of data. And for each category a specific encoding of information could be found.

The given time series data are not cyclical, the data might fall in a category of trend time series. However the trend of global decrease is not interesting as long as it is common for all patterns found in a dataset. Some samples have the characteristic of rapid increase and after a culmination point it is either slowly or rapidly decreasing, while others contain a wave shape which could be considered as just one period of some goniometric function. The final remarkable category of patterns are signals with exponential decrease.

There are basically three major approaches to performing clustering of time series data. You can either work directly with raw data, indirectly with features extracted from the raw data, or indirectly with models built from the raw data [8]. In the following sections we would like to compare the first two approaches.

3.1 Raw Time Series Data

Abassi et al. [9] described some biological patterns found in their data. For the clustering of impedance profiles a hierarchical clustering of all data points was used. A bottom-up approach used in agglomerative clustering works well for grouping similar responses if they are aligned in time and all sample have the same length. We applied the very same approach to a test dataset. The resulting clustering is quite good (see Figure 2a), however using all data points as an input for a clustering algorithm is not just computationally expensive but also does not represent patterns well.

3.2 Model Based Encoding

Since the impedance values change over time in a smooth fashion, we wish to fit our data to a curved function. Thus, we assume that the data can by represented by a general model:

$$m_t = f(t_t) + \varepsilon_t, \quad t = 1, 2, \dots T$$

where the ε are the errors modelled by a Gaussian distribution $N(0, \sigma^2)$ and m_t is a value of a model in time t. While doing curve fitting we try to minimize the root mean square error, defined as:

$$E_{RMSE} = \sqrt{\frac{1}{n} \sum_{i=1}^{T} (y_i - f(t_i))^2}$$

The obtained parameters of the model are further used in the clustering process. A general polynomial function is defined as follows:

$$p_n(t) = a_n t^n + a_{n-1} t^{n-1} + \dots + a_1 t_1 + a_0 \tag{1}$$

where a_n to a_0 are parameters. To fit the parameters to the data the Levenberg-Marquardt [10] method was used, which is a special type of the Newton method.

In order to get comparable parameters without missing values models must be simple and universal. Otherwise curve fitting might not converge or the RMSE would be too huge.

As an input for the clustering algorithm we used parameters from n equations. With higher degrees of polynomials the number of inputs grows and is given by formula $(n+1)(n+2)/2-1$. Polynomials of higher degrees fit more precisely to various types of data. Experimentally we found that polynomials with degree between 4 to 6 give a good trade-off between the number of input attributes and precise representation of inputs. By leaving out the last parameter of the polynomial we can easily get rid of the vertical transition of curve.

Another model is base on an exponential function defined as follows:

$$y(t) = a \cdot e^{-bt} + c \tag{2}$$

Many processes in biology have an exponential trend, also in our case some categories could be easily fitted to this model.

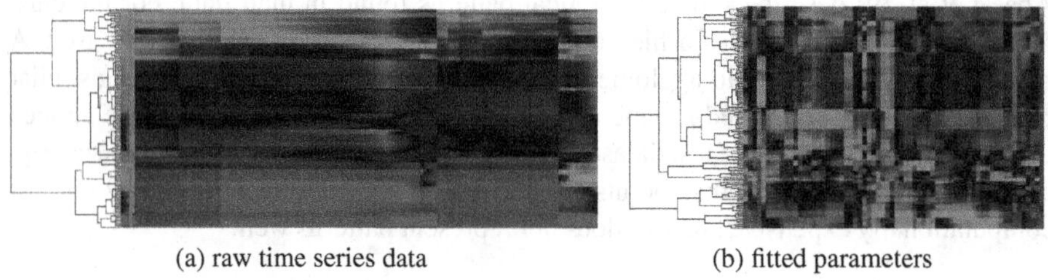

(a) raw time series data (b) fitted parameters

Fig. 2. Visual representation of clustering results

4 Experiments

In order to have a precise evaluation metric we have created a dataset with samples from different experiments and afterwards an expert classified the data into 5 categories (see Figure 6). To understand better the way how the experts analyse the data we run a forward selection algorithm on the dataset with labels. Surprisingly not many attributes are needed to decide such a task when you have the advantage of having labels. With only two attributes, one from the beginning and other from the end of the time series data was enough to obtain 98% classification precision (similarly the decision tree only uses 6 time points from 3 different parts of the time series for classification – see Figure 3). This result might suggest splitting the analysis of the signal into multiple parts. The selected time points are almost equally distributed which signifies that all parts of the measurement are important and we can not draw any conclusion from the analysis of just one subsequence.

Any approach based on a similar selection (or aggregation) of specific attributes might suffer from overrating absolute values of samples. This decision tree would fail on a dataset which contained similar but either horizontally or vertically shifted samples

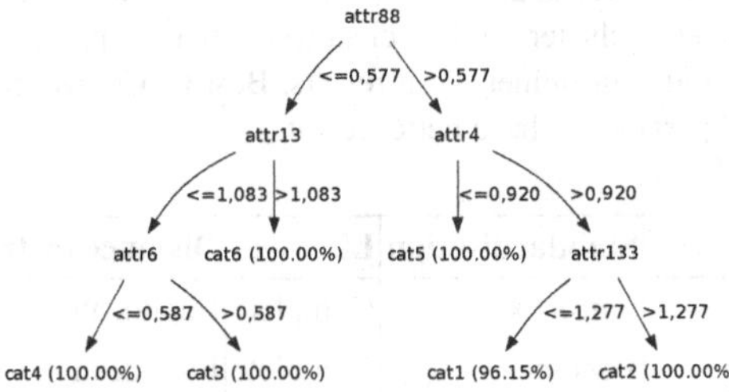

Fig. 3. Decision trees are able to decide the classification of time series with only very few attributes. At least one attribute is taken from the beginning of measurement and other from the end.

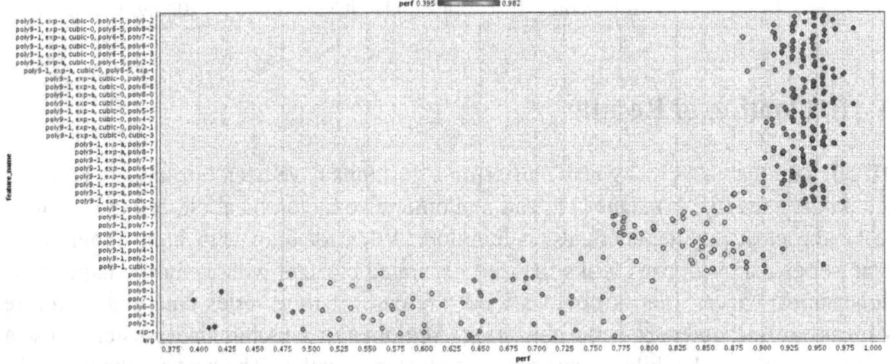

Fig. 4. Classification performance of forward selection on the testing dataset. Algorithm was given a set of 66 attributes to choose from, with 3 attributes we can get the precision above 90%, with 4 and 5 attributes the precision is even better. However using more than 5 attributes would lead to over fitting (worse results).

(where a similar effect appears either sooner or later in time). Therefore we would like to introduce an approach which is resistant to this shift and capable of finding similar patterns.

It is quite clear that including polynomials with increasing degree will introduce some duplicate information. To avoid this issue we used forward selection again with 10 fold cross-validation, to see which attributes are more significant. As you can see from Figure 4 at least three attributes are needed to obtain 90% precision in classification. This proves that the information included in our estimated parameters is more general that information in raw time series data. One of the chosen parameters is the a coefficient of an exponential model (see (2)). This parameter does characterise well the rapid increase or decrease at the beginning of curve, so including this one does make sense. Another parameter signifies horizontal movement of the curve, which seems to be important for the assignment to a category.

Table 1. Comparison of CPCC for agglomerative hierarchical clustering with different settings. CPCC closer to 1 means better clustering. It is clear to see that preprocessing of data and chosen distance metric have a significant influence on results. Best CPCC was achieved with raw data, however in average fitted parameters have better results.

Input	Standardisation	Linkage	Distance metric	CPCC
raw time series	min-max	Complete	Euclidean	0.811
	z-score	Complete	Euclidean	0.652
	maximum	Complete	Euclidean	0.806
	z-score	Average	Canberra	0.944
fitted parameters	min-max	Complete	Euclidean	0.747
	maximum	Average	Euclidean	0.778
	z-score	Average	Canberra	0.894
	z-score	Complete	Canberra	0.889

5 Evaluation of Results

Throughout the years many new clustering algorithms have been introduced, however the oldest ones like k-means [11] and agglomerative hierarchical clustering [12] tend to be the most popular methods in literature. When dealing with high dimensional time series data, k-means looks for well-separated clusters with rounded shape in n-dimensional space. That is obviously not the case of time series data and therefore k-means fails to find reasonable clustering. Agglomerative hierarchical clustering use a bottom-up approach while merging closest clusters together. The best results were obtained when average linkage was used, on the other hand the worst results were obtained with the single linkage.

Unsupervised learning is a challenging field mainly due to the lack of a universal evaluation criterion. To deal with this problem we have chosen a semi-supervised approach.

From traditional evaluation metrics we used the Cophenetic correlation coefficient (CPCC) [13] which is one of many evaluation metrics that can be used as optimization criteria. The value of the CPCC should be maximized, however a higher value of the CPCC does not guarantee that user would prefer this result to another clustering. Also should be noted that the number of clusters does not influence the value of CPCC.

By fitting parameters we manage to lower the dimensionality of input data. Clustering is less sensitive to noise and therefore more stable. However, some attributes are more important than others. Clustering produced directly by hierarchical clustering corresponds to counting area below a curve. Therefore patterns with rapid decrease are always together. This might not be the case of the other approach, nevertheless this is just a question of proper weighting and selection of input attributes, which should be done by user.

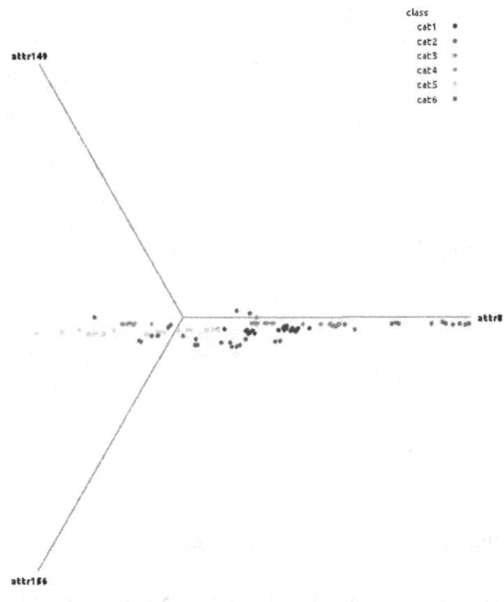

(a) raw time series data

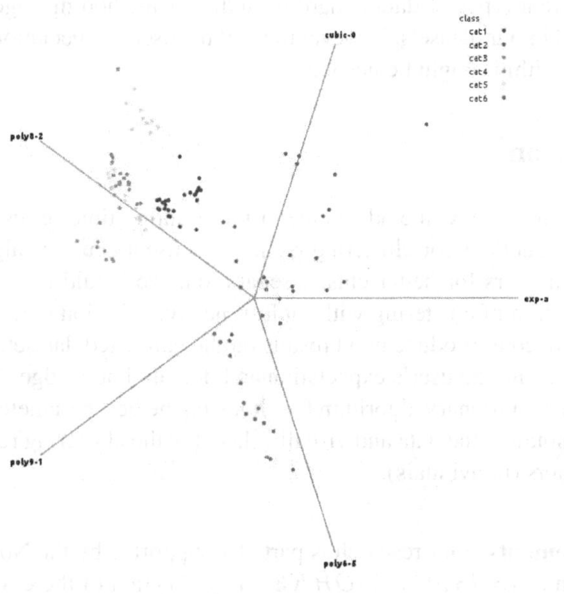

(b) fitted parameters

Fig. 5. Visualization of important attributes for both approaches, which were selected by an evolutionary optimization. It is clear to see that raw time series data does not form separable nor compact clusters. A linear projection was used for both visualizations.

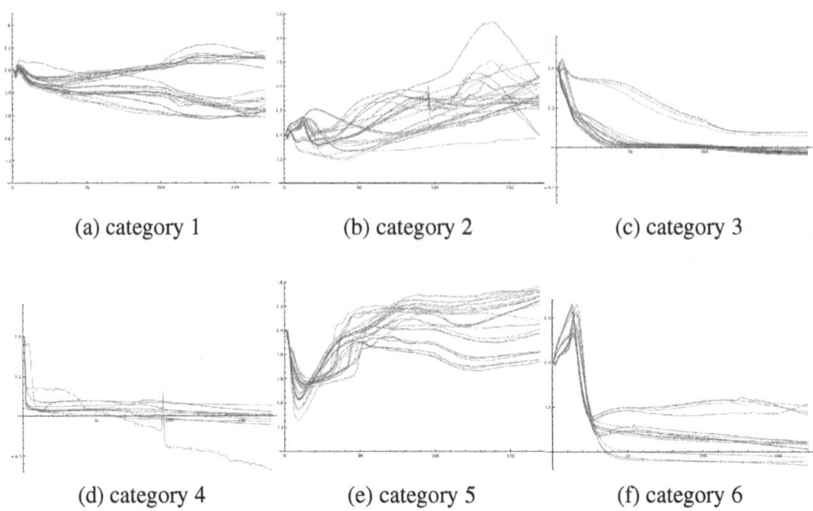

(a) category 1 (b) category 2 (c) category 3

(d) category 4 (e) category 5 (f) category 6

Fig. 6. Manual classification of patterns into 5 categories which was done by an expert

Using similar models is problem specific, and a generalized approach that would work on different types of data is hard to find. Our method manages to capture some patterns found in our datasets, however to fulfil the user's expectations some training in clustering algorithms might be needed.

6 Conclusion

In this contribution show the advantage of representing time series signals by parameters of fitted functions for clustering results. We use a heuristic algorithm to select a subset of parameters for better cluster separation. We would like to focus on the interactive evolution of clustering with multi-objective criterion optimization. The fitted parameters proved to produce good results on the annotated dataset, however, we need to take into account the user's expectation and domain knowledge. This could be done by running an evolutionary algorithm for choosing the best parameters while the user is iteratively annotating the data and visually checking the clustering results on a different set of parameters (individuals).

Acknowledgements. Our research is partially supported by the Novel Model Ensembling Algorithms ($SGS10/\ 307/OHK3/3T/181$) grant of the Czech Technical University in Prague.

We would like to thank Petr Bartůněk, Ph.D. and Antonio Pombinho, M.Sc. from the IMG CAS institute for supporting our research, providing the data and letting us publish all details of our work.

References

1. Han, J., Kamber, M.: Data mining: concepts and techniques. Morgan Kaufmann Publishers Inc., San Francisco (2000)
2. Bifulco, I., Fedullo, C., Napolitano, F., Raiconi, G., Tagliaferri, R.: Global optimization, meta clustering and consensus clustering for class prediction. In: Proceedings of the 2009 International Joint Conference on Neural Networks, IJCNN 2009, pp. 1463–1470. IEEE Press, Piscataway (2009)
3. Caruana, R., Elhawary, M., Nguyen, N., Smith, C.: Meta clustering. In: Proceedings of the Sixth International Conference on Data Mining, ICDM 2006, pp. 107–118. IEEE Computer Society, Washington, DC (2006)
4. Guyon, I., Luxburg, U.V., Williamson, R.C.: Clustering: Science or art. In: NIPS 2009 Workshop on Clustering Theory (2009)
5. Golay, X., Kollias, S., Stoll, G., Meier, D., Valavanis, A., Boesiger, P.: A new correlation-based fuzzy logic clustering algorithm for fmri. Magnetic Resonance in Medicine 40(2), 249–260 (1998)
6. Möller-Levet, C.S., Klawonn, F., Cho, K.-H., Wolkenhauer, O.: Fuzzy Clustering of Short Time-Series and Unevenly Distributed Sampling Points. In: Berthold, M., Lenz, H.-J., Bradley, E., Kruse, R., Borgelt, C. (eds.) IDA 2003. LNCS, vol. 2810, pp. 330–340. Springer, Heidelberg (2003)
7. Wilpon, J.G., Rabiner, L.R.: A modified k-means clustering algorithm for use in speaker-independent isolated word recognition. The Journal of the Acoustical Society of America 75, S93 (1984)
8. Liao, T.W.: Clustering of time series data – a survey. Pattern Recognition 38(11), 1857–1874 (2005)
9. Abassi, Y.A., Xi, B., Zhang, W., Ye, P., Kirstein, S.L., Gaylord, M.R., Feinstein, S.C., Wang, X., Xu, X.: Kinetic cell-based morphological screening: prediction of mechanism of compound action and off-target effects. Chem. Biol. 16(7), 712–723 (2009)
10. Marquardt, D.W.: An algorithm for least-squares estimation of nonlinear parameters. Journal of the Society for Industrial and Applied Mathematics 11(2), 431–441 (1963)
11. MacQueen, J.B.: Some methods for classification and analysis of multivariate observations. In: Cam, L.M.L., Neyman, J. (eds.) Proc. of the Fifth Berkeley Symposium on Mathematical Statistics and Probability, vol. 1, pp. 281–297. University of California Press (1967)
12. Lance, G.N., Williams, W.T.: A general theory of classificatory sorting strategies. The Computer Journal 9(4), 373–380 (1967)
13. Sokal, R.R., Rohlf, F.J.: The comparison of dendrograms by objective methods. Taxon 11(2), 33–40 (1962)

Pattern Recognition in EEG Cognitive Signals Accelerated by GPU

Pavel Dohnálek, Petr Gajdoš, Tomáš Peterek, and Marek Penhaker

Department of Computer Science and Department of Measurement and Control, FEI, VSB - Technical University of Ostrava,
17. listopadu 15, 708 33, Ostrava-Poruba, Czech Republic
{pavel.dohnalek,petr.gajdos,tomas.peterek,marek.penhaker}@vsb.cz

Abstract. Analysing of Electroencephalography (EEG) cognitive signals becomes more popular today due to availability of essential hardware (EEG headsets) and sufficient computation power of common computers. Fast and precise pattern matching of acquired signals represents one of the most important challenges. In this article, a method for signal pattern matching based on Non-negative Matrix Factorization is proposed. We also utilize short-time Fourier transform to preprocess EEG data and Cosine Similarity Measure to perform query-based classification. The recognition algorithm shows promising results in execution speed and is suitable for implementation on graphics processors to achieve real-time processing, making the proposed method suitable for real-world, real-time applications. In terms of recognition accuracy, our experiments show that accuracy greatly depends on the choice of input parameters.

Keywords: Electroencephalography, cognitive signal, pattern matching, graphics processors, matrix factorization.

1 BCI Implementation Accelerated by GPU

For more than twenty years graphics cards have been used purely to accelerate 3D graphics computations. The motivation was simple - 3D computations are time consuming to such extent that classic general purpose processing units (CPUs) can no longer perform well enough to accommodate both application logic and displayable data. Graphics card took upon themselves the burden of computing this displayable data, eventually evolving into processing units based on massive parallelization. Today, somewhat ironically, graphics cards are capable of general purpose computing, taking the load from CPUs even more and performing much better than any current CPU could.

There are several software technologies that enable programmers to harness the parallel computing power of modern GPUs, among which most notable are OpenCL by Khronos Group, CUDA by NVIDIA and DirectCompute by Microsoft. In our research we decided to focus our efforts on the CUDA technology.

In order to massively utilize parallelism in various tasks, CUDA uses a structure of threads. These threads are organized into three-dimensional blocks, which

Á. Herrero et al. (Eds.): Int. JointConf. CISIS'12-ICEUTE'12-SOCO'12, AISC 189, pp. 477–485.
springerlink.com

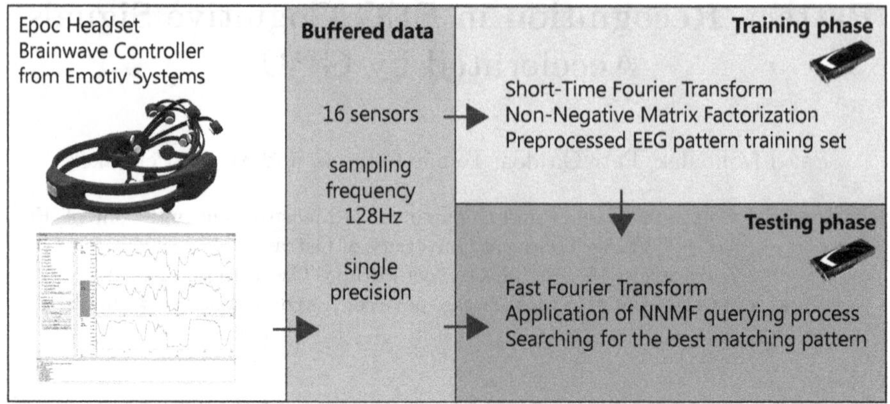

Fig. 1. The GPU accelerated BCI schema

are in turn organized into two-dimensional grids. While blocks can contain at most 512 threads, it is not defined which dimension should contain them. It is legal to distribute the threads into the three-dimensions without limitation as long as the total number of threads in a single block does not exceed 512. While it is perfectly legal to create a block of 16 x 16 x 1 threads, one cannot specify a block whose dimensions are 32 x 32 x 1 as the total number of threads in this case is 1024. Grids do not constrain the number of blocks they can support as long as neither of the two dimensions exceed the value of 65,535. Overall, single multiplication makes it clear that any grid can contain up to 2,198,956,147,200 threads. That does not, however, mean that all these threads could be computed all at once. For details on how threads are distributed for computing and how they are executed, we encourage the reader to further study the matter in [3].

1.1 The CUBLAS Library

As massive parallel computing gained on popularity, several libraries computing linear algebra problems were created. Very well known is the freely available BLAS library whose massively parallel version designed for CUDA is called CUBLAS. CUBLAS is, like its non-parallel counterpart, structured into three levels. The first level contains basic functions to perform scalar and vector operations. Level 2 is a set of functions designed to compute matrix-vector operations while the purpose of Level 3 functions is to perform various matrix-matrix operations.

For NNMF computations, Level 3 functions will be the most significant as NNMF factor matrix update rules rely heavily on matrix-matrix multiplication. It is obvious that the efficiency of matrix-matrix multiplication algorithm greatly influences the overall performance of NNMF computation. It turns out CUBLAS provides good-performance algorithms while maintaining very simple API, thus greatly simplifying the implementation.

All matrices to be processed by CUBLAS library must be stored in the column-major format. Full CUBLAS API along with brief but still sufficient descriptions of individual functions can be found in CUDA Toolkit 4.0 CUBLAS Library by NVIDIA Corporation.

1.2 Non-Negative Matrix Factorization

Due to technological advances in the field of data collection in various application fields ranging from molecular research to deep space exploration, it has become increasingly difficult to process all collected data. There are many techniques being used for automated data analysis like image and speech recognition, natural language processing, document classification, signal processing and many others. One of them is Non-negative Matrix Factorization, commonly abbreviated as NNMF, which specializes in discovering and learning underlying features of an input matrix. In our research we utilize NNMF as the cornerstone of our EEG pattern recognition algorithm, accelerating the process using CUDA technology.

The NNMF Problem. Many algorithms (and their versions) exist to efficiently compute NNMF, two of the arguably most used ones originally proposed by Lee and Seung in [4]. First, let us formally state the NNMF problem:

Given the input matrix $V_{M \times N}, M \geq 0, N \geq 0$, and the rank $K \ll min(M, N)$, minimize the error matrix E in the equation

$$V = WH + E, \tag{1}$$

where W and H, subjects to non-negativity constrains $W \geq 0, H \geq 0$, are factor matrices with dimensions $M \times K$ and $K \times N$. The algorithms proposed by Lee and Seung use multiplicative update rules to iteratively update the factor matrices, minimizing an error function. One such function is the Frobenius Norm:

$$\|V - WH\|^2. \tag{2}$$

In [4] Lee and Seung prove that the Frobenius Norm is non-increasing if the factor matrices are updated using the following multiplicative update rules:

$$H_{xa} \leftarrow H_{xa} \frac{(W^T V)_{xa}}{(W^T WH)_{xa}}, \tag{3}$$

$$W_{bx} \leftarrow W_{bx} \frac{(VH^T)_{bx}}{(WHH^T)_{bx}}, \tag{4}$$

where a denotes a row of matrix H and a column of matrix W, x represents a column of matrix H and y is the row index of matrix W. We utilize these update rules to compute the factor matrices used later to recognize patterns belonging to particular actions in our EEG signal. Many more algorithms to compute NNMF along with other versions of the NNMF problem such as Sparse NNMF or Large-scale NNMF can be found in [2]

1.3 Action Recognition

Once we have obtained the factor matrices from our training data set using the equations (3) and (4), we compute the Moore-Penrose pseudoinverse matrix of W using Singular Value Decomposition. This pseudoinverse matrix is then used to convert any incoming query vector $\mathbf{q}$ to the required dimensionality and vector space by the following formula:

$$\mathbf{h_q} = W^{\dagger}\mathbf{q}, \tag{5}$$

where $\mathbf{h_q}$ denotes the query vector converted into the vector space of matrix H while $W^{\dagger}$ is the Moore-Penrose pseudoinverse of W [1].

Once we have obtained vector $\mathbf{h_q}$, we can then compare it to all vectors (columns) of matrix H and measure its similarity with them. To measure the similarity, we use Cosine Similarity Measure defined as

$$\frac{\mathbf{h_q^T} \cdot \mathbf{h_x}}{||\mathbf{h_q}||\,||\mathbf{h_x}||}, \tag{6}$$

where $\mathbf{h_x}$ denotes a column vector of matrix H. We seek the maximum of the above expression iterating over all column vectors $\mathbf{h_x}$, of which there are N, obtaining the final formula of the action recognition step:

$$\max_{0 \le x < N}\{\frac{\mathbf{h_q^T} \cdot \mathbf{h_x}}{||\mathbf{h_q}||\,||\mathbf{h_x}||}\}. \tag{7}$$

2 Dataset

There are two types of data we need to use in our algorithms. For NNMF we require a training dataset, that is a set we use to determine the factor matrices H and W. The second set serves us to test the recognition results and thus is called the test dataset. Both the training and testing data are obtained using Emotiv EPOC neural headset. The data was collected during approximately 20 minutes of scanning EEG waves for two actions, called Neutral and Push. The Neutral action represents the state when a headset user, the subject of the measuring, clears his mind and thinks of nothing. The Push action is recorded by the headset when the user thinks of pushing an imaginary object away.

During the recording, 203060 different values were collected from one of the EEG sensors and stored for brainwave pattern recognition. There is a total of 14 relevant EEG sensors in the headset, yielding 14×203060 values to process. When collecting EEG data, the headset also creates many descriptors for the collected data called data chunks. Each data chunk stores information about which data it belongs to, described as a starting index and the number of values the chunk covers, as well as the action the data represents. Each sensor provides different data for a given chunk, but the chunk describes the same amount of data regardless of sensor. That is, if the first chunk has the length of 16, the chunk describes first 16 values collected by each of the sensors. Any two different chunks, however, can have different lengths.

When the data is obtained, it is then preprocessed using Short-Time Fourier Transform (STFT). This transformed data is then used to assemble a training matrix in the following manner: every column represents data processed in one STFT window for each sensor. Let M be the number of rows in the matrix, x be the amount of computed Fourier coefficients and y be the number of EEG sensors used to collect the data. M is then determined as $(\frac{x}{2}+1) \times y$. We only store $(\frac{x}{2}+1)$ Fourier coefficients since the rest is duplicate and carries no new information.

As mentioned above, the length of a data chunk may be different from any other chunk. To construct the matrix, we need to unify the number of values to be taken for STFT. We choose to find out which chunk is the smallest and pick its length. In our experiments the smallest data chunk had the length of 12. If we choose to compute 64 Fourier coefficients, we receive a matrix with $33 \times 14 = 462$ rows. The amount of columns, N, is the number of STFT windows computed. In our dataset there are 2026 computed STFT windows (rougly 10% of the whole dataset), which are properly stored in a matrix and later used as training data in NNMF. The rest is then used as test vectors.

Having constructed the training matrix, its resulting size being 462×2026, we only need to choose the number of features we are looking for by NNMF. Let K be this number. We learned that for given data the best results in terms of both speed of computation and accuracy in pattern recognition are obtained when K is given the value of 10.

3 Experiments

We conducted several experiments to determine the performance of our accelerated implementations. There are several significant parameters whose alteration can influence both execution and accuracy performance. For each experiment we adjusted one of these parameters and measured the performance. Results of the experiments are summarized in tables below, with the abbreviations having the following meaning:

- **TMD** - Training matrix dimension
- **# test vectors** - Number of test vectors
- **STFT** - Short-Time Fourier Transform
- **NNMF** - Non-negative Matrix Factorization
- **LP** - Learning Process (STFT+NNMF)
- **CT** - Classification time for the whole data set
- **CTR** - Classification time for a single input vector
- **ACC** - Classification accuracy

All time values are expressed in milliseconds. Accuracy is measured simply as a ratio between total number of test vectors and correctly classified ones. Experiments were performed on a machine equipped with NVIDIA GeForce 9600GT graphics card, Intel Core i5 750 processor and 8 GB DDR3 RAM.

Table 1. Computation times given in milliseconds showing the performance dependency on the number of Fourier coefficients being computed for the input data

	Number of FFT coefficients				
	64	**128**	**256**	**512**	**1024**
TMD	462 x 2026	910 x 2026	1806 x 2026	3598 x 2026	7182 x 2026
# test vectors	3656				
STFT	5975	5226	5413	5351	5569
NNMF	1264	1295	2371	4352	8065
LP	7239	6521	7784	9703	13634
CT	19235	21044	23525	30154	178714
CTR	5.26	5.76	6.43	8.25	48.88
ACC	64.21%	59.74%	51.37%	41.96%	44.26%

Table 2. Computation times given in milliseconds showing the performance dependency on the size of the window used in STFT. Note that input matrix dimensions are not dependent on the window size. The input matrix is bigger for certain window sizes because for these sizes it was necessary to adjust the number of FFT coefficients computed.

	Size of the STFT window				
	50	**100**	**150**	**200**	**250**
TMD	462 x 2026	910 x 2026	1806 x 2026	1806 x 2026	1806 x 2026
# test vectors	3656	1828	1219	914	732
STFT	5975	5366	5570	5382	5460
NNMF	1264	1326	2293	2356	2340
LP	7239	6692	7863	7738	7800
CT	19235	10186	9052	5257	3993
CTR	5.26	5.57	7.42	5.75	5.46
ACC	64.21%	60.40%	47.15%	42.84%	41.94%

Table 3. Computation times given in milliseconds showing the performance dependency on the number of features that NNMF looks for

	Number of features					
	10	**20**	**40**	**80**	**160**	**320**
TMD	462 x 2026					
# test vectors	3656					
STFT	5975	5547	5211	5367	5507	5273
NNMF	1264	1202	1513	2215	3806	7442
LP	7239	6749	6724	7582	9313	12715
CT	19235	19719	20185	20640	22073	25538
CTR	5.26	5.39	5.52	5.65	6.04	6.99
ACC	64.21%	61.74%	53.12%	57.48%	44.17%	45.70%

Table 4. Computation times given in milliseconds showing the performance dependency on the number of features that NNMF looks for

	Number of NNMF iterations					
	1	10	100	1000	10000	100000
TMD	462 x 2026					
# test vectors	3656					
STFT	5413	5257	5975	5258	5367	5444
NNMF	390	437	1264	5741	53944	535986
LP	5803	5694	7239	10999	59311	541430
CT	19640	19595	19235	19312	19734	19484
CTR	5.37	5.36	5.26	5.28	5.40	5.33
ACC	29.24%	57.26%	64.21%	61.83%	59.14%	67.59%

Table 5. Computation times given in milliseconds showing the performance dependency on the number of NNMF iterations

	Training dataset size (% of the dataset)					
	10	20	40	50	70	80
TMD	462 × 2026	462 × 4056	462 × 8116	462 × 10146	462 × 14206	462 × 16236
# test vectors	3656	3250	2438	2032	1220	814
STFT	5975	10249	20389	25053	35303	40731
NNMF	1264	1357	2216	2808	3759	3978
LP	7239	11606	22605	27861	39062	44709
CT	19235	17612	13463	11278	6615	4337
CTR	5.26	5.42	5.22	5.55	5.42	5.33
ACC	64.21%	55.87%	52.06%	52.70%	56.78%	53.91%

Table 6. NNMF computation times given a square matrix with the number of desired features equal to half the dimension of the matrix

	Matrix dimension				
	256	512	1024	2048	4096
GPU	577	1326	6833	49530	392200
CPU	5912	57751	503107	6544050	not tested

Table 7. NNMF computation times given a square matrix with the number of desired features being exactly 200

	Matrix dimension				
	256	512	1024	2048	4096
GPU	811	1450	3416	11747	41449
CPU	9844	38875	167289	979518	not tested

Table 1 shows the computation times for different numbers of FFT coefficients. It should be noted that the coefficients themselves compose the training matrix, hence the increase in one of the matrix dimensions. While there is a significant increase of classification time for 1024 coefficients, the time is still well within acceptable boundaries.

As shown in table 2, we learned that window size used for STFT has no significant impact on the performance. Note that training matrix dimensions are not dependent on STFT window size - the increase is caused by the change in the number of FFT coefficients that was necessary for certain window sizes.

Table 3 expresses results similar to 2. It is obvious that while the increase in the number of features certainly impacts performance, the impact is of no great import. While it might be interesting to measure the performance for even more features, 320 features is arguably the reasonable maximum number of features to look for in our application. Experiments have shown us that 10 features is the best choice for both performance and recognition accuracy, as stated in chapter 2.

Number of NNMF iterations (table 4) seem to have the greatest impact on the performance, but it should be noted that NNMF converges well within 100 iterations and therefore, once again, number of iterations have very little impact on practical usefulness of our solution. It should be noted that although after approximately 30 iterations the cost function converged on its local minima, there is still significant variance in recognition accuracy. This is due to the fact that the factor matrices are initialized randomly.

Table 5 shows how computation times depend on the amount of training data. This is the only limiting factor we have encountered, seeing 40% of the whole dataset as the reasonable maximum for real life application. The limitation, however, applies only for the training process, with classification times for a single vector constantly hovering between 5 and 6 milliseconds, which was to be expected. Usually, in real life applications, the training phase of the algorithm can be allowed to run longer as it is usually done only once. Therefore, it does not need to be done in real time. As we expected, the time needed to compute Fourier transform increases linearly with the dataset size, adding approximately 5 seconds of computation time to every 10% dataset size increment. One thing to note is that classification accuracy is not strictly increasing with the training dataset size. In fact, the best accuracy was achieved with the smallest training dataset we experimented with.

Considering NNMF to be the cornerstone of our implementation, with data preprocessing and classification methods being subjects to change in future research (especially due to not the best recognition accuracy), we also performed experiments on NNMF computation times alone for different training matrix dimensions. The results can be seen in tables 6 and 7. While part of the GPU version of the algorithm was also computing the pseudoinverse matrix, it was not part of the CPU version, giving the CPU version a significant edge. Also, the CPU version was multithreaded to utilize the full computing power of the processor. Still, it can be clearly seen that our GPU implementation of the NNMF algorithm is far superior to its CPU counterpart with the worst case of GPU performance being 10 times better than that of the CPU.

4 Future Research and Conclusion

Both the performance and price of graphics processing units show great promise for use in the BCI field. Our next efforts will focus on creating BCIs with more advanced EEG devices, increasing the performance by utilizing multiple GPUs and improving the recognition accuracy, which still remains a great challenge to solve. We presented a method of creating a BCI capable of real-time pattern recognition in brainwaves using a low-cost hardware, resulting in a very cost-efficient way of solving the problem, and we certainly feel this method to be worth further elaboration.

Acknowledgement. This work was supported by the Bio-Inspired Methods: research, development and knowledge transfer project, reg. no. CZ.1.07/2.3.00/ 20.0073 funded by Operational Programme Education for Competitiveness, co-financed by ESF and state budget of the Czech Republic.

References

1. Benetos, E., Kotti, M., Kotropoulos, C.: Musical instrument classification using non-negative matrix factorization algorithms and subset feature selection. In: Proceedings of 2006 IEEE International Symposium on Circuits and Systems, ISCAS 2006, vol. 5, p. 4 (2006)
2. Cichocki, A., Phan, A.H., Zdunek, R.: Nonnegative Matrix and Tensor Factorizations: Applications to Exploratory Multi-way Data Analysis and Blind Source Separation. Wiley, Chichester (2009)
3. Kirk, D.B., Hwu, W.-M.W.: Programming Massively Parallel Processors: A Hands-on Approach, 1st edn. Applications of GPU Computing Series. Morgan Kaufmann (February 2010)
4. Lee, D.D., Sebastian Seung, H.: Algorithms for Non-negative Matrix Factorization. In: Leen, T.K., Dietterich, T.G., Tresp, V. (eds.) Advances in Neural Information Processing Systems 13, pp. 556–562. The MIT Press, Cambridge (2001)

On the Potential of Fuzzy Rule-Based Ensemble Forecasting

David Sikora[1], Martin Štěpnička[2], and Lenka Vavříčková[2]

[1] Department of Informatics and Computers, Faculty of Science, University of Ostrava, 30. dubna 22, Ostrava, Czech Republic, David.Sikora@osu.cz

[2] Centre of Excellence IT4Innovations – Division of the University of Ostrava – Institute for Research and Applications of Fuzzy Modeling 30. dubna 22, Ostrava, Czech Republic {Martin.Stepnicka,Lenka.Vavrickova}@osu.cz

Abstract. It is a well-known fact that there is no individual forecasting method that is generally for any given time series better than any other method. Thus, no matter how efficient a chosen method is, there always exists a danger that for a given time series the method is highly inappropriate. To overcome such a problem and avoid the danger of choosing an inaccurate method, distinct ensemble techniques that combine more individual forecasting methods are designed. These techniques construct a forecast as a (linear) combination, i.e. a weighted average, of forecasts by individual methods. It is important to stress that even very sophisticated approaches to determine appropriate weights often fail to outperform so called equal weights approach, i.e. a simple arithmetic mean of individual forecasts.

This contribution attempts to construct a novel ensemble technique that determines the weights based on time series features such as trend, seasonality, kurtosis or stationarity. The knowledge how to determine weights comes from the regression analysis. However, in order to capture the desirable issues of robustness and mainly of interpretability, the knowledge how to combine individual methods is encoded in a linguistic description, i.e. in a specific fuzzy rule base that uses linguistic evaluative expressions. Thus, the mechanism of determination of particular weights is perception-based logical deduction – a unique fuzzy inference technique that is designed for linguistic descriptions. An exhaustive experimental justification is provided in order to confirm the promising potential of the given direction of research. The suggested ensemble approach based on fuzzy rules demonstrates both, lower forecasting error and higher robustness in comparison to individual methods as well as to the equal weights ensemble.

1 Introduction

Time series forecasting is an important tool for support of individual and organizational decision making. The main goal is to predict a behavior of a given

Á. Herrero et al. (Eds.): Int. JointConf. CISIS'12-ICEUTE'12-SOCO'12, AISC 189, pp. 487–496.
springerlink.com

phenomenon based solely on the past patterns of the same event [20]. Formally, we are given a finite sequence $y_1, y_2, \ldots, y_T$ of real numbers that is called a time series and our task is to forecast the next values $y_{T+1}, y_{T+2}, \ldots, y_{T+h}$, where h denotes the so called *forecasting horizon.*

Distinct mainly statistical time series forecasting methods have been designed and are nowadays widely used in practice, let us recall e.g. decomposition, ARIMA or exponential smoothing methods [20]. However, no method that would be superior to any other has been ever found. It is an obvious and well-known fact that for each forecasting method there exist time series on which the method fails to outperform all the other forecasting methods. Thus, relying on a single method is a highly risky strategy.

It is interesting that even searching for methods that outperform any other for narrower specific subsets of time series has not been successful yet. Let us recall e.g. Armstrong et al. [3]:

"Although forecasting expertise can be found in the literature, these sources often fail to adequately describe conditions under which a method is expected to be successful".

1.1 Motivation for Ensemble Techniques

Based on the above mentioned empirical observations concerning the non-existence of a unique optimal forecasting methods so called *ensemble techniques* have started to be designed and successfully applied. The main idea of *"ensembles"* consists in an appropriate combination of more forecasting methods in order to avoid the risk of choosing a single unappropriate one. Typically, ensemble techniques are constructed as a linear combination of the individual ones. Formally, let us assume that we are given a set of M individual methods and for a given time series and a given forecasting horizon h, j-th individual method provides us with the following prediction:

$$\hat{y}^{(j)}_{T+1}, \hat{y}^{(j)}_{T+2}, \ldots, \hat{y}^{(j)}_{T+h}, \quad j = 1, \ldots, M.$$

Then the ensemble forecast is given as follows

$$\hat{y}_{T+i} = \frac{1}{\sum_{j=1}^{M} w_j} \cdot \sum_{j=1}^{M} w_j \cdot \hat{y}^{(j)}_{T+i}, \quad i = 1, \ldots, h \tag{1}$$

where $w_j \in \mathbb{R}$ is a weight of the j-th individual method. Usually, these weights are supposed to be normalized, that is $\sum_{j=1}^{M} w_j = 1$.

Let us mention that it was Bates and Granger [5] who first showed significant gains in accuracy through combinations of more methods. Let us also recall another work [21] where the authors combined various time series forecasts and showed that for set of forecasts, their linear combination could constitute a forecast with an error variance smaller than the one of the individual forecasts. They found out that combining procedures did produce an overall forecast superior to individual forecasts on the majority of tested time series. Also other sources confirm an approach using the combination given by (1) enables to obtain a forecast that exceeds the individual ones in accuracy as well as robustness, see e.g. [4].

Nevertheless, how to combine individual forecast, i.e., how to determine appropriate weights, is still relatively an open question. Although one would expect sophisticated approaches to rule, Makridakis et al. [8] showed that taking a simple average outperforms taking a weighted average. This could be interpreted that the so called "equal weights approach" [9], that is an arithmetic mean, is a hardly beatable benchmark and that trying to find appropriate non-equal weights leads rather to a random damage of the main averaging idea that is behind the robustness and accuracy improvements provided by equal weights.

1.2 Motivation for a Fuzzy Rule Based Approach

Nevertheless, although the equal weights performs as accurately as mentioned above, there are works that promisingly show the potential of more sophisticated approaches. We would like to highlight e.g. Lemke and Gabrys in [18] that is based on distinct features of time series such as: measure of the strength of trend, measure of the strength of seasonality, standard deviation, skewness or kurtosis. Given time series considered as elements of the feature space and clustered by k-means algorithm. Individual methods were ranked according to their performance on each cluster, three best methods for each cluster were determined. For a given new time series, closest cluster is determined and then the given the three best methods are combined.

It should be stressed that this approach performed very well on a sufficiently big set of time series. For us, the publication of Lemke and Gabrys [18] is among two main motivations because if nothing else, it demonstrate, that there is a dependence between time series features and accuracy of some forecasting method that allows to efficiently determine weights for ensemble techniques.

The second major motivation comes from the so called *Rule-Based Forecasting* developed by Collopy and Armstrong [3,9]. These authors came up with linguistically given (IF-THEN) rules that encode an expert knowledge "how to predict a given time series". Once more we recall in the above emphasized observation from [3] that the forecasting expertise that one may find in the literature usually fails in describing conditions under which certain methods do work properly and thus, is not practically applicable.

Most of the rules from [3,9] do not set up weights however, they often set up rather a specific model parameter, e.g. the smoothing factors of the Brown's exponential smoothing with trend. Moreover, in antecedents the rules very often use properties that are not crisp but rather vague, recall e.g. expression such as: "last observation is unusual; trend has been changing; unstable recent trend; recent trend is long" etc., see [9]. For such cases, using crisp rules that are either fired or not and nothing between, seems to be less natural than using fuzzy rules. Similarly, the use of crisp consequents such as: "add 10% to the weight; subtract

0.4 from beta; add 0.1 to alpha" etc. [9], seems to be less intuitive then using vague expression that are typical for fuzzy rules.

By following the main ideas of rule-based forecasting by Collopy and Armstrong and of using time series features (meta-learning) by Lemke and Gabrys, we aim at obtaining an interpretable and understandable model that besides providing an unquestionable forecasting power helps to understand "when what works", nonchalantly said.

2 Methodology

2.1 Forecasting Methods

Having in mind the goal of our investigation, we have chosen the most often used forecasting methods that are at disposal to the widest community of users. These are: Decomposition Techniques (DT), Exponential Smoothing (ES), seasonal Autoregressive Integrated Moving Average (ARIMA), Generalized Autoregressive Conditional Heteroscedasticity (GARCH) models, Moving Averages (MA) and finally, Random Walk process (RW) and Random Walk process with a drift (RWd). Since it is by far out of the topic of the paper to introduce details related to the chosen methods, we only refer to the relevant literature [7,14,20].

In order to avoid any bias from a naive implementation of the above listed methods, we adopted implementations of these methods by professional software packages such as ForecastPro® (for ES, ARIMA and MA), Gretl® (GARCH and RWd) and NCSS® (DT). These tools executed fully automatic parameter selection and optimization which made possible to concentrate the investigation purely on the combination technique. Moreover, their arithmetic mean (AM), that represents the equal weights ensemble, was also determined and used as a valid benchmark.

Remark 1. We are fully aware of the potential of recent computational intelligence methods for the time series forecasting task. For instance, neural networks [11], evolutionary computation [10], fuzzy techniques [25] or their combinations [24] have been demonstrated to have a great potential. Furthermore, in [4] the authors claim that ensemble of computational techniques perform even better than ensembles of standard statistical techniques.

However, the implementation of such techniques is not sufficiently standardized yet and readers have much worse access to such software packages which makes the use of ensembles of such techniques much less feasible. In order to keep the main direction of the paper that is the ensemble technique and not the individual method setting and optimization, we leave this direction for future research.

2.2 Time Series Features

For our purposes, some important features need to be extracted from given time series. Particularly, we used the following features: *Strength of trend, Strength of seasonality, Length of the time series, Skewness, Kurtosis, Coefficient of variation, Stationarity, Forecasting horizon and Frequency.*

Since most of them are standard and well-known, we only briefly describe some of them. Coefficient of variation is a relative measure of data variability that is obtained by the standard deviation divided by a mean:

$$CV = \frac{S_y}{\mid \bar{y} \mid} = \frac{\sqrt{\frac{1}{T-1}\sum_{t=1}^{T}(y_t - \bar{y})^2}}{\frac{1}{T}\sum_{t=1}^{T} y_t}$$

which is a correct formula because all time-series contained only positive values. Stationarity assumes that the mean and the autocovariances of a time series do not change in time and it may be verified by many tests. We employ the Augmented Dickey-Fuller test [12] that under the null hypothesis assumes that a given time series is non-stationary, under the alternative hypothesis that time series is stationary. Value $(1-P)$ where P stands for the obtained P-value serves as the measure stationarity of a given time-series. Strength of trend and seasonality are similarly determined on such tests. Forecasting horizons are different for different time series frequencies (annual, quarterly and monthly time series) and they are adopted from the M3 competition instructions because this was a source of our data-sets, see the latter Section. Length of the time series is determined by the number of observations of the time series.

Since the range of some features can significantly vary, it is crucial to normalize the range of features to the interval [0,1].

2.3 Datasets and Evaluation

To develop and validate the model we have used data from the well-known M3 data-set repository[1] that serves as a general benchmark database provided by the authority of the International Institute of Forecasters. We have chosen 198 time series from the fields Microeconomics, Industry, Macroeconomics, Finance and Demography. This selected data-set was divided into 2 sets (training set and testing set) both containing 99 time series. The training set was used in order to identify our model, that is for a fuzzy rule base identification. The testing set was used in order to test whether the determined knowledge encoded in the fuzzy rules works generally also for other time series out of the training set.

The global performance of a forecasting model is evaluated by an error measure. Historically, *Mean Absolute Error* or *(Root) Mean Squared Error* are very popular error measures. However, Mean Square Error is too sensitive to outliers [2] and furthermore, both methods are scale-dependent measures, i.e. every single time series has a different impact on the overall results [1] and hence, these measures may hardly be used for a comparison across more time series. *Symmetric Mean Absolute Percentage Error* (SMAPE) given by [16]:

[1] The repository contains 3000 time series from the M3 time series forecasting competition [19].

$$\text{SMAPE} = \frac{1}{h} \sum_{t=T+1}^{T+h} \frac{|e_t|}{(|y_t| + |\hat{y}_t|)/2} \times 100\%, \tag{2}$$

where $e_t = y_t - \hat{y}_t$ for $t = T+1, \ldots, T+h$ is a scale independent error measures that can be easily used to compare methods across different time series and thus, will be used in this paper.

3 Fuzzy Rule-Based Ensemble Design

In this section, we briefly describe the main points of the realization of the investigation. We describe how we reach the goal – an identification of a fuzzy rule base that serves as a flexible model that determines ensemble weights purely on the chosen time series features and thus, describing the dependence of forecasting efficiency of individual time series based on the features.

3.1 Fuzzy Rule Base Identification

The fuzzy rule base may be identified by distinct approaches. Because of the reason of missing reliable expert knowledge mentioned above, from the very beginning we omit the identification by an expert and we focus on data-driven approaches.

There are three natural data-driven approaches to identify a fuzzy rule base for our purposes. The very first one and up to now the only approach that led to provided experimentally approved results is the regression analysis. Therefore, we will devote separate Subsection 3.2 to this topic. The other two approaches that are still under construction focus on the identification of fuzzy rules with help of fuzzy clustering and with help of linguistic associations mining that is based on GUHA method [15] and its fuzzy variant [17,23]. Since these approaches are not experimentally approved yet, we avoid describing the details how they are implemented in the fuzzy rule base identification process. However, it is important to mention them here because all three approaches are supposed to be combined.

3.2 Regression Analysis

By the linear regression one can explore the dependence of expected SMAPE errors obtained by each method on the measured features of a given time series. It is obvious that ensemble weights should be (proportionally) higher if a given method is supposed to provide lower SMAPE error and vice-versa. Thus, it is natural to put

$$w_j = (1 - \text{SMAPE}_j) \tag{3}$$

where SMAPE_j denotes the SMAPE error of the j-th method, and to estimate directly the weight of the j-th method w_j instead of the SMAPE value.

So, we use the multiple regression analysis represented by the following system of linear equations in a matrix form:

$$\mathbf{w} = \mathbf{F} \cdot \boldsymbol{\beta} + \boldsymbol{\varepsilon} \tag{4}$$

that may be expanded as follows:

$$\begin{pmatrix} w_{1,j} \\ w_{2,j} \\ \vdots \\ w_{I,j} \end{pmatrix} = \begin{pmatrix} 1 & f_{1,1} & \cdots & f_{1,K} \\ 1 & f_{2,1} & \cdots & f_{2,K} \\ \vdots & \vdots & \ddots & \vdots \\ 1 & f_{I,1} & \cdots & f_{I,K} \end{pmatrix} \cdot \begin{pmatrix} \beta_0 \\ \beta_1 \\ \vdots \\ \beta_K \end{pmatrix} + \begin{pmatrix} \varepsilon_1 \\ \varepsilon_2 \\ \vdots \\ \varepsilon_I \end{pmatrix}$$

where $w_{i,j} = (1 - \mathrm{SMAPE}_j)$ for the i-th time series; $f_{i,k}$ is the k-th measured feature value of the i-th time series, $\boldsymbol{\beta} = (\beta_0, \ldots, \beta_K)^{\mathrm{T}}$ is a vector of unknown regression coefficients, $\boldsymbol{\varepsilon} = (\varepsilon_0, \ldots, \varepsilon_I)^{\mathrm{T}}$ is an error vector. Symbol I stands for the number of time series and K stands for the number of features thus, in our case $I = 99$.

In order to estimate relationship between features of the time series and the desired SMAPE values (and thus to determine the weights) for each individual forecasting method, the vector of the linear regression parameters has been determined using the least square criterion minimization. In this way we obtain seven regression models – one for each forecasting method.

For the sake of clarity and simplicity of the models, only significant features were considered. For the choice of statistically significant features we applied the forward stepwise regression. This algorithm begins with no variables (features) in a model, then it selects a feature that has the largest R^2 value. This step is repeated until all remaining features are not significant. Only statistically significant features were used in the multiple regression which estimated the unknown parameters. Furthermore, only in the case that a P-value of an parameter estimated in the multiple regression was smaller than a certain threshold of significance we rejected the null hypothesis and the estimated parameter was found statistically significant. Note, that for each method, different features played the significant role.

The obtained regression models were sampled and in this way we obtained nodes in the reduced features spaces with only significant features and weights. For example, in the case of the ARIMA method only the *Strength of Trend* and *Variation coefficient* were found significant and thus we obtained a regression model in the form

$$Weight_{ARIMA} = \beta_0 + \beta_1 \cdot \textit{Strength of Trend} + \beta_2 \cdot CV$$

which after its sampling provided gave us with a set of nodes

$$[Weight_{ARIMA,\,\ell}, \textit{Strength of Trend}_\ell, CV_\ell], \quad \ell, \ldots, L$$

where L denotes the number of sample points.

These nodes served as learning data for the so called *Linguistic Learning Algorithm* [6] that automatically generates linguistic descriptions, i.e. fuzzy rule bases with linguistic evaluating expressions, that jointly with a specific fuzzy inference method *Perception-based Logical Deduction* [22] may derive conclusions based on imprecise observations. Due to this, we obtain seven linguistic descriptions – each of them determining weights of a single individual method based on transparent and interpretable rules, such as:

IF *Strength of Trend* **is** *Big* **AND** *CV* **is** *Small* **THEN** $Weight_{ARIMA}$ **is** *ExBig*.

No matter the origin (method of identification) of fuzzy rules, such an ensemble technique will be naturally called *"Fuzzy Rule-Based Ensemble"* (FRBE).

4 Results, Conclusions and Future Work

In order to judge its performance, the fuzzy rule-based ensemble was applied on the 99 time series from the testing set. Table 1 shows that in both the average and the standard deviation of SMAPE values over all testing set, the suggested ensemble outperforms all the individual methods. Moreover, the equal weights method (AM), that is highly suggested in the literature, has been outperformed as well.

Table 1. Average of the SMAPE forecasting errors and standard deviation of the SMAPE

Methods	Average Error	Methods	Error Std. Deviation
DT	21.59	DT	24.52
GARCH	17.27	GARCH	21.22
RWd	15.95	RWd	20.62
RW	15.26	RW	19.43
MA	15.11	MA	19.27
ARIMA	14.44	ARIMA	20.31
ES	14.43	ES	18.39
AM	14.40	AM	18.42
FRBE	**14.18**	FRBE	**18.03**

Although one might consider the improvement to be less than required, it is evident that fuzzy rule-based approach performs very well even against the equal weights method. The fact that the victory has been reached not only in the accuracy but also in the robustness (standard deviation of the error) is also worth noticing.

One should also note that this is a preliminary study that opens this topic for further steps. The use of other techniques for fuzzy rule base identifications (fuzzy cluster analysis, linguistic associations mining etc.) are supposed to be used in order to improve the results shortly. Furthermore, deep redundancy [13] and/or consistency analysis of obtained fuzzy rule bases is highly desirable in order to improve the knowledge that is linguistically encoded in them, is also highly desirable.

Thus, it may be concluded that in this introductory study, we have clearly stated the motivations and main ideas and mainly we have demonstrated the promising potential of the fuzzy rule-based forecasting that entitles us to continue in the foreshadowed future work.

Acknowledgements. This work was supported by the European Regional Development Fund in the IT4Innovations Centre of Excellence project (CZ.1.05/1.1.00/02.0070). Furthermore, we gratefully acknowledge partial support of projects KONTAKT II - LH12229 of MŠMT ČR, 01798/2011/RRC of the Moravian-Silesian region and SGS11/PřF/2012 of the University of Ostrava.

References

1. Armstrong, J.S.: Evaluating methods. In: Armstrong, J.S. (ed.) Principles of Forecasting: A Handbook for Reasearchers and Practitioners. Kluwer, Boston (2001)
2. Armstrong, J.S., Collopy, F.: Error measures for generalizing about forecasting methods: Empirical comparisons. International Journal of Forecasting 8, 69–80 (1992)
3. Armstrong, J.S., Adya, M., Collopy, F.: Rule-Based Forecasting Using Judgment in Time Series Extrapolation. In: Armstrong, J.S. (ed.) Principles of Forecasting: A Handbook for Reasearchers and Practitioners. Kluwer, Boston (2001)
4. Barrow, D.K., Crone, S.F., Kourentzes, N.: An Evaluation of Neural Network Ensembles and Model Selection for Time Series Prediction. In: Proc. of 2010 IEEE IJCNN 2010. IEEE, Barcelona, Spain (2010)
5. Bates, J.M., Granger, C.W.J.: Combination of Forecasts. Operational Research Quarterly 20, 451–468 (1969)
6. Bělohlávek, R., Novák, V.: Learning Rule Base of the Linguistic Expert Systems. Soft Computing 7, 79–88 (2002)
7. Box, G., Jenkins, G.: Time Series Analysis: Forecasting and Control. Holden-Day, San Francisco (1976)
8. Makridakis, S., Andersen, A., Carbone, R., Fildes, R., Hibon, M., Lewandowski, R., Newton, J., Parzen, E., Winkler, R.: The Accuracy of Extrapolation (Time-Series) Methods - Results of a Forecasting Competition. Journal of Forecasting 1, 111–153 (1982)
9. Collopy, F., Armstrong, J.S.: Rule-Based Forecasting: Development and Validation of an Expert Systems Approach to Combining Time Series Extrapolations. Management Science 38, 1394–1414 (1992)
10. Cortez, P., Rocha, M., Neves, J.: Evolving Time Series Forecasting ARMA Models. Journal of Heuristics 10, 415–429 (2004)
11. Crone, S.F., Hibon, M., Nikolopoulos, K.: Advances in forecasting with neural networks? Empirical evidence from the NN3 competition on time series prediction. International Journal of Forecasting 27, 635–660 (2011)
12. Dickey, D.A., Fuller, W.A.: Distribution of the Estimators for Autoregressive Time Series with a Unit Root. Journal of the American Statistical Association 74, 427–431 (1979)
13. Dvořák, A., Štěpnička, M., Vavříčková, L.: Redundancies in systems of fuzzy/linguistic IF-THEN rules. In: Proc. EUSFLAT 2011, pp. 1022–1029 (2011)

14. Hamilton, J.D.: Time Series Analysis. Princeton University Press, New Jersey (1994)
15. Hájek, P.: The question of a general concept of the GUHA method. Kybernetika 4, 505–515 (1968)
16. Hyndman, R., Koehler, A.: Another look at measures of forecast accuracy. International Journal of Forecasting 22, 679–688 (2006)
17. Kupka, J., Tomanová, I.: Some extensions of mining of linguistic associations. Neural Network World 20, 27–44 (2010)
18. Lemke, C., Gabrys, B.: Meta-learning for time series forecasting in the NN GC1 competition. In: Proc. of 2010 FUZZ-IEEE 2010. IEEE, Barcelona (2010)
19. Makridakis, S., Hibon, M.: The M3-Competition: Results, Conclusions and Implications. International Journal of Forecasting 16, 451–476 (2000)
20. Makridakis, S., Wheelwright, S., Hyndman, R.: Forecasting methods and applications, 3rd edn. John Wiley & Sons, USA (2008)
21. Newbold, P., Granger, C.W.J.: Experience with Forecasting Univariate Time Series and Combination of Forecasts. Journal of the Royal Statistical Society Series a-Statistics in Society 137, 131–165 (1974)
22. Novák, V.: Perception-Based Logical Deduction. In: Reusch, B. (ed.) Computational Intelligence, Theory and Applications (Advances in Soft Computing), pp. 237–250. Springer, Berlin (2005)
23. Novák, V., Perfilieva, I., Dvořák, A., Chen, Q., Wei, Q., Yan, P.: Mining pure linguistic associations from numerical data. International Journal of Approximate Reasoning 48, 4–22 (2008)
24. Štěpnička, M., Donate, J.P., Cortez, P., Vavříčková, L., Gutierrez, G.: Forecasting seasonal time series with computational intelligence: contribution of a combination of distinct methods. In: Proc. of EUSFLAT 2011, pp. 464–471 (2011)
25. Štěpnička, M., Dvořák, A., Pavliska, V., Vavříčková, L.: A linguistic approach to time series modeling with the help of the F-transform. Fuzzy Sets and Systems 180, 164–184 (2011)

Recognition of Damaged Letters Based on Mathematical Fuzzy Logic Analysis

Vilem Novak and Hashim Habiballa

Centre of Excellence IT4Innovations, division of the University of Ostrava
Institute for Research and Applications of Fuzzy Modeling,
30. dubna 22, 701 03 Ostrava 1, Czech Republic
{Vilem.Novak,Hashim.Habiballa}@osu.cz

Abstract. In this paper, we present an alternative method which turned out to be quite successful in recognizing letters that can be seriously damaged almost up to the level when even people have problems to recognize them correctly. The method is based on application of mathematical fuzzy logic with evaluated syntax.

1 Introduction

A problem quite often encountered in image analysis is to recognize letters that are more or less damaged. The damage can be caused by wrong lightning, damaged original model, or by subsequent modifications made in the image. The situation in all cases is that we face a torso of an original letter and our goal is to guess with high certainty what kind of letter we see. The problem also relates to recognition of handwritten letters. People have amazing abilities to recognize even partial torsos of letters but the task is difficult for a computer. There are several recognition techniques which work with high accuracy. These are applications of, usually, neural networks or feature extraction techniques. Both have specific problems which are solved with various success.

In this paper we will present a special method which is based on mathematical fuzzy logic calculus with evaluated syntax. This calculus was initiated in [4] in propositional version and further developed in first order version in [1] and crowned in the book [2]. The pattern recognition method was originally described in [3]. Recently, however, we had to solve the problem of recognizing letters stamped on hot iron. The letters are badly visible because of the hot background and, moreover, they can be damaged already in the stamping moment. The original method was modified accordingly and proved to be quite effective in the task. The recognition rate, of course, depends on the image preprocessing but once the letter is somehow extracted from the image, the recognition rate is close to 100%.

2 A Formal Logical Theory of Patterns

The proposed method is based on the theory of mathematical fuzzy logic with evaluated syntax Ev_{L}. The idea is to split the image into a grid of parts, each of

Á. Herrero et al. (Eds.): Int. JointConf. CISIS'12-ICEUTE'12-SOCO'12, AISC 189, pp. 497–506.
springerlink.com

which can consists of one or more pixels. We suppose to be able to characterize the given part more specifically by a certain logical formula. Its meaning can be, for example that the given part is "a house, a pipe, a tree, a branch, a steep curve", etc. or, of course, simply "a pixel". This opens the possibility to evaluate similarity of images on the basis of content of its parts. Since the the basic concept of $\mathrm{Ev}_{\text{Ł}}$ is that of *evaluated formula* a/A where A is a formula and a is its evaluation being element of the algebra of truth degrees, we have a natural means at disposal which can be used for the described task; simply A represents the content of the given part of the image and a is its degree of truth. The procedure was first published in [3] as a method using which we could realized some preliminary recognition of images to select in some sense "suspicions" ones that could later on be passed to a more detailed analysis using other sophisticated methods. In this paper we present a simplified application which still turned out to be very effective for recognition of even severely damaged letters.

2.1 Fundamental Notions

First, we consider a special language J of first-order $\mathrm{Ev}_{\text{Ł}}$. We suppose that it contains a sufficient number of terms (constants) $t_{i,j}$ which will represent *locations* in the two-dimensional space (i.e., selected parts of the image). Each location can be whatever part of the image, including a single pixel or a larger region of the image.

The two-dimensional space will be represented by matrices of terms taken from the set of closed terms M_V:

$$M = (t_{i,j})_{\substack{i\in I\\ j\in J}} = \begin{pmatrix} t_{11} & \dots & t_{1n} \\ \vdots & \vdots & \vdots \\ t_{m1} & \dots & t_{mn} \end{pmatrix} \tag{1}$$

where $I = \{1, \dots, m\}$ and $J = \{1, \dots, n\}$ are some index sets. The matrix (1) will be called the *frame of the pattern*. In other words, the frame of the pattern is the underlying grid of parts of the given image. The pattern itself is the letter which we suppose to be contained in the image and which is to be recognized. A vector $t_i^L = (t_{i1}, \dots, t_{in})$ is a *line* of the frame M and $t_j^C = (t_{1j}, \dots, t_{mj})$ is a *column* of the frame M.

The simplest content of the location is the *pixel* since pixels are points of which images are formed. A *pixel* is represented by a certain designated (and fixed) atomic formula $P(x)$ where the variable x can be replaced by terms from (1), i.e., it runs over the locations. Another special designated formula is $N(x)$. It will represent "nothing" or also "empty space". We put $N(x) := \mathbf{0}$.

The algebra of truth values is the standard Łukasiwicz MV-algebra

$$\mathcal{L} = \langle [0, 1], \vee, \wedge, \otimes, \rightarrow, 0, 1\rangle \tag{2}$$

where

$$\wedge = \text{minimum}, \qquad \vee = \text{maximum},$$
$$a \otimes b = \max(0, a+b-1), \qquad a \to b = \min(1, 1-a+b),$$
$$\neg a = a \to 0 = 1-a.$$

Formulas of the language J are *properties of the given location* (its content) in the space. They can represent whatever shape, e.g., circles, rectangles, hand-drawn curves, etc. As mentioned, the main concept in the formal theory is that of *evaluated formula.* It is a couple $^{a}/A$ where A is a formula and $a \in [0,1]$ is a syntactic truth value. In connection with the analysis of images, we will usually call a *intensity* of the formula A. Note that $^{0}/\mathbf{0}$, i.e., "nothing" has always the intensity 0.

Let M_Γ be a frame. The *pattern* Γ is a matrix of evaluated formulas

$$\Gamma = \left(^{a_{ij}}/A_x[t_{ij}]\right)_{\substack{i \in I_{M_\Gamma} \\ j \in J_{M_\Gamma}}} \tag{3}$$

where $A(x) \in \Sigma(x)$, $t_{ij} \in M_\Gamma$ and I_{M_Γ}, J_{M_Γ} are index sets of terms taken from the frame M_Γ.

A *horizontal component* of the pattern Γ is

$$\Lambda_i^H = \left(^{a_{ij}}/A_x[t_{ij}] \in \Gamma \mid t_{ij} \in t_i^L\right), \qquad j \in J_{M_\Gamma} \tag{4}$$

where t_i^L is a line of M_Γ. Similarly, a *vertical component* of the pattern Γ is

$$\Lambda_j^V = \left(^{a_{ij}}/A_x[t_{ij}] \in \Gamma \mid t_{ij} \in t_j^C\right), \qquad i \in I_{M_\Gamma} \tag{5}$$

where t_i^C is a column of M_Γ. When the direction does not matter, we will simply talk about *component.* The *empty component* is

$$E = (^{0}/\mathbf{0}, \dots, ^{0}/\mathbf{0}).$$

Hence, a component is a vertical or horizontal line selected in the picture which consists of some well defined elements represented by formulas.

Dimension of the component $\Lambda = (^{a_1}/A_1, \dots, ^{a_n}/A_n)$ is $\dim(\Lambda) = k^{max} - k^{min} + 1$, $1 \leq k^{min}, k^{max} \leq n$, where

$$k^{min} = \min_{i=1,\dots,n} \{i \mid \vdash_a \neg(A_i(x) \Leftrightarrow \mathbf{0}), a > 0\}, \tag{6}$$
$$k^{max} = \max_{i=1,\dots,n} \{i \mid \vdash_a \neg(A_i(x) \Leftrightarrow \mathbf{0}), a > 0\} \tag{7}$$

This definition takes into account that there may be "holes" in the component which, however, should be considered to contribute to the component itself and hence to its dimension. On the other hand, the outer empty parts of the component are *not* included in the computation of its dimension.

A pattern is *horizontally (vertically) normalized* if Λ_1^H (Λ_1^V) contains at least one evaluated formula $^{a}/A$ such that $\vdash \neg(A \Leftrightarrow \mathbf{0})$. A pattern is *normalized* if it is both horizontally and vertically normalized. This definition means that there is at least one location in the given line where the truth evaluation of the respective formula A is equal to 1 (for example, at least one pixel with the intensity 1).

Intensity of a pattern is the matrix

$$Y_\Gamma = (a_{ij})_{\substack{i \in I_{M_\Gamma} \\ j \in J_{M_\Gamma}}} . \tag{8}$$

The intensity $Y_{\Lambda_i^H}$ ($Y_{\Lambda_j^V}$) of a horizontal (vertical) component is defined analogously.

Maximal intensity of a pattern Γ is

$$\check{H}_\Gamma = \max\{a_{ij} \mid a_{ij} \in Y_\Gamma \text{ and } \vdash_{a_{ij}} \neg(A_x[t_{ij}] \Leftrightarrow \mathbf{0})\}.$$

Minimal intensity $\hat{H}_\Gamma$ of a pattern Γ is defined analogously. Intensity of a pattern Γ is said to be *normal* if $\check{H}_\Gamma = 1$. The pattern with normal intensity is called *normal*.

2.2 Comparison of Patterns

To be able to *compare* two patterns

$$\Gamma = (a_{ij}/A_x[t_{ij}])_{\substack{i \in I_{M_\Gamma} \\ j \in J_{M_\Gamma}}}$$

and

$$\Gamma' = \left(a'_{ij}/A'_x[t_{ij}]\right)_{\substack{i \in I_{M_{\Gamma'}} \\ j \in J_{M_{\Gamma'}}}} ,$$

their frames must have the same dimension. This can always be assured because we can embed two given frames in bigger frames of equal size. Thus, without loss of generality, we will assume in the sequel that $I_{M_\Gamma} = I_{M_{\Gamma'}}$ and $J_{M_\Gamma} = J_{M_{\Gamma'}}$.

The patterns will be compared both according to the content as well as intensity of the corresponding locations. Hence, we will consider a bijection $f : M_\Gamma \longrightarrow M_{\Gamma'}$ $f(t_{ij}) = t'_{ij}$, $i \in I, j \in J$ between the frames M_Γ and $M_{\Gamma'}$.

Let two components, $\Lambda = (a_1/A_1, \ldots, a_n/A_n)$ and $\Lambda' = (a'_1/A'_1, \ldots, a'_n/A'_n)$ be given. Put $K_1 = \min(k^{min}, k'^{min})$ and $K_2 = \max(k^{max}, k'^{max})$ where k^{min}, k'^{min} are the corresponding indices defined in (6) and k^{max}, k'^{max} are those defined in (7), respectively. Note that K_1 and K_2 are the left-most and right-most indices of some nonempty place which occurs in either of the two compared patterns in the direction of the given components. Furthermore, we put

$$n^C = \sum\{b_i \Vdash_{b_i} A_i(x) \Leftrightarrow A'_i(x), a/A_{i,x}[t] \in \Lambda, a'/A'_{i,x}[f(t)] \in \Lambda', K_1 \leq i \leq K_2\}, \tag{9}$$

$$n^I = \sum\{b_i = a_i \leftrightarrow a'_i \mid a_i/A_{i,x}[t] \in \Lambda, a'_i/A'_{i,x}[f(t)] \in \Lambda', K_1 \leq i \leq K_2\}. \tag{10}$$

The number n^C represents the total degree in which the corresponding places in both patterns tally *in the content.* This extends the power of the procedure as from the formal point of view, A'_i may differ from A_i but they still may represent the same object; at least to some degree b_i. The number n^I is similar but it reflects the compared intensity of the objects residing in the respective locations.

The components Λ and Λ' are said to *tally in the degree* q if

$$q = \begin{cases} \frac{n^C+n^I}{2(K_2-K_1+1)} & \text{if } K_2 - K_1 + 1 > 0 \\ 1 & \text{otherwise.} \end{cases} \tag{11}$$

We will write

$$\Lambda \approx_q \Lambda'$$

to denote that two components Λ and Λ' tally in the degree q. When $q = 1$ then the subscript q will be omitted.

It can be demonstrated that if all formulas A in ${}^a/A$, for which it holds that $\vdash \neg(A \Leftrightarrow \mathbf{0})$ and which occur in Λ and Λ', are the same then we can compute q using the following formula:

$$q = 1 - \frac{1}{2} \frac{\sum_{K_1 \le i \le K_2} |a_i - a'_i|}{K_2 - K_1 + 1}. \tag{12}$$

Theorem 1. *Let Λ and Λ' be components. Then*

$$\Lambda \approx \Lambda' \quad \textit{iff} \quad \vdash A_i \Leftrightarrow A'_i \quad \textit{and} \quad a_i = a'_i$$

holds for every ${}^{a_i}/A_i \in \Lambda$ and ${}^{a'_i}/A'_i \in \Lambda'$.

This theorem says that the components Λ and Λ' tally in the degree 1 iff they are the same, i.e. their content as well as their intensity coincide.

The pattern Γ can be viewed in two ways:

(a) From the *horizontal view,* i.e., as consisting of horizontal components

$$\Gamma = \left(\Lambda^H_i\right)_{i \in I} = (\Lambda^H_1, \dots, \Lambda^H_m). \tag{13}$$

(b) From the *vertical view,* i.e., as consisting of vertical components

$$\Gamma = \left(\Lambda^V_j\right)_{j \in J} = (\Lambda^V_1, \dots, \Lambda^V_n). \tag{14}$$

If the distinction between horizontal and vertical view of the pattern is inessential, we will simply use the term *pattern* in the sequel.

A *subpattern* (horizontal or vertical) $\Delta \subseteq \Gamma$ of $\Gamma = (\Lambda_1, \dots, \Lambda_p)$ is any connected sequence

$$\Delta = (\Lambda_{j_1}, \dots, \Lambda_{j_k}), \qquad 1 \le j_1, j_k \le p \tag{15}$$

of components (horizontal or vertical, respectively) from Γ. If $\Lambda_{j_1} \ne E$ and $\Lambda_{j_k} \ne E$ then Δ is a *bare subpattern* of Γ and the number k is its *dimension.*

Δ is a *maximal bare subpattern* of Γ if $\bar{\Delta} \subseteq \Delta$ for every bare subpattern $\bar{\Delta}$. The dimension of a maximal bare subpattern of Γ is the *dimension of* Γ and it will be denoted by $\dim(\Gamma)$.

Recall from our previous agreement that, in fact, we distinguish horizontal ($\dim_H(\Gamma)$) or a vertical dimensions ($\dim_V(\Gamma)$) of the pattern depending on whether a pattern is viewed horizontally or vertically. Note that both dimensions are, in general, different.

Suppose now that two patterns Γ and Γ' are given and let $\Delta \subseteq \Gamma$ have a dimension k. The following concepts of maximal common subpattern and the degree of matching of patterns are inspired by the paper [5].

Let q_0 be some threshold value of the the degree (11) (according to experiments, it is useful to set $q_0 \approx 0.7$). We say that Δ *occurs in* Γ' with the degree q_0 (is a q_0-*common subpattern* of both Γ and Γ') if there is a subpattern $\Delta' \subseteq \Gamma'$ of dimension k for which the property

$$\Lambda_i \approx_q \Lambda'_i, \qquad q \geq q_0 \tag{16}$$

holds for every pair of components from Δ and Δ', $i = 1, \ldots, k$, respectively. If $q_0 = 1$, then Δ is a common subpattern of Γ and Γ'.

A q_0-common subpattern of Γ and Γ' is *maximal* if every subpattern $\bar{\Delta} \subseteq \Gamma$ such that $\bar{\Delta} \supseteq \Delta$ is not a q_0-common subpattern of Γ and Γ'.

Lemma 1. *Let $q_0 > 0$ and $\Delta, \bar{\Delta}$ be different maximal q_0-common subpatterns of Γ and Γ'. Then $\Delta \cap \Delta' = \emptyset$.*

Let $\Delta_1, \ldots, \Delta_r$ be all maximal q_0-common subpatterns of Γ and Γ'. The *degree of matching* of Γ and Γ' is the number

$$\eta(q_0) = \frac{1}{2}\left(\frac{\sum_{j=1}^{r} \dim(\Delta_j)}{\dim(\Gamma)} + \frac{\sum_{j=1}^{r} \dim(\Delta_j)}{\dim(\Gamma')}\right) \tag{17}$$

where $\dim(\Delta_j) \geq 2$ for all $j = 1, \ldots, r$.

We must compute (17) separately for horizontal and vertical view so that we obtain horizontal η_H as well as vertical η_V degrees of matching, respectively. Then the *total degree of matching* of the patterns Γ and Γ' is the number

$$\bar{\eta}(q_0) = \frac{\eta_H(q_0) + \eta_V(q_0)}{2}. \tag{18}$$

Theorem 2.

$$1 = \bar{\eta}(1) = \eta_H(1) = \eta_V(1) \qquad \textit{iff} \qquad \vdash A_i \Leftrightarrow A'_i \quad \textit{and} \quad a_i = a'_i$$

holds for every $a_i / A_{i,x}[t] \in \Gamma$ and $a'_i / A'_{i,x}[t] \in \Gamma'$.

This theorem generalizes Theorem 1. Namely, it says that two patterns match in the degree 1 iff they are the same. More precisely, if we put the threshold $q_0 = 1$ then both patterns match in a degree $\bar{\eta} = 1$ if and only if they are virtually the same. This justifies our belief that the formal apparatus above works also in more general case when only similarity can be estimated.

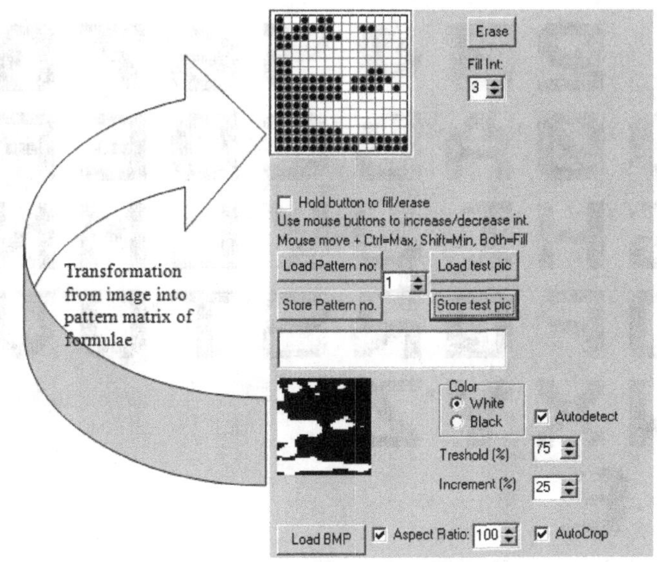

Fig. 1. Transformation of the image into the pattern Γ which is the matrix of evaluated formulas. These represent, in our case, just full circles with various intensities (as depicted in the figure).

3 Experimental Results

In this section we will present results of an experimental program PrePic developed on the basis of the theory described above. The program transforms each image into the component Γ. Then, using the proposed method a pattern in a given image can be recognized. Images in our experiment were both BW with simple bitmap or shades of gray as well as color. The patterns are both numbers and letters consisting of circles because they were printed on metal using a special printing head. Therefore, we finally decided not to distinguish more complex formulas and, in fact, took only atomic formulas $P(x)$ representing full circles into account. This was sufficient for our purpose though the full power of the above described method was not used. But, as can be seen from the figures, the results are still fairly convincing.

Dimensions of all patterns is 16 so that $I_{M_\Gamma} = J_{M_\Gamma} = \{1, \ldots, 16\}$. Each pattern is thus

$$\Gamma = \left(a_{ij}/P_x[t_{ij}]\right)_{\substack{i\in\{1,\ldots,16\}\\ j\in\{1,\ldots,16\}}}$$

where a_{ij} represent various intensities of the full circles. Example of transformation of a given image (namely the letter 'E') into a pattern is depicted in Figure 1. Because patterns are very simple, we can use formula (12) to compute the degree q.

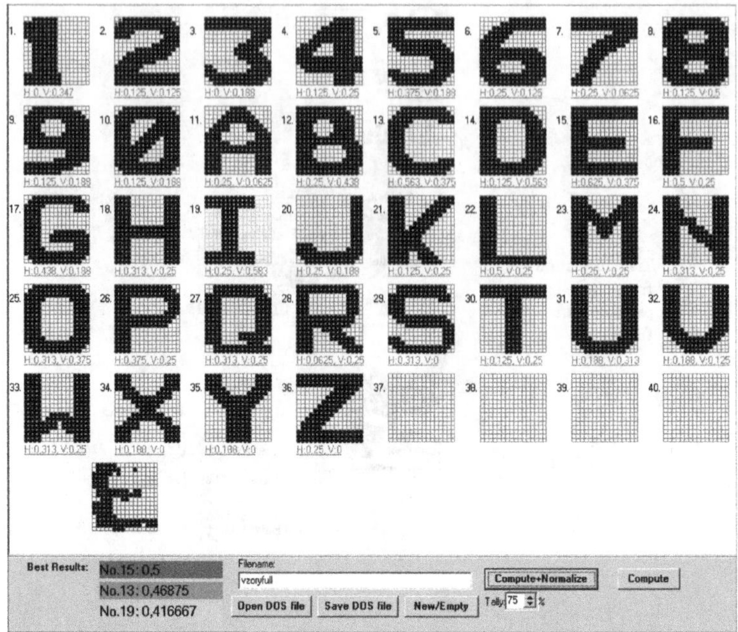

Fig. 2. List of patterns. Below is a pattern Γ of a damaged letter 'E' and three closest recognized letters together with their total degrees of matching, namely 'E' (1, degree = 0.5), 'C' (2, degree=0.47) and 'I' (3, degree=0.42). The threshold value $q_0 = 0.75$.

Figure 2 contains a list of patterns (numbers and letters) to be recognized. Below is example of a given pattern (the damaged letter 'E') and the list of the first three best matching letters with the corresponding total degree of matching $\eta(0.75)$ given in (18). One can see that the algorithm recognized the letter 'E' with the highest degree of matching equal to 0.5.

In Figure 3, few further results of recognition of letters are shown. The algorithm always marks the first 3 best results. These can be found on the left of each image together with the corresponding total degrees of matching (in %) (below). One can see that our method is indeed very robust.

Let us remark that the full potential of our method was not utilized because the patterns are quite simple and so, only atomic formulas $P(x)$ representing full circles were considered. In a more complicated problem we might introduce a formal logical theory characterizing properties of the respective locations in the image and then,

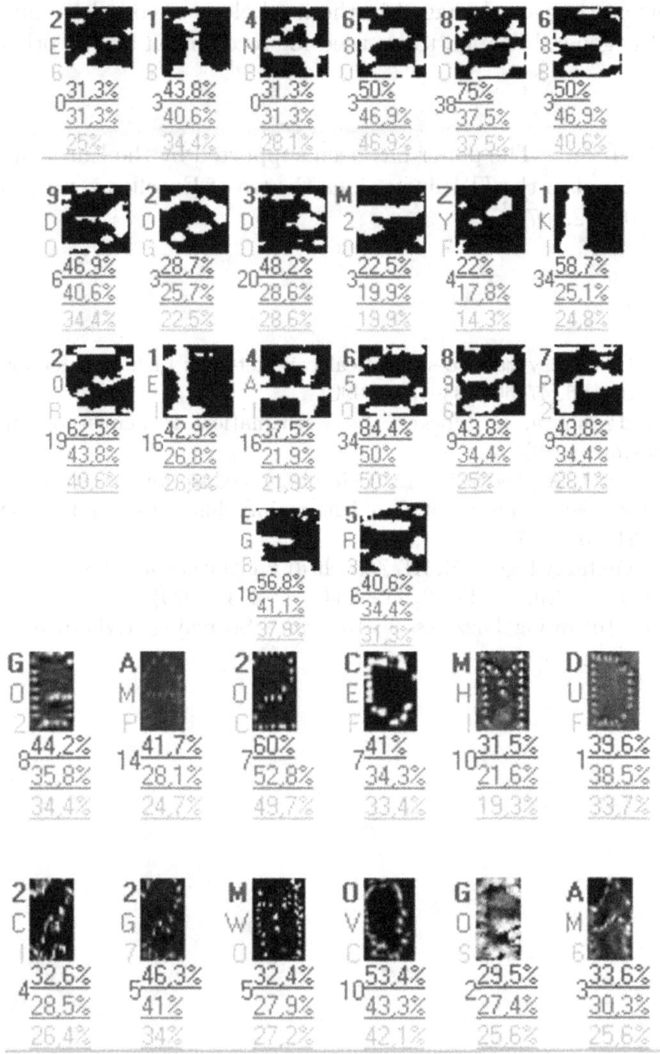

Fig. 3. Results of recognition of numbers and letters

4 Conclusion

In this paper we presented an original method for recognizing letters using a method based on the theory of mathematical fuzzy logic with evaluated syntax. The practical results confirm that the method is very robust and enables to recognize letters that are damaged in the level close to complete unreadability. Our method is general enough to be used for recognition also of other patterns than letters.

Acknowledgement. The paper has been supported by the European Regional Development Fund in the IT4Innovations Centre of Excellence project (CZ.1.05/1.1.00/02.0070).

References

1. Novák, V.: On the syntactico-semantical completeness of first-order fuzzy logic I, II. Kybernetika 26, 47–66, 134–154 (1990)
2. Novák, V., Perfilieva, I., Močkoř, J.: Mathematical Principles of Fuzzy Logic. Kluwer, Boston (1999)
3. Novák, V., Zorat, A., Fedrizzi, M.: A simple procedure for pattern prerecoginition based on fuzzy logic analysis. Int. J. of Uncertainty, Fuzziness and Knowledge-Based Systems 5, 31–45 (1997)
4. Pavelka, J.: On fuzzy logic I, II, III. Zeitschrift für Mathematische Logik und Grundlagen der Mathematik 25, 45–52, 119–134, 447–464 (1979)
5. Schek, H.J.: Tolerating Fuzziness in Keywords Searching. Kybernetes 6, 175–184 (1977)

Image Reconstruction with Usage of the F-Transform

Pavel Vlašánek[1] and Irina Perfilieva[2]

[1] Department of Informatics and Computers, University of Ostrava, 30. dubna 22, 701 03 Ostrava, Czech Republic
[2] Centre of Excellence IT4Innovations - Division of the University of Ostrava - IRAFM, 30. dubna 22, 701 03 Ostrava, Czech Republic
{Pavel.Vlasanek,Irina.Perfilieva}@osu.cz

Abstract. The article is focused on image reconstruction with usage of the F-transform. The results will be demonstrated on the picture of the Lena damaged by three kinds of the masks. We will compare reconstructed images based on the 2D F-transform with reconstructed images based on Gaussian radial basis function.

1 Introduction

By image reconstruction we understand a restoration of the damaged image. By damage we mean everything which does not belong to the original image or, in other words, which we want to replace in it. For testing purposes used in this article we will use the damage which is defined by predefined mask. We have a damaged image and a respective mask of this damage as inputs, and we produce a reconstruction. The damaged pixels are not used for calculations, their reconstructed values are obtained by computation. We will use the image in Fig. 1 as original one and three kinds of masks in Fig. 2 which characterize damages that appear in experiments. Their respective combinations are in Fig. 3 so that all of them will be used as our inputs.

There is a wide range of image reconstruction algorithms. We will demonstrate two examples of algorithms. First algorithm uses the radial basis functions [3]. It has been selected because the first author has some previous experiences with this type of reconstruction. Second algorithm uses the F-transform and it is our novel contribution to the existed methods. Outputs will be demonstrated on a single input image damaged in three different ways. The first kind of damage is overlaying the input image by some text. The text can be additional information added by camera as date and/or time or some watermark for example. The second type is inpaint which means some scratches or physical damage like fold. The last type is a noise. We will see that a noise can be removed by applying the F-transform with a proper chosen partition, the details are also in [5]. Every damage is chosen in advance by the corresponding to it mask. Let us remark, that if not known beforehand, the mask can be created manually according to a visual appearance of a damage. Thus, the requirement on having mask at disposal is not very much restrictive. Computations will be done on the gray scale image of the known benchmark Lena with the size 512×512. The CPU and sequential way of processing will be used. The output image will be compared with the non-damaged one.

Á. Herrero et al. (Eds.): Int. JointConf. CISIS'12-ICEUTE'12-SOCO'12, AISC 189, pp. 507–514.
springerlink.com

Fig. 1. Undamaged image for our testing

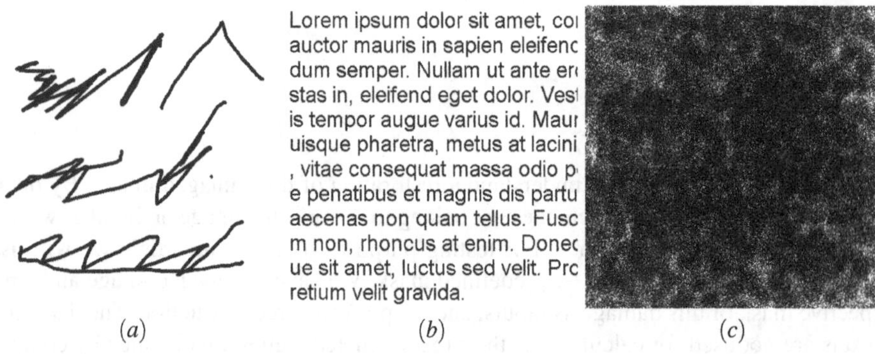

Fig. 2. (a) inpaint mask; (b) text mask; (c) noise mask

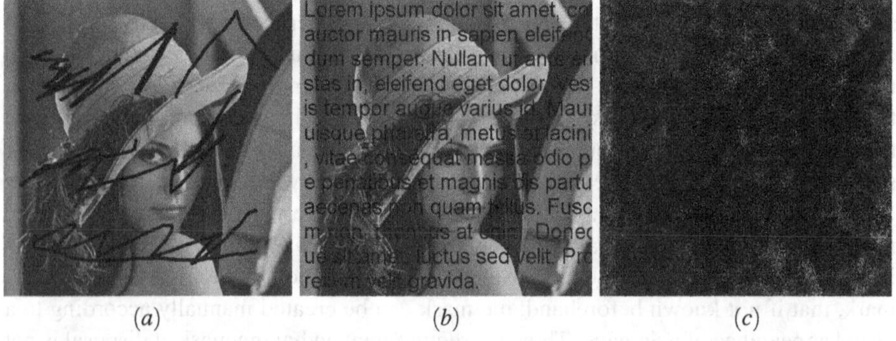

Fig. 3. Completed images used as input - (a) inpaint damage; (b) text damage; (c) noise damage

2 Radial Basis Functions for Image Reconstruction

The radial basis functions (RBF) are useful tool for image reconstruction. As a first step we have to choose some circular function ϕ which is symmetric around a center $[x_i, y_i]$. Computation with usage of the RBF is defined as

$$f(x,y) = \sum_{i=1}^{n} \lambda_i \phi(r) \tag{1}$$

$$A\overline{\lambda} = \overline{f} \tag{2}$$

where r is $\sqrt{(x-x_i)^2 + (y-y_i)^2}$ and (2) is linear form of (1). We consider (1) at non-damaged points (x,y) where points (x_i, y_i) are taken from non-damaged area as well. We can say, that from geometrical point of view we identify centers of the RBF with the non-damaged pixels use all of them to compute $\overline{\lambda}$. Thus, (1) leads to a system of linear equations (2) where $\overline{\lambda}$ is the vector of unknowns $\lambda_1, \ldots, \lambda_n$, $\overline{f}$ is the vector at left-hand sides. To guarantee a solvability of (2) it has been proposed to modify the system (2) and considered it in the following extended form

$$\begin{vmatrix} A & P \\ P^T & 0 \end{vmatrix} \begin{vmatrix} \overline{\lambda} \\ \overline{\gamma} \end{vmatrix} = \begin{vmatrix} \overline{f} \\ \overline{0} \end{vmatrix} \tag{3}$$

where $A, \overline{\lambda}, \overline{f}$ are taken from (2), $\overline{\gamma}$ is a vector of unknown coefficients of a polynomial function [2], i.e., $\gamma_0 + \gamma_1 x + \gamma_2 y$ and P is matrix of the x, y coefficients from that polynomial function. The solution of (3) contains the solution $\overline{\lambda}$ of (2). We extend the applicability of equation (1) to the damaged area and consider the new function

$$\hat{f}(x,y) = \sum_{i=1}^{n} \lambda_i \phi(r) \tag{4}$$

where the point (x,y) belongs to the whole area (damaged and non-damaged) and λ_i are components of the solution $\overline{\lambda}$. The important remark is that the points (x,y) from the non-damaged area functions f and $\hat{f}$ are equal. To conclude, $\hat{f}$ is the reconstructed

(a) (b) (c)

Fig. 4. Reconstructed images by method RBF - (a) inpaint mask; (b) text mask; (c) noise mask

image with the help of radial basis function ϕ. For purposes of the article we will use ϕ in form of Gaussian function $e^{-(\varepsilon r)^2}$ with $\varepsilon = 5$ [7] computed in 5×5 neighborhood of the non-damaged pixels. The results of the reconstruction are in Fig. 4.

3 Fuzzy Transformation

We are going to describe the F-transform (see [4,5] for more details) on discrete examples. Due to the space limitations, we introduce the F-transform for one-variable functions. Let us introduce the notion of fuzzy partition defined for $[1,n]$. We have to define set of basic functions $A_1,...,A_n$ where $A_k : [1,n] \to [0,1], k = 1,...,n$. For our cases we will use triangles defined by

$$A_1(x) = \begin{cases} 1 - \frac{(x-x_1)}{h}, & x \in [x_1,x_2], \\ 0, & \text{otherwise}, \end{cases}$$

$$A_k(x) = \begin{cases} \frac{|x-x_k|}{h}, & x \in [x_{k-1},x_{k+1}], \\ 0, & \text{otherwise}, \end{cases}$$

$$A_n(x) = \begin{cases} \frac{(x-x_{n-1})}{h}, & x \in [x_{n-1},x_n], \\ 0, & \text{otherwise}. \end{cases}$$

where x_i are nodes of $[1,n]$ and h is a distance between two of them. The F-transform is defined as follows

$$F_k = \frac{\sum_{j=0}^{l-1} f(p_j)A_k(p_j)}{\sum_{j=0}^{l-1} A_k(p_j)}$$

where F_k are components which will be used for approximation of the related values, l is number of the input points and p_j is point itself. We will use the inverse F-transform for computation of the approximated values.

$$f(p_j) = \sum_{k=1}^{n} F_k A_k(p_j).$$

3.1 1D Reconstruction

Let us illustrate how the F-transform is used in image reconstruction. Due to our presentation for one-variable functions, we illustrate the technique on a single line of a two-dimensional image. Some pixels are damaged (unknown points) and will be recomputed with the help of their neighbor (non-damaged points). The illustration is in Fig. 5 where circles ($\cdot$) are non-damaged points and crosses ($+$) are approximated ones.

The damaged points of Fig. 5 are marked in Fig. 6 by yellow color. The reconstruction of the whole line is shown in Fig. 7. It is important that the components of the F-transform are computed from the non-damaged points only. The inverse F-transform is applied to points of the whole area.

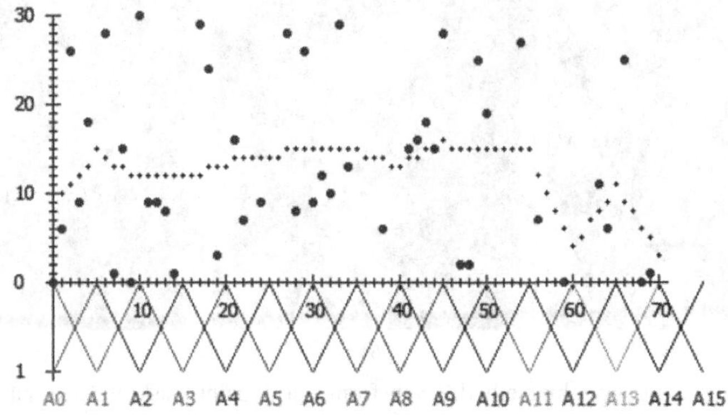

Fig. 5. The screenshot of 1D reconstruction

Fig. 6. Input line with non-damaged and damaged pixels

Fig. 7. Approximation based on fuzzy computation

Fig. 8. Final line with pixels

The result of the reconstruction will be original line with usage of the computed pixel instead of damaged ones. The final image is in Fig. 8.

3.2 2D Reconstruction

In the two dimensional image, every pixel have a position defined by axes x and y and intensity which is a value in the chosen range $[0,1]$ or $[0,255]$. From implementation point of view we will store and use an image as $M \times N$ array. The difference between one dimensional and two dimensional computation is usage of the second set of the basic functions $[B_1,...,B_n]$ for partition of the domain on the y axis. We have to compute components of the F-transform. The components are used in inverse computation [4] and are defined as follows

$$F_{k,l} = \frac{\sum_{j=1}^{M}\sum_{i=1}^{N} f(p_i,q_j)A_k(p_i)B_l(q_j)}{\sum_{j=1}^{M}\sum_{i=1}^{N} A_k(p_i)B_l(q_j)}.$$

(a) (b) (c)

Fig. 9. Reconstructed images by method F-transform - (a) inpaint mask; (b) text mask; (c) noise mask

where $f(p_i, q_j)$ is an image intensity value at the respective pixel, $A_k(p_i)$ is membership of the position p_i measured by function A_k and similarly q_j by function B_l. Inverse F-transform is defined as

$$f_{p_i,q_j} = \sum_{k=1}^{n} \sum_{l=1}^{m} F_{k,l} A_k(p_i) B_l(q_j)$$

where f_{p_i,q_j} is computed intensity of the pixel. There are results of the reconstruction in Fig. 9.

4 Results

The following root mean square deviation is used for comparison of a reconstructed image with the original one. It takes distances between values of original and reconstructed pixel and used is as follows

$$RMSD = \sqrt{\frac{\sum_{i=1}^{N \cdot M} (x_r - x_o)^2}{N \cdot M}}$$

where x_r is intensity of the reconstructed pixel, x_o intensity of the original pixel, N is width of the image and M is height. Results are in Table 4 and some details on the parts of the Lena in Fig. 10.

Table 1. The results by RMSD

Mask	RBF	FT
Noise	12.47	11.72
Inpaint	5.23	4.87
Text	4.62	4.56

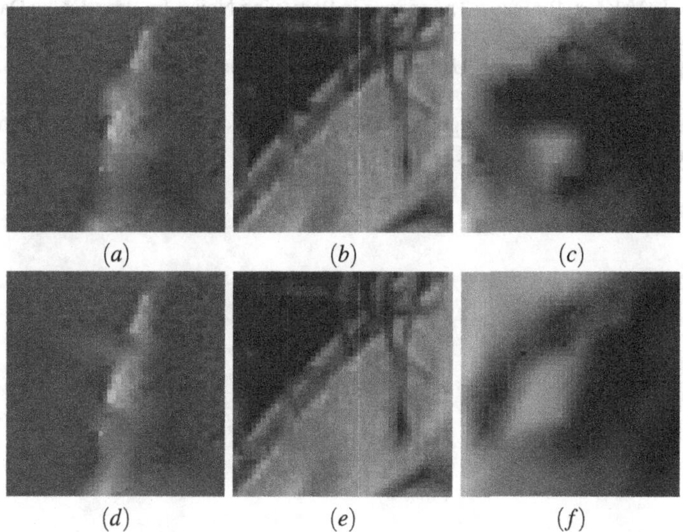

Fig. 10. Details of the reconstruction by F-transform where the damage is caused by (a) inpaint; (b) text; (c) noise. Details of the reconstruction by RBF where the damage is caused by (d) inpaint; (e) text; (f) noise.

5 Conclusion

F-transform were not primarily designed for image reconstruction, but there is shown that we can use it for this purpose. Moreover this approach shows slightly better results than the alternative method via RBF with basic settings. Future work will follow the path of the automatic detection of the damaged pixels. Another topic will be deep comparison with radial basis functions in general.

Acknowledgement. This work was partially supported by the European Regional Development Fund in the IT4Innovations Centre of Excellence project (CZ.1.05/1.1.00/02.0070) and by the project sgs12/PřF/2012.

References

1. Baxter, B.J.C.: The interpolation theory of radial basis functions, Dissertation thesis, Trinity College, 134 p (1992)
2. Zapletal, J., Vaněček, P., Skala, V.: Influence of essential parameters on the rbf based image reconstruction. In: Proceedings of SCCG 2008, Slovakia, pp. 178–185 (2008)
3. Uhlíř, K., Skala, V.: Radial basis function use for the restoration of damaged images. Computer Vision and Graphics, 839–844 (2006)
4. Di Martino, F., Loia, V., Perfilieva, I., Sessa, S.: An image coding/decoding method based on direct and inverse fuzzy transforms. Int. J. Approx. Reasoning 48, 110–131 (2008)

5. Perfilieva, I., Valášek, R.: Fuzzy Transforms in Removing Noise. In: Proc. Fuzzy Days, Berlin, pp. 225–234 (2005)
6. Ronovský, A., Vlašánek, P.: Using Radial-Basis Functions on Image Reconstruction. WOFEX, Ostrava (2011)
7. Fornberg, B., Driscoll, T.A., Wright, G., Charles, R.: Observations on the behavior of radial basis functions near boundaries. Comput. Math. Appl. 43, 473–490 (2002)

Image Compression with Artificial Neural Networks

Stephane Kouamo* and Claude Tangha

University of Yaounde I, Department of Computer Science,
P.O. Box 812 Yaounde, Cameroon
{skouamo,tangha}@gmail.com

Abstract. In this paper, we make an experimental study of some techniques of image compression based on artificial neural networks, particularly algorithm based on back-propagation gradient error [5]. We also present a new hybrid method based on the use of a multilayer perceptron which combines hierarchical and adaptative schemes. The idea is to compute in a parallel way, the back propagation algorithm on an adaptative neural network that uses sub-neural networks with a hierarchical structure to classify the image blocks in entry according to their activity. The results come from the *Yann Le Cun* database [7], and show that the proposed hybrid method gives good results in some cases.

Keywords: neural networks, image compression and coding, back-propagation.

1 Introduction

For many years, there has been a steady growth in the needs of digital images (fixed or animated) in multiple areas : telecommunications, broadcasting multimedia, medical diagnostics, meteorology, robotics, etc. Yet this type of data represents an enormous mass of informations difficult to transmit and store with current means [1]. Besides the problem of storage, if this information has to be transmitted over a network, the time of transmission is often too long. To counter these problems, the compression of these images becomes a necessary operation. Image compression is the operation which passes from one representation to another with the aim of reducing the number of bits used for the coding. The decompression returns the image to a less compact representation, but more able to be manipulated or shown. At present, the main core of image compression technology consists of three important processing stages: pixel transforms, quantization and entropy coding [5]. Several techniques are used to realize the compression of fixed images or movies. Among the best known are : the Karhunen-Loeve Transformation (KLT), compression by *Discrete Cosine Transformation* (DCT), compression by *Discrete Wavelet Transformion* (DWT), the *Fractal compression method* and then compression by artificial

* Corresponding author.

Á. Herrero et al. (Eds.): Int. JointConf. CISIS'12-ICEUTE'12-SOCO'12, AISC 189, pp. 515–524.
springerlink.com

neural networks [5,8,11,12]. The approaches using Artificial Neural Networks (ANN) for intelligent processing of the data seem to be very promising, this being essentially due to their structures offering possibilities of parallel calculations and the use of the training process to allow the network to adapt itself to the data to be treated.

This paper presents with tests to support, an experimental study of some techniques of back propagation algorithm involved in image compression. We also present a hybrid method, that sometimes produces better results.

The rest of the document is organized in four sections. The first section presents direct neural network development for image compression. The second section presents the proposed method. The third section is reserved for the implementation of studied methods (we shall limit ourselves to neuronal techniques). The fourth section, presents results obtained from all studied methods. Then, we end with a conclusion.

2 Compression by Artificial Neural Networks

An Artificial Neural Network (ANN) is a computation model whose design is inspired schematically by the functioning of real neurons (human or not), it can be considered as a set of interconnected cells that communicate with each other and with the outside through linking known as *synaptic weights* [1,3]. There are several techniques of image compression using neural networks. We shall limit ourselves to the use of a multilayer perceptron and the algorithm of back-propagation gradient error. Therefore we have : basic back propagation, hierarchical back propagation and adaptative back propagation.

2.1 Basic Back Propagation

Back propagation is one of the neural networks which are directly applied to image compression [4,5]. The neural network used here is a multilayer perceptron with three layers. The network structure can be represented by an *input layer*, one *hidden layer* and an *output layer* as shown in Fig.1 below. The input and output layer have the same size and are fully connected to the hidden layer. The number of neurones of the hidden layer is less than (strictly) the previous two. Compression is achieved at the hidden layer by designing the value of K (the number of neurones at the hidden layer). Lets N the number of neurones of the input layer, w_{ji} the connexion weights between the input and the hidden layer which can be described by a matrix of order $K \times N$, and w'_{ij} the connexion weights between the hidden layer and the output layer which can also be described by a matrix of order $N \times K$. The input image is split up into blocks or vectors of 4×4 and 8×8. Image compression is achieved by training the network in such a way that the coupling weights w_{ji} scale the input vector of N dimension into a narrow channel of K dimension ($K < N$) at the hidden

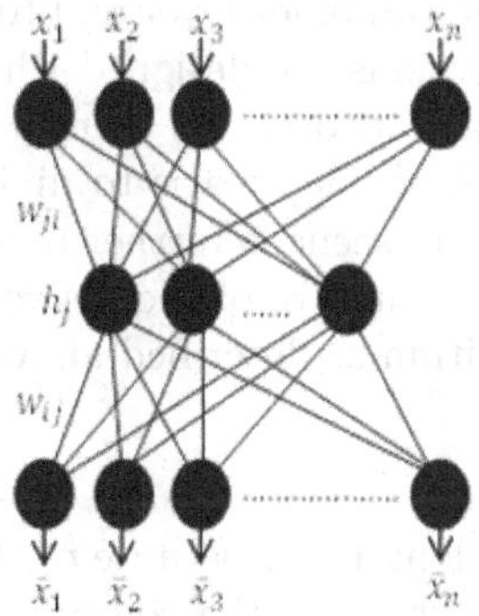

Fig. 1. Structure of a neural network with basic back propagation

layer and produce the optimum output value which makes the quadratic error between input and output minimum [5].

2.2 Hierarchical Back Propagation

The structure of the neural network with basic back propagation can be extended to construct a *hierarchical neural network* by adding two more hidden layers into the existing network [5,10]. Therefore, we have an input layer, an output layer and one hidden layer containing a *combiner layer* connected to the input layer, a *decombiner layer* connected to the output layer and a *compressor layer* connected to the *combiner* and the *decombiner* layer as shown in Fig.2 below.

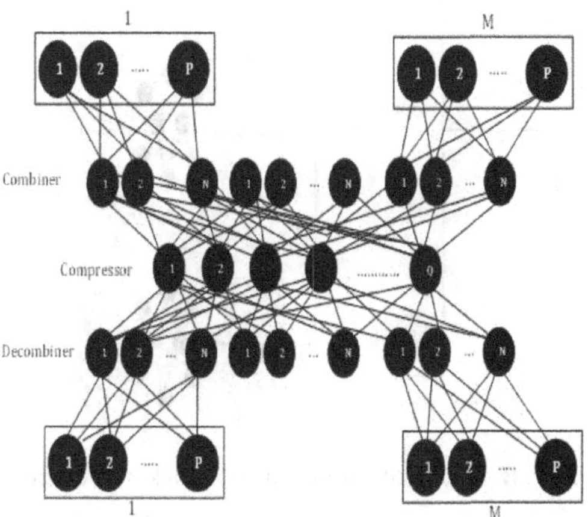

Fig. 2. Hierarchical neural network structure

From the input layer to the combiner layer and from the decombiner layer to the output layer, local connections are designed which have the same effect as M fully connected neural sub-networks. As seen in Fig.2, all the three hidden layers are fully connected. The basic idea is to divide an input image into M disjoint sub-scenes and each sub-scene is further partitioned into T pixel blocks of size $p \times p$ [10,5]. After the structure of the network is created, we can apply basic back propagation algorithm as described above.

Remark 1

- training stage take a too long time, because of the hidden layer struture;
- the system converges faster when the compression ratios are low enough, which entails a bigger capacity to operate on large-sized images.

2.3 Adaptive Back Propagation

By exploiting the basic back propagation, *Carrato* proposes in his article [3] a number of combinations based on the principle that different neural networks are used to compress the image blocks with different levels of complexity. N neural sub-networks are then trained and the input image is divided into N subsets, each subset corresponding to a level of complexity. The general structure for the adaptive schemes can be illustrated in Fig.3 in which a group of neural networks with increasing number of hidden neurones (h_{min}, h_{max}) is designed. Using an

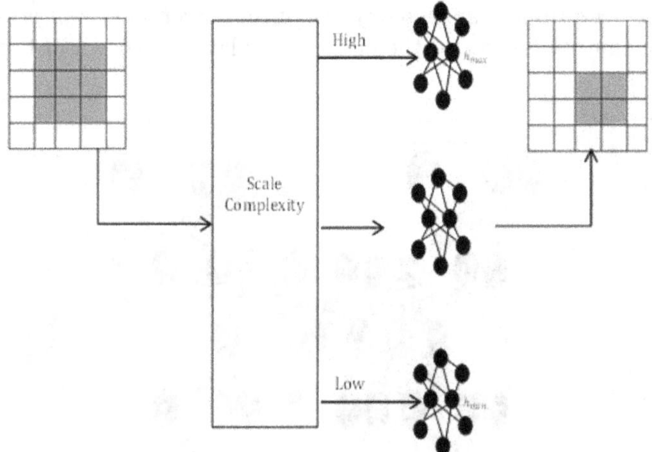

Fig. 3. Adaptative neural network structure

adaptative back propagation to train a neural network to do image compression consists of making this classification on the input image. After this classification, we can apply the basic back propagation on the differents subsets.

Remark 2

- the algorithm permits one to deal better with the quality of the reconstructed image;
- the decompression time are reduced by using this algorithm.

Observations

- the hierarchical and adaptive neural network techniques derive from basic back propagation and try to improve its performances;
- the main difference between hierarchical and adaptive neural network is that hierarchical uses a 3 hidden layers to makes compression while adaptive makes image compression by taking into account the noise between the blocks of nearby pixels of the input image.

3 Hybrid Method Proposed

By exploiting the advantages of hierarchical and adaptive neural network techniques, we proposed a new method called *hybrid method* to optimize at most the profits obtained. The idea is to construct a multilayer perceptron that has the structure of an adaptive neural network but which uses sub-networks to classify image blocks in entry according to their levels of activity, with the structure of a hierarchical neural network. The structure of hybrid method is shown by the Fig.4 below. When the structure of the network is built we can then apply

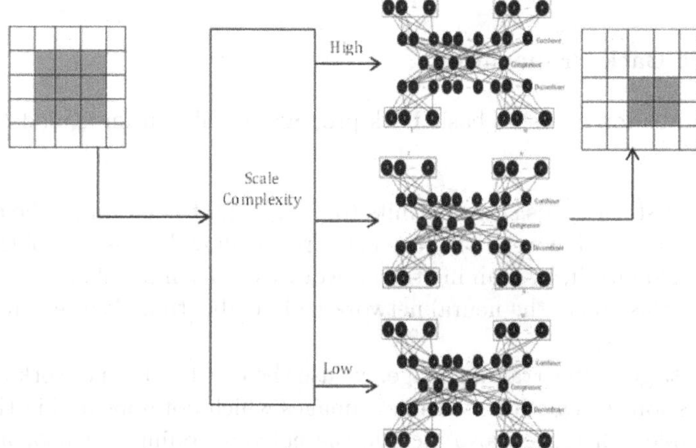

Fig. 4. Hybrid neural network structure

the back propagation algorithm in a basic way as described above. So the image in entry will be split up into N subsets corresponding to the sub-networks, and each subsets will be trained by its corresponding sub-network which has a hierarchical structure.

Remark 3

- the system deals better the quality of the reconstructed image but take a long time to make training stage ;
- the decompression time is reduced ;

To have a greater reduction in compression and decompression time, we compute this algorithm in a parallel way.

4 Implementation of Studied Techniques

4.1 Sample Preparation and Architecture

We use the cross-validation protocol to choose the samples of training and validation data. Then, we used 90% of data for the training stage and 10% for the using stage. The data result from the *yann leCun* database [7], and the working environment used is:

- experiment conditions : 2 PC intel Celeron processor 900, 320*GB HDD*, 4*GB* Memory ;
- operating system : *Ubuntu* 11.04 (*linux* distribution);
- programming language : *GNU Octave version* 3.2.4, the *language C* and the *openmpi* library.

4.2 Basic Back Propagation

Image compression by using basic back propagation algorithm is made into two steeps :

1. **training stage** : this stage permits to design and to determine the architecture (the connexion weight) of the network which will takes care of the image compression itself, by training the network with a sample of data reserved to it. After this stage, the neural network will be able to make the compression on new images.
2. **using stage** : after training stage, we use the constructed network to realize compression (using stage) with new images which not appeared in the training sample. This steep also include the entropy coding of vector h_j at the hidden layer (it is a binary coding which permits to obtain the size of the compressed image in bits). In the case of the training stage ends successfully, the codes of these coupling weights will be required to adapt the networks with the input images which were not met yet.

During the training stage, the input image is split up into blocks or vectors of size 8×8 or 4×4. After an operation of normalization of the grayscale pixels of the input image such that they are all included in the interval $[0, 1]$, the training image blocks are used to make compression and decompression on a linear network as follow:

$$h_j = \sum_{i=1}^{N} w_{ji} x_i \quad 1 <= j <= k$$

for the encoding and,

$$\bar{x} = \sum_{j=1}^{K} w'_{ij} h_j \quad 1 \leq i \leq N$$

for the decoding, until the mean square error between the expected outputs and that obtained is minimal. That means:

$$\sum (x_i - \bar{x}_i) \leq \alpha$$

Where α is a real very near to 0.

The reason for using normalized pixel values is due to the fact that neural networks can operate more efficiently when both their inputs and outputs are limited to a range of $[0, 1]$. Good discussions on a number of normalization functions and their effect on neural network performances can be found in [1,4,5].

4.3 Hierarchical Back Propagation

In accordance with the hierarchical neural network structure shown in Fig.2, the principle can be described like this : the image in entry is split up into M sub-scenes of size $p \times p$, then we apply an *outer loop neural network* ($OLNN$) by taking only the *input*, the *combiner* and the *output* layer out of the network. After it, we make an *inner loop neural network* ($ILNN$) by selecting the three hidden layers (*combiner*, *compressor* and *decombiner*). Finally we reconstruct the network with connection scale weights obtained after each step [5]. Once we have completly assembled the entire network, we can apply the basic back propagation as detailed above. After both training stages, 4 sets couples weights are conceived and can be attributed in the following way : both sets couples of weights obtained after the stage 1 (OLNN) are attributed to the external layers like this : *input* – *combiner* and *output* – *decombiner* ; and both sets couples obtained after the stage 2 (ILNN) are attributed to the internal layers as *combiner* – *compressor* and *decombiner* – *compressor*. After the training stage, the network is ready for the compression.

4.4 Adaptive Back Propagation

The adaptive neural network structure is illustrates as shown in Fig.3. After the network structure has been defined, training is applied as follows : a block of the image taken as input is passed to all sub-networks, and then we determine its complexity level by using the ratio *signal/noise* (SNR or $PSNR$). This rate will be used for rough classification of image blocks in the same number of subsets that sub-networks of neurones, each subset corresponding to a level of complexity. After this initial coarse classification is completed, each neural network is then further trained by its corresponding refined sub-set of training blocks through back propagation algorithm proposed above.

4.5 Hybrid Method

The general structure of the hybrid network can be observed in Fig.4 above. The principle is the same like the adaptive back propagation but, sub-networks use to classify image blocks in entry according to their levels of complexity, have the structure of a hierarchical neural network. After the classification is completed, each neural network is trained through basic back propagation algorithm.

5 Results

The training stage was conducted with successive compression rate of 30%, 50% and 90%. The images used come from the $Yann\ LeCun$ database [7], from the file called "$train - images$", these images are successively split up into blocks of 4×4 and 8×8 and we use 5 bits to encode one pixel of the input image. The using stage is made with 10% of the sampling data planned for that purpose. The quality of the reconstructed images is evaluated by both the $PSNR$ and the EQM.

$$EQM = \frac{1}{T}\sum_{i=1}^{T}(\bar{x}_i - x_i)^2$$

$$PSNR = 10\log_{10}\left(\frac{255^2}{EQM}\right)$$

With T: image size, x_i: i^{th} pixel of the original image, $\bar{x_i}$: i^{th} pixel of the reconstructed image.
The compression rate (TC) and the number of bit(s) per pixel (*bit rate*) will be used to assess the performance of compression in terms of saving space and are determined by the equations below :

$$TC = \frac{M}{N}$$

$$bitrate = \frac{\log_2(N)}{(N/P)}$$

Where N represents the number of output neurones, M that of the hidden neurones, and P is the bits number used to codify one neurone. Obtained results are grouped into the following Table 1.

Table 1. Different studied techniques

	Hybrid method			Basic back propagation (4 × 4)			Hierarchical back propagation			Adaptative back propagation			Parallel hybrid method		
Compression rate (in %)	30	50	90	30	50	90	30	50	90	30	50	90	30	50	90
Training time (in min)	29	40	47	30	38	40	29	41	48	35	42	50	19	25	29
Compression time (in second)	5.06	5.09	5.11	5.02	5.04	5.07	5.07	5.11	5.12	5.06	5.09	5.11	3.12	3.14	3.16
Decompression time (in min)	8	12	13	9	13	13	12	15	18	8	12	13	5	7	8
PSNR (in dB)	30.03	32.84	34.80	28.42	30.83	33.10	29.45	31.01	33.85	29.93	32.10	34.14	30.03	32.84	34.80
Bit rate (*bit/pixels*)	0.58	2	3	0.58	2	3	0.58	2	3	0.58	2	3	0.58	2	3

Table 1 above, shows that :

(1) The performances in terms of the quality of the reconstructed images and the bit(s) used to code a pixel increases when the number of images used for the training stage increases.
(2) The low rate compression allow to reduce considerably the number of bit(s) used for the coding of pixels, but are followed by a visual degradation of the quality of the reconstructed image.
(3) Adaptive back propagation provides good results in terms of quality of the reconstructed image and the number of bit(s) used to encode a pixel.
(4) The hierarchical back propagation produces good learning time compared to other compression techniques based on back propagation when rates compression are low.
(5) Hybrid method proposed produces better results than those obtained with back propagation techniques.
(6) The use of parallelism to compute hybrid method proposed contributes to reduce by almost the half the computation (training, compression and decompression) times.
By using vectors blocks of 4×4 compared to vectors blocks of 8×8 pixels, a significant gain of 4.01 dB on average can be achieved in the quality of reconstructed images. But in return, an average decrease of 1.2 $bit/pixel$ is recorded. By adding the advantages of hierarchical back propagation and those of the adaptive back propagation, we improve the results provided by the techniques based on back propagation algorithm, which are already acceptable.

6 Conclusion

In this paper, after briefly describing the various techniques of image compression, we present the neural networks techniques based on the use of back propagation algorithm involving image compression. Obtained results show that training and decompression time decrease gradually as the compression rate is down for all the studied method. Decrease of computation time is followed by a visual degradation of the reconstructed image quality, a loss of 4.46 dB can be observed. Implementation tests done on the hybrid method proposed by using the strengths of compression techniques based on hierarchical and adaptive back propagation, give good results that the other studied method and is very promising. One possible way of improvement is to try to compute our algorithm by using the schemes of the KLT method and apply it to images with bigger sizes than 16×16, and use SSIM to evaluate the quality of reconstructed images.

References

1. Benamrane, N., Benhamed Daho, Z., Shen, J.: Compression des images medicales par reseaux de neurones. USTO, Traitement Du Signal 6, 631–638 (1998)
2. Benbenisti, et al.: New simple three-layer neural network for image compression. Opt. Eng. 36, 1814–1817 (2000)
3. Carrato, S.: Neural networks for image compression. In: Neural Networks: Adv. and Appli., 2nd edn., vol. 2, pp. 177–198. Gelende Pub. North-Holland, Amsterdam (1992)
4. de Bodt, E., Cottrell, M., Verleysen, M.: Using the Kohonen algorithm for quick initialisation of simple competitive learning algorithm. In: European Symposium on Artificial Neural Networks (2001)
5. Jiang, J.: Images compression with neural networks, A Survey. Signal Processing: Image Communication (1998)
6. Karayiannis, N.B., Pai, P.I.: Fuzzy vector quantization algorithm and their application in image compression. IEEE Trans. Image Process. 4(9), 1193–1202 (2008)
7. Le-Cun, Y.: A competitive learning method for asymmetric threshold network. In: COGNITIVA 1985, Paris, June 4-7 (1985)
8. Mallat, S.G.: A Wavelet Tour of Signal Processing, pp. 145–150. Academic Press (1999)
9. Mallat, S.G.: A theory for multiresolution signal decomposition: the wavelet representation. IEEE Trans. Pattern Anal. Mach. Intell. 11(7), 674–693 (1989)
10. Namphol, A., Chin, S., Arozullah, M.: Image compression with a hierarchical neural network. IEEE Trans. Aerospace Electronic Systems 32(1), 326–337 (1996)
11. Rabbani, M., Jones, P.W.: Digital Image Compression Techniques. Tutorial Texts. SPIE Optical Engineering Press (1991)
12. Ramel, M.J.Y., Agen, F., Michot, J.: Fractal compression, Jacquin methods, triangular subdivisions and Delaunay. EPUT, Depts.-Info (2004)
13. Zhang, L., et al.: Generating and conding of fractal graphs by neural network and mathematical morphology methods. IEEE Trans. Neural Networks 7(2), 400–407 (1996)

Image Compression Methodology Based on Fuzzy Transform

Petr Hurtik and Irina Perfilieva

Centre of Excellence IT4Innovations division of the University of Ostrava Institute for Research and Applications of Fuzzy Modeling
30. dubna 22, 701 00, Ostrava, Czech Republic
{petr.hurtik,irina.perilieva}@osu.cz

Abstract. The main objective of our work is to develop an effective algorithm for image compression. We use both lossy and non-lossy compression to achieve best result. Our compression technique is based on the direct and inverse fuzzy transform (F-transform), which is modified to work with dynamical fuzzy partition. The essential features of the proposed algorithm are: extracting edges, automatic thresholding, histogram adjustment. The article provides a comparison of our algorithm with the image compression algorithm (JPEG) and other existing algorithms [1, 7] based on fuzzy transform.

1 Introduction

By image compression we mean a reduction in size of the image with the purpose to save space and by this, a transmission time. Digital images are usually identified with their intensity functions which, being measured in the interval $[0,1]$, can be represented by fuzzy relations. Therefore, in the literature on fuzzy sets and their applications, a continuously growing interest to the problems of image compression was expected. However, this was not the case. Below, we will give a short overview of main ideas which influenced a progress in image compression on the basis of fuzzy sets.

A pioneering publication of Lotfi A. Zadeh [10] discussed the issue of data summarization and information granularity. It has been noticed that a max – min - composition with a fuzzy relation works as a summarization/compression tool. Then in a series of papers (see [2, 3]), the idea to associate image compression with the theory of fuzzy relation equations was intensively investigated. The correspondence between a quality of reconstruction and a t-norm in a generalized max $-t$ - composition with a fuzzy relation was analyzed in [3, 4]. A new idea which influences a further progress in fuzzy based image compression came with the notion of F-transform [5]. In [1], it has been shown that the F-transform based image compression is better than the best possible fuzzy relation based one. However, the former was still worse than JPEG technique. A certain improvement of the F-transform based image compression was announced in [6].

A new wave of interest to the discussed problem came with more sophisticated applications of the F-transform to image processing, especially to the problem of edge detection [9]. It has been noticed that the quality of reconstructed image strongly depends on the quality of reconstructed edges. This idea is elaborated in details in the proposed contribution.

Á. Herrero et al. (Eds.): Int. JointConf. CISIS'12-ICEUTE'12-SOCO'12, AISC 189, pp. 525–532.
springerlink.com

2 Compression

Image compression means a reduction in size of the image. By image we mean a discrete function f with two variables which is defined on the domain $[1,N] \times [1,M]$ and takes values from [0, 255]. The value $f(x,y)$ characterizes intensity of the gray level of the pixel whose coordinates are (x,y). Below, we will refer to f as to intensity function or image. By compression we mean a certain transformation of f which results in a new image function f' defined on $[1,N'] \times [1,M']$ where $N' < N, M' < M$. A compression is characterized by its ratio CR which is equal to $N'M'/NM$. We have to solve two problems: reduce size of compressed image and obtain decompressed image most similar to original one.

We propose a compression alorithm which is based on the discrete F-transform in combination with saving sharp edges. This algorithm consists of the following steps: find and store information about gradient (section 2.1); compute range of intensity over an image block and make a desicion regarding further partition of this block (section 2.2); compress by the F-transform (section 2.3) and store histogram of the original image (section 2.4).

Let us make a short overview of some contemporary techniques used for compression. The idea to partition an image area into blocks according to respective ranges of the intensity function is taken from png graphics format. Representation of a compressed image by the result of a certain transform is usual for the JPEG format. Modern compression algorithms use several transforms: discrete wavelet transform, discrete cosine transform, Burrows-Wheeler transform and many others.

In our approach, we combine both lossy and nonlossy compression - gradient pixels are stored by nonlossy format, areas by lossy F-transform. We propose decompression of an image after compression. Decompression is the inverse transformation with respect to compression, it means that we transform $N' \times M'$ back into $N \times M$.

2.1 Image Gradient Separation

Gradient separation (or edge detection) is the first step of the image compression algorithm. The notion of edge is informally characterized as an area where a significant change of intensity occurs. In practice, this characterization connects edges with ares where first or second derivative of intensity function f attains its extremal value. In our approach, we take the above given characterization literally and propose to classify an edge area on the basis of a difference g between maximal and minimal values of intensity function f over it:

$$g(x,y) = max(f(x',y')) - min(f(x',y')) \tag{1.1}$$

$$x' \in \{x-1,x,x+1\}; y' \in \{y-1,x,y+1\}.$$

The area with high values of the difference g is not a subject of compression. Due to this fact, a sharpness of a reconstructed image is as good as in the original one. The proposed approach is sensitive to noise, more than if partial derivatives are computed by e.g., Sobel operators. The result of the gradient separation algorithm is shown in fig 1. We propose fixed mask matrix 3×3 pixels. In order to reduce that kind of sensitivity,

Fig. 1. Left: original picture. Right: pixels with hight difference value

we propose to use a dynamic threshold T for selecting high values of the difference g. Due to a space limitation, we will skip a detailed description of choosing T.

2.2 F-Transform

Below, we shortly recall the basic facts about one-dimensional the F-transform [5]. For simplicity, we apply it to a function f of one variable defined on $[a,b]$: let $x_1 < ... < x_n$ be fixed nodes within $[a,b]$. We say that fuzzy sets $A_1,...,A_n$, identified with ther memership functions $A_1(x),...,A_n(x)$ defined on$[a,b]$ form a fuzzy partition of $[a,b]$ if they fulfil the following conditios for $i = 1,...,n$ are fulfilled:

1. $A_i : [a,b] \to [0,1], A_i(x_i) = 1$;
2. $A_i(x) = 0$ if $x \notin (x_{i-1}, x_{i+1})$, where we assume $x_0 = x_1 = a$ and $x_{n+1} = x_n = b$;
3. $A_i(x)$ is a continuous function on $[a,b]$;
4. $A_i(x)$ stricly increases on $[x_{i-1}, x_i]$ for i= $2,...,n$ and stricly decreases on $[x_i, x_{i+1}]$ for $i = 1,...,n-1$.
5. $A_i(x)$ stricly increases on $[x_{i-1}, x_i]$ for i= $2,...,n$ and stricly decreases on $[x_i, x_{i+1}]$ for $i = 1,...,n-1$.

$$F_i = \frac{\sum_{j=1}^{m} f(p_j)A_i(p_j)}{\sum_{j=1}^{m} A_i(p_j)} \tag{1.2}$$

for $i = 1,...,n$. Shapes of basic functions are not prederminited, so that we use triangular membership functions due to simplicity of coding. Following inverse F-transform is then defined by:

$$f_{F,n}(p_j) = \frac{\sum_{i=1}^{n} F_i A_i(p_j)}{\sum_{i=1}^{n} A_i(p_j)} \tag{1.3}$$

In our case, we are using two-dimensional F-transform described in [9]. Let us remark that in the above given characterization of a fuzzy partition, we did not use the Ruspini condition. By this, we obtain a certain flexibility in choosing a partition.

2.3 Evaluation of Intensity Range in Area

Image compression algorithm is usually applied to smaller subareas. The main problem is finding size of those areas. For instance, if we have large area of one color, but with some small detail of different color, we have two options: we can compress it as one area, but the detail will be lost. Or we can divide the are with the detail into smaller areas in order tol keep that small detail. In the last case, we have to memorize many small areas of one color. We propose to solve this problem by using the F-transform with a non-uniform partition which is chosen on the basis of the following procedure. Each area E is characterized by its width E_w and its height E_h; At the beginning of algorithm we set $E_w = N; E_h = M$. The values of function g are computed on the basis of (1) where $x' \in [max(0, x - E_w), x]; y' \in [max(0, y - E_h), y]$. If $g(x,y) \leq D$, $D \in [0,255]$ and means user defined threshold for control algorithm power we choose the respective area E as an element of the partition of the F-transform. Owerthise, we divide area E into four symmetrical subareas and continue recursively. Dividing is terminated if the condition of minimal difference D is true, or the condition of minimal area is true:

$$(E_w \leq S \vee E_h \leq S) \vee g(x,y) \leq D. \tag{1.4}$$

In (4) S means minimal size (of width, or height) of basic funcions and D means threshold of minimal intensity difference. These two values S and D are defined by a user, and both of them influence power the of compression algorithm. The result of the divide algorithm with red colored borders of an areas is shown in fig 2.

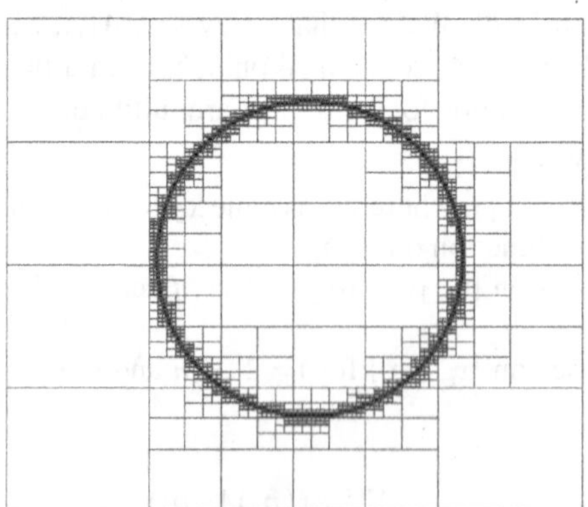

Fig. 2. Example of dividing the area

2.4 Image Histogram

The F-transform based compression is lossy and therefore the histogram of an original image is changed after decompression. We propose to store the histogram of the original image and apply it to obtain a better reconstruction of compressed image. We compute the cumulative distribution function, say C as a characteristic of the histogram and store the respective vector of values of C. This additional vector is added to the stored information about the compressed image. As a results, the quantity of the stored information increases so that the proposed earlier compression ratio does not fully characterize a size of the stored information. To take into account the actual a size of the stored information we increase the compression ratio by the respective quantity. For example: if an image has the dimension 512×512 px then we increase compress ratio value by 0.0009.

3 Decompression

The decompression is a transform from $M'N'$ space back to MN space. We propose the decompression algorithm based on the inverse F-transform (3). Because an application of the direct and inverse F-transform leads to the lossy decompression, our goal is to minimize data loss. We propose to minimize the loss by decompression of the stored gradient pixels (chap. 3.1) and histogram restore (chap. 3.2).

3.1 Decompression of Gradient Pixels

The area with high values of the difference g (see Section 2.1) is added to the image reconstructed by the inverse F-transform. We have to put pixels from this area into their own layer above the currently decompressed layer. After that we can merge layers hierarchicly.

3.2 Histogram Restore

After applying the F-transform and its inverse the range of intensity changes. In order to restore the range of intensity of an original image in the reconstruction, we use the

Fig. 3. Left: without histogram restore, PSNR = 29dB. Right: with histogram restore, PSNR = 30dB.

Fig. 4. Left: original; right: proposed reconstruction, CR=0.08, PSNR=29dB

Fig. 5. Left: proposed reconstruction, CR=0.25, PSNR=37dB; right: proposed reconstruction, CR=0.44, PSNR=43dB

Fig. 6. Left: JPEG, CR=0.25, PSNR=39dB; right: JPEG, CR=0.43, PSNR=46dB

stored information about the cumulative distribution function C. This step allows to increase the quality of reconstruction. For example, there is figure 3 for comparison between image with and without histogram restore.

4 Estimation of a Quality of Reconstruction

The following criterion is used for estimation of a quality of a reconstructed image. *PSNR* (Peak Signal to Noise Ratio) measures a similarity between an original image and its reconstruction after. Higher value of *PSNR* means better quality of result.

$$PSNR = 20log\left(\frac{max(f)}{\sqrt{MSE}}\right)[dB] \tag{1.5}$$

$$MSE = \frac{1}{M \cdot N}\sum_{x=0}^{M-1}\sum_{y=0}^{N-1}(f(x,y) - q(x,y))^2$$

By $max(f)$ we mean the is maximum value of the intensity of the original image f. By q we mean the means intensity value of the decompressed image.

5 Experiments

In Figures 4 and 5, we demonstrate the results of the proposed technique. In Figure 6, we show the results of the JPEG algorithm with the same compression ratio (*CR*). For the chosen benchmark "Cameraman" from the Coral Gallery, the JPEG is slightly better. However, for the created by us picture in Fig. 2, the proposed algorithm shows better results that JPEG algorithm. In order to [1] you can see, that the results of the proposed algorithm are slightly better that the previous one.

6 Conclusion

We have proposed a new compression method on the basis of the F-transform. In comparison with the previous one [7], the newly proposed compression uses the following improvements: edge extraction, dynamic area division described by non-uniform partition and histogram adjustment. Our next research will be focusing on large-size images, color images, estimation of a speed of our algorithm and detailed comparison with the JPEG technique.

Acknowledgement. This work was parially supported by the European Regional Development Fund in the IT4Innovations Centre of Excellence project (CZ.1.05/1.1.00/02.0070) and the project SGS12/PrF/2012.

References

1. Di Martino, F., Loia, V., Perfilieva, I., Sessa, S.: An image coding/decoding method based on direct and inverse fuzzy transforms. International Journal of Approximate Reasoning 48, 110–131 (2008)
2. Hirota, K., Pedrycz, W.: Fuzzy relational compression. IEEE Trans. Syst., Man, Cybern. 29(pt. B), 407–415 (1999)
3. Nobuhara, H., Pedrycz, W., Hirota, K.: Fast Solving Method of Fuzzy Relational Equation and Its Application to Lossy Image Compression/Reconstruction. IEEE Trans. Fuzzy Syst. 8(3), 325–334 (2000)
4. Pedrycz, W., Hirota, K., Sessa, S.: A Decomposition of Fuzzy Relations. IEEE Trans. Syst., Man, Cybern. 31(pt. B), 657–663 (2001)
5. Perfilieva, I.: Fuzzy transforms. Fuzzy Sets and Systems 157, 993–1023 (2006)
6. Perfilieva, I., De Baets, B.: Fuzzy Transform of Monotonous Functions with Applications to Image Processing. Information Sciences 180, 3304–3315 (2010)
7. Perfilieva, I., Pavliska, V., Vajgl, M., De Baets, B.: Advanced image compression on the basis of fuzzy transforms. In: Proceedings of the Conference on IPMU 2008, Torremolinos, Malaga, Spain (2008)
8. Perfilieva, I.: Fuzzy transforms of monotone functions with application to image compression. Information Sciences 180, 3304–3315 (2010)
9. Perfilieva, I., Hodakova, P., Hurtik, P.: F1-transform Edge Detector Inspired by Cannys algorithm. IPMU, Italy (accepted, 2012)
10. Zadeh, L.: Fuzzy sets and information granularity. In: Gupta, M., Ragade, R., Yager, R.R. (eds.) Advances in Fuzzy Set Theory and Applications, pp. 3–18. North Holland, Amsterdam (1979)
11. Ziv, J., Lempel, A.: A Universal Algorithm for Sequential Data Compression. IEEE Transactions on Information Theory 23(3), 337–343 (1977)

Implementation of Background Knowledge and Properties Induced by Fuzzy Confirmation Measures in Apriori Algorithm

Iva Tomanová and Jiří Kupka

Centre of Excellence IT4Innovations – Division of the University of Ostrava – Institute for Research and Applications of Fuzzy Modeling
30. dubna 22, Ostrava, Czech Republic
{Iva.Tomanova,Jiri.Kupka}@osu.cz

Abstract. This work contributes to so-called association analysis. Its goal is to search for dependencies (called associations) between attributes in large scale data sets. Recently the authors theoretically studied some properties of fuzzy confirmation measures and possible application of background (resp. expert) knowledge into associations mining process. In this work we implement our recent results into well-known Apriori algorithm. Despite of the fact that the presented algorithm allows us to mine linguistic associations, i.e., associations interpretable in natural language, basic ideas of this algorithm can be easily extended to less specific model of fuzzy sets.

1 Introduction

We contribute to so-called fuzzy association analysis. Our first motivation comes from the paper [9] where a method for searching for linguistic associations, i.e., for associations in the form of natural language sentences, was introduced. Linguistic association can have a form of IF-THEN rule (for example, *"IF the area of the base of a cylinder is roughly big AND the height of this cylinder is more or less big THEN the volume of this cylinder is big."*). Such associations are easily interpretable even for somebody who is not an expert in data analysis.

The method from [9] was partially based on the GUHA method (see [4], [5] or [11]) which is the first method used for mining associations from data sets. But there exists another algorithm (Apriori algorithm from [1]) which made association analysis popular worldwide. In both methods some crisp confirmation measures can be used for checking validity of mined associations.

In this contribution we work with fuzzy confirmation measures whose existence was justified in [2]. Recently in [7] and [8], the authors studied validity of several properties of fuzzy confirmation measures and also possible implementation of background (resp. expert) knowledge into the mining process. In the latter case we were motivated by good interpretability of mined linguistic associations because possible direct cooperation with the user of the mining process is one of our main motivations.

Á. Herrero et al. (Eds.): Int. JointConf. CISIS'12-ICEUTE'12-SOCO'12, AISC 189, pp. 533–542.
springerlink.com

In this paper we demonstrate how to apply results from [7] and [8], and some specific properties of the model of evaluative linguistic expressions (see [6] and [9]) into known Apriori algorithm. Our algorithm might be computationally more complex than the original one as we use more complex model of linguistic expressions. But we also suggest several ways allowing us to possibly reduce number of tested associations. Some of them are based on properties of fuzzy confirmation measures and, consequently, can be used in the original algorithm as well. Additionally, this is the first algorithm where linguistic associations can be mined without transforming to the non-fuzzy case. Additionally, the proposed algorithm is the first case where hierarchical structure of evaluative linguistic expressions is taken into consideration and this feature can substitute some preprocessing steps. It should be emphasized that the same ideas can be used in other models of the same kind.

The paper consists of six parts and its organization is as follows - in Section 2 some basic notions and the task of linguistic associations mining are introduced. Three pairs of fuzzy confirmation measures are introduced in Section 3. In Section 4 we mention some properties of fuzzy confirmation measures. And finally, a part of a modified Apriori algorithm and concluding remarks are in Sections 5 and 6, respectively.

2 Basic Notation

In this section we introduce several simple mathematical concepts that are necessary for mining of linguistic associations. As there is not enough space in this contribution, for basic notions from fuzzy mathematics (e.g., membership degree, t-norm, t-conorm, implication operator etc.) we refer to [10] and references therein. We work with the product, minimum and Łukasiewicz t-norm, respectively. Further, we work with fuzzy sets S of the form $S : X \to I$ where $X \subseteq \mathbb{R}$ and $I = [0,1]$. The set of all convex fuzzy sets on X is denoted by $\mathcal{F}(X)$.

In the following text we were inspired by the theory of *evaluative linguistic expressions, resp. predications* ([9]). We work with *evaluative linguistic expressions*, i.e., with *atomic evaluative expressions Big (Bi), Medium (Me)* and *Small (Sm)* which can be modified by linguistic hedges (introduced by L. A. Zadeh), for example, *Very (Ve), More or Less (ML)* etc. Thus, one can obtain evaluative linguistic expressions *Ve Sm, ML Me, Bi* etc. Further, *evaluative linguistic predications* are evaluative linguistic expressions considered in a certain context. Since contexts can be mathematically expressed as closed intervals, one can represent evaluative linguistic predications as fuzzy sets.

There are more mathematical models of evaluative linguistic predications. Namely, the original one from [9] and the novel one ([6]) we work with here. Thus, linguistic predications can be represented by fuzzy covering $\mathcal{P} := \{S_k\}$ of a chosen universum U (for details we refer to [6]). The latter model allows us to work with specifying evaluative linguistic expressions containing the formula "*but not*". In this work, we use only an example of evaluative linguistic predications with linguistic hedges *more or less* or *very* – see Example 1 and Figure 1. Note that later $|X|$ denotes a *number of fuzzy sets* S_k from $\mathcal{P}$.

Example 1. In this contribution we consider the attribute (resp. property, variable) X on certain interval $[c,d]$ whose covering $\mathcal{P}$ contains 9 fuzzy sets $\{S_k\}$, (i.e., $S_k : [c,d] \to [0,1]$ and $|X| = 9$)

$$\begin{array}{lll} S_1 \sim Ve\ Sm, & S_2 \sim Sm \text{ but not } Ve\ Sm, & S_3 \sim ML\ Sm \text{ but not } Sm, \\ S_4 \sim Ve\ Me, & S_5 \sim Me \text{ but not } Ve\ Me, & S_6 \sim ML\ Me \text{ but not } Me, \\ S_7 \sim Ve\ Bi, & S_8 \sim Bi \text{ but not } Ve\ Bi, & S_9 \sim ML\ Bi \text{ but not } Bi. \end{array}$$ □

For every attribute X we can define sets $P^1(X) := \{S_k\}$ and, for various indexes i

$$P^i(X) = \left\{S \in \mathcal{F}([c,d]) \mid S = \mathtt{OR}_{j=1}^i S_j, \text{ where } S_j \in P^1(X)\right\},$$

where OR can be represented by a relevant t-conorm (see also Remark 1). Further, for every context we distinguish subsystems representing *small, medium* and *big* values, respectively, i.e., we can have $P^i_{Sm}(X), P^i_{Me}(X), P^i_{Bi}(X)$ for various indexes i. Finally, we put $P(X) = \bigcup_i P^i(X)$.

Example 2. For Example 1 it makes sense to consider only $i = 1, 2, 3$. Then $P^1_{Sm}(X) = \{S_1, S_2, S_3\}$, $P^1_{Me}(X) = \{S_4, S_5, S_6\}$ and $P^1_{Bi}(X) = \{S_7, S_8, S_9\}$, where each $S_k \in \mathcal{P}$ represents a suitable evaluative linguistic predication. For instance, for *small* values we have

$$\begin{array}{ll} S_1 \ \mathtt{OR}\ S_2 \sim Sm \in P^2_{Sm}(X), & S_1 \sim Ve\ Sm, \\ S_2 \ \mathtt{OR}\ S_3 \sim ML\ Sm \text{ but not } Ve\ Sm \in P^2_{Sm}(X), & S_2 \sim Sm \text{ but not } Ve\ Sm, \\ S_1 \ \mathtt{OR}\ S_2 \ \mathtt{OR}\ S_3 \sim ML\ Sm \in P^3_{Sm}(X), & S_3 \sim ML\ Sm \text{ but not } Sm. \end{array}$$

For *medium* and *big* values we construct P^i_{Me}, P^i_{Bi} analogously (for details see [6]). □

It is possible to construct either simpler or more comlex mathematical models than the one from Examples 1 and 2, but in this contribution we work only with this one as it is the most simple nontrivial mathematical model of evaluative linguistic predications.

A *specificity ordering* $\preceq$ is an ordering of fuzzy sets interpreting evaluative linguistic predications. We denote by $S' \preceq S$ the fact that, for each $x \in X$, $S'(x) \leq S(x)$ holds.

Example 3. Let S, S' denote fuzzy sets from Examples 1 and 2. If $S' \sim Ve\ Sm$ and $S \sim Sm$ then $S' \preceq S$. □

In our task we consider a numerical data set in the form of two-dimensional table $\mathcal{D}$

$$\mathcal{D} := \begin{array}{c|cccc} & X_1 & X_2 & \dots & X_m \\ \hline o_1 & e_{11} & e_{12} & \dots & e_{1m} \\ o_2 & e_{21} & e_{22} & \dots & e_{2m} \\ \vdots & \vdots & \vdots & \ddots & \vdots \\ o_n & e_{n1} & e_{n2} & \dots & e_{nm} \end{array}$$

where an element of table is a real number $e_{ij} \in \mathbb{R}$ ($e_{ij} = [o_i]_j$), it is a value of jth *attribute* (property) X_j measured on ith *object* (observation, transaction) o_i. Let $\mathcal{D}_o$ denote the set of rows (resp. objects) of $\mathcal{D}$.

Now contexts of all attributes must be specified. Mathematically, for $j = 1, 2, \ldots, m$, a *context* of any attribute X_j is a closed interval $[c_j, d_j]$. Any context should be set by the expert (user) which is more natural. When contexts are specified, one can work with fuzzy sets $P(X_j)$ introduced above.

Example 4. Consider 10 objects in an attribute *Age* with values $\{28, 45, 67, 32, 56, 70, 43, 73, 33, 72\}$. Then the context of the attribute *Age* must be given by the expert as $[0, 110]$. □

Our goal is to search for dependencies between given disjoint sets of attributes $\{Y_l\}_{l=1}^p, \{Z_k\}_{k=1}^q \subseteq \{X_j\}_{j=1}^m$. We look for unknown *linguistic associations* of the form

$$A(\{Y_l\}_{l=1}^p) \Rightarrow B(\{Z_k\}_{k=1}^q), \tag{1}$$

($A \Rightarrow B$ in short) where A, B are conjunctive evaluative linguistic predications, i.e., predications of the form

$$A = \mathtt{AND}_{l=1}^p (Y_l \ is \ S_l), \ S_l \in P(Y_l), \tag{2}$$

and $\Rightarrow$ denotes a relationship between A and B. This relationship can be given by chosen confirmation measures introduced in the next section (for more details we refer to [2]). Let us remark that the left side of (1) is called *antecedent* and the right side of (1) is called *succedent* or *consequent.*

Below we also work with so-called itemsets. A *k–itemset* T is a set of k ordered pairs (l, S_l) where any $l \in \{1, 2, \ldots, m\}$ denotes an index for which $S_l \in P(Y_l)$. Clearly, see the next example, there exists a one-to-one correspondence between linguistic predications (2) and p–itemsets. Consequently, we can identify p–itemsets with relevant linguistic predications - for instance, the cardinality of a p–itemset T can be considered as a cardinality of A, (1) can be thought as $T \Rightarrow R$ where T is a p–itemset and R is a q–itemset, respectively, and so on.

Example 5. Assume that linguistic predications are defined in all attributes. Then an expression "X_2 is *very small* AND X_5 is *big but not very big*" can be represented by 2–itemset $\{(2, S_2), (5, S_5)\}$, where $S_2 \sim$ *Ve Sm* and $S_5 \sim$ *Bi but not Ve Bi*, respectively. □

Similarly, it is easy to see that specificity ordering can be extended to the set of itemsets in a very natural way. Namely, for a p–itemset $T = \{(i, S_i)\}_{i \in I}$ and q–itemset $R = \{(j, E_j)\}_{j \in J}$ we denote $T \preceq R$ if $I \subseteq J$ and $S_i \preceq E_i$ for any $i \in I$. Finally, we can specify an operator C representing cardinality of a given p–itemset (resp. (2)). For any $o \in \mathcal{D}_o$ we get

$$\mathtt{C}(A)(o) = \mathtt{AND}_{l=1}^p S_l([o_i]_l). \tag{3}$$

Note that mathematical representation of AND's is mentioned in the next subsection (see Remark 1).

3 Fuzzy Confirmation Measures

Association analysis offers many ways how to test the validity of mined association (i.e., hypothesis). In this subsection we introduce so-called (fuzzy) confirmation measures (called quantifiers in GUHA method) which are used for this purpose.

Shortly, the authors of [2] justified using of three pairs of fuzzy confirmation measures. Below $\otimes$ denotes a t-norm, $\cdot$ is an ordinary product and $\rightarrow$ represents any *generalized implication*. So for associations of the form (1) we specify three *support measures*. Namely, a *t-norm-based support measure*

$$supp_t(A \Rightarrow B) := \sum_{o \in \mathcal{D}_o} A(o) \otimes B(o), \tag{4}$$

a *minimum-based support measure*

$$supp_m(A \Rightarrow B) := \sum_{o \in \mathcal{D}_o} \min\{A(o), B(o)\}, \tag{5}$$

and an *implication-based support measure*

$$supp_c(A \Rightarrow B) := \sum_{o \in \mathcal{D}_o} A(o) \cdot (A(o) \rightarrow B(o)). \tag{6}$$

Finally, for any of support measures (4), (5) and (6), its *confidence measure* is defined by

$$conf_p(A \Rightarrow B) := \frac{supp_p(A \Rightarrow B)}{\sum_{o \in \mathcal{D}_o} A(o)}, \text{ for } p = t, m, c. \tag{7}$$

Usually, the user specifies which pair of confirmation measures is used (see remark below) and specifies support and confidence thresholds. We say that a given rule $A \Rightarrow B$ is *valid* if its support and confidence degrees are greater than or equal to those thresholds.

Remark 1. For completeness, we have to specify how AND's and OR's are mathematically represented. For (4), (7) or (5), (7) or (6), (7) conjunctions are represented by t-norms or minimum or ordinary product, respectively. Disjunctions are represented by relevant t-conorms.

In the rest of this section we mention properties (see [8]) proved for fuzzy confirmation measures introduced above. Below a symbol $\vdash$ means that if the left side of the rule is valid then the right side is valid as well.

Lemma 1. *([8]) Let A, B, C, D are evaluative linguistic predications then for measures* (4), (5), (6) *and relevant* (7)

$$(A \Rightarrow B),\ (A \Rightarrow C) \vdash (A \Rightarrow (B \text{ OR } C)).$$

Lemma 2. ([8]) *For measures* (4), (5) *and relevant* (7)

$$(A \text{ AND } B) \Rightarrow (C \text{ AND } D) \vdash (A \text{ AND } B \text{ AND } D) \Rightarrow C.$$

4 Background Knowledge and Special Properties

As a *background knowledge* (resp. *expert knowledge*) we call prior information or experience about a given data set $\mathcal{D}$ which can be specified by the user. Background knowledge can be specified as a set $\mathcal{B}$ of associations (denoted by $\Rightarrow^*$) for which we assume that they are fully valid in $\mathcal{D}$. Here the full validity of $(E \Rightarrow^* F) \in \mathcal{B}$ means that $conf(E \Rightarrow^* F) = 1$ for chosen confirmation measure. In [8], the authors clarified that it can be useful to apply the set $\mathcal{B}$ into the process of mining associations. The following result is valid only in this case.

Lemma 3. ([8]) *For measures* (4), (5), (6) *and relevant* (7)

$$A \Rightarrow B,\ B \Rightarrow^* C \vdash A \Rightarrow C.$$

Lemma 4. *For measures* (4), (5), (6) *and relevant* (7)

$$A \Rightarrow B',\ B \Rightarrow^* C \vdash A \Rightarrow C,\ \text{whenever } B' \preceq B.$$

Proof. We consider t-norm-based support measure (4). We obtain $supp_t(A \Rightarrow B') = \sum_{o \in D_o} A(o) \otimes B'(o)$. According to [8], $B \Rightarrow^* C$ implies that $B(o) \leq C(o)$ for any $o \in \mathcal{D}_o$. From $B' \preceq B$ it holds $B'(o) \leq B(o) \leq C(o)$. Consequently, we obtain directly from (4) and (7) that $supp_t(A \Rightarrow B') \leq supp_t(A \Rightarrow C)$ (resp. $conf_t(A \Rightarrow B') \leq conf_t(A \Rightarrow C)$).

As minimum-based confirmation measures are a special case of t-norm-based ones, it remains to finish this proof for implication-based confirmation measures (6). We obtain $supp_c(A \Rightarrow B') = \sum_{o \in D_o} A(o)(A(o) \rightarrow B'(o))$. As above we have $B'(o) \leq B(o) \leq C(o)$. Consequently, we obtain from (6) and (7) that $supp_c(A \Rightarrow B') \leq supp_c(A \Rightarrow C)$ (resp. $conf_c(A \Rightarrow B') \leq conf_c(A \Rightarrow C)$). □

5 Modified Algorithm Apriori

The Apriori algorithm is one of the best known algorithm used for searching for associations. In the first step of this algorithm frequent itemsets are discovered. Then candidate associations are generated and tested by chosen confidence measure. In this section we demonstrate how to implement our mathematical model into this algorithm. The computational complexity of the proposed algorithm is higher, however our algorithm allows to adapt mined association to the data set and hence, in some way, substitutes some preprocessing steps.

The aim of this section is twofold. Firstly, we demonstrate how the properties described above and background knowledge can be implemented into the Apriori algorithm (e.g., [1]). Secondly, we suggest an implementation of our model of evaluative linguistic expressions.

The proposed algorithm is the following:

INPUT:

Data description - notation:

n ... the number of objects,
m ... the number of attributes,
$\mathcal{D}_o$ ... the set of objects,
X_j ... the jth attribute $j = 1, \ldots, m$,
e_{ij} ... the value of jth attribute measured on ith object.

What is specified by the user:

$supp_p$ ... the support measure (p is one of t, m, c),
α ... minimal support threshold,
γ ... minimal confidence threshold and a suitable *linguistic description*,
$[c_j, d_j]$ ... the context of attribute X_j, for any $j = 1, 2, \ldots, m$,
$\mathcal{P}(X_j)$ ... fuzzy covering $\{S_{jk}\}_{k=1}^{|X_j|}$ on X_j consisting of fuzzy sets S_{jk}, where $|X_j|$ is the number of fuzzy sets $P^1(X_j) := \mathcal{P}(X_j)$ (e.g., see Example 2).

Other symbols:

C_r ... a set of candidate r–itemsets, (e.g., $C_1 = \{(j, S_{jk}) \mid \forall j = 1, \ldots, m;\ S_{jk} \in P^1(X_j)\}$) (We start with $C_r = \emptyset$.),
L_r ... a set of large r–itemsets (We start with $L_r = \emptyset$.),
$\mathcal{A}$ ... a set of found associations (We start with $\mathcal{A} = \emptyset$.).

OUTPUT: The set of linguistic associations $\mathcal{A}$.

STEP 1. Construct a set $C_1 := \{(j, S_{jk}) \mid S_{jk} \in P^1(X_j),\ j = 1, 2, \ldots, m\}$ of all 1–itemsets and, for each $t \in C_1$, compute (see (3))

$$count(t) = \sum_{o \in D_o} \mathtt{C}\,(t)(o). \tag{8}$$

STEP 2. Check the $count(t)$ of each $t \in C_1$:

(a) If $count(t) \geq \alpha$ then put t into L_1.
(b) If $count(t) < \alpha$ and $t \in P_Q^1(X_j)$[1] (Q is one of Sm, Me, Bi), then check all $t' \in P_Q^2(X_j)$, satisfying $t \preceq t'$. If $count(t') \geq \alpha$ for such t', put $t' \in L_1$. Otherwise do the same with 1–itemsets from $P_Q^3(X_j)$.

STEP 3. Set $r = 1$.

STEP 4. As in the original algorithm, to generate C_{r+1} from large r–itemsets, i.e., use r–itemsets from L_r. The only difference is that we have to deal with linguistic expressions which can be ordered by specificity ordering. In order to keep cardinalities of r-subitemsets, every generated $(r + 1)$–itemset the most broad expressions mentioned in r–itemsets.

Example 6. Let only pairs $\{t_1, t_2\}$, $\{t_1, t_3\}$, $\{t_1, t_4\}$ and $\{t_2, t_3\}$ be in L_2. Then only $\{t_1, t_2, t_3\} \in C_3$ while $\{t_1, t_2, t_4\} \notin C_3$ (resp. $\{t_2, t_3, t_4\}$, $\{t_1, t_3, t_4\}$) because $\{t_2, t_4\} \notin L_2$ (resp. $\{t_2, t_4\}$, $\{t_3, t_4\} \notin L_2$).

[1] Here and below we use the fact that k–itemsets can be identified with elements of fuzzy covering.

STEP 5. For any $t \in C_{r+1}$ do the following:

(a) Compute *count*(t) by (8).
(b) If *count*$(t) \geq \alpha$, put t in L_{r+1}.
(c) If *count*$(t) < \alpha$ we have to consider "broader" linguistic expressions in every attribute as in STEP 2(b). For example, if

$$t = \{(u_1, S_{u_1}), (u_2, S_{u_2}), \ldots, (u_i, S_{u_i}), \ldots, (u_{r+1}, S_{u_{r+1}})\} \tag{9}$$

and (u_i, S_{u_i}) is such that $S_{u_i} \in P_Q^h(X_j)$, $Q \in \{Sm, Me, Bi\}$ and $h \leq 2$, then instead of S_{u_i} we take all $S'_{u_i} \in P_Q^{h+1}(X_j)$. Thus, we check (8) for $r+1$–itemset

$$t' = \{(u_1, S_{u_1}), (u_2, S_{u_2}), \ldots, (u_i, S'_{u_i}), \ldots, (u_{r+1}, S_{u_{r+1}})\}.$$

If *count*$(t') \geq \alpha$ then we do not check $\tilde{t} \in C_{r+1}$ for which $t' \preceq \tilde{t}$ and put $t' \in L_{r+1}$. But we have to check other possible combinations of "broader" expressions in this step.
(d) For any $t \in L_{r+1}$ we may assume that elements of t are ordered by their cardinalities. I.e., for (9) we assume

$$count\{(u_i, S_{u_i})\} \leq count\{(u_k, S_{u_k})\} \text{ whenever } u_k \leq u_i.$$

STEP 7. If $L_{r+1} = \emptyset$ and $r \geq 2$ then follow the next steps.
STEP 8. Set $w = 1$.
STEP 9. Choose element $t \in L_r$ of the form

$$t = \{(u_1, S_{u_1}), (u_2, S_{u_2}), \ldots, (u_i, S_{u_i}), \ldots, (u_r, S_{u_r})\}. \tag{10}$$

The r–itemset t can be decomposed into $t'(w)$ and $t''(w) := t \setminus t'(w)$ where w denotes that t' consists of w elements. For instance,

$$t'(w) = \{(u_1, S_{u_1}), (u_2, S_{u_2}) \ldots, (u_w, S_{u_w})\},$$

$$t''(w) = \{(u_{w+1}, S_{u_{w+1}}), (u_{w+2}, S_{u_{w+2}}) \ldots, (u_r, S_{u_r})\}.$$

Clearly, such decomposition defines an association $a(w) := t'(w) \Rightarrow t''(w)$. For all w–itemsets $t'(w)$ we do the following steps.
STEP 10. If $conf_p(a(w)) < \gamma$, then

(a) $t'(w)$ is replaced by w–itemset $\tilde{t}$ possessing "broader" expressions (i.e., $(i, S_i) \in t'(w)$ implies that there exists $(i, \tilde{S}) \in \tilde{t}$ such that $S_i \preceq \tilde{S}$) an association $a(w) := \tilde{t} \Rightarrow t''(w)$ is checked instead of $a(w)$. (As in STEP 5 (c) - all possible combinations of "broader" expressions have to be considered here.).

If none association $\tilde{t} \Rightarrow t''(w)$ can be constructed then choose different $t'(w) \subseteq t$ and repeat this step. If this is not possible and $w < r$, put $w := w+1$ and repeat this step with another $t'(w) \subseteq t$. If $w = r$, then $L_r := L_r \setminus t$ and go to STEP 9 whenever $L_r \neq \emptyset$. In the latter case, $r := r - 1$ and go to STEP 8.
STEP 11. If $conf_p(a(w)) \geq \gamma$ then

(a) put $a(w)$ into set $\mathcal{A}$.

For confirmation measures (4) and (5) and relevant (7) we can use the following reduction tools.

(b) It follows from Lemma 2 that, for z={w+1, ..., r-1}, $a(z)$ is valid if the antecedent of $a(z)$ contains $t'(w)$.
(c) It follows from Lemma 1 that all associations $t'(w) \Rightarrow \tilde{t}$ are valid whenever $t''(w) \preceq \tilde{t}$.
(d) Elements of $t'(w)$ can be replaced by elements of t with lower cardinality and the validity is not corrupted.

Example 7. We consider 3–itemset $\{t_1, t_2, t_3\}$ where $\mathtt{C}(t_3) \leq \mathtt{C}(t_2) \leq \mathtt{C}(t_1)$ and we obtain $a(1) := (t_1 \Rightarrow t_2 \texttt{ AND } t_3) \in \mathcal{A}$, then $(t_2 \Rightarrow t_1 \texttt{ AND } t_3)$ and $(t_3 \Rightarrow t_1 \texttt{ AND } t_2)$ are also valid associations. □

For all confirmation measures from Section 3:

(e) Lemmas 3 and 4 and associations from $\mathcal{B}$ might be applied in this step as well.

STEP 12. If all associations $a(w)$ generated from t were checked and $w < r$, put $w := w + 1$ and go back to STEP 9. If $w = r$, then $L_r := L_r \setminus t$ and go to STEP 9 whenever $L_r \neq \emptyset$. In the next, $r := r - 1$ and go to STEP 8. In the latter case $r = 1$ then it means the end of the algorithm.

Since there is not enough space, we could not carefully explain STEPs 10 and 11. Neither we could provide an analysis of the complexity of our algorithm.

6 Conclusions

In this paper we provide a demonstration of a novel algorithm for mining of linguistic associations. The number of linguistic associations which must be checked is definitely higher, so we have suggested some steps which can be used for reductions of mined associations (STEP 11 (b), (c) and (d)). We realize that there is still enough work to be finished (e.g., to estimate complexity of the proposed algorithm, to specify how to use the background knowledge, to reduce the set of mined linguistic associations in a reasonable way etc.).

However, the proposed algorithm contains several things which are novel. They are already mentioned in the introductory section - for instance, we search for linguistic associations in a "fuzzy" way, we implemented some properties of fuzzy confirmation measures in order to reduce the number of associations which must be checked, and since we work with mathematical model of fuzzy sets which are hierarchically ordered, this kind of association analysis need not require so precise preprocessing. There are currently many papers contributing to this task, for example see [3] or [12]. Finally, it must be mentioned, that ideas implemented in our algorithm are valid in general and can be easily extended to other mathematical models.

Acknowledgements. This work was supported by the European Regional Development Fund in the IT4Innovations Centre of Excellence project (CZ.1.05/1.1.00/02.0070). Furthermore, we gratefully acknowledge support by the grant 01798/2011/RRC of the Moravian-Silesian region.

References

1. Agrawal, R., Srikant, R.: Fast Algorithms for Mining Association Rules, vol. 1215, pp. 487–499. Citeseer (1994)
2. Dubois, D., Hüllermeier, E., Prade, H.: A systematic approach to the assessment of fuzzy association rules. Data Mining and Knowledge Discovery 13, 167–192 (2006)
3. Fu, H.: Cluster analysis and association analysis for the same data, pp. 576–581. University of Cambridge, UK (2008)
4. Hájek, P.: The question of a general concept of the guha method. Kybernetika, 505–515 (1968)
5. Hájek, P., Havránek, T.: Mechanizing hypothesis formation. Mathematical foundations for a general theory. Springer, Heidelberg (1978)
6. Kupka, J., Tomanová, I.: Some extensions of mining of linguistic associations. Neural Network World 20, 27–44 (2010)
7. Kupka, J., Tomanová, I.: Some dependencies among attributes given by fuzzy confirmation measures. In: Proc. of the LFA-EUSFLAT 2011, France, pp. 498–505 (2011)
8. Kupka, J., Tomanová, I.: Dependencies among attributes given by fuzzy confirmation measures. Expert Systems with Applications 39(9), 7591–7599 (2012)
9. Novák, V., Perfilieva, I., Dvořák, A., Che, Q., Wei, Q., Yan, P.: Mining pure linguistic associations from numerical data. International Journal of Approximate Reasoning 48(1), 4–22 (2008)
10. Novák, V., Perfilieva, I., Močkoř, J.: Mathematical principles of fuzzy logic. Kluwer Academic Publishers, Boston (1999)
11. Rauch, J.: Logic of association rules. Applied Intelligence 22, 9–28 (2005)
12. Tsay, Y.J., Chang-Chien, Y.W.: An efficient cluster and decomposition algorithm for mining association rules. Information Sciences 160, 161–171 (2004)

Author Index